EVALUATION OF FIRE SAFETY

EVALUATION OF FIRE SAFETY

D. Rasbash
formerly of Edinburgh University, UK

G. Ramachandran
Universities of Manchester and Leeds, UK

B. Kandola
Parsons Brinkerhoff Technologies Limited, UK

J. Watts
Fire Safety Institute, Vermont, USA

M. Law
Arup Fire (Ove Arup), London, UK

John Wiley & Sons, Ltd

Telephone (+44) 1243 779777

Email (for orders and customer service enquiries): cs-books@wiley.co.uk
Visit our Home Page on www.wileyeurope.com or www.wiley.com

Other Wiley Editorial Offices

John Wiley & Sons Inc., 111 River Street, Hoboken, NJ 07030, USA

Jossey-Bass, 989 Market Street, San Francisco, CA 94103-1741, USA

Wiley-VCH Verlag GmbH, Boschstr. 12, D-69469 Weinheim, Germany

John Wiley & Sons Australia Ltd, 33 Park Road, Milton, Queensland 4064, Australia

John Wiley & Sons (Asia) Pte Ltd, 2 Clementi Loop #02-01, Jin Xing Distripark, Singapore 129809

John Wiley & Sons Canada Ltd, 22 Worcester Road, Etobicoke, Ontario, Canada M9W 1L1

Wiley also publishes its books in a variety of electronic formats. Some content that appears in print may not be available in electronic books.

Library of Congress Cataloging-in-Publication Data

Evaluation of fire safety / D. Rasbash ... [*et al.*].
p. cm.
Includes bibliographical references and index.
ISBN 0-471-49382-1 (Cloth : alk. paper)
1. Fire protection engineering. 2. Fire prevention. I. Rasbash, D.
TH9145 .E94 2004
628.9′22–dc22

2003023868

British Library Cataloguing in Publication Data

A catalogue record for this book is available from the British Library

ISBN 0-471-49382-1

Typeset in 9.5/11.5pt Times by Laserwords Private Limited, Chennai, India
Printed and bound by CPI Antony Rowe, Eastbourne

CONTENTS

Preface

PART I STRUCTURE OF THE FIRE PROBLEM 1

1 The place of fire safety in the community 3
- 1.1 The nature of the fire hazard 3
- 1.2 Interaction between fire hazard and other hazards 3
- 1.3 Major fire hazard areas 4
- 1.4 The total cost of fires 5
- 1.5 Prescriptive and functional approach to fire safety 5
- 1.6 Purpose and outline of this book 6
- 1.7 Definitions 7
- Symbols 9
- References 9

2 The fire safety system 11
- 2.1 Basic questions of fire safety 11
- 2.2 Fire safety objectives 12
- 2.3 Steps in fire safety design 13
- 2.4 Sources of fire safety data 16
- 2.5 Subsystems 17
  - 2.5.1 Fire occurrence and fire prevention 17
  - 2.5.2 Fire development and fire control 19
  - 2.5.3 Harmful effects 19
  - 2.5.4 Direct detriment 20
  - 2.5.5 Consequential effects and fire accommodation 20
  - 2.5.6 Fire safety design and management 21
- 2.6 Contribution of fire safety engineering 21
- 2.7 Approaches to quantitative evaluation of fire safety 22
- 2.8 Other systems approaches 26
- 2.9 Risk management 28
- 2.10 Trade-off, equivalency, cost benefit, and cost effectiveness 29
- 2.11 How safe is "safe enough?" 30
- Symbols 33
- References 33

3 Review of some major fire & explosion disasters 35
- 3.1 Introduction 35
- 3.2 Major disasters in buildings involving sudden rapid spread of fire 35
  - 3.2.1 Summerland Leisure Center, Isle of Man, 1973 35

3.2.2 The Stardust Dance Club fire, Dublin, 1981 38
3.2.3 King's Cross Underground Station, London, 1987 41
3.2.4 Bradford City Football Ground, 1985 41
3.2.5 Interstate Bank, Los Angeles, 1988 43
3.2.6 Dupont Plaza Hotel, Puerto Rico, 1986 45
3.3 Fires in which extensive spread of smoke was a major factor 51
3.3.1 Beverly Hills Supper Club, Kentucky, 1977 51
3.3.2 Fairfield Home, Nottinghamshire, 1974 53
3.4 Fires associated with explosions 54
3.4.1 MV Betelgeuse, Ireland, 1979 54
3.4.2 Ronan Point, London, 1968 56
3.4.3 Piper Alpha Platform, North Sea, 1988 60
References 61

4 Requirements from public and private authorities for fire safety 63

4.1 Introduction 63
4.2 US regulatory environment 63
4.2.1 Codes and standards 64
4.2.2 Consensus codes and standards 64
4.2.3 Limitations of existing codes 65
4.2.4 Code adoption and enforcement 66
4.3 National Fire Protection Association 66
4.3.1 NFPA codes and standard-making process 67
4.3.2 Performance option 67
4.3.3 Life safety code 68
4.3.4 NFPA building code 68
4.4 Model building and fire codes 68
4.4.1 Model code writing organizations 69
4.4.2 International Code Council (ICC) 70
4.5 Other nonprofit organizations 70
4.5.1 Product standards and testing 70
4.5.2 Underwriters laboratories 71
4.5.3 Engineering societies 71
4.5.4 Insurance organizations 72
4.6 US federal agencies 72
4.6.1 Workplace fire safety 73
4.6.2 Consumer products 73
4.6.3 Transport 73
4.6.4 Nonregulatory activities 73
4.7 Canadian regulations 74
4.7.1 National Research Council 74
4.7.2 Canadian Commission on Building and Fire Codes 75
4.7.3 Objective-Based Codes 76
4.7.4 Institute for Research in Construction 77
4.7.5 Canadian Standards 77
4.8 United Kingdom 78
4.8.1 Buildings 78
4.8.2 Industrial and process hazards 79

4.8.3 Consumer fire safety 80
4.8.4 Transport 81
4.8.5 Fire safety audits and checklists 81
References 82

PART II QUANTIFYING FIRE SAFETY **85**

5 Physical data 87

5.1 Introduction 87
5.2 Burning and ignition 87
5.2.1 Mechanisms of burning 87
5.2.2 Properties of flaming combustion of liquids and solids 88
5.2.3 Ignition 91
5.2.4 Identification and power of ignition sources 95
5.3 Spread of fire 95
5.3.1 Radiation from flames 96
5.3.2 Convection from flames 100
5.3.3 Measurements of heat transfer from flames 101
5.3.4 Radiation from hot gas layers 102
5.3.5 Convection from hot gas layers 103
5.3.6 Fire spread along surfaces 105
5.4 Circumstances favorable to rapid fire spread 106
5.5 Sudden massive flaming in buildings 106
5.6 Sudden massive flaming during fires in process industries 108
5.7 Production and movement of smoke and toxic gases 109
5.8 Postflashover fires in buildings 112
5.9 Interaction between fire and structures 114
5.10 Defenses against smoke 117
5.11 Fire detection 118
5.12 Fire suppression 119
5.13 Interaction between fire and people 121
5.14 Explosions 122
5.15 Fire scenarios in fire safety design 123
Nomenclature 125
References 127

6 Sources of statistical data 133

6.1 Introduction 133
6.2 Fire departments and brigades 134
6.2.1 United Kingdom 134
6.2.2 United States of America 136
6.2.3 Japan 138
6.2.4 Other countries 138
6.3 Insurance organizations and fire protection associations 139
6.3.1 United Kingdom 139
6.3.2 United States of America 140
6.3.3 Other Countries 141

6.4 Special databases 141
6.4.1 United Kingdom 141
6.4.2 United States of America 142
6.4.3 International 143
6.5 Other data sources 143
6.5.1 Minor databases 143
6.5.2 Research studies 144
6.6 Ancillary statistics 145
6.6.1 Population data 145
6.6.2 Building stock 146
6.6.3 Economic data 146
6.6.4 Other ancillary statistics 147
6.6.5 International sources 147
6.7 Discussion 147
6.7.1 Use of statistics 147
6.7.2 Limitations in the use of national statistics 148
6.7.3 Gaps in national fire statistics 149
6.7.4 International comparisons 150
Acronyms 151
References 152

7 Occurrence and growth of fire 153

7.1 Introduction 153
7.2 Probabilistic approach 154
7.3 Probability of fire starting 154
7.4 Probable damage in a fire 158
7.5 Extent of spread 162
7.6 Fire growth rate 164
7.7 Fire severity 169
7.8 Special factors 170
Symbols 172
References 173

8 Life loss 175

8.1 Introduction 175
8.2 Characteristics of fire casualties 176
8.3 Location of casualties 177
8.4 Nature of injuries 178
8.5 Materials first ignited 180
8.6 Casualty rate per fire 180
8.7 The time factor 182
8.8 Evacuation model 184
8.9 Multiple-death fires 186
8.10 Other measurements of life risk 190
8.11 Impact of product choices on life risk – US method 192
Symbols 195
References 195

9 Property damage 197

9.1 Introduction 197
9.2 Probability distribution 197
9.3 Extreme value theory 203
9.3.1 Introduction 203
9.3.2 Extreme order distributions 204
9.3.3 Estimation of extreme order parameters 204
9.3.4 Behavior of large losses 207
9.3.5 Return period 209
9.3.6 Average and total loss 209
9.4 Multiple regression 215
Appendix : Properties of extreme order statistics 217
9A.1 Basic properties 217
9A.2 Estimation of extreme order parameters 217
9A.3 Variation in sample size 219
References 220

10 Performance of fire safety measures 221

10.1 Introduction 221
10.2 Fire prevention measures 221
10.2.1 Introduction 221
10.2.2 Dwelling fires – house-to-house visits 222
10.2.3 Chip or fat pan fires – television advertisement 222
10.2.4 Space heater fires – television advertisement 225
10.3 Automatic detectors 227
10.3.1 Performance 227
10.3.2 Effectiveness – property protection 228
10.3.3 Effectiveness – life safety 231
10.3.4 Reliability 232
10.3.5 Addressable systems 233
10.3.6 False alarms 233
10.4 Sprinklers 235
10.4.1 Performance 235
10.4.2 Effectiveness – property protection 237
10.4.3 Effectiveness – life safety 239
10.5 Structural (passive) fire protection 239
10.5.1 Fire resistance 239
10.5.2 Compartmentation 243
10.5.3 Means of escape facilities 243
10.5.4 Fire door 245
10.6 Effectiveness of other fire protection devices 246
10.6.1 Fire extinguishers 246
10.6.2 Ventilation systems 246
10.7 Interactions 247
10.7.1 Automatic detectors and fire brigade 247
10.7.2 Sprinklers and fire brigade 248

10.7.3 Sprinklers and structural fire protection 248
10.7.4 Sprinklers and ventilation systems 249
References 251

PART III METHODS OF MEASURING FIRE SAFETY 253

11 Deterministic fire safety modeling 255

11.1 Introduction 255
11.2 Models of enclosure fires 256
11.2.1 Theory and concept of zone models 257
11.3 Dynamics of enclosure fires 257
11.3.1 Heat release 257
11.3.2 Fire-generated flows 259
11.3.3 Stratification 261
11.3.4 Mixing 263
11.3.5 Heat transfer and mass loss to surfaces 265
11.3.6 Flow through openings 267
11.4 Zone modeling of preflashover enclosure fires 270
11.4.1 Conservation of mass and energy 270
11.4.2 The hot layer 272
11.4.3 Flame and burning objects source terms 273
11.4.4 Fire plume source terms 276
11.4.5 The hot-layer source terms 277
11.4.6 Heat conduction source terms 279
11.4.7 Convective heat flux source terms 280
11.4.8 Radiative heat flux source terms 280
11.4.9 Flows through vertical vents source terms 280
11.4.10 Products of combustion source terms 285
11.5 One-zone modeling of postflashover fires 289
11.5.1 Theoretical model 289
11.5.2 Numerical solution procedure assumptions 293
11.6 Field modeling of enclosure fires 294
11.6.2 Method of solution 300
11.7 Evacuation modeling 302
11.7.1 Evacuation model – EXITT 305
11.7.2 Evacuation model – EGRESS 306
11.7.3 Evacuation model – SIMULEX 307
Nomenclature 309
Greek symbols 313
References 314

12 Model Validation 319

12.1 Introduction 319
12.2 Validation of zone models 320
12.3 The Harvard V zone model (first) 320
12.3.1 Experimental setup 321
12.3.2 The fire 321
12.3.3 Measurements 321
12.3.4 Comparison of results 321

12.4 Harvard VI zone model 324
12.4.1 Experimental setup 325
12.4.2 The fire 325
12.4.3 Measurements 325
12.4.4 Discussion of results 325
12.4.5 Conclusions 328
12.5 FAST zone model 328
12.5.1 Experimental setup 330
12.5.2 Comparison of results 330
12.5.3 Another validation of FAST 331
12.6 CFAST 333
12.6.1 CFAST validation 334
12.6.2 Discussion of results 334
12.7 FAST and HARVARD VI 336
12.7.1 Experimental setup 336
12.7.2 Discussion of results 336
12.8 Validation of field models 339
12.9 JASMINE field model 339
12.9.1 Enclosure fires with natural ventilation 340
12.9.2 Enclosure fires (forced ventilation) 342
12.9.3 Ceiling jet for confined and unconfined ceilings 345
12.9.4 Tunnel fires 346
12.10 CFDS-FLOW3D field model 350
12.10.1 Single-enclosure fires with natural ventilation 350
12.10.2 Multiroom enclosure fires with natural ventilation 353
12.10.3 Large single-cell enclosure fire 356
12.11 Conclusions concerning validation of zone and field models 359
Symbols 359
References 360

13 Point systems – a single index 363

13.1 Introduction 363
13.2 Concepts 364
13.2.1 Defensibility 365
13.2.2 Heuristic models 365
13.2.3 Multiattribute evaluation 365
13.2.4 Application 366
13.3 Examples of established point systems 366
13.3.1 Gretener method 366
13.3.2 Dow fire and explosion index 367
13.3.3 Fire Safety Evaluation System (FSES) 367
13.3.4 Multiattribute evaluation examples 369
13.4 Multiattribute evaluation 372
13.4.1 Attributes 373
13.4.2 Attribute weighting 375
13.4.3 Attribute ratings 377
13.4.4 Scoring methods 382
13.5 Criteria 384

13.6 Summary 386
Glossary 386
Nomenclature 387
References 388

14 Logic trees 391

14.1 Introduction 391
14.2 Management of fire safety 392
14.3 Logic trees 392
14.4 Decision tree 393
14.5 Event tree 395
14.6 Fault tree 397
References 398

15 Stochastic fire risk modeling 399

15.1 Introduction 399
15.2 General model 400
15.3 Markov model 400
15.3.1 Mathematical representation 400
15.3.2 Markov chains 402
15.3.3 Markov process 402
15.4 State transition model – spread within a room 404
15.5 State transition model – spread from room to room 407
15.6 Network models 408
15.6.1 Elms and Buchanan 408
15.6.2 Platt 410
15.6.3 Ling and Williamson 411
15.7 Random walk 414
15.8 Percolation process 416
15.9 Epidemic theory 417
Acknowledgment 418
References 419

16 Fire safety concepts tree and derivative approaches 421

16.1 Introduction 421
16.2 Fire safety concepts tree 421
16.2.1 Configuration of the tree 421
16.2.2 Applications 423
16.3 The GSA Goal-Oriented System approach 425
16.3.1 Components of the goal-oriented systems approach 426
16.3.2 Quantitative aspects 427
16.3.3 Limitations of the GSA approach 431
16.3.4 Summary of the GSA approach 432
16.4 Modified approach 432
16.4.1 Postulates of fire spread 433
16.4.2 Stress-strength models 434
16.4.3 Probability distributions 437
16.4.4 Modified approach 437

16.4.5 Example 441
16.4.6 Limitations of the modified approach 442
16.4.7 Summary of the modified approach 443
16.5 WPI engineering method 443
16.5.1 Framework of the WPI engineering method 443
16.5.2 Event trees 445
16.5.3 Quantification 445
16.5.4 Limitations of the WPI engineering method 448
16.5.5 Summary of the WPI engineering method 448
References 448

17 Fire safety assessment in the process industries 451

17.1 Introduction 451
17.2 Legislation for control of major hazards 451
17.3 Point schemes used in the process industry 453
17.3.1 Dow chemical points scheme 453
17.3.2 The mond fire, explosion and toxicity index 455
17.4 Instantaneous Fractional Annual Loss (IFAL) 455
17.5 Hazard and operability studies 457
17.6 Failure Mode and Effects Analysis (FMEA) 457
17.6.1 Failure Mode Effects and Criticality Analysis (FMECA) 458
17.6.2 Reliability of fire protection systems 459
17.7 Inherent safety 460
17.8 Stages in safety design for new plant 460
17.9 Examples of logic tree analysis in process industries 461
References 461

Index 463

PREFACE

David Rasbash began publishing and teaching about the evaluation of fire safety in the 1970s. The accumulation of contributions to the subject over the succeeding years, by himself and others, reached a stage where a textbook was clearly needed, and David's colleagues managed to persuade him that he was the ideal person to prepare such a book. Having agreed, he planned the book's structure and enlisted 'Ram' Ramachandran, Baldev Kandola, and Jack Watts to contribute a number of the chapters. During the final stages of his illness, David could not complete the task and he was happy to accept the suggestion that Margaret Law might take over and bring the book to completion. Margaret has filled in the gaps in David's chapters and has consulted the other authors about any tricky parts in their work. She found the process to be tiring but completely fascinating.

An engineering approach to the evaluation of safety is not, of course, a new subject. However, what is new about this book is that it brings together data, information, and techniques that are particularly relevant to evaluating fire safety. The authors hope that not only students but also practising engineers will want to dip into its pages many times.

PART I STRUCTURE OF THE FIRE PROBLEM

1 THE PLACE OF FIRE SAFETY IN THE COMMUNITY

1.1 The nature of the fire hazard

The hazard of fire is the consequence of uncontrolled, exothermic chemical reactions, especially between organic materials and air. It is particularly associated with combustible materials and energy sources used by people in everyday life. Although fire threatens both life and property and its control occasions much expenditure, the hazard must be set against the benefit gained from these resources so that a balanced view can be obtained. Moreover, living standards are highly dependent on the use of buildings. The extra danger when fires occur in an enclosure, with the heat and smoke being trapped rather than moving relatively harmlessly upward, needs to be set against the intrinsic value of using buildings. It follows that one cannot, in general, eliminate fire hazard, although one can reduce it to an acceptably low level by suitable design procedures.

1.2 Interaction between fire hazard and other hazards

Fire takes its place alongside many other hazards in living. These include health hazards such as epidemics and sickness, industrial transportation and domestic accidents, as well as natural hazards such as earthquakes, floods, hurricanes, and so on. The fire hazard can of course be reduced by a severe restriction in the use of energy and combustible materials, but this could bring in its wake suffering and cost in excess of any alleviation of the fire problem. It could even give rise to conditions that prompt other hazards, particularly health hazards. There is a tendency for people who specialize in fire safety to look at the fire problem in isolation. One must be careful not to lose perspective in so doing, for example, with regard to the benefits that might ensue using a material or process that might incidentally impose an increased fire hazard.

This point is illustrated diagrammatically in Figure 1.1 (Rasbash, 1974). Risks are associated with the act of living. Some risks have to be taken, while others are taken voluntarily. Risks are taken to obtain a benefit, of which perhaps a notional measure might be denoted by A. Amongst the risks, there are those with fire, which may inflict a penalty of "detriment" of fire damage and hurt because of fire occurrence. These may be assigned a notional value of f_d ("d" for detriment). The fire danger requires a fire safety programme that inflicts a cost of f_c ("c" for cost). In the same way, other hazard scenarios inflict detriment h_d, and safety programme costs of h_c. The

Evaluation of Fire Safety D. Rasbash, G. Ramachandran, B. Kandola, J. Watts and M. Law
 ISBN: 0-471-49382-1

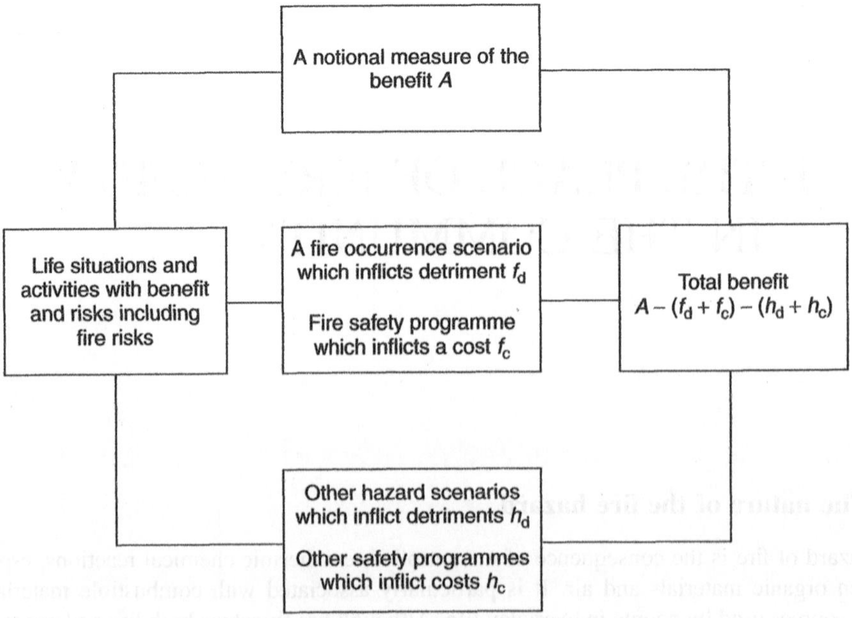

Figure 1.1. Fire safety in the community

object of any rational programme toward controlling fire safety should be to maximize the total benefit: $[A - (f_d + f_c) - (h_d + h_c)]$.

Two examples serve to illustrate this point. Insulation in houses saves energy and would thus increase A. Insulation, particularly on the inside surfaces of a room, is also known to increase the rate of fire spread even if the insulation is not combustible. The introduction of such insulation would therefore serve to increase f_d. Would $[A - f_d]$ be increased by the introduction of insulation? Many effective insulating materials are in themselves highly flammable. This tends to rule out their use on interior walls. It is normal in these circumstances to introduce a noncombustible layer on the inside wall with extra cost f_c. In this case, the relevant benefit is the change in value of $[A - (f_d + f_c)]$.

The provision of smoke stop doors is common in buildings, particularly in the United Kingdom. These of course occasion a certain cost that contributes to f_c. As long as they can be opened when necessary by people escaping a fire, such doors reduce the risk of death in the event of a fire and thus reduce f_d. But an extra cost that tends not to be brought into the equation is the inconvenience of having these doors scattered about buildings, particularly to those who have a physical handicap. There is a consequent reduction of the general benefit factor A, although in this case the reduction is difficult to quantify. This factor usually manifests itself by the doors being propped open much of the time, thus nullifying much of the reduction in f_d. Again, this can be overcome by having such doors held open and closed only following automatic detection of fire. This substantially increases the cost f_c.

1.3 Major fire hazard areas

Fires causing loss and damage can occur wherever human activity occurs. Perhaps the most frequent location for fires is within buildings. These include both domestic and nondomestic premises, and the latter can extend to a wide range of occupancy, such as factories of various kinds, buildings where there are special risks to the public, including places of public assembly and places where people sleep, such as hotels and hospitals. Industrial occupancies extend beyond buildings to include mines, process plant housed in the open, offshore installations, agricultural crops, and forestry. Finally, there is a whole range of facilities for road, rail, marine, and air transport even extending in recent times to satellites and space modules. For most of these hazard areas, a considerable and costly fire occurrence background has built up over the years and has given rise to extensive requirements for fire safety. In the world of fire insurance, specific hazard areas are often called "risks."

1.4 The total cost of fires

The total cost of a fire to a community may be represented by the sum $(f_d + f_c)$ for all the fire risks in the community; this would include all buildings, plant, processes, means of transport, and so on. Many items contribute to the sum. With regard to f_d, the detriment produced by fires, we have, of course, the direct toll of life and injury and the actual financial losses caused by fire. There are indirect or consequential effects due to disruption of facilities, loss of trade, and employment. There is also public concern and anxiety, particularly following major disasters and the cost of any inconvenience caused. The cost of fire safety measures f_c includes costs aimed at preventing fires, controlling them when they occur, and mitigating their direct and indirect effects. They include the cost of services such as the fire brigade, fire insurance, and a substantial part of building control or other regulating procedures.

Information on the direct financial loss due to fires has been available in the United Kingdom since World War II. However, it was realized in the 1950s that this direct financial loss was only the tip of the iceberg since it is necessary to be concerned with the total cost of fire. An early exercise to deal with this matter was made by Fry (1964). He found that the direct fire loss in the United Kingdom when corrected for rising prices had remained relatively constant until 1957, although there were indications of an increase after that date. During the whole of the period covered, the direct financial fire loss represented about 0.2% of the gross national product. However, when some other costs of fire relevant to f_c were included, particularly incremental building costs and the costs of fire services and insurance, the total cost of fire to the nation was found to approach about 1% of the gross national product.

In an analysis for 1976, Rasbash (1978) added estimates of costs of indirect loss, fatalities, injuries, and inconvenience to f_d and of fire prevention to f_c. This increased the total value of $(f_d + f_c)$ relative to the gross national product by 50%. The fire precaution costs were about twice as great as the cost of losses and hurt. This points to the necessity of being sufficiently discerning in fire safety design to ensure that the increase in the cost f_c brings about a comparable reduction of the expected detriment f_d. The estimated detriment in the Rasbash analysis did not include the cost of public anxiety, which is a major factor following the occurrence of fire and explosion disasters.

Since about 1980, Wilmot has collected data that provide a continuous overview of costs of fire precautions and fire detriment for a number of countries. These are summarized in Section 6.7.4.

1.5 Prescriptive and functional approach to fire safety

In the past, and indeed for the most part in the present as well, the provision of fire safety has been through enactments that have been prescriptive. This may be regarded as the traditional approach to fire safety. More recently, as test methods for performance of items of fire defense have become available, the entirely prescriptive approach to fire safety has become modified, in requiring that items of fire defense fulfill a performance standard. Moreover, there has been a move in recent years from prescriptive to functional, that is, what is proposed can be shown to bring about sufficient safety from fire. This recognizes the multifaceted approach to fire safety and the demand for obtaining cost-effective fire safety. To achieve this it is necessary to specify not only the objective of the fire safety activity but also the degree of fire safety aimed for. There is a tendency for official legislation, at least in the United Kingdom, to be somewhat open ended in this matter. Thus, the Health and Safety legislation generally aims for the level of hazard to be "as low as reasonably practicable" (ALARP) while recognizing risk levels that are either negligible or intolerable. "Not reasonably practicable" may be defined as incurring costs in bringing about a reduction in risks that are seriously out of proportion to the benefits achieved by the reduction in risk (Royal Society, 1983). The relative value of f_c to f_d referred to in the previous section, would indicate that, at least for the United Kingdom as a whole, the level of fire safety reaches this standard. Building Regulations (England and Wales) now aim for some requirements to be for "appropriate levels of safety." Nevertheless, insofar as the requirements are functional rather than prescriptive, the detailed way in which these aims are accomplished is left to the designer.

The difference between the prescriptive and the functional approach is that in the latter it is necessary to quantify the elements of fire safety, particularly how much "fire" can be expected, how much "safety" is being installed, and at how much cost. This helps ensure that money is spent on safety where it is most needed and the least costly regime of precautions capable of providing sufficient safety is put in place. It also helps to give flexibility to designers and to demonstrate that solutions to fire safety for a given risk are equitable and fair. This aspect will assume increasing importance as harmonization is sought on fire safety design between countries with different traditional approaches to fire safety. It has been the practice in the past to follow fire and explosion disasters by lurches of requirements for fire defense. A quantitative fire safety design procedure for complex plant and building hazards would be a major step in avoiding disasters in the first place. Currently, there is a move toward the functional approach to fire safety in buildings by defining the constituent elements to be expected of fire control and fire safety needs in buildings of a given hazard type and setting up performance standards for each of these elements (Bukowski and Tanaka, 1991). It is visualized that these performance standards would not require special expertise for supervision by a control authority.

There is a tendency, particularly in the reports of public inquiries following disasters, for a detailed range of prescriptive measures to be laid down to ensure the disaster "never happens again." Much of this tends to become embodied in prescriptive requirements. However, this need not necessarily be the case. An example of a recommended scheme following a disaster, where the object was to give flexibility of design and management, is given in the Keane report into the inquiry into the Stardust disaster in Ireland (1982). This report indicated the way the hazard in public assembly buildings might be assessed and appropriate fire safety introduced to fit the hazard.

1.6 Purpose and outline of this book

The last few decades have seen the development of methodologies that will allow a designer to accomplish the change from a prescriptive approach to a functional approach to fire safety. It is the purpose of this book to provide a description of these methodologies. Part I deals with the structure of the fire problem and, in addition to this introductory chapter, contains in Chapter 2 a description of the fire safety system. This will outline the constituent and interdependent components of the system, particularly precautions for prevention, protection, and accommodation, concepts of fire safety design and management and the place of quantitative objectives in dealing with fire safety. The major input into fire safety are the lessons of disasters, lessons we continue to have to learn. Chapter 3 gives summaries of some recent fire and explosion disasters that have been studied in detail and those lessons that are currently being absorbed into fire safety requirements. A range of prescriptive requirements for fire safety has been inherited from the past and will be outlined in Chapter 4. An appreciation of these is an important part of the functional approach to fire safety since usually the levels of safety they represent form a basis against which functional approaches to fire safety can be judged. Part II will be devoted to the data that are available for a quantitative functional approach to fire safety. Although Chapter 5 will outline recent physical experimental data, particularly on fire behavior and control, Part II will deal mainly with data from statistical sources on various aspects of fire safety. Part III – Methods of measuring fire safety – will describe the methods currently being developed to pursue the functional approach, particularly methods to quantify fire safety and measure it against objectives. This will feature deterministic, probabilistic, and stochastic methods as well as the use of logic diagrams in fire safety evaluation. The book does not discuss economic aspects. Topics such as cost-benefit analysis, consequential losses, value of human life, decision analysis, and application of Utility Theory, all in relation to fire, are discussed elsewhere (Ramachandran, 1998).

1.7 Definitions

It is desirable to set down the meaning of a number of terms that will be used frequently in this book.

First the word "fire." Fires occur because sources of ignition come into contact with or develop within combustible materials. Most fires, of course, are wanted fires, since they are the most widespread way of making energy available for general use. As far as the context of this book is concerned, fires are mainly of interest where they extend beyond the point of origin to cause hurt, damage, expense, or nuisance. This would exclude wanted fires, unless they fall into the above category, and indeed those unwanted fires that do not extend beyond the point of origin to cause detriment in the above way. But the term is wider than those "fires that result in a call to the fire brigade," which is often taken as a definition of the term "fire."

The word "risk" has been defined as the potential for realization of unwanted negative consequences of an event or process (Rowe, 1977) or the chance of injury or harm (Cassell, 1974). Following this, "fire risk" may be stated as being the chance for injury or harm associated with the occurrence of fire, as defined above. It will be a major concern of this book to quantify the "chance" or "potential for realization" of the risk by characterizing the expected frequency of its occurrence against the severity of the consequences. The words "risk" and "hazard" are interchangeable in general usage. However, in recent years it has become accepted in the professional engineering world that the word "hazard" should cover descriptive definition of the dangerous situation and that the word "risk," a quantification or estimation of the hazard. Thus the nomenclature of the Institution of Chemical Engineers (Jones, 1992) defines "hazard" as

> "a physical situation with a potential for human injury, damage to property, damage to the environment or some combination of these."

"Risk" is defined as

> "the likelihood of a specified undesired event occurring within a specified period or in specified circumstances. Risk may be either a frequency (the number of specified events occurring in unit time) or a probability (the probability of a specified event following a prior event) depending on the circumstances."

More briefly, a glossary of terms associated with fire (British Standard 4422, 1984) defines fire risk as the probability of a fire occurring and fire hazard as the consequences of the event if fire occurs. It will be noted that there is a lack of coincidence between these two pairs of definitions. The latter pair also masks the fact that a fire, if it occurs, can have a whole gamut of possible effects ranging from a call to the fire brigade without damage to the destruction of a city. In this book, the assessment and quantification of fire risk will usually be visualized as the product of the frequency (F) with which fire occurs with each product of the probabilities (p_i) relevant to specific harmful effects (Ha_i) that may follow.

$$\text{Fire risk} = F(p_1 Ha_1 \ldots p_i Ha_i \ldots p_n Ha_n) \quad [1.1]$$

Equation [1.1] embraces both the above pairs of definitions for n harmful effects under consideration. It may not be possible to sum these harmful effects directly for two reasons. Firstly, they may not be expressible in similar terms, for example, number of deaths, direct loss due to damage, and public anxiety. Secondly, the specified harmful effects may overlap, for example, the chances that area damaged may exceed 100 m^2 and 1000 m^2. Where the harmful effect is readily expressible as a mean value, particularly financial loss or areas damaged, then the fire risk can also be expressed as the product of frequency and the mean effect.

The above differentiation between hazard and risk will generally be followed in this book, but it will not be followed slavishly since, in the fire safety world, particularly the insurance world, there is an inherited tendency to use the words "risk" and "hazard" interchangeably and to use the word "risk" for a specific hazard area. The term "risk agent" is the name given to entities, particularly people, exposed to the risk.

The term "major hazard" has come into use to describe an activity, process, or a situation in which the consequences of an incident may be disastrous or catastrophic. The likelihood of such a disaster may be very small, although the public perception of the risk may be influenced by the catastrophic consequence. It is possibly as a counter to this that the professional engineering world has sought to discourage the use of the word "risk," particularly in this situation, except as a quantitative statement of likelihood.

"Safety" is regarded in this book as the inverse, the complement or the antithesis of risk, that is, the lack of potential for unwanted negative consequences of an event, process, or activity. Assuming that air exists everywhere or cannot be rigorously excluded, there is a fire hazard and consequent risk wherever combustible material is present. There are thus very few situations indeed in which one can say that there is a complete absence of fire risk and that fire safety is complete. The quantification of safety will be approached through the quantification of fire risk associated with processes and activities. These may be said to be "fire safe" when a sufficiently low fire risk is associated with them. It should be noted that in this sense the word "safe" covers both a description of the harmful effects arising from the hazard and a quantification of freedom from these effects. For a given harmful effect Ha_1, and assuming that F is substantially less than one per year, which is generally the case for frequency of fires in buildings attended by the fire brigade (Chapter 7), the safety for this harmful effect may be expressed as

$$\text{Safety}(Ha_1) = 1 - Fp_1 \quad [1.2]$$

This is the probability in a year that the harmful effect by fire will not occur.

An alternative definition of safety is

$$\text{Safety}(Ha_1) = 1/Fp_1 \quad [1.3]$$

This is the expected time interval between fires that brings about the harmful effect.

In the fire safety world, one frequently comes across the terms "fire prevention," "fire protection," "fire safety design," and "fire safety management." There is as yet no general consensus on the meaning of these terms, particularly the first two of them. Thus, the term "fire protection" is often implied to cover all of the above terms. This is apparent in the activities of many organizations in this field known as Fire Protection Associations or Organizations. The term "fire prevention" is often used by Fire Services to cover all aspects of fire safety other than direct fire-fighting actions carried out by themselves. The British Standard Glossary of fire terms (British Standard 4422, 1984, Part 1) defines fire prevention as

> "measures to prevent the outbreak of a fire and/or to limit its effects"

and fire protection as

> "design features, systems, equipments, buildings or other structures to reduce dangers to persons and property by detecting, extinguishing or containing fires."

It will be noted that the second part of the definition of fire prevention overlaps heavily with the definition of fire protection. The IChemE nomenclature (loc.cit) defines fire prevention as

> "measures taken to prevent outbreaks of fire at a given location."

and fire protection as

> "design features, systems or equipment which are intended to reduce the damage from a fire at a given location."

Specific meanings for these terms as used in this book, which are in line with the IChemE nomenclature, will emerge in Chapter 2, which will introduce the concept of fire safety as a system. The term "fire safety" itself is comparatively recent. It is used to cover all aspects of safety from fire. It is finding increasingly widespread use in this sense, although it is sometimes limited to safety of life only.

"Fire Safety Engineering" is a relatively new term used to describe the discipline concerned with the design and management of Fire Safety for situations in which hazards exist. Traditionally, the terms "Fire Protection Engineering" in the United States and "Fire Engineering" in the United Kingdom have been used. The term "Fire Safety Engineering" was adopted by the author, who found after inquiries that it was less confusing to lay people than "Fire Engineering."

Symbols

A	A notional measure of benefit associated with risk situations
f_c	A fire safety programme that inflicts a cost
f_d	A fire occurrence scenario that inflicts a detriment
h_c	Safety programme other than fire, which inflict costs
h_d	Safety programme other than fire, which inflict detriments
F	Frequency with which fire occurs
p_1, p_i, p_n	Probabilities of specific harmful effects associated with fire
Ha_1, Ha_i, Ha_n	Harmful effects associated with fire

References

British Standards Inst (1984). BS 4422 Glossary of Terms Associated with Fire, Part I.
Bukowski, R W and Tanaka, T (1991). Towards the goal of a performance fire code. *Fire and Materials*, **15**, 175.

Cassell (1974). *Cassell's English Dictionary*, 4th Edition, Cassell, London.

Fry, J F (1964). The cost of fire. *Fire International*, **1**, Part 5, 36–45.

Jones, D (1992). Nomenclature for Hazard and Risk Assessment in the Process Industry, Institution of Chemical Engineers, Rugby.

Keane, Mr Justice, R (1982). *Report on the Tribunal of Inquiry on the Fire at Stardust, Artane, Dublin on the 14th February 1981*. The Stationery Office, Dublin.

Ramachandran, G (1998). *The Economics of Fire Protection*, E. & F. N. Spon, London.

Rasbash, D J (1974). *The fire problem. Symposium on Cost Effectiveness and Fire Protection*, Building Research Establishment.

Rasbash, D J (1978). The place of fire safety in the community. *Fire Engineers Journal*, **38**(111), 7.

Royal Society (1983). *Risk Assessment*, A study group report, p. 23, 161.

Rowe, W D (1977). *An Anatomy of Risk*, Wiley, New York.

2 THE FIRE SAFETY SYSTEM

2.1 Basic questions of fire safety

The efficient design of fire safety for a unit such as a building or a plant where there is a fire hazard depends on obtaining answers to three questions. First, how much is the fire hazard? The answer can generally be divided as follows:

1. The likelihood that a fire with unwanted effects will occur.
2. Given that such fires do occur, the ways in which they can develop and be controlled.
3. The potential for the harmful effects produced by these fires to cause detriment, particularly hurt to people and damage to property and processes.

The second question that arises is whether the level of fire safety from the fire hazard so evaluated is acceptable. The acceptability of the hazard will depend on how safe is "safe enough." If the safety level is not high enough, the third question is what further different safety measures need to be introduced. The acceptability of the measures will depend on their cost, the latter including both the financial cost and any deleterious effect they would have on the function of the unit concerned.

The fire safety system considers these questions as an integrated whole. Here, we follow Beard's (1986) definition of a system as an entity, conceptual or physical, which consists of interdependent parts. In the present context, the system is the concept of the fire safety of the whole unit concerned. The model of the fire safety system that will be put forward in this chapter follows the line of the above questions and is based primarily on a series of suggested steps in the evaluation and design of fire safety for specific hazard areas (Rasbash, 1977, 1980). There will also be reference to other systemic approaches to fire safety, particularly the General Services Agency (GSA) and the National Fire Protection Association (NFPA) systems approaches, and the risk management approaches in insurance and industry. However, the dimension of a fire safety system can extend beyond single-hazard areas to cover collections of buildings, ships, aircraft, plants, and so on and indeed whole cities and communities. In large measure, these will tend to be a summation of the fire safety systems of the individual units manifesting the fire hazard.

Evaluation of Fire Safety D. Rasbash, G. Ramachandran, B. Kandola, J. Watts and M. Law
 ISBN: 0-471-49382-1

2.2 Fire safety objectives

Many organizations are concerned with the assessment, preparation, and dissemination of requirements for fire safety. Their contribution will be summarized in some detail in Chapter 4. The activity of such organizations represents the traditional and generally accepted way of achieving fire safety. In their deliberations, the procedures they follow must necessarily be influenced at least by the fundamental questions posed in Section 2.1.

In the activities of the above organizations, fire safety objectives are, in general, stated in one of three ways:

(a) To protect life

(b) To protect property

(c) To ensure that a disaster, which has caused the fire safety activity, "must never happen again."

The first two of the above objectives are open to quantitative definition, although rarely explicitly stated. The third objective reflects the impact of major disasters, particularly those with multiple fatality, fire, and explosion, on fire safety requirements and legislation. In absolute terms, it is unachievable in that even if all the precautions recommended for fire safety are put in hand and managed faithfully, there would still remain a remote chance of a fire disaster of similar dimensions happening. This especially follows when it is recognized that all safety measures are subject to human and/or mechanical failure. However, the third objective does imply a major additional objective in its own right, in that the disaster has caused such shock, concern, and anxiety to the public to bring about a demand for this objective to be pursued. Society as a whole has become involved in such a major way that the third objective may be regarded as a manifestation of societal risk or concern.

There is, in addition, another major objective of fire safety that individuals need to bring into focus, and that is maintenance of function. Whatever harm the fire may do, it is necessary for people to carry on and awareness is needed of the circumstances in which the occurrence of a fire would make this very difficult. Jeopardy of the functioning of an organization may arise particularly as a result of the destruction of certain specific key assets in a fire. The objectives of fire safety may therefore be extended to five areas as indicated in Table 2.1.

Most fire deaths and hurt occur near the point of fire origin and result from fires in items such as clothing, furniture, and heating equipment. Objective 1 is therefore, for the most part, the province of consumer legislation and public education on fire safety matters. Objective 2 is usually the prime objective of requirements for public buildings and certain industrial processes. The requirements have been framed particularly in response to disasters that have highlighted major hazards to life, or to the anticipation of such disasters. Objective 3 is usually the province of the management of the enterprise concerned. Objective 4 is deeply embedded in fire safety legislation, although rarely stated overtly at the present time. It can occur where a fire on one person's property involves that of another and it is of particular importance where a fire can grow to involve a whole city or part of a city. This was a relatively frequent occurrence until a hundred or so years ago, but because of fire safety requirements built into city design, it is infrequent in

Table 2.1. Major objectives of fire safety

1. Life safety of individuals
2. Life safety where there is a major societal concern
3. Loss prevention of individual premises and assets
4. Loss prevention of premises and assets where there is a major societal concern
5. Maintenance of function

Table 2.2. Life safety objectives for fires in buildings

1. Protect life (and limb) of individual users or occupants from fires (and explosions) that result from activities for which they (or their immediate family) are responsible.
2. Protect life of individual users of the building from fire that results from activities of: (1) owner or manager of the premises, or provider of services to premises and (2) other users.
3. Protect life of building users from fire that arises from activities of people outside the building.
4. Protect nonusers of building from fire that occurs within the building.
5. Protect life of people called to deal with emergencies, especially firefighters.

Western society today. Objective 5 is normally covered by insurance, which allows for financial cover of assets that could be destroyed, although industrial and commercial organizations may need to take special steps to take into account this aspect of fire safety.

Carelessness on the part of one party causing hurt to another is also a factor in 1 and 2 above. In multiple fatality disasters, many of those killed and hurt are likely to be completely innocent parties. However, fires that involve one or two deaths rarely become matters of major concern, unless a number of incidents of a similar type come before the public. Nevertheless, according to the degree of responsibility of those who may suffer the hurt, different levels of fire safety might be called for. With this in mind, it has been suggested that Objective 1 in Table 2.1 could be extended for fire hazards in buildings, as indicated in Table 2.2 (Rasbash, 1980).

The majority of fire deaths in buildings are in the ambit of Item 1 in Table 2.2. Typically, the fire and the exposed are in the same dwelling or even in the same room. Ignition of clothing due to carelessness, of beds or armchairs due to smoking and misuse of heating and electrical appliances are major causes in this item. In Items 2 and 3, smoke and toxic gas from a fire often move to surround individuals concerned and hinder their escape. Explosions also cause collapses that kill people away from the explosion source. In Item 4, fire spreads into a building from an outside ignition source or another burning building. An explosion, as a consequence of a leak into the building from outside, also comes into this category. Item 5 covers the spread of fire from the building to other buildings or to collapse of buildings onto people outside due to fire or explosion. With regard to Item 6, firemen can protect themselves against normal smoke hazards but are endangered by sudden increases in the flame size, by collapse of the building without warning or by the release of exceptionally toxic fumes.

It is important when attempting to obtain a rational approach to fire safety not only to recognize the relevant objectives that are being pursued but also to give them quantitative definition. It is reasonable to do this in financial terms for Objectives 3 and 4 in Table 2.1, and in so doing, to pursue a further possible objective of optimizing total fire safety costs to obtain a minimum value of $(f_d + f_c)$ (Section 1.2 and 2.10). This is more difficult for Objectives 1 and 2. Having defined the extent and frequency of hurt that one might be prepared to tolerate, it is possible to approach the optimization of fire safety procedures by looking for a minimum cost and inconvenience for measures specified to bring about the desired degree of safety.

2.3 Steps in fire safety design

Given a specific fire hazard area, for example a specified dwelling, factory, ship, or railway tunnel, the process of designing fire safety has been broken down into a number of steps (Rasbash, 1977, 1980), which, with some minor modifications, have been reproduced in Table 2.3. The steps logically follow each other in the sequence indicated. These steps may all be regarded as

Table 2.3. Steps in the evaluation of fire safety

1.	Define the fire hazard area.
1a.	Identify people, property, and processes at risk from fire and explosion incidents within the fire hazard area.
2.	Define the fire safety objectives.
3.	Assess materials that can burn.
4.	Assess sources of ignition.
5.	Assess the conditions of fire spread that would lead to an established fire.
6.	Assess agents that cause fire (i.e. that bring 3, 4, 5 together).
7.	Estimate the probability of fires being caused.
8.	Assess the means available of limiting fire, (1) active means (2) passive means.
9.	Estimate the courses of fire behavior.
10.	Assess the harmful agents produced by fires and their capacity to harm people and property
11.	Estimate the production and range of action of harmful agents produced by fires.
12.	Assess methods of protection against the harmful agents.
13.	Estimate the direct detriment to people and property that may be caused by fires.
14.	Assess available methods of protecting people and processes from the indirect effects caused by direct detriment.
15.	Estimate indirect detriment.
16.	Judge whether estimated direct and indirect detriment comply with fire safety objectives. If Step 16 shows that the objectives of fire safety are not met, then carry out the following steps.
17.	Postulate changes in the fire safety situation, for example in the precautions taken.
18.	Estimate the effect of changes on achievement of fire safety objectives.
19.	Define an acceptable method of achieving objectives, taking into account cost and convenience.
20.	Formulate and express fire safety requirements.

component parts of a fire safety system for the hazard area concerned. They interconnect in the manner shown in Figure 2.1, which may be regarded as a diagram of the fire safety system. It will be seen that the steps concerned are mainly squares representing data acquisition steps or circles representing data processing steps. Except for Step 1, each data acquisition step feeds into at least one data processing step.

Steps 1, 1a, and 2 in Table 2.3 are introductory steps and provide basic information concerning the risk. It is necessary to first define the type of hazard area and the occupancy as described in Section 1.3. This action will give access to relevant legislation literature and fire codes based on previous experience with the type of hazard area concerned (Chapter 4) and to comparative information available in many statistical compilations on fire safety (Chapter 6). Guided by such information and for a given specific hazard area, this step leads to Step 1a, which identifies who and what may be hurt by fire or explosion within the area. This includes the numbers, the nature, and likely location of people both inside and outside the specific hazard area that could be exposed to the effects of an incident within the area and the material items such as stock, equipment, plant and buildings that could be put at risk. In recent years, there has also been concern on the way a fire can damage the environment, particularly by pollution of air caused by smoke from the fire and ground contamination by toxic materials in run-off fire-fighting water. Beyond this, there are processes necessary for maintenance of functions that may be affected if people are hurt or items are damaged or destroyed. These processes include manufacture, servicing, and business processes associated with the enterprise as a whole. Steps 1 and 1a are therefore essential information gathering steps of the sort of fire and explosion experience that may be expected and what may be endangered by such experience.

The definition of objectives as required by Step 2 would, if put forward in quantitative terms, differentiate the systemic approach to fire safety from the traditional, empirical approach of

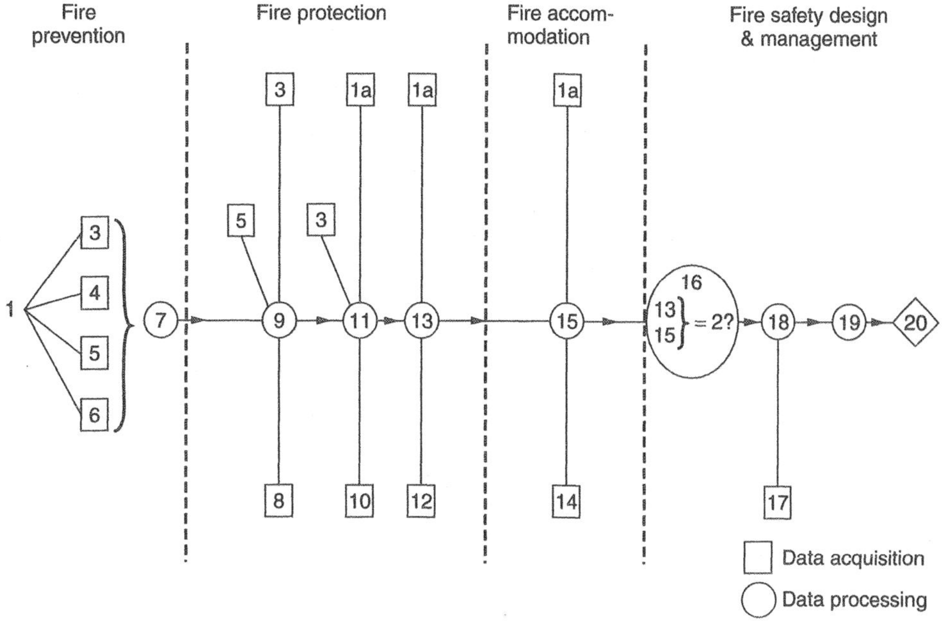

Figure 2.1. Steps in the evaluation of fire safety (see Table 2.3 for description of the steps)

regulatory authorities. The objective could also be an optimum financial balance of cost of precautions and residual risk or a minimum cost necessary to achieve a safety level that may not be expressed in financial terms. The steps from 3 to 16 seek to quantify the hazard to allow comparison with objectives. Data in the acquisition steps (Figure 2.1) are obtained from a detailed study of the specific hazard area concerned. In general, people and property are involved in direct detriment caused by fire and it is estimated in detail by the time Step 13 is reached. Insofar as air or water pollution may cause environmental damage, this should be included in direct detriment. Processes become involved in indirect detriment, in consequential loss and interruption losses, and people who need to recover from direct detriment. These should be accounted for by the time Step 15 is reached. The societal concern associated with Objectives 2 and 3 in Table 2.2 may also be regarded as coming into this category, the process disturbed being the smooth running of society as a whole. If Step 16 shows that the fire safety of the hazard area does not meet the objectives, then it is necessary to carry out the fire safety design process implied in Steps 17 to 19 to ensure that it does.

Finally Step 20, "formulate and express requirements," may be regarded as a fire safety management step and it is an integral part of the fire safety management process, which requires that fire safety measures be applied in practice and kept under constant review (Section 4.8.5).

The fire safety system as illustrated in Figure 2.1 has the potential of becoming highly complex. It will tend to be characterized over a period of time as the relevant data are acquired. However, it is possible to regard the clusters of factors as leading to separate data processing points in Figure 2.1, that is, to Steps 7, 9, 11, 13, 16, 18, and 19 as subsystems. One could also regard certain continuous groups of processing points as, for example, 7 + 9 or 9, 11 + 13 as enlarged subsystems. These subsystems, which are listed in Table 2.4, may be associated with specific limited fire safety objectives. Names are suggested for these subsystems, some of which have been featured in Figure 2.1. It is necessary to feed information into such subsystems appropriate to the data input leading to it. Thus, information on the probability and location of the occurrence

Table 2.4. Fire safety subsystems

Subsystem designation	Data processing steps (see Figure 2.1)	Area of application	Processed data input needed or assumed	Suggested name for subsystem
(i)	7	Occurrence of fire, fire prevention methods	–	Fire occurrence or fire prevention
(ii)	9	Fire growth, fire size, fire extinction	Fire occurrence	Fire development or fire control
(iii)	7 + 9	Total amount of fire	–	Amount of fire
(iv)	11	Amount of harmful effects	Amount of fire	Harmful effects
(v)	13	Direct detriment. Safety from harmful effects	Amount of harmful effects	Direct detriment
(vi)	11 + 13	Direct detriment. Safety from harmful effects	Amount of fire	Main safety
(vii)	9 + 11 + 13	Direct impact of fire. Fire protection methods	Fire occurrence	Fire impact or fire protection
(viii)	7 + 9 + 11 + 13	Total direct cost of fire	–	Total direct cost
(ix)	16	Consequential detriment	Direct detriment	Consequential detriment or fire accommodation
(x)	7 + 9 + 11 + 13 + 16	Total cost of fire	–	Total cost of fire
(xi)	18 + 19	Designing acceptable fire safety	Present situation + change + objectives	Fire safety design

of established fire associated with Step 7 would need to be fed into a subsystem based on $9 + 11 + 13$ in which the objective is to estimate direct damage to assets or hurt to people. It is usually in dealing with one of these subsystems or even part of a subsystem that much of the quantitative approach to fire safety has been pursued up to the present time. Section 2.5 will consider these subsystems in some detail. However, before this, it is necessary to outline sources of fire safety data available for the system.

2.4 Sources of fire safety data

The main source of information on fire safety is accounts of fire and particularly fire and explosion disasters that have occurred in the past. Over the course of history, the lessons that have been learnt from past fire disasters have been assimilated in requirements and legislation and into accepted fire safety design. Thus, extensive city fires that are so much a feature of the history of fire are very rare nowadays in Western society because of basic steps in fire safety. These include fire separation of buildings either by party walls or space across streets, noncombustible exteriors to buildings, and organization of fire brigades. Fire disasters in theaters, which occurred particularly toward the end of the nineteenth century, have been countered by the statutory introduction of fire safety measures such as separation of the stage area from the auditorium by the safety curtain and protected means of escape from the auditorium. Also disasters that occurred because of the rapid spread of fire through combustible linings, draperies, and furnishings in a public place, as in the Coconut Grove disaster, have been countered by control of the performance in fire of such items. However, learning from fire disasters is a continuous process and is still a major input into the improvement of fire safety. For this reason, Chapter 3, which gives accounts of some recent fire and explosion disasters, has been introduced as an indication of where we stand currently on this matter. Fire in high-rise buildings, leisure and transport facilities, and industrial processes handling flammable fluids and dangerous substances feature in this summary.

The bulk of recent fire experience in many countries is encapsulated in the form of fire statistics. These provide a major input of data into the fire safety system and into the processes of fire safety design and management. Data based on fire statistics can be fed into specific parts of the system, particularly as lead information into the various subsystems. Statistical information will be considered in detail in Part II of the book.

The third major branch of fire safety information is provided by experimental observation and scientific interpretation of fire processes and methods of countering fire. The present generation has seen a major increase of data in this area, such that it is possible now to describe in quantitative terms major areas of ignition, fire and harmful agent development and control that previously had not been possible. A broad survey of such information as it exists at present will also be given in Part II (Chapter 5). This information also extends to the behavior of people in fire situations.

Throughout Table 2.3, the term *identify and quantify* has been used in association with data acquisition steps. It is of course essential to identify a specific need for data before those data can be quantified. The identification process is aided greatly by experience of past fires together with a detailed examination of the hazard situation. However, the process of quantification still leaves many gaps. In many of the areas where quantification is called for, there is a dearth of objective data. This lack may not only cover intrinsic properties that are measurable but also ways of making use of these properties to predict what needs to be estimated. This is particularly so in the case of the development of fire and the spread of harmful effects. Even where statistical and experimental data are available, doubts can arise as to the relevance of such data to a real hazard. It is inevitable that engineers under such conditions, particularly when working under time constraints, will supplement objective data with subjective data based on their own and other people's judgments. The quality of such data inevitably depends on the experience of the people involved. Indeed, in certain approaches to quantitative fire safety evaluation for recognizable types of hazard, the experience and judgment of a group of people may be deliberately and systematically harnessed to provide necessary quantitative data.

2.5 Subsystems

2.5.1 FIRE OCCURRENCE AND FIRE PREVENTION

Steps 3 to 7, in Subsystem (i), is associated with the prediction of the tendency of fires to start, that is, the expected frequency of fire (F in equation 1.1). The position of this subsystem has been indicated in Figure 2.1. Traditional knowledge of fire ordains that fire will occur when three constituent factors are brought together, namely, combustible materials, heat, and air. Bearing in mind that air is always present unless it is deliberately made absent, it does not need special consideration in the present context. However, two further items are necessary for fire, particularly when it is defined as in Section 1.7. First is the ability of a combustion zone to spread from the point of ignition sufficiently to form what may be called an *established fire* or a fire that can be specifically recognized as following a definition, for example, causing damage or a call to the fire brigade. Second is the agent or agents that bring the conditions for an established fire together. Steps 3 and 4 respectively identify and quantify the materials that can burn and the sources of ignition, and embrace the first two essential components of fire. These are invariably highlighted in methods of risk assessment. Much statistical information has been published, and even more unpublished information exists, tabulating these factors for different occupancies. However, it is desirable in exercises that seek to quantify safety that these steps should go further than the normal pinpointing of materials that can burn and potential sources of ignition. They should include the quantity of heat required to produce fire conditions, the heat that can be produced by the fire itself, that is, the fire load, and indeed, in certain instances, the properties of the reaction itself. The latter is implicitly present in data such as flammability limits, fundamental burning velocity, and conditions for ignition and extinction. The power and the potential for ignition of different sources also need to be classified. Some further detail on these matters is given in Chapter 5. Step 5 covers the ability of the combustion zone to spread from the point of ignition, and is the factor that probably most controls whether there is or is not a fire.

In most hazard areas, extensive amounts of combustible material are present as well as many potential sources of ignition and even many conditions that will allow fire to spread and perhaps even spread rapidly. Yet, fire is a very rare condition, since a causative agent is needed to bring these constituents of fire physically together and induce fire. This is covered by Step 6. The most important of these causative agents are

1. human beings,
2. failure of mechanical and electrical and other forces under human control,
3. natural forces.

They may operate either by introducing an ignition source, for example, smoking materials to where there is flammable material, or vice versa, for example, spillage of flammable liquid near an electrical source, or removing a barrier between ignition source and fuel, for example, fire guard. Human failure includes deliberate, careless, or unintentional introduction of combustible materials to sources of ignition and vice versa. In addition, mistakes made in design, manufacture, and operation of machinery or plant could have the same effect. The natural forces under Step 3 are mainly gravity and wind, although lightning, earthquakes, and tremors are also candidates. Some information on these items is available in certain national statistics. However, in carrying out Step 6, direct experience within a hazard area and knowledge of management attitudes count a great deal.

Information on Steps 3, 4, 5, and 6 is basically what is needed to estimate likelihood of fire occurrence. For most risks, it is difficult to carry this out quantitatively as an exercise in its own right, chiefly because of the widespread use of combustible materials of different kinds, the availability of sources of ignition, and the absence of quantified information on human behavior referred to under Step 6. However, it is possible to make a shortcut to this point by taking figures on the frequency of fire occurrence from statistical information covering a class of similar hazard area. A statistical figure may be adjusted taking into account information in Steps 3 to 6, which suggests a departure, either beneficial or otherwise, from those average conditions for which statistics may be assumed to apply. Experience here of many hazard areas is a useful background in making such a judgment. The precautions taken to prevent fires in the hazard area play a major part in the management of fire safety. They include items such as management, education, and training of staff and other risk agents, housekeeping, design, and maintenance of power and plant equipment, record keeping and follow-up of hazardous occurrences. These may be classified as *fire prevention* measures and the extent to which they are in operation would be relevant to establishing the probability of fires occurring in a given hazard area.

2.5.2 FIRE DEVELOPMENT AND FIRE CONTROL

Subsystem (ii) covers the evaluation of the ability of fires to grow and be controlled. Steps 3 and 5, as well as being input into the fire occurrence subsystem, are major inputs into this system as well in governing what may burn and how rapidly combustible material may become involved. The quantification of rapidity of fire spread is a major objective of fire safety science. Given the existence of combustible materials, some of the common factors that may give rapid fire spread are considered in Chapter 5. The factors particularly depend on the geometry of the combustible materials in relation to the environment and the potential ignition sources. Situations that may give rise to disastrously rapid and extensive fire spread are described in Chapter 5.

Step 8 is a survey of the installed fire protection methods, which may be divided into (1) active methods and (2) passive methods. Active methods include the means of detection, control, and extinction of fire, the availability and effectiveness of the fire brigade, and the extent to which people on the premises have been trained to recognize and cope with fires. Passive methods include the control of the fire-spread conditions that might cause a small fire to become a big one and the means of compartmentation, segregation, and separation against fire within the hazard area. The passive methods are therefore complementary to the fire spread factors included in Step 5. In all situations in which protective requirements play a part, the reliability of the measures taken, and therefore their maintenance, is an essential component of the information needed. With addition of information available in earlier steps, particularly Steps 3 and 5, it is theoretically possible to estimate the courses that fires can take when initiated in various ways, and in various parts of the hazard. This is covered in Step 9. There is a large chance element depending on, for example, the availability and manifestation of various mechanisms of fire spread, the spatial distribution of items to which fires can spread, factors that control burning rate as well as fire spread such as extraneous wind conditions, the time and extent of window shatter, and the probability and effectiveness of functioning of active and passive fire safety measures. A probabilistic distribution of the fire sizes and size/time histories can be more meaningful than an average or maximum fire size, and approaches to this are being developed, which will be described in Part III.

The specific courses of fires through a specified fuel arrangement within the hazard area are often referred to as *fire scenarios* (Chapter 5). These fire scenarios may even postulate the heat output of a fire as a function of time within the hazard area. A statistical approach is also now available for providing the mean and variance of expected growth rates for specific types of hazard areas (Chapter 7).

2.5.3 HARMFUL EFFECTS

Step 1a has served to identify what is at risk. Steps 10 and 11 and Subsystem (iv) define and quantify the amount of the relevant harmful effects that may be associated with the fires defined by Subsystem (ii), particularly specific fire scenarios designated as representing fire development within this subsystem. The major harmful effect to property produced by a fire is heat, but under some conditions, particularly explosions, pressure effects and missiles may become the dominant causes of harm. All these can influence plant or buildings. Where people are concerned, smoke and toxic products also present a dominant hazard. Occasionally, the corrosive nature of the combustion products may cause harm to property. There has also been concern when carbon fiber products are involved in fire that the fibers released may harm electronic equipment (Fiskel and Rosenfield, 1982). Radioactive materials and toxic materials, particularly in industrial plants, although not created by fire, may be dispersed by a fire or explosion. Heat pressure and missiles produced by a fire or explosion may also give rise to the formation or release of other harmful effects, for example, the collapse of a building produces falling masonry, the breaking of a tank may release toxic, corrosive, or flammable materials. This may allow fire to be started in other areas.

For each item or person at risk, and for each kind of harmful effect, there is a critical value of the effect above which harm may be done and a relationship between the value of the effect and the amount of harm done. For heat, smoke, and toxic products, the time of exposure is an important aspect of these damage relationships. Within this subsystem, these critical values are defined and the possible range of distance and time at which they may operate.

2.5.4 DIRECT DETRIMENT

Having defined the potential damaging power of the harmful effects in Subsystem (iv), the actual damage and hurt will depend on the protective methods available to protect the risk agents from these harmful effects. These include methods of escape, smoke control installations, the assistance of emergency services, explosion relief, blast walls, and salvage methods during fire-fighting operations. Fire resistance and distance of separation are major factors in protecting risk agents from heat and in this respect, some of the factors in Step 8(2) are relevant. Dispersion mechanisms for heat, smoke, and toxic products would include buoyancy forces from the fire itself, wind, and imposed air currents. However, these dispersing agencies may also be the means of dispersing dangerous substances such as radioactive and highly toxic materials over a wide area and causing extensive environmental damage. The protective methods may involve the sheltering of the risk agent while in the hazard area. Such data provide the major input into estimating the direct detriment that may be caused by fire covered in Subsystem (v). In practice, it is often difficult to separate the potential of manifestation of harmful effects from the extent to which they may actually bring about damage and it is therefore usually convenient to combine data processing Steps 11 and 13 into one subsystem, namely, Subsystem (vi). This has been called *main safety* subsystem. Given postulated or calculated fire scenarios, this subsystem covers the amount of direct hurt or damage these scenarios can bring about. Subsystem (vii), which also includes the development of fire, has been labeled as the fire protection subsystem since it includes consideration of all direct fire protection methods as distinct from fire prevention methods. Its position is indicated in Figure 2.1. Within this subsystem, values of $p_i\ Ha_i$ (equation 1.1) are estimated insofar as they cover direct detriment.

As far as the occupancy type is concerned, the estimation of direct detriment at this stage would allow a check across to available statistics. Information on damage to people and property by fire, particularly people, is well documented in routine fire reports. If the intention of the exercise is to proceed only with further steps in the analysis, one could feed in details of expectation at this point obtained basically from available statistics but modified according to the information received from previous steps.

2.5.5 CONSEQUENTIAL EFFECTS AND FIRE ACCOMMODATION

Subsystem (ix) deals with the consequential effects that may stem from direct detriment (see also Figure 2.1). The process of dealing with these effects has been called *accommodation* of fire as distinct from prevention and protection against fire described earlier. However, the word does not meet universal approval because of an implication of tolerance and "contingency planning" has been suggested. In industry and commerce, plant, equipment, stocks or data vulnerable to fire may be essential components of processes or operations. Damage to a single small item may affect the condition of a whole process and may even go beyond the hazard area under consideration. Such losses cover indirect business interruption or consequential losses. There is very limited statistical information available on this aspect of risk. It varies in detail from one hazard area to another, and can in general be ascertained only by direct observation and inquiry. As far as people are concerned, death and injury will generally give rise to compensation demands and hospitalization services. There could, in addition, be trauma that may be long lasting, as well as societal concern referred to earlier.

A great deal of detriment and hurt following direct loss and hurt is usually covered by insurance, which is thus a planned accommodation to the loss. Moreover, in addition to the protection methods outlined in the earlier steps to protect risk agents from the direct effects of fire, specific facilities may be available to protect processes threatened by the consequential effects. These may take the shape of duplication of sensitive items, the dispersion of facilities, or of contingency plans for dealing with an emergency (Woolhead, 1976). Taking account of these should allow the expectation of direct detriment estimated in Step 13 to be extended to cover the extra risk agents concerned with indirect detriment giving the total expectation of detriment. An assessment of total expectation of loss taking into account all data processing Steps from 7 to 16 is contained in Subsystem (x) that is entitled *Total Cost of Fire*. The objective of this subsystem is to obtain a total cost of fire risk as covered in equation [1.1] for both direct and indirect harmful effect.

2.5.6 FIRE SAFETY DESIGN AND MANAGEMENT

If the safety estimated for risk agents does not come up to the required standard, changes that could improve matters might be postulated. This may be regarded as the fire safety design process and is covered in Subsystem (xi) (see also Figure 2.1). This would address changes to particular factors that occur in earlier steps and cover fire prevention and protection methods, fire fighting, and contingency plans. A possible change might be to move certain items away from the hazard area, which would modify Step 1a. Alternatively, it might be decided to modify objectives in Step 2. The costs and effectiveness of the changes considered would be major inputs into this subsystem and its objective would be the definition of an acceptable fire safety design.

The steps in Table 2.3 were put forward originally as a method of assessing the safety for a hazard area that was already in existence. For a postulated new building or facility with its accompanying fire hazards, it is very desirable that fire safety design is incorporated early in the design process and certainly well before the building or facility is completed or even seeking legislative approval. The earlier steps, particularly 3 to 13, would cover this initial design process and would tend to be assimilated with later steps 17 to 19. However, these later steps would accommodate later changes called for by interested bodies or a changing environment.

2.6 Contribution of fire safety engineering

The processing of data for the steps in Table 2.3 relevant to a real situation requires a substantial understanding of fire safety engineering. Indeed, the methodology of obtaining the content of these steps for hazard areas of different kinds may be taken as encompassing the bulk of the subject. Many of the steps imply specialization in their own right and even clusters of specialization. If the objectives of the system of precautions are stated in probabilistic terms, it is necessary to feed in data that contain probabilistic expressions of the phenomena concerned. This might occur in any of the data steps.

Malhotra (1991) has put forward a list of fire safety measures that need to be considered by a fire safety engineer when designing fire safety for a building.

1. Fire prevention
2. Fire detection/alarm

3. Fire growth/control
4. Means of escape
5. Smoke control
6. Structural stability
7. Fire-spread control
8. Fire extinction
9. Fire fighting
10. Fire safety management.

Fire prevention is the broad objective of Subsystem (i) and fire safety management follows the fire safety design process in Subsystem (xi) in formulating the requirements of the design process and monitoring and auditing their application (Chapter 4). There are other measures that require quantification in the growth of fire or methods of protection against the harmful effects of fire (Steps 8 and 12 of Table 2.3). Thus, measures 2, 3, 6, 7, 8, and 9 above would generally be considered in Step 8 and measures 2, 4, and 6 in Step 12. Fire detection and alarm plays a major part in both Step 8 and Step 12: in the former by setting in train active measures of fire defense and in the latter by warning people of danger and expediting their escape. Moreover, insofar as fire occurrence is recognized by Step 5 as going beyond an incipient fire to one that is in line with some form of definition of fire (Section 1.7), then fire detection and alarm may also be regarded as contributing to the fire prevention subsystem. Insofar as the specified measures can influence stages in the fire safety system, as estimated by Steps 9, 11, and 13 in Table 2.3, they may be regarded as subsystems of the subsystem involving these stages.

2.7 Approaches to quantitative evaluation of fire safety

It will be apparent from the above comments on the fire safety system that even within a part of the system a wide range of factors may be present, which can influence the objectives of this system or even the partial objectives of a subsystem. This is illustrated in a study of fire safety effectiveness statements (Watts *et al.* 1979) that was addressed particularly to life safety in buildings. Watts lists 66 variables that affect life safety. Of these, 10 described the occupants (Data Step 1), 17 the features of the building (Data Steps 5, 8) 11 the means of egress (Data Step 12), 12 the means of detection, alarm, and extinguishment (Data Step 8), 9 the means of smoke control (Data Step 12), and 6 the properties of the potential fuel (Data Steps 3, 5, 8). In Watt's approach, one can thus recognize an integration of factors occurring within the broader range of disciplines given by Malhotra above. All but four of the 66 variables could be regarded as occurring in the fire protection subsystem, the four exceptions being in the fire occurrence subsystem.

It is necessary for any quantitative approach to the evaluation of fire safety to not only recognize the relevant factors but also quantify and order them in such a manner as to allow their contributions to fire harm and fire safety to be assessed. In general, there are two quite different ways of doing this, which may be respectively termed as *point schemes* and *mathematical models*.

With point schemes following the identification of the relevant factors, a methodology is developed for assessing their importance in achieving or hindering the stated objective, particularly safety of life or property or its converse risk to life or property. The methodology usually involves the systematic harnessing of knowledge and experience of a group of relevant experts. The main object is to develop a system of points according to recognizable levels of the variables involved, which could be processed in a simple manner to give the necessary level of safety or risk. In its application, no detailed knowledge of the way, and in which part of the system the factors contribute to fire safety, is assumed. It is however necessary to calibrate point schemes against an acceptable standard, usually buildings or processes that are regarded as sufficiently safe. Point schemes are also referred to as risk or safety rating schemes, index systems, and numerical grading. With mathematical models, the processes contributing to the safety objectives are directly modeled, particularly through the involvement of quantitative data in one or more of the data processing steps listed in Figure 2.1.

Mathematical models are basically of three kinds; deterministic, probabilistic, and stochastic (Kanury, 1987). However, there is a great deal of overlap between these different types of models. Deterministic models rest on the assumption that the behavior of the factor involved known quantitative relationships with time and space. Elements of fire safety relating particularly to the spread and growth of fire, the formation and movement of harmful agents, and the movement of people have been modeled in this way. Answers to objectives are provided in the form of a "yea" or "nay" because of the assumed certainty of knowledge of the processes. However, the data input into individual factors of a deterministic model can be cast into a statistical form if the likely nature of this input is known to vary. Thus, items such as response to fire brigade, wind direction, and expected fuel load in given premises over time would be expected to manifest variability. Moreover, a basically deterministic model can be applied to a wide range of similar units, for example, retail premises or office buildings in which perhaps a basic fire growth model is served with data representing a wide range of premises. Probabilistic models take into account the contribution of a number of factors by ordering the factors in a logical way, assessing their likelihood of coming into play. The performance of the system as a whole is then estimated by compounding the probabilities. The answers are provided in the form of probability of achieving objectives.

There are difficulties in probabilistic models in dealing with elapse of time. Stochastic models may be regarded as intermediate between deterministic and probabilistic models and apply particularly when random elements involving time and movement are associated with deterministic processes. These models are useful in characterizing the movement of hazardous conditions, for example, flammable vapors, fire, or smoke through time and space. They may also find use in modeling movement of people as they seek to gain access to a safe place.

Mathematical models have found their major use so far within Subsystems (ii), (iii), and (v) (Table 2.4), particularly those aspects that deal with the growth of fire, the emission and movement of smoke, and the movement of people and their egress to a safe area. The elapsed time following onset of fire plays a fundamental part in these processes and the focus of the calculations is to estimate if the people will have enough time to escape before their way of escape becomes blocked. The time (T_f) taken for dangerous conditions due to fire and smoke to spread through a building following the onset of an established fire will depend primarily on the position of the fire and the geometry and fire safety properties of the building. The time taken for a successful egress by people will depend on the time they receive a warning of fire (T_p), the time they respond to warning (T_a), the time it takes to achieve relative safety (T_{rs}), and the time to safe egress in the open air (T_s).T_p is dependent on the fire detection system in place but T_a, T_{rs}, and T_s are highly dependent on the nature of the people at risk. The total of these times needs to be less than the elapsed time, T_f, from ignition for the fire to develop untenable environmental conditions. Marchant (1980) has reviewed components of an escape route system that influence these times and has classified the importance of these components on a scale of 1 to 5, 1 being the most important influence. Factors that need to be taken into account in developing models of safe egress during a fire are given in Table 2.5.

The use of logic trees plays a major part in setting up mathematical models of fire safety. Indeed, Figure 2.1 itself may be regarded as a simple form of logic tree in that it illustrates how specific items of data feed into various points to control safety. The most widely used logic trees

Table 2.5. Selected variables influencing life safety and egress

Variables that describe the occupants
1. Physiological/psychological condition
2. Sociological orientation
3. Previous training
4. Familiarity of the building
5. Egress leadership
6. Alertness
7. Irrational actions/behavior
8. Occupant load
9. Density in corridors/exit ways
10. Ratio of immobile to mobile occupants
Variables that describe the features of the building
11. Height of building
12. Construction class of building
13. Fire resistance of structural members
14. Compartmentation
15. Fire resistance of exit way enclosure
16. Fire resistance of vertical shafts
17. Fire resistance of separation of hazardous areas
18. Protection of openings in fire resistant enclosures
19. Heat actuated automatic closing devices
20. Exposure protection
21. Exterior fire spread
22. Windows
23. Electrical system
24. Mechanical system
25. Elevators
26. Centrally located watch desk
27. Ignition prevention measures
Variables that describe the means of egress
28. Exit way dimensions
29. Egress capacity
30. Remoteness/independence of exit ways
31. Dead end exit ways
32. Lighted exit ways
33. Obvious/identified exit ways
34. Operation of exit way doors
35. Vertical exit way design
36. Heliport on roof
37. Exterior fire escape
38. Balconies
39. Rescue by Fire Department
Variables that describe the means of detection, alarm, and extinction
40. Automatic detection system
41. Manual alarm system
42. Distinctive audible alarms
43. Public address system

Table 2.5 (*continued*)

44. Emergency control system
45. Automatic notification of the Fire Department
46. Automatic extinguishing system
47. Standpipe system
48. Portable fire extinguishers
49. Systems maintenance
50. Suppression by the Fire Department
51. Suppression by the in-house fire brigade
Variables describing the means of smoke control
52. Structural smoke control
53. Pressurization of adjacent compartment
54. Manual HVAC shutdown
55. Separate shaft for exhaust
56. Exit ways used as return air plenum
57. Automatic shaft vents
58. Compartmented stairway
59. Opening protection for smoke partitions
60. Smoke-actuated automatic closing devices
Variables that describe the properties of the potential fuel
61. Probability of ignition
62. Energy load
63. Rate of energy release
64. Duration of the fire
65. Toxicity of the combustion products
66. Light attenuation by the combustion products

are event trees and fault trees. With event trees, the outcome of a critical event is mapped. Thus, a critical event may be the occurrence of an "established fire" and the tree follows an input of factors as exemplified in Subsystems (ii) to (ix). Another common critical event is the occurrence of a leak of flammable fluid in a process industry. The event tree would follow the history of this leak until it encounters an ignition source and produces a fire or explosion. A fault tree specifies a certain fault and moves backward from the immediate causes of the fault to elemental causes that are responsible for it. Thus, the occurrence of fire itself may be regarded as a fault and Subsystem (i) is a first step to setting up a fault tree that aims at predicting the likelihood of a fire occurring. On the other hand, the occurrence of a major fire disaster itself may be recognized as a fault and the contributing factors leading to that disaster can be identified and quantified by using a fault tree, or as a possible outcome in an event tree. In recent years, there has been an average of less than one major fire disaster per annum (more than 10 people killed) in buildings in the United Kingdom, even though there are about 100,000 fires per annum to which the fire brigade is called, and possibly ten times as many to which the fire brigade is not called. A common factor in many fire disasters in buildings is the sudden change in the fire situation from a small unthreatening fire to a frightening extensive fire. It is important to recognize properties of the building and contents and possible defects in management that may bring this about. Taking cognizance of this in the fire safety design process has at least as important an effect in reducing the likelihood of fire disaster as efforts addressed to reduce the frequency of fires. This matter will be dealt with in Chapters 3 and 5.

Other logic trees known as *success trees* and *decision trees* are also in use. These aim at predicting success of objectives and modeling the outcome of decisions in the fire safety design process. The above methods for the quantification of fire safety will be developed in Part III of the book.

2.8 Other systems approaches

The GSA systems approach to fire safety was developed in the early 1970s by Nelson and may be regarded as the earliest systems approach to fire safety. It covered particularly Subsystem (ii), the fire development subsystem in Table 2.4 as applied to specific federal buildings in the United States. Given the occurrence of an "established fire," it sought to model and estimate the probability of fire spread from the compartment of origin of a fire to the whole floor containing the compartment and thence to the entire building. This was then compared with preset objectives limiting probabilities of fire spread throughout the building. Later modifications of this approach (Nelson, 1977) extended to cover Subsystem (iv) harmful effects, particularly insofar as they affect life safety and maintenance of function (Subsystem (ix)). The GSA system will be covered in detail in Part III, Chapter 16.

The NFPA has developed a systems approach to fire safety based on a logic tree of the success-tree kind, in that the aim of the approach is success in achieving fire safety objectives (NFPA, 1980). This tree, which forms the basis of a number of models of fire safety developed in the United States, will be dealt with in greater detail in Chapter 16. However, at this stage, the parallels between the NFPA system and the systemic approach developed above need to be pointed out. Thus, success in achieving objectives is stated to be obtained in one of two ways: (1) prevent fire ignitions (2) manage fire impact. The first of these is aligned with Subsystem (i), that is, fire occurrence and prevention in Table 2.4. The second may be aligned with Subsystem (vii), fire impact and fire protection. However, there are differences in approach in the structure of the two branches of the NFPA tree. Thus, "prevent fire ignition" is achieved by (1) control heat energy source, or (2) control heat energy transfer, or (3) control fuel response. There is no specific mention of a fire-spread characteristic as postulated in Step 5, although this may be presumed to be present in the factor "control fuel response." There is also no specific mention of the agents that bring the components of fire together as in Step 6, although there is substantial cover of agents that contribute to "control heat energy source." Since the agent that brings the other fire occurrence factors together is often highlighted as "the cause of fire," this perhaps is a limitation of the NFPA tree. The fire safety system represented in Figure 2.1 does not have a specific step dealing with heat energy transfer. However, Step 3 is presumed to include the knowledge of heat transfer necessary to ignite materials and Step 4 is presumed to include the heat transfer that ignition sources are capable of providing.

On the "manage fire impact" fire, the contributing factors are stated to be "manage fire" and "manage exposed." These are covered in Subsystem (ii) (fire development) and Subsystem (vi) (main safety) in Table 2.4 respectively. "Managing exposed" is stated to be achieved by either limiting the amount exposed or safeguarding the exposed and the latter by "defending in place" or by "moving the exposed." Defending while moving is a necessary requirement of "move exposed" and this is provided in the NFPA system by a factor called *provide protected path* as a necessary part of "move exposed." All these are factors that would be part of the data of Subsystem (vi) to be considered either during the initial fire safety assessment or part of the safety design process in Subsystem (xi).

A feature of the NFPA system is the manner in which "prevent ignition," "manage fire" and "manage exposed" are postulated as alternatives to achieving success in fire safety. In practice, it is very rarely possible to rely completely on any one of these, and fire safety design almost

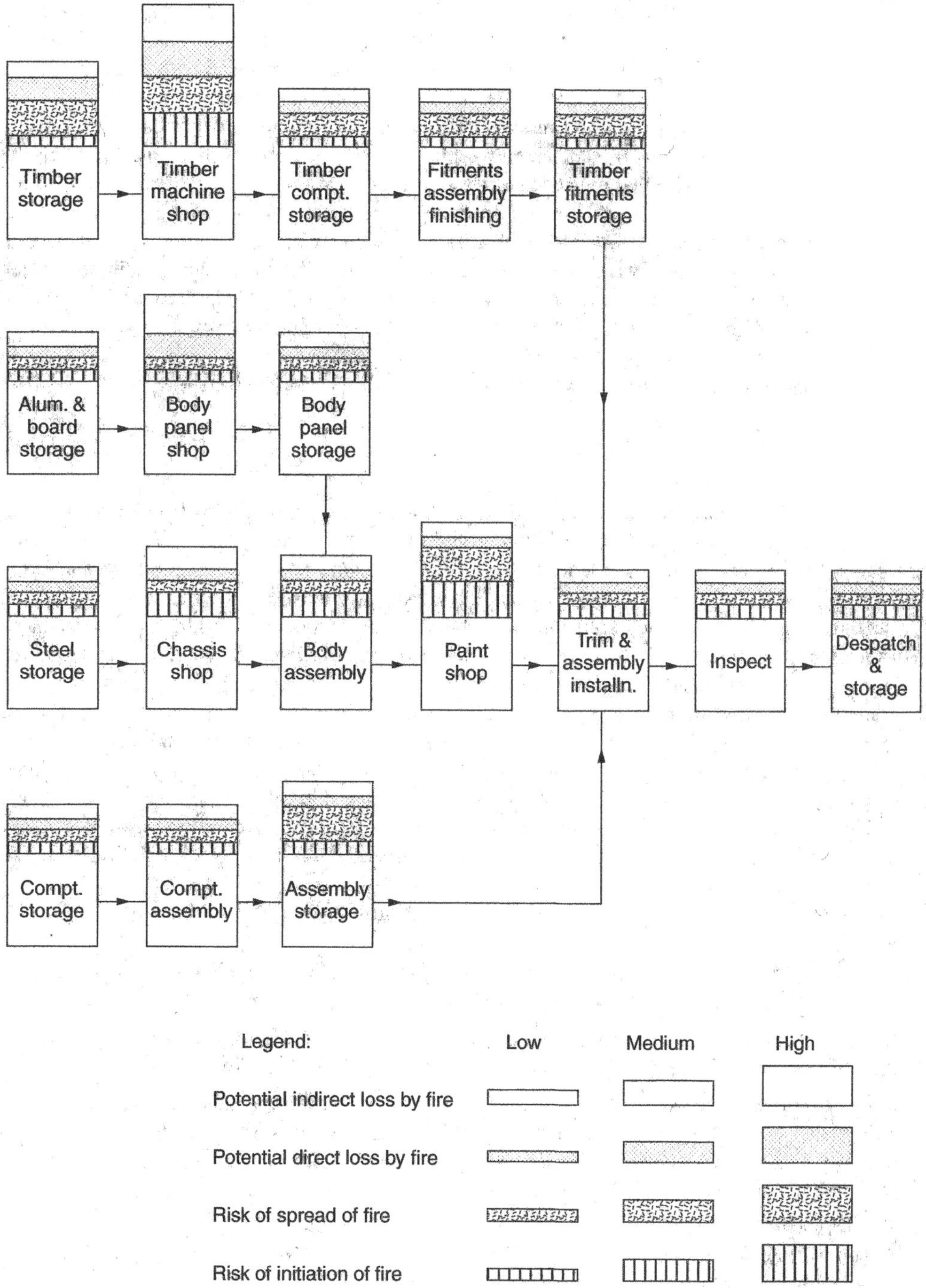

Figure 2.2. Example of representation of integrated fire hazard for industrial premises

invariably depends on an amalgam of all three. The NFPA system does not extend to cover consequential effects of fire and the range of precautions needed to cover these effects. However, the tree is very useful as a detailed indicator of components of fire prevention and fire protection that contribute to fire safety, and where they play their part in the system.

Another early systems approach to fire safety is covered by a document entitled "Management Strategy for Fire," produced by the UK Fire Protection Association. This was focused particularly on industrial premises where there was concern for potential indirect loss from fire, that is, where the "maintenance function" objective was important. The operations for a factory were divided into units and each unit was examined with a view to identifying four components of hazard, risk of initiation of fire (Subsystem (i)), a rudimentary method of quantifying them was made by judging whether the hazard component was low, medium, or high. This would then lead to an overall view of the total fire hazard that could be represented diagrammatically in a manner exemplified in Figure 2.2. Thus, the four components of hazard followed in the sequence given imply a similarity of approach with the systems outlined in Section 2.3 above.

2.9 Risk management

In recent years, an activity known as risk management has grown up within insurance and industrial organizations (Crockford, 1980). This activity is concerned with the identification and handling of a wide range of risks that is inherent in the operation of an industrial organization. These risks may be due to many causes; there are however substantial similarities in the procedures for dealing with them. Fire and explosion risk is but one of a number that might give rise to major disasters. Wind, storm, earthquakes, and floods are also risks of this kind. There is a whole range of accident risks associated with safety of individuals. There are technical risks associated with new processes, marketing risks associated with inadequate monitoring of the market and change of market habits, labor risks with availability and control of staff, liability risks resulting from inadvertent damage to third parties, particularly by products being manufactured, and political and social risks from nationalization, government intervention and so on. Finally, there are the everyday security risks associated with criminal activities of various kinds. There is an increasing tendency for management of such risks in industry to fall within the responsibility of a risk management group or adviser.

Four common components of these risks may be identified

1. *The threat or the hazard.* These are the factors that could produce an adverse result. Many have been enumerated in the previous paragraph.
2. *Resources.* These are the assets, people, processes, and earnings that could be affected by these threats. In the stepwise fire safety system, these are identified in Step 1.
3. *Modifying factors.* These are features, both internal and external, that tend to increase or reduce the probability of the threat becoming a reality, or the severity of the consequences if it does. As far as fire is concerned, these would find expression within the range of data acquisition steps of Figure 2.1.
4. *The consequences.* This is the manner in which the threat manifests its effects upon the resources. For the stepwise fire safety system, this is pinpointed in the data processing steps, particularly 13 and 16.

In general, the items concerned, particularly the modifying factors, are monitored by checklists. An important part of risk management is stated to be the measurement of risk for each of the threats and for each of the resources and, with knowledge of the modifying factors, the estimation

Table 2.6. Statement of loss expectancy used in insurance industry

Estimated maximum loss (EML): Usually expressed as percentage of value of unit under consideration. The fraction is likely to be charged in a serious conflagration.
Maximum possible loss: Financial loss that would occur under catastrophic or extremely unfavorable conditions (Failure of two or more protective systems – active and passive).
Maximum probable loss: Maximum financial loss under normal conditions, for example one protective system failing.
Normal loss expectancy: Financial loss under average operating conditions – all protective systems functional.

of the probability of the threat materializing and the consequences that occur. Usually, these at present are stated to be of high, low, or medium probability with low, medium, high, and possibly catastrophic consequences. However, in some cases, a disciplined methodology is followed on the lines of equation 1.1 to calculate the expectation of loss (Hauan, 1980, Munday *et al., 1980*).

In the insurance industry, it is customary to use estimates of expected loss under different conditions in data for estimating premiums. Some of the definitions of loss expectancy are listed in Table 2.6. The association, as indicated in Table 2.6, of the loss with the failure of items of fire safety defense, would allow a quantification of the probabilities of the loss occurring.

Having identified the risk, a number of methods of handling the risk are available. These methods may be roughly equated with the subsystems of prevention, protection, and accommodation associated with Subsystems (i), (vii), and (ix) respectively of Table 2.4. Thus, the risk may be avoided or eliminated, or the probability of its occurrence reduced. This aspect of handling is known as *risk reduction* and may be equated to prevention. Risk protection is a second method of handling risk and may be identified directly with the objectives of protection mentioned earlier, particularly in the reduction in the effects if the hazard materializes. A method called *transfer* is the means of reducing the vulnerability of a particular risk by arranging for someone else to carry part or the entire burden. This is normally done by insurance and may be regarded as part of "accommodation." The expense of all these items in the fire safety system would be regarded as contributing to the fire costs f_c (Figure 1.1). Finally, there is "financing" or "retention" of risk in which one recognizes the risk and carries it oneself. This may also be regarded as an accommodation, but in this case, it would form part of the fire penalty or detriment costs. A contingency plan may be set up within the organization to cover the part of the risk that is not covered by insurance. In general, a small frequent risk may be quantified without too much difficulty and can be carried by the firm. The difficulty arises with very infrequent risks that can cause large losses. Major fire disaster is typical of these.

In general, the requirements for fire safety design and management of any specific hazard mean that precautions are needed in the three domains of prevention, protection, and accommodation. The balance between these three will depend on the understanding of the hazard and the degree of benefit associated with the presence of the hazard. When there is no benefit associated with a hazard, as for example with disease, then as understanding of the hazard increases, there will be a tendency for the management to become dominated by prevention. Where, as in fire, a considerable benefit accrues in situations that give rise to the hazard (Section 1.2), then management will generally consist of precautions in all three domains.

2.10 Trade-off, equivalency, cost benefit, and cost effectiveness

There are many factors that can influence fire safety that would find their place in one or more of the data acquisition steps in the fire safety system. A common activity in fire safety design is to seek to trade-off a particular design feature, which is being called for by regulations, but which is proving difficult or expensive to implement, with a less expensive or less inconvenient feature. What is called for in these circumstances is that the new approach should provide "equivalent" fire safety. If the fire safety system lends itself to a quantitative approach, the fire safety objective could normally be expressed within one of the data processing Steps 7, 9, 11, 13, and 16, particularly, 9, 13, and 16. It is important, however, that the subsystem chosen should be large enough to accommodate the factors that are being considered. If there is no interaction between these factors and other factors in the fire safety system, then estimation of trade-off can be quite straightforward, that is, by how much does each of the factors in the absence of the others improve the fire safety. Complications may arise if one or more of the factors also affects the performance of other components of the system.

A common situation is trade-off between sprinklers and fire resistance. This can be accommodated in Subsystem (ii). In this case, the objective can be defined in Step 9, in that, for example, the fire proceeding beyond a compartment has a certain limited probability. However, if sprinklers, fire resistance, or improvement in fire properties of combustible linings is to be traded off with certain aspects of means of escape, then this would need to be done within the larger Subsystem (vii), since the objective would need to be expressed in Step 13 and the relevant data concerning means of escape are not fed in until Step 12. On the other hand, if the activities, precautions, and hazard areas concerned are covered by a recognized points scheme, equivalence may be regarded as achieved by balancing the allocated points appropriately.

By far the most frequent trade-off calculation compares standard fire protection methods with insurance costs. The standard fire protection methods may be assigned a cost that is part of f_c (see Figure 1.1). The insurance company presumes that this would result in a lower value of f_a and a lower value of insurance cost is charged such that the total cost of fire precautions f_c met by the insured may be reduced. The actual trade-off of a lower value of f_c with a lower insurance premium is, however, carried out by the insurer.

In general, fire safety design tends to be a trade-off between increased cost of fire prevention, protection, and accommodation methods f_c and a lower expected cost of fire detriment f_d. This is the essence of the cost-benefit approach to fire safety. It may be found by investigating the loss and effect of different levels of fire precautions and the resultant effect on f_c that there is an optimum value where the sum $(f_p + f_c)$ is a minimum. However, this is not invariably the case (Rasbash, 1980), as it depends on the rate at which expected fire losses are reduced as fire precaution costs are increased. This is illustrated in Figure 2.3 (a), (b), and (c). (a) indicates a situation in which the rise of precaution costs is less than the initial effect these have on the fire losses. Under these conditions, an optimum is possible. (b) indicates a situation in which the increasing cost of precautions is always more than the reduction in fire losses – no optimum is possible. (c) indicates a situation in which there is a certain high minimum precautions cost. An optimum will appear in total cost, this optimum may still be higher than total cost, in the absence of the precautions considered. A situation of this kind may arise in protecting a risk with precautions that have substantial real basic cost, for example, sprinklers.

Trade-off or equivalency exercises may also indicate a set of precautions where the total cost f_c is a minimum given that f_p is a constant. This particular approach, which is a cost effectiveness approach, needs to be adopted where it is difficult to express f_d in financial terms, for example where the major objectives are associated with life safety, that is, either 1 or 2, in Table 2.2.

2.11 How safe is "safe enough?"

As indicated earlier, absolute fire safety is unobtainable and in fire safety design one is inevitably aiming at a level of fire safety that may be regarded as "safe enough." What should this be? (Given

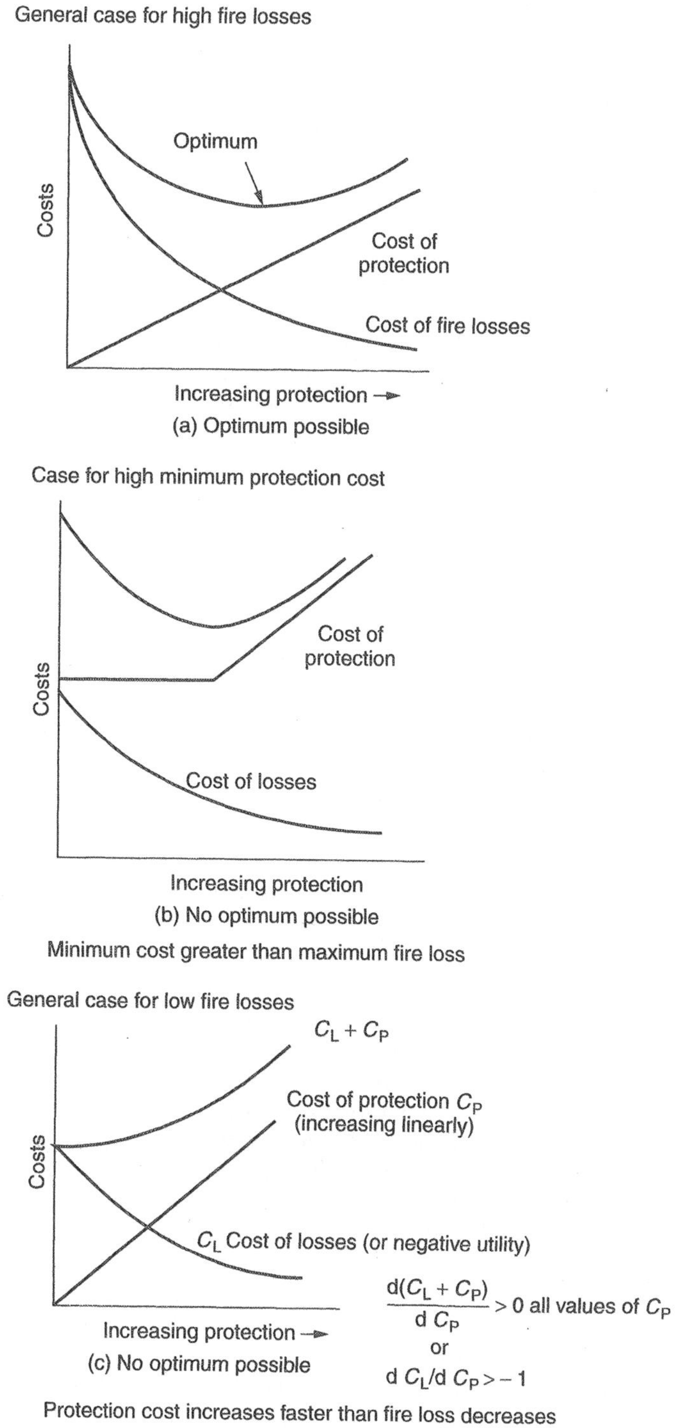

Figure 2.3. Cost effectiveness of fire protection. (a) Optimum possible; (b) No optimum possible – Minimum cost greater than maximum fire loss and (c) No optimum possible – protection cost increases faster than fire loss decreases

the assessment of risk, this is the question that needs to be faced in judging "risk evaluation" (Section 1.7)). In recent years, this question has come to be considered on a much broader scale related to how safe man-made enterprises should be in general, particularly enterprises such as industrial and nuclear power plants that have a potential of producing a catastrophe (CIRIA, 1981, Royal Society, 1983). The answer that is emerging is that the level of safety that is acceptable, particularly for life risk, should be at least that which has been acceptable for risks of a similar kind in the past, having in mind not only the nature of the risk but also the characteristics of the population bearing the risk (Rowe, 1977). A great deal of statistical and anecdotal information of man-made and natural disasters of different kinds has been collected (Rasmussen, 1975, Nash, 1977), which forms a background to this approach. A difficulty arises when one is concerned with an enterprise that gives rise to hazards of a kind that have not been experienced before. Risks from the fallout of radioactive material following a disaster in a nuclear power plant or development of malignant species, if control is lost in genetic engineering enterprises, are candidates for such concern. The perception of risk plays an important part on what is or is not acceptable. A detailed review of this aspect of safety evaluation is available (Royal Society, 1983).

A discussion document of the UK Health and Safety Executive (HSE, 1987) explored in depth the tolerability of risk from nuclear power stations. They suggested that it would be intolerable if a member of the public were exposed to a risk of death of 1 in 10^4 per annum from any large-scale industrial hazard. The risk would be broadly acceptable if it was below 1 in 10^6 per annum. Between these two criteria, the principle of "as low as reasonably practical (ALARP)" should operate. The chance of an accident at a nuclear installation that would bring about more than 100 deaths by cancer should be less than 1 in 10^6. More recently, somewhat more stringent criteria, particularly for individual risk, have been recommended for land-use planning near major industrial hazards (HSE, 1989).

With fire safety, one is dealing with a hazard that is well known to mankind and for which exists a long history of disasters followed by regulation and sufficient safety. Moreover, in recent years, many countries have taken to collecting comprehensive statistics on the occurrence and effects of fire. The potential quantitative measure exists, therefore, of the current levels of fire safety within a community. Assuming such levels are acceptable, this information can be analyzed to produce benchmarks for safety that can be used in a quantitative approach to fire safety design. On this basis, Rasbash (1984) has put forward criteria for acceptability for death by fire to an individual and for multiple fatality fire disasters. For fire risk to an individual, target acceptable probabilities of 10^{-5} to 10^{-7} per annum were suggested according to the nature of the person at risk and the benefit obtained from the risk activity. A summary of recommendations for multiple fatality fire disasters for specific buildings is given in Table 2.7, which is based on the frequency of such fires in Western countries mainly during the period 1946–1982. Such criteria may be used in a manner similar to the use of criteria in quantitative risk assessments for industrial processes. However, the requirement would appear to be more stringent than those suggested above for industrial nuclear installations. Thus, instead of 100 deaths, a target probability of fire risks of 1×10^{-6} per annum is associated with the occurrence of more than five deaths. This target is the product of Fp_i (equations 1.1, 1.2, 1.3), where F is the frequency of fire occurrence and p_i is the probability that given a fire a harmful effect of more than five deaths will occur. The value of F for buildings considered in Table 2.7 is on the order of once in 10 to 100 years (Chapter 7). This implies that given a fire, one is looking for a value of p_i of about one in 10,000 to 100,000 to achieve an acceptable level of safety.

In practice, a difficulty arises in quantitative fire safety design for buildings in general. It is due primarily to limited control and widespread potential for fire, which means there is a dearth of information on many of the factors contributing to fire safety, particularly human factors. (There may be less difficulty for hazardous industrial processes under strict control). As a result, it is

Table 2.7. Target probabilities for fire risks in buildings

Maximum number at risk in building	N (number of fatalities)			
	>5	>15	>100	>500
Less than 15	5×10^{-7}	–	–	–
15–100	1×10^{-6}	3×10^{-7}	–	–
100–500	2×10^{-6}	5×10^{-7}	6×10^{-8}	–
Greater than 500	4×10^{-6}	8×10^{-7}	1×10^{-7}	5×10^{-8}

comparatively rare to calculate reliable quantitative estimates of risk against which benchmarks, such as those indicated above, can be readily applied. What tends to occur in quantitative models is that where judgment is used to fill the gap in knowledge, it is made to err on the safe side. As a result, final estimates of risk, particularly those based on an input of a large number of contributing factors, tend to exceed targets of acceptability such as those put forward in Table 2.7. In general, the approach that is usually adopted in quantitative models of fire safety is comparing estimates of fire safety involving novel elements with those obtained with similar situations in which prescriptive requirements inherited from the past apply and are deemed to be sufficient.

The practise that is developing, therefore, is to calibrate the quantitative approach against a risk situation of a similar kind that is deemed to be acceptable. This particularly covers risks, for example, those in buildings, for which there is a major inheritance of prescriptive legislation (Chapter 4) and for which the act of following this legislation may be regarded as safe enough. Thus, given clear and identifiable objectives, a comparison can be made between a traditional approach favored in one country to another favored elsewhere or between a traditional accepted approach and an alternative approach or a novel approach developed by a new technology. This method can be used both for mathematical models for fire safety and for point schemes. Indeed, it is part of the process of the development of point schemes to calibrate them against known acceptable risks of similar kind. However, one must be wary in all quantitative approaches of this kind against stretching the bounds of similarity beyond credibility. One must be mindful that at least two major elements are similar in comparisons of this kind, namely, a credible fire scenario and a people behavior scenario.

Symbols

f_c	a fire safety programme that inflicts a cost
f_d	a fire occurrence scenario that inflicts a detriment
F	frequency with which fire occurs
Ha_i	harmful effects associated with fire
P_i	probability of specific harmful effect associated with fire.
T_a	period of response to warning
T_f	elapsed time from ignition to develop untenable conditions
T_p	elapsed time from ignition to receipt of warning of fire
T_{rs}	period to achieve relative safety
T_s	period to achieve egress to open air

References

Beard, A N (1986). Towards a systemic approach to fire safety. *Proc. 1st Int. Symposium on Fire Safety Science*, Hemisphere Publishing Co. New York, p. 943.

CIRIA (1981). *Rationalisation of Safety Serviceability – Factors in Structural Codes*, Construction Industry Research and Information Association.

Crockford, N (1980). *An Introduction to Risk Management*, Woodhead Faulker, Cambridge.

Fiskel, J and Rosenfield, D B (1982). Probabilistic methods for risk assessment. *Risk Analysis*, **2**(1), 1.

Hauan, O (1980). *Classification of Fire Risks in Industrial Plants*, Seminar on Industrial Safety, Det Norske Veritas.

HSE (1987). *The Tolerability of Risk from Nuclear Power Stations*, HMSO.

HSE (1989). *Risk Criteria for Land-use Planning in the Vicinity of Major Industrial Hazards*, HMSO.

Kanury, A M (1987). On the craft of modelling in engineering and science. *Fire Safety Journal*, **12**(1), 65.

Malhotra, H L (1991). The role of a fire safety engineer. *Fire Safety Journal*, **17**, 131.

Marchant, E (1980). *Modelling Fire Safety and Risk. Fires and Human Behavior*, D Canter (Ed.), John Wiley & Sons, Chichester, Chapter 16.

Munday, G *et al* (1980). *Instantaneous fractional annual loss – a measure of the hazard of an industrial operation. 3rd Int. Symposium, Loss Prevention and Safety Promotion in the Process Industries*, Basle, Switzerland.

Nash, J R (1977). *Darkest Hours*, Pocket Books, New York.

Nelson, H E (1977). *Directions to Improve Application of Systems Approach to Fire Protection Requirements for Buildings*, SFPE Technology Report 77-8, Boston, USA.

NFPA (1980). Fire Safety Concepts Tree, NFPA Standard 550.

Rasbash, D J (1977). The definition and evaluation of fire safety. FPA, *Fire Prevention Science and Technology*, (16) Supplement to Fire Prevention (Suppl. 118).

Rasbash, D J (1980). Analytical approach to fire safety. *Fire Surveyor*, **9**(4).

Rasbash, D J (1984). Criteria for acceptability for use with quantitative approaches to fire safety. *Fire Safety Journal*, **8**, 141–158.

Rasmussen, N C (1975). An Assessment of Accident Risks in US Commercial Power Plants, WASH 1400.

Rowe, W D (1977). *An Anatomy of Risk*, John Wiley & Sons.

Royal Society (1983). *Risk assessment – a study group report*, Royal Society, London.

Watts, J, Milke, J A and Bryan, J L (1979). *A Study of Fire Safety Effectiveness Statements*, College of Engineering, University of Maryland, College Park.

Woolhead, F E (1976). Philosophy of risk management. *Fire Engineers Journal*, **36**(102), 5.

3 REVIEW OF SOME MAJOR FIRE & EXPLOSION DISASTERS

3.1 Introduction

Fire and explosion disasters and the concern and investigation that follow them probably provide the major input into requirements for fire safety. In this chapter, a number of disasters that have occurred in the last three decades will be reviewed, with particular emphasis on important lessons that have been learned that are currently being incorporated into fire precautions. A common feature in a number of disasters in buildings is a sudden spread of fire from an apparently small and limited fire to one that is highly threatening and disastrous. There have been fires in which the rapid spread of smoke was the main cause of the disaster. A major feature of disasters, particularly in industrial and transport processes, has been an explosion, which was either solely responsible for the disaster or worsened the fire situation greatly. Special consideration is given here to disasters with these features.

3.2 Major disasters in buildings involving sudden rapid spread of fire

3.2.1 SUMMERLAND LEISURE CENTRE, ISLE OF MAN, 1973

A picture of the Summerland complex is shown in Figure 3.1, and a diagram of a part of the premises called the *Main Solarium Level* is shown in Figure 3.2 (Government Office, 1974). This level was the one at which people entered the premises. It contained a solarium in which shows were given, an amusement arcade 32 m long by 17 m wide and 4 m high that opened to the solarium, and a restaurant. Above the latter two areas there were three further stories in which various activities were carried out and which overlooked the solarium. There was also an open-air public area called *novelty golf* outside the amusement arcade and solarium.

Three extensive areas of flammable surfaces were incorporated into the building and played a major part in the fire disaster. Firstly, the roof and most of the wall, shown in Figure 3.1, were constructed of large panes of Oroglas (poly methyl methacrylate). Those concerned with the design of the building were not fully conversant with the flammability properties of this material. Secondly, abutting onto the Oroglas and stretching upwards from the rear part of the amusement arcade to the top of the building was a wall of Galbestos cladding. This was a steel

Evaluation of Fire Safety D. Rasbash, G. Ramachandran, B. Kandola, J. Watts and M. Law
 ISBN: 0-471-49382-1

Figure 3.1. External view of the Summerland leisure complex

sheet surfaced on both sides with resin, and it had passed a standard fire test for roofing systems, which exposes specimens to 12 kW/m^2 radiation. Thirdly, between the sheet steel outer wall and the amusement arcade there was a cavity 0.3 m wide and 12 m long, the inner wall of which was Decalin, a form of fiberboard (class 4, which is the lowest grading in the BS 476, Part 7 Spread of flame test).

The fire occurred while a concert was in progress in the solarium. Following warning of a fire and signs of smoke at the far end of the amusement arcade, the first rapid spread of fire took place along the length of the amusement arcade. This produced a flame above the front of the arcade that involved the upper stories, and above the side, which involved the Oroglas wall. The fire then spread rapidly to the whole of the wall and the roof. Figure 3.3 shows a picture of the fire at this stage.

The cause of the fire was a plastic booth that had been set on fire by children at point X (see Figure 3.2). This ignited the resin on the sheet steel and the flame soon spread to the resin within the cavity and to the fiberboard material and supporting wooden struts on the other side. The fire on the outside wall of the building was controlled by extinguishers. Within the confines of the cavity, to which air had limited access, the fire persisted and a fuel-rich fire developed over a period of about 20 min before the fiberboard wall burned through. The place of access of the air to the cavity was at the corner of the building where there was an incomplete seal and the point of breakthrough into the amusement arcade was near this point at the closed end of the arcade. At this point, flame and fuel-rich gas was ejected into the arcade, followed by a continuous flame

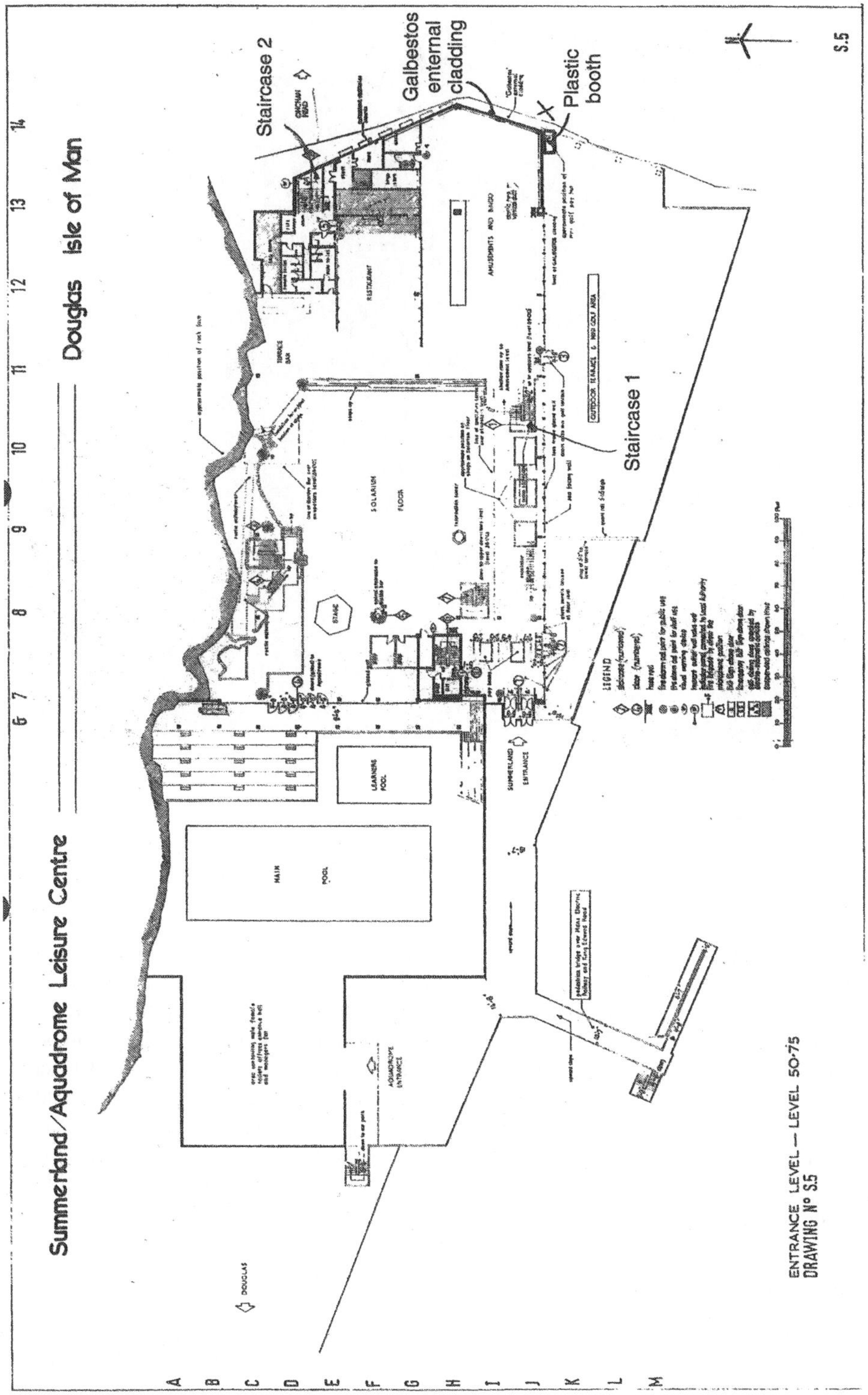

Figure 3.2. Main solarium level, Summerland

Figure 3.3. Fire spread involving the Oroglas wall and roof of Summerland

from inside the cavity. This could have acted as a powerful ignition source for the combustible wall surfaces that were present in the arcade. The flames spread along the length of the arcade in 1 to 2 min.

There were 3000 people in the building at the time, of whom 50 persons – men, women, and children – perished in the fire. They were found in different parts of the premises. Many people died on an open staircase (staircase 1 in Figure 3.2) that people were using to escape from the upper floors. It was fully exposed to the flames that poured out from the open end of the amusement arcade, and to flames from the nearby fire on the Oroglas wall. A number of bodies were also found on stairway 2. This was intended to be a protected stairway but it served also as a service stairway and a permanent opening was made between one of the floors and the stairway through which much smoke must have passed during the fire. A number of exit doors from the main solarium level were locked at the time of the fire and there was no disciplined evacuation procedure. One reason for delay was the need for parents to look for their children who were in different parts of the leisure center.

3.2.2 THE STARDUST DANCE CLUB FIRE, DUBLIN, 1981

Forty-eight young people died in this incident. There were about 800 people in the dance hall when the fire occurred (Keane, 1982). In the early stages, the fire involved the West Alcove (Figure 3.4), measuring 17 m by 10 m, with a floor that sloped upward toward the back where the

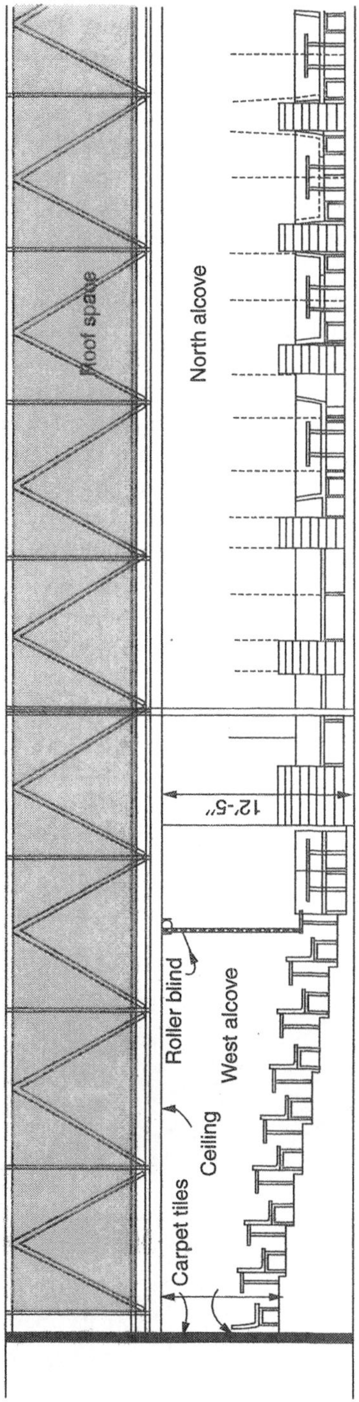

Figure 3.4. Stardust club interior (section) showing west alcove

height was 2.4 m. The alcove, which was empty and partially cut off from the main body of the dance hall by a roller blind (Figure 3.5), had rows of seats each measuring 0.9 m long with 4-cm-thick PVC covered polyurethane foam seating and backrests. The back row was installed against the back wall, which was lined with carpet tiles; these gave a class 3 to 4 performance in the standard spread of flame test. Mineral fiber insulating tiles, which were almost noncombustible, formed the ceiling of the alcove above which there was a roof space. Some mechanism caused the ignition on the back row of seats so that a line fire developed. The people in the hall were aware of there being a small fire in this area for some 6 to 7 min, during which time the roller blind was raised. The flame impinged on the carpet tiles and the back wall rapidly became involved. Then, within tens of seconds, all the seats in the alcove began to burn, and flames and smoke spread very quickly to the main body of the ballroom.

Experiments carried out during the investigation showed that when one of the 0.9-m seats was ignited along its whole length it could produce a heat transfer of 100 kW/m^2, on the back wall. The combination of burning seats plus carpet tiles produced a burning rate at the back of the alcove of about 800 kW/m run, sufficient to produce extensive flaming under the ceiling and high levels of radiation down on the seats in front. Full-scale tests showed that the heat transfer from the flames just ahead of the back wall rose rapidly to a peak of 250 kW/m^2. The result was that the tops of the seats well ahead of the back wall were exposed to a heat transfer in excess of 60 kW/m^2, sufficient to produce spontaneous ignition in a few seconds. The array of seats below the ceiling acted as an extensive surface that could respond to the very high heat transfer rates.

A noteworthy point is that a part of the ceiling in the alcove collapsed, which led to the shattering of the roof and the venting of some of the heat and smoke. Had venting not occurred, it is likely that many more of the 800 people in the hall would have been killed.

Figure 3.5. Stardust club interior – alcove behind roller blind

3.2.3 KING'S CROSS UNDERGROUND STATION, LONDON, 1987

The fire at King's Cross caused the deaths of 31 people. The origin of the fire was a wooden escalator about 40 m long leading from railway platforms to the booking hall (subsurface). It was the left-hand escalator of a group of three running in a semicircular shaft 8 m in diameter. The side balustrade and handrail of this escalator was separated from the wall and ceiling of the escalator shaft by a combustible horizontal surface 0.3 m in width. Surfaces on the escalator were class 3 or 4 spread of flame; these included the risers and treads of the steps, the balustrade, and probably also the laminated surfaces separating escalators from each other and the wall. In addition, the first meter of the wall and ceiling was covered by an advertising hoarding, which, because of the presence of varnish and plastic covered advertisements, was the most flammable item associated with the escalator shaft. There was also evidence that the ceiling had been recently painted in such a way that if it had been subjected to the spread of flame test, it would probably have given a class 3 or 4 performance. When a sample that included the plaster and concrete base was exposed to a heat transfer rate of 75 kW/m^2 in a cone calorimeter, a number of the layers of the paint delaminated in 2 to 3 s, ignited in 5 s and produced a heat output of between 200 and 450 kW/m^2, with much smoke, for about 20 s.

The Inquiry (Fennell, 1988) found that the fire was ignited by a lighted match falling through a gap at the edge of the escalator into residues of lubricating grease that had been allowed to collect beneath the escalator. This led to what appeared as a comparatively small fire in the middle of the escalator. Suddenly, after a period of about 15 min, fire shot up the escalator and quickly involved the upper part of the escalator shaft as well as the combustible material in the booking hall above. People in the booking hall were confronted with a wall of flame and black smoke moving rapidly toward them.

During the inquiry, different opinions were expressed concerning the mechanism of sudden rapid flame spread, including a mechanism that involved delamination of the thick paint on the ceiling. A field modeling study carried out at Harwell, which was reported late in the investigation, indicated that flames within the trench of the escalator did not move vertically upwards, as had been assumed up to that time. It was likely that the flames were confined to the trench itself for a significant time, thus promoting flame spread along the escalator. Tests on a one-third-scale escalator system, carried out by the Health and Safety Executive, showed that this could indeed have happened. Heat transfer rates within the escalator trench of about 150 kW/m^2 were measured, and were found to be predominantly convective rather than radiative. It was decided by the Inquiry that this was the mechanism of rapid fire spread.

However, the opinion was also expressed (Rasbash, 1991) that the spread of fire out of the trench on to the advertising hoarding, which was also observed, might have rapidly produced a line source heat transfer to the ceiling of the order of 100 kW/m^2. A consequent spread up the hoarding/ceiling combination and under the ceiling to the booking hall could take about 20 s. There could thus have been a double blow awaiting those in the ticket hall. Most of the people who were killed were in the ticket hall and many had been sent there from the underground station through a separate escalator as a way of escaping the fire.

3.2.4 BRADFORD CITY FOOTBALL GROUND, 1985

At about 15.40 on 11 May, 1985, during a football match, a fire started in the main stand of the Bradford City Football Ground. The stand was some 90 m long, and within 7 min of the first sign of fire the stand was completely alight. Fifty-six people lost their lives. A full report is given by Popplewell (1985).

The stand was set on the side of a hill. It was divided into two approximately equal longitudinal sections by a wooden fence 1.5 m high. Above the fence, spectators were provided with timber seats affixed to timber frame; below the fence there were polypropylene seats fixed to concrete. Access to and from the seating sections was from a long corridor extending along the length of the back of the stand. The corridor was located at the highest point of the stand, next to the perimeter wall that contained exit doors and turnstiles leading to and from the outside road.

Because of the natural slope of the hill, there was a void underneath the wooden floor of the stand, which varied in depth from 0.21 to 0.75 m. There were gaps in the flooring and between the seats through which, over the years, combustible rubbish had been allowed to fall and which had accumulated to a depth of about 0.2 m. There was also a close-boarded roof to the stand that was covered throughout its length with roofing felt.

The fire started at 15.40 in the void beneath the wooden seats just above the wooden fence, near one end of the stand. People nearby felt heat and looking down through a gap, could see flames. Figure 3.6 shows a photograph taken early in the fire indicating smoke coming up through the gaps in the floor and with flame underneath seat J141. Flames spread upward through the void under the floor and appeared through gaps above the floor at 15.43. Within a minute, the area of the fire visible above the floor was several square meters in extent and increased to about 10 m^2 in a further minute, at which time the flame height extended to the roof. By 15.46, there was a serious spread of fire involving the roof, and by 15.47 the whole of the stand, roof, and seats were completely alight as shown in Figure 3.7.

Most of the people who died were trying to escape through the back corridor. In the first few minutes of the fire, there was no major effort on the part of spectators to escape, as the fire did not appear to be threatening. However, once the fire covered several meters above the floor, then a rush to escape started, particularly upwards to the corridor at the back of the seats. It was likely that at this time the corridor was beginning to become smoke logged. Although there were exits from the corridor onto the street, most were closed and a number of them, particularly the turnstiles, were locked. The exits could not accommodate the number of people attempting

Figure 3.6. Early stage: fire under the seating, Bradford City

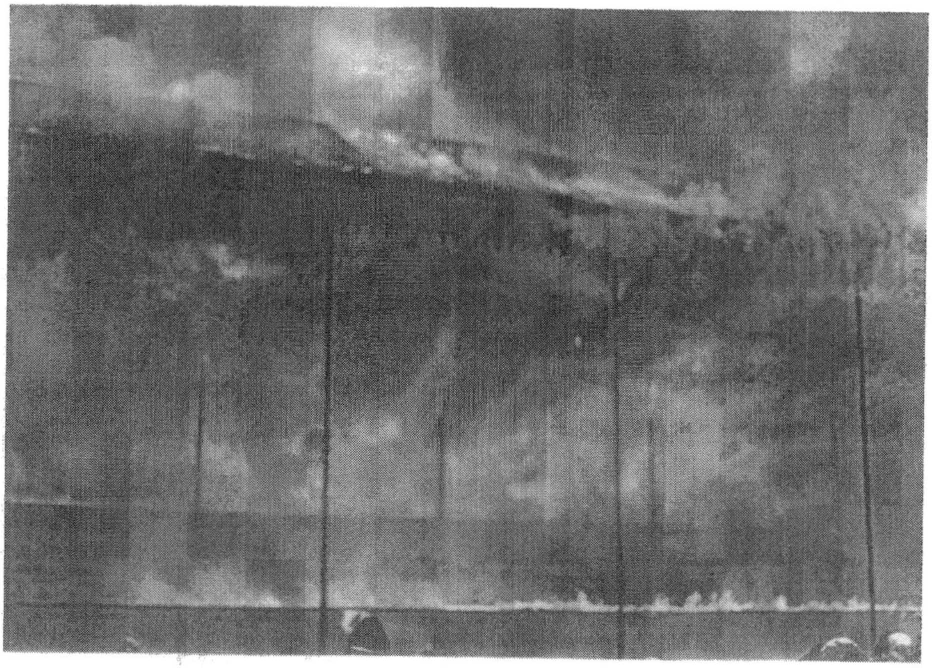

Figure 3.7. Complete involvement of roof and stand, Bradford City

to use them in the minute or two that was available for escape and the majority of people who died did so near exits at the center of the corridor and toward the end of the corridor where the fire started.

3.2.5 INTERSTATE BANK, LOS ANGELES, 1988

On Wednesday, 4 May and continuing into 5 May, 1988, fire spread from floor to floor through exterior openings of this 62-story tower building in downtown Los Angeles. The fire destroyed four floors and damaged a fifth. It claimed one life, injured approximately 35 occupants and 14 fire personnel, and resulted in a property loss of over \$200 million. Details of the building and the fire are given in reports by Routley (1989) and Klem (1989).

Built in 1973, the building contains approximately 1625 m^2 net area of office space per floor, built round a central core (see Figure 3.8). Below the ground there is a basement accommodation, a garage, and a pedestrian tunnel. Approximately 4000 people work in the building. They are mainly staff of the banking corporation, but tenants occupy several floors. The tower contains four main stairways. The two northern stairs are enclosed within a common shaft while the southeastern stair has a pressurized vestibule separating each floor area from the stair shaft. The building has a structural steel frame, protected by a sprayed-on fire protective coating, with steel floor pans and lightweight concrete decking. The exterior curtain walls are glass and aluminum.

There were extensive smoke detector installations, including detectors connected to magnetic hold-opens for doors at each end of the elevator lobby to the 12th floor (G + 11), which was the floor of fire origin. However, it was reported that these detectors were not connected to the fire alarm system. A sprinkler system was being installed and was virtually complete, but the valves between the standpipe riser and the system on each floor were not yet open.

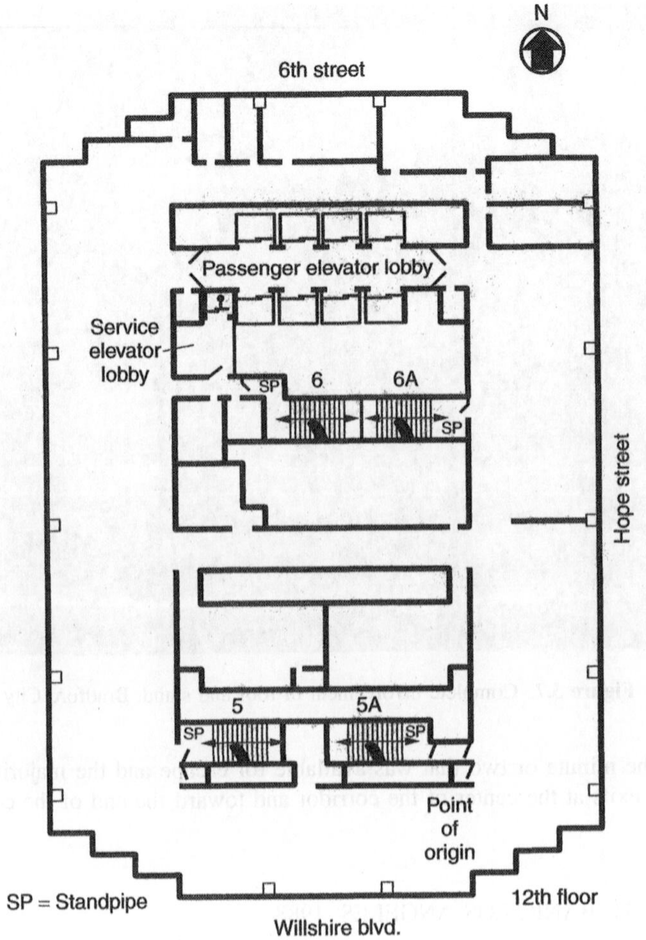

Figure 3.8. Typical floor area in Interstate Bank building (Nelson, 1987)

At approximately 10.25 p.m. on the night of the fire, employees of the sprinkler system contractor heard glass falling and saw smoke at the ceiling level on the fifth floor (G + 4). A manual alarm was pulled but sounded for only a few seconds. It is believed that security personnel on the ground floor silenced the alarm. The fire originated in an open-plan office area in the southeast quadrant of the 12th floor (see Figure 3.8). The area of origin contained modular furniture with numerous personal computers and terminals. The cause is thought to be electrical in origin, but the precise source was not determined. The fire extended to the entire open area and several office enclosures to fully involve the 12th floor, except for the passenger elevator lobby, which was protected by the automatically closing fire doors.

Fire also extended to floors above, primarily via the outer walls of the building. Windows broke and released large flames. Flames also penetrated through gaps between the curtain wall panels and the ends of the floor slab. The curtain wall construction was such that it did not butt the edge of the slab. The external flames gave heavy exposure to the windows on successive floors as the fire extended upward from the 12th to the 16th (G + 15) floor. The flames were estimated to be lapping 9 m up the face of the building. The curtain walls on these floors, including windows, spandrel panels, and mullion, were almost completely destroyed by the fire. The fire extended at a rate estimated at 45 min per floor and burned intensely for approximately 90 min on each level. This resulted in at least two floors being heavily involved during most of the fire. Fire fighting prevented extension beyond the 16th floor.

Firefighters faced unusual challenges. The steel structural frame interfered with radio communications. Access to upper floors was by stairwells that were filled with heat, smoke going up, and waters cascading down. Two "airborne engine companies" were deployed via helicopter. Problems were encountered with pressure reducing valves on standpipes and overpressure caused several hose ruptures and made hand lines difficult to control. The heat of the fire caused several aluminum alloy valves in the occupant hose cabinets to fail creating high-pressure water leaks. These leaks took water from the supply that was available for hand lines and caused additional water damage on floors below the fire. Thirty-two attack companies used approximately 20 hand lines on the fire floors. The estimated water flow through hand lines at the height of the fire was 150 L/s (2400 US gallons per minute).

The floors below the fire received massive water damage. Those above were heavily damaged by heat and smoke. In spite of the total burnout of four-and-a-half floors, there was no damage to the main structural members, and only minor damage to one secondary beam and a small number of floor pans. Although there was concern for structural integrity during the incident, postfire analysis indicated that there was no danger of major or minor structural collapse. It was noticed that quality control in the application of the sprayed-on fire protection was unusually good.

The bank's disaster plan went into effect immediately and proved to be well worth the many person-hours of planning. If this fire had occurred during normal working hours, many more lives would have been at risk. However, with people occupying the 12th floor, there could have been earlier alarm and possibly first aid fire fighting. The barrier weakness at the junctions of curtain walls and floor slabs is now well recognized and easily remedied. If the automatic sprinkler system had been in operation, the risk to both life and property would have been significantly reduced.

Nelson (1989) was able to reconstruct important aspects of the fire growth and behavior, by making an engineering analysis of the fire. One of the models used in his analysis was an early version of FPETOOL (Nelson, 1990) that included ASET (Available Safe Egress Time). The rate of heat release curve used to simulate the burning of the initial fire was based on measurements for similar computer workstations. It is shown in Figure 3.9. The calculated growth before flashover is shown in Figure 3.10 and is compared with the fast and moderate fire growth curves of NFPA 72E (1990). Nelson (1989) extended the study to examine the potential for flashover. Figure 3.11 shows how ceiling height and floor area affect the rate of heat release needed to produce flashover. (The three curves represent different ceiling heights – 1.8, 2.7, 3.7 m. The abscissa is floor area, ranging from 50 to 1400 m^2, and the ordinate is the rate of heat release. Triangles indicate floor area dividing points below which smoke temperatures cannot exceed 600 °C because of insufficient oxygen.)

3.2.6 DUPONT PLAZA HOTEL, PUERTO RICO, 1986

On 31 December, 1986, a mid-afternoon fire at the Dupont Plaza Hotel and Casino in San Juan, Puerto Rico, resulted in 97 fatalities, over 140 injuries, and property losses involving millions of dollars. The fire department arrived approximately 5 min after notification of the fire, but it took nearly five hours to achieve final extinction because of the severity and magnitude of the fire that confronted them and the complexity of the rescue problems. Reports by USFA and Polaris Research Development (1987) and Klem (1987) give comprehensive descriptions of the fire.

Built in the early 1960s, the 22-story hotel was L-shaped on plan, with the main entrance at one story above grade, accessed by a large ramp. There was a ballroom at ground level, the Casino

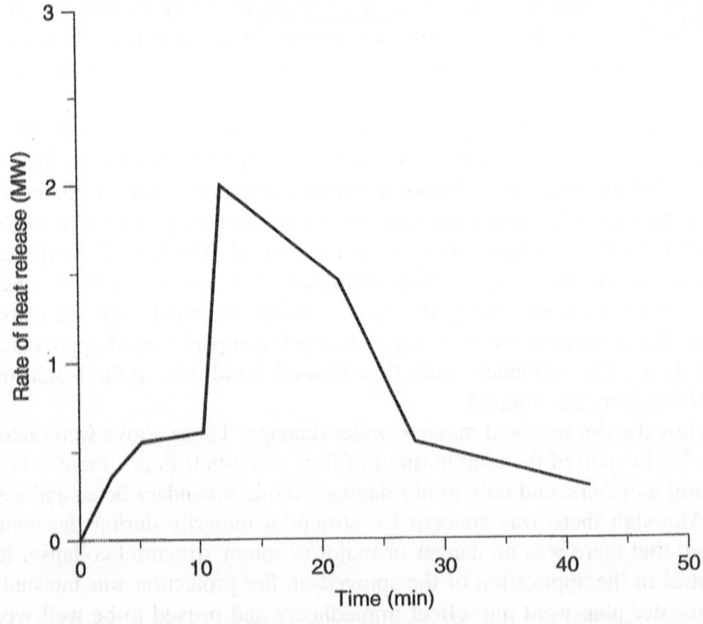

Figure 3.9. Rate of heat release for typical computer work station (Nelson, 1987)

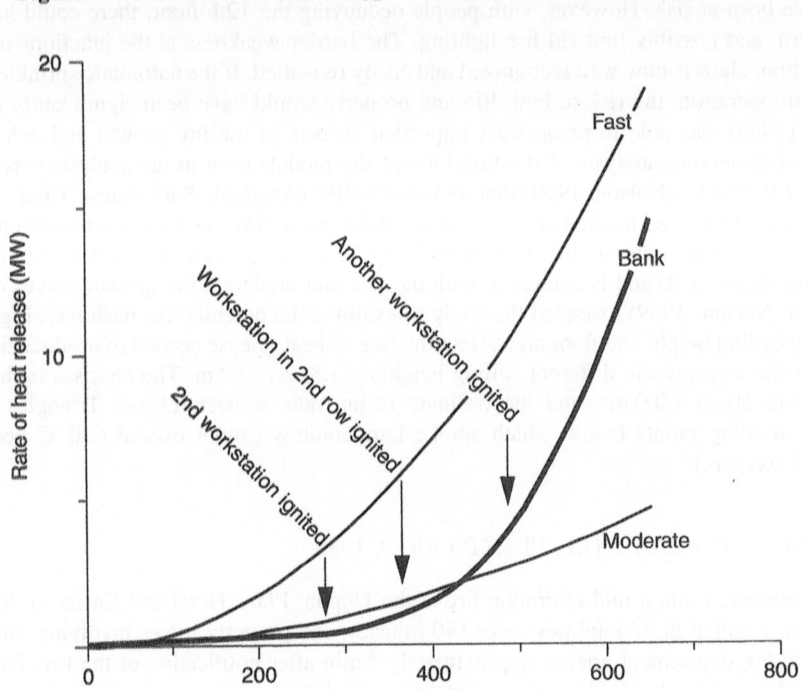

Figure 3.10. Estimated rate of heat release on 12th floor of Interstate Bank building (Nelson, 1987)

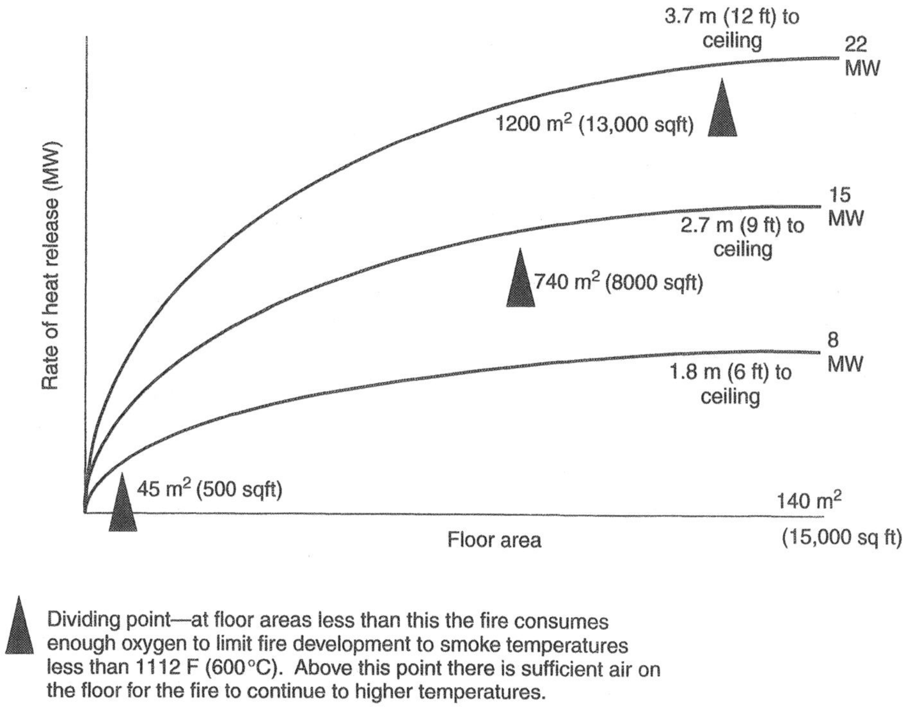

Figure 3.11. Maximum fire size in a closed space as a function of floor area and distance from fire to ceiling (Nelson, 1987)

at entrance level, and there were various shops, restaurants, and conference rooms surmounted by a 17-story tower with 423 guestrooms.

The construction in the ballroom area included unprotected noncombustible and some combustible material. The casino, lobby area, and the high-rise tower contained fire-resistive construction. The ballroom can best be described as a general-function, conference-type room, which could be divided by single free-hanging panels. A portion of the ballroom was being used for storage at the time of the fire. Both the casino and the ballroom had suspended ceilings, creating large voids below the structural ceiling. Three sides of the casino had floor-to-ceiling windows, as did the atrium that connected the ballroom and casino. The casino had two means of egress (exit ways): on the ballroom side there was a pair of free-swinging tempered glass doors, and at the opposite end there was a solid-core inward-opening wood door.

There were no automatic detection and alarm systems. There was a local-only, manual fire evacuation alarm system installed in the high-rise tower. There were no sprinklers. There was a standpipe and hose system in the tower.

The fire was discovered at approximately 3.22 p.m. in the south part of the ballroom, which was unoccupied. The fire department was notified at about 3.45 p.m. The authorities later determined that the fire was deliberately set in a large stack of recently delivered guestroom furniture, temporarily stored in the ballroom. Subsequent spread of the fire is shown in Figure 3.12. The fire developed quickly after ignition. It involved the stored materials, the combustible interior – including cardboard and wood packing – the carpeted walls, stacked chairs, and a combustible, removable partition that separated the south ballroom, where the fire started, from the adjoining north ballroom. While the south ballroom was approaching full involvement, the fire

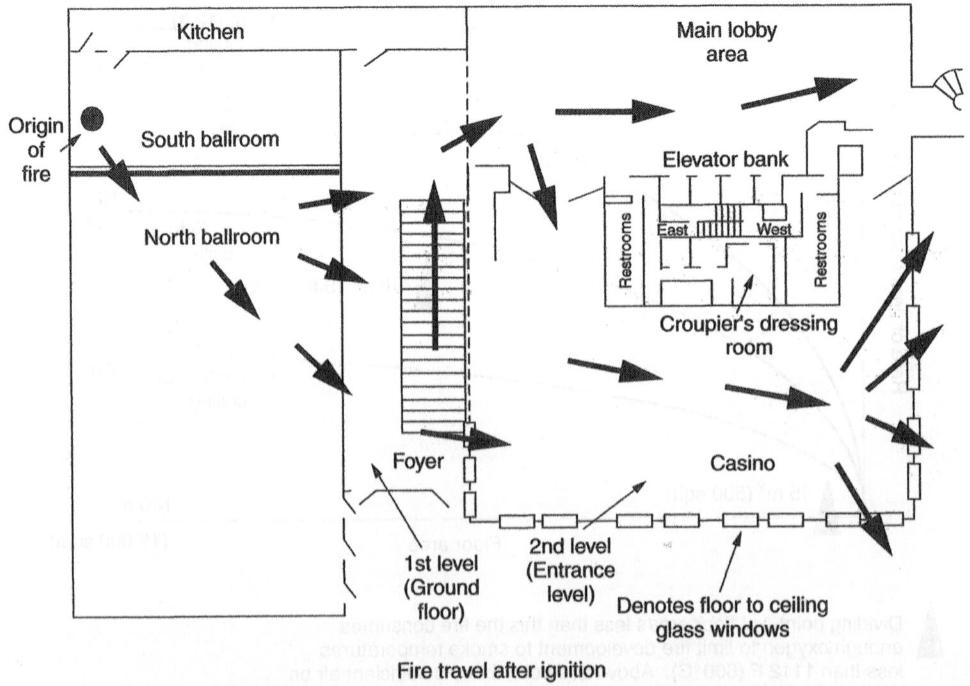

Figure 3.12. Fire travel in the Dupont Plaza Hotel (Nelson, 1989)

penetrated the combustible partition, spreading products of combustion to the north ballroom and to a foyer connecting the ballroom level (ground floor) to the entrance level directly above. There was limited extent of fire growth and damage in the north ballroom, but it is believed that fire reentered the upper portion of the north ballroom at some time after 10 min, probably upon breakage of the glass partitions at the lobby level of the foyer. This would explain the damage found in the balcony area and on the ballroom ceiling.

Witness accounts indicate that smoke emerged into the foyer from an open door of the north ballroom at approximately seven minutes. This was the first time that smoke entered the view of the general public, and initially it was cool and thin. The foyer contained the main staircase to the lobby. Flashover occurred in the south ballroom at about 10 min after established burning. It involved large portions of combustible wall material, and ignited portions of the partition between north and south, significant parts of the wood flooring, and the stacked chairs. Intense heat from the flashover broke the glazed partition between the south ballroom and the foyer so that a massive quantity of smoke and flames was released into the foyer.

Once the fire had reached casino level, smoke spread to the high-rise tower by several means, including service and passenger elevators, the hotel's HVAC system, toilet exhausts for guestrooms, the exterior of the building, and through stairways connecting the casino level to each guest room floor. Occupants of the high-rise became aware of the fire and many moved to the roof of the building and waited for helicopter rescue, while others stood on guestroom balconies awaiting rescue.

Between 200 and 250 people were estimated to have been in the casino at the time of the fire. They began to leave the casino once smoke was spotted through the glass walls overlooking the atrium and ballroom. Most of them moved towards the west portion of the casino, which led to the main hotel lobby and to an exterior spiral stairway leading to ground. Not all of them were able to safely reach this exit before the flame front vented through the lobby and out of the opening to the spiral stairway. When this occurred, the main exit path for the remaining occupants of the casino became blocked. As the fire grew in intensity and as more people became aware of the fire, they rushed to the only available exit – the wooden door at the far end of the casino. Because two simultaneous actions were needed to unlatch and open the door, which unfortunately opened inward as well, it was difficult in the crowded conditions to exit by this route and people began looking for alternative means of escape. Some people began to break the exterior glass walls at the rear of the casino, and then jumped to the ground 5 m below.

Ninety-three victims were found dead at the scene and four died later in hospital. Fatalities were mainly clustered near the west exit of the casino. Other fatalities were located in the lobby area, in a passenger elevator, and one fatality was found in a guestroom on the fourth floor (G + 3). All but eight of the fatalities at the hotel were burned beyond recognition. Analysis of blood samples indicated that carbon monoxide alone or combined with hydrogen cyanide played a major role in the deaths of the nonburned victims but was probably not significant for those who were burned (Clark *et al.*, 1990).

Table 3.1. Fire characteristics calculated in engineering analysis of Dupont Plaza Hotel fire (Nelson, 1987)

Mass burning rates

- Preflashover (free) burning of initially ignited fuel package
- Postflashover burning of materials in the south ballroom
- Burning of foyer ceiling after flashover of south ballroom but prior to entry of fire into casino

Rates of heat release

- Preflashover (free) burning of initially ignited fuel package
- Rate of energy flow from south ballroom to north ballroom prior to flashover in south ballroom
- Rate of energy flow from north ballroom to foyer prior to flashover in south ballroom
- Postflashover energy flow from south ballroom
- Energy release from the foyer ceiling after flashover of south ballroom but prior to entry into casino

Smoke temperatures

- South ballroom
- North ballroom
- Foyer prior to flashover in the south ballroom
- Foyer following flashover in the south ballroom

Other fire characteristics

- Smoke layer depth
- Velocity of smoke/fire front
- Mass product in smoke layer
- Oxygen concentration in smoke layer
- Visibility in smoke layer
- Flame length (extension)
- Flame spread
- Potential response of sprinklers
- Potential response of smoke detectors
- Fire duration

Nelson (1987) conducted an engineering analysis of the development and growth of the fire and its effects that led to a description of the course of this fire, fitting the available information. For example, he concluded that approximately 40 s after flashover in the south ballroom, a deep, hot, toxic smoke front traversed the lobby, forcing the occupants to flee and blocking the exits from the casino to the lobby. Additionally, a wall of flame lapped out of the ballroom up to the wooden ceiling of the foyer and across a major portion of that ceiling. Very quickly, this added fuel to the fire. At about 2 min after the flashover, most of the glazing surrounding the foyer had failed and a sudden flow of hot gases and unburned fuel traversed from the foyer into the casino. It quickly changed to a flame front that swept the length of the casino in about 20 s. A large body of flame then broke through the windows of the west wall of the casino. Table 3.1 lists the specific fire characteristics described in the analysis. Figure 3.13 shows predicted smoke layer temperature in the north and south ballrooms and in the foyer. These are consistent with the patterns of damage observed after the fire. Nelson's report illustrates the value of analytical, engineering methods, and how they should be used to evaluate not only potential fire hazards but also the impact of different circumstances and different fire protection measures on the course of a fire.

It was believed that the fire was set intentionally because of labor dissatisfaction. Inevitably, the extensive litigation that ensued included much independent analysis to determine behavior, response, and performance of any materials or structural assemblies that might have contributed to the resulting fatalities. Owing to the adversarial nature of litigation, most of this work was cloistered until the time of legal disclosure. A panel was convened nine years after the fire to discuss some of the research outcomes from the diverse analyses of this fire (Lund, 1995). The

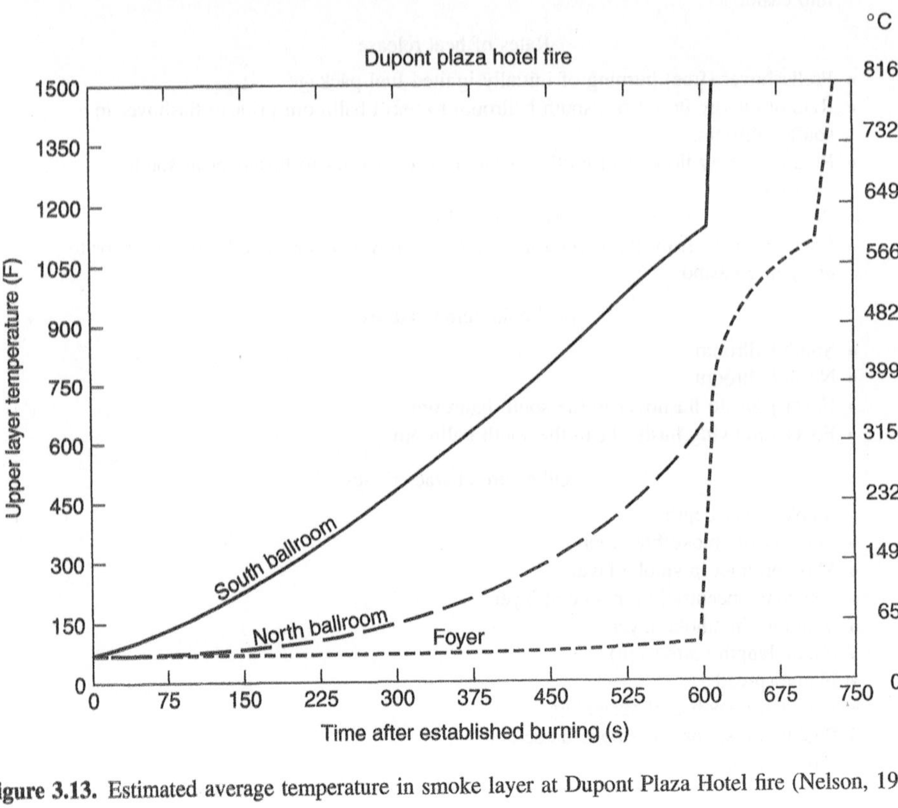

Figure 3.13. Estimated average temperature in smoke layer at Dupont Plaza Hotel fire (Nelson, 1989)

reported accomplishments included a new intermediate scale calorimeter, more advanced models of glass breakage in fires, further development of scale modeling of smoke flow in buildings, and enhanced integration of Computational Fluid Dynamics (CFD) models with computer animation.

3.3 Fires in which extensive spread of smoke was a major factor

3.3.1 BEVERLY HILLS SUPPER CLUB, KENTUCKY, 1977

Around 8.45 p.m. on the evening of 28 May, 1977, a fire occurred at the Beverly Hills Supper Club at Southgate, Kentucky. In addition to the total destruction of the Club, the fire resulted in the death of 164 persons. Best (1978) and Lawson (1984) give detailed information about this tragic fire.

The Club had dining facilities, a nightclub with live entertainment, lounges, and a number of rooms used for private parties. A rebuilt club was opened in 1972, but there had been numerous addition and alterations since then, the latest being in 1976. The total floor area had become approximately 5000 m^2 by the time of the fire (see Figure 3.14).

The building was mainly unprotected, noncombustible construction. The ground floor of a small two-story front section contained the main entrance, foyer, main dining room, main bar area, the Zebra Room, office areas, coat-check room, and part of the Viennese Room. The upper story contained a number of small party rooms, lavatories, and dressing rooms. The remainder of the building was single story. It contained the other part of the Viennese Room, the kitchen, various utility and storage areas, the Cabaret Room, the Empire Room, and the Garden Rooms.

The interior finishes were primarily wood or dense fiberboard panels on the walls and carpet on the floors. There were suspended ceiling assemblies with noncombustible ceiling tiles, which supported recessed lighting fixtures. In the oldest part of the building where the fire started, evidence was found of combustible tiles in ceiling assemblies installed earlier and left in place when the noncombustible assembly was installed beneath. Furnishings consisted of tables, tablecloths, and other dining and entertainment accessories including padded, vinyl-clad chairs throughout the various rooms.

There were ten exits from the building, including an employee exit from the kitchen. A main corridor connected all ground floor dining areas and entertainment areas to the main entrance. This corridor had no smoke or fire partitions. There were no internal fire division walls anywhere in the building, no smoke or fire detectors, no fire alarm system, and no automatic sprinkler system.

When the fire occurred, there were about 3500 patrons and 250 employees in the club. All the major rooms and most of the smaller rooms were occupied. Investigation established that the fire started in the unoccupied Zebra Room, at the front of the building (Figure 3.14). The most probable cause was electrical and combustibles located there, which would have fed the fire. The concealed combustible tiles and wooden supports provided fuel for continued fire spread in the concealed spaces. Evidence indicated that the fire burned for a considerable time prior to its discovery at 8.45 p.m. The county police-fire communications center received notice of the fire at 9.01 p.m. Despite attempts at extinction, flashover occurred in the Zebra room. The fire subsequently broke out of the room through the double doors at the north end and then spread rapidly throughout the building.

The biggest crowd, about 1000 persons, was in the Cabaret Room, some 45 m away from the room of origin, down the main corridor (Figure 3.14). Most of the victims were occupants of the Cabaret Room. After a bus boy warned patrons in the room that there was a fire and indicated the emergency exits, some exiting began. Soon after, smoke came into the Cabaret Room through its main entrance (from the main corridor). This main entrance provided the exit used normally by patrons (in nonfire conditions). Two other exits were available. One, at the northwest corner, led

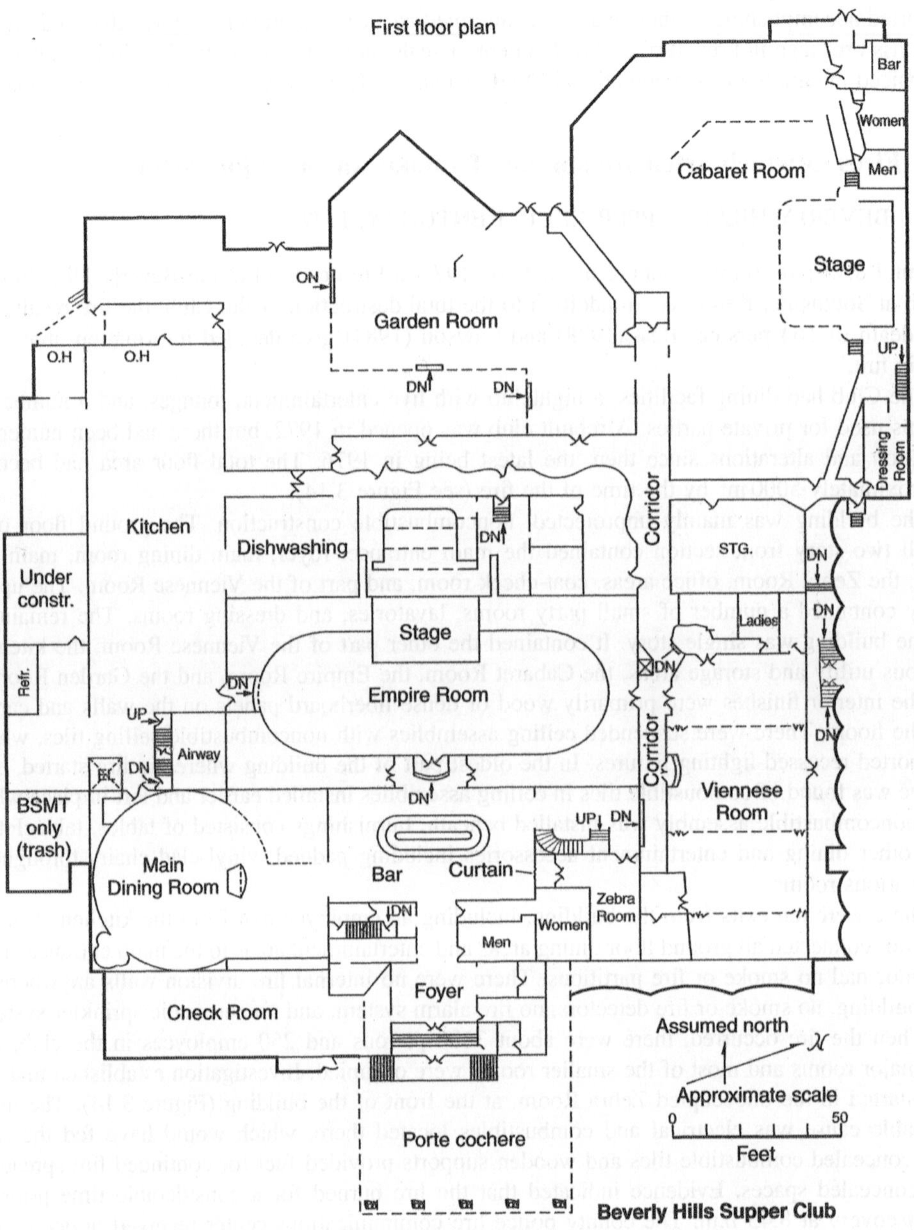

Figure 3.14. Floor plan of Beverly Hills Supper Club (Emmons, 1983)

through double doors to a service bar area and then to another set of double doors to the outside. The other, at the northeast corner, led through a door and across a short corridor to a single door to the outside.

Bright (1977) undertook a qualitative analysis of fire spread from the Zebra Room to the Cabaret Room. He concluded that the rapidity of the spread down the main corridor, somewhere between 2 and 5 min was undoubtedly a factor in the large loss of life in the Cabaret Room.

The corridor had combustible wall linings and a combustible carpet assembly on the floor, which would have assisted rapid spread. Emmons (1983) applied principles of fluid dynamics to derive numerical estimates that would explain the fire behavior. There was an apparent discontinuity, because the rate of fire and smoke spread was minimal for about 15 min after discovery and then it suddenly increased in the long corridor, extending about 45 m in less than 5 min. He hypothesized that smoke migration was minimal initially because with all doors closed there was no significant force to move it. There was a strong movement of smoke in a northerly direction once the doors in the Garden Room were opened, which is consistent with reports that the prevailing wind was from the south. Emmons estimated a rate of smoke travel, and Figure 3.15 shows a plot of smoke volume against time. This was based on principles of conservation of energy and mass in a fluid network, supplemented by eyewitness information. He also developed an estimate of the rate at which the flame spread down the north–south corridor and he concluded that there would have been rapid spread regardless of the combustibility of the linings.

3.3.2 FAIRFIELD HOME, NOTTINGHAMSHIRE, 1974

In December 1974, a fire took place at the Fairfield Home, Edwalton, Nottinghamshire, UK, in which 19 people died. There was a Committee of Inquiry that reported it in July 1975. (DHSS, 1975). This fire was of importance, in that it was one of those that had taken place in a CLASP building, and there had been concern about these buildings for some five years.

CLASP means Consortium of Local Authorities Special Programme. This organization had brought together a method of construction that was a light pin-jointed steel frame on a thin slab

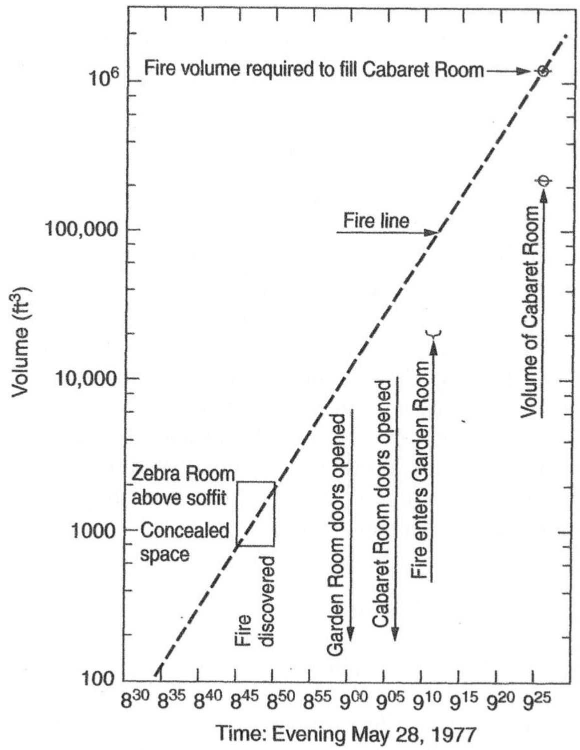

Figure 3.15. Fire growth in the Beverly Hills Supper Club (Emmons, 1983)

foundation having structural floor and roof diaphragms. The building was designed to be used in areas where there might be subsidence and where it could give in in various directions and not crack. The design, however, involved the use of a large void or cavity in which the pin structure was erected. This cavity extended across the whole area of the building with the ceiling of the lower floor as the lower surface of the cavity and the floor of the upper floor, or the roof, as the upper surface. The structure could be a number of stories high and in these buildings it was postulated when they were designed first that the ceiling and floor acted as a membrane protection to the steel frame, thus giving it the necessary measure of fire resistance between the floors.

The dominant factor in this type of property was the presence of voids that could cause both the spread of smoke and the spread of fire. In many instances, the upper surface of the void was of combustible materials, usually timber but occasionally plywood; the lower surface of the void might have been constructed from fiber insulating board or perforated board. Voids extended horizontally across the whole level of the building.

The Fairfield Home was a single story building made up of a central area and five outlying dormitory houses, all connected to the central area by corridors. A single undivided roof covered the whole building from end to end. The fire occurred at about 2 a.m. in a bedroom in house no. 1 and probably burned for some time before effective action was taken. The Committee of Inquiry decided that it was substantially the early travel of smoke through the voids that caused many of the fatalities. Nine people died in house no. 1 in which the fire started. In addition, smoke traveled through the void into house no. 2 and in this way was thought to bypass fire doors and lobbies leading to the corridor between the two houses. Nine people died in house no. 2, most of them in bedrooms with the bedroom door shut. Also the fire traveled to many other parts of the building, particularly house no. 5, which was substantially destroyed. The ceiling of the building was plasterboard, 3/8 inch thick, dropped into aluminum tees that were supported by mild steel straps. A glass fiber insulating quilt was laid on the ceiling but did not cover it all directly. The ceiling in the area where the fire started was found to be much distorted and disturbed. The underside of the roof was wood but there was bitumen in the roof structure as well. In certain areas of the communicating corridors, the plasterboard suspended ceiling was perforated. The door of the bedroom in which the fire started was open so there was a plentiful supply of air for fire development there, and smoke evolved could either leak through into the void directly or pass into a corridor where perforations conducted the smoke into the void. Smoke could also pass into closed bedrooms by penetrating the ceiling at the outer edges.

The dominating factor in this incident was the deep inadequacy of fire safety in the building itself due to the presence of the extensive void. The building was of a popular design, and a large number of such buildings had been built before sufficient experience was accumulated to appreciate the hazard. In this way, it has similarities with other manufactured or built items, which have a design fault that does not show up until an extensive amount of capital has been invested. This left a major problem of what one did to improve the safety of the many other CLASP buildings so that at least the hazard could be lived with even though it was impossible to eliminate. The judicious combination of fire detection, fire stopping, and subdivision of the voids served to reduce the risk to a tolerable level.

3.4 Fires associated with explosions

3.4.1 MV BETELGEUSE, IRELAND, 1979

MV Betelgeuse was a tanker of 61,776 gross registered tonnage, carrying a cargo of 75,000 tonnes of Arabian heavy crude and 40,000 tons of Arabian light as well. It was off-loading at a jetty about 400 m off Whiddy Island in Bantry Bay, Ireland, when the first of several explosions

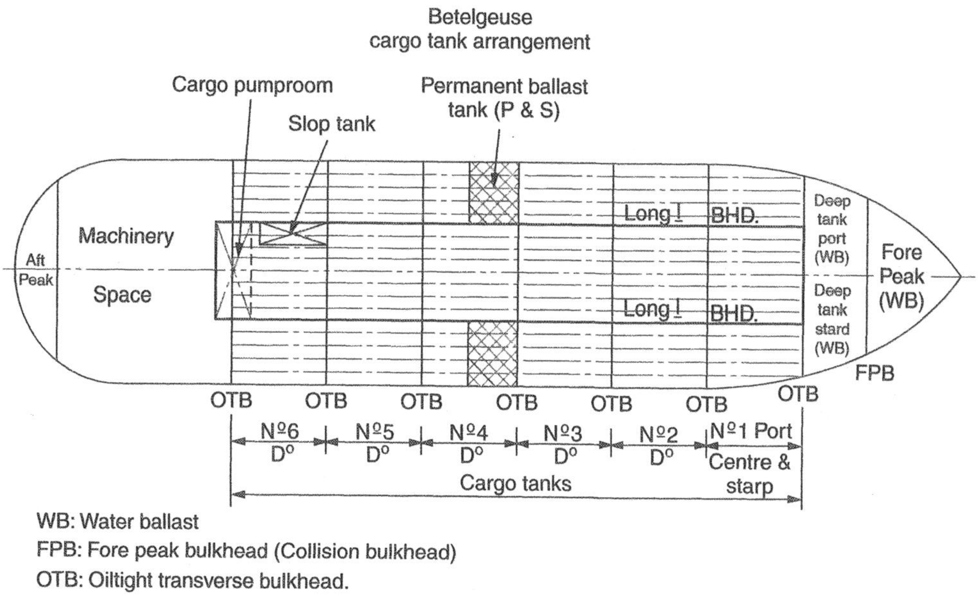

Figure 3.16. MV Betelgeuse as tied up at the jetty

occurred. An account is given in Costello (1980). Figure 3.16 is a diagram of the ship as it was tied up at the jetty showing the arrangements of the main tanks of the ship. There were 18 tanks in 6 groups of 3, each group of 3 consisting of a center tank, and port and starboard wing tanks. Number 4 wing tanks were further divided to provide two ballasting tanks 4(a), 4(b).

At about 00.30 on 8 January, it had already off-loaded its heavy crude from number 1 and 6 tanks and the central tanks, prior to off-loading the light crude from wing tanks 2, 3, 4(b), and 5. Certain water ballasting activities were being carried out, mainly involving the 2 to 5 center tanks that had discharged cargo. A muffled explosion took place in tanks 4(a) that were very near the center of the ship. This broke the back of the ship just aft of number 3 tanks and caused a release of light crude from wing tanks 3. A fire was started near the center of the ship, but as crude oil was released the fire soon found its way around the side of the ship. At 1.06 a violent explosion occurred in the rear part of the ship in all three number 6 tanks and in the center tank of number 5. The back of the ship was further broken and there was a further major release of light crude and an intensification of the fire. There was no connecting passageway from the jetty to the island, and 50 people who were on the ship and the jetty died as a result of the fire. Figure 3.17 shows the ship broken into three parts (central part under water) some 12 h later.

Evidence given at the public inquiry indicated that the primary cause of the disaster was the failure of the structure of the ship as a result of the stresses caused by the discharging and ballasting operations. Parts of the ship from which the cargo had been discharged were subject to an upward force, and parts in which cargo or ballast was present were subject to a downward force. Over the period of operation of the tanker there had been substantial corrosion on the structural members of the ship leading to wastage of these members. Calculations based on the load ballast conditions of the ship at the time of the disaster showed that the longitudinal structural members near the deck of number 4 ballast tanks would have buckled under the strain. There was strong evidence that this buckling took place and that it occurred prior to the explosions in the 4(a) tanks. It was postulated that this buckling caused structural failure between the vapor space of

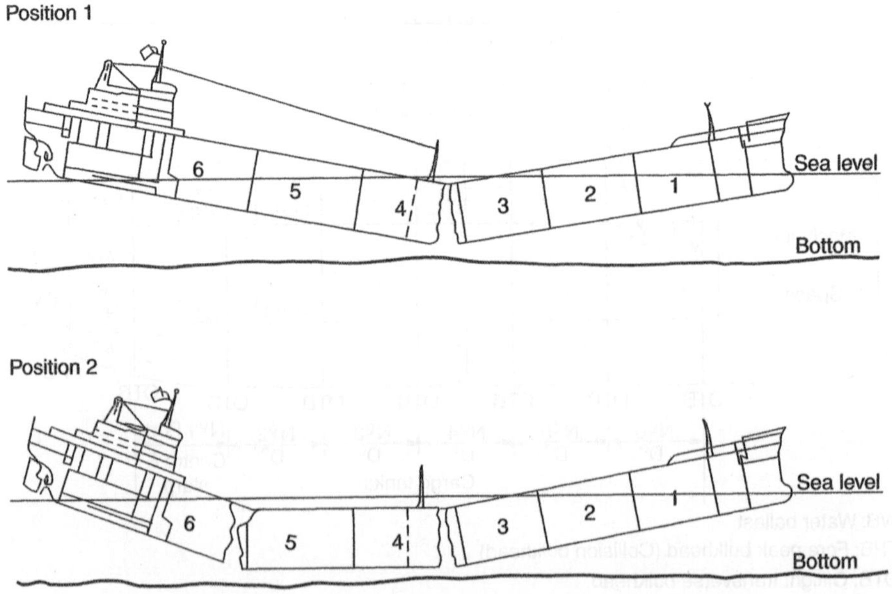

Figure 3.17. MV Betelgeuse some 12 hours after the explosion

the number 3 wing tanks containing light crude and the number 4 ballast tanks, allowing vapor to enter and an explosive mixture to develop within the latter. Moreover, rubbing or impact between steel parts caused by the buckling process could have established sparks sufficiently incendiary to ignite the gas mixture. The resulting explosion in the ballast tanks caused the bottom plates of the ship to fracture. This would have resulted in a massive release of light crude on both sides of the ship.

Among a number of recommendations, the tribunal strongly supported the use of inert gas to displace air in spaces above tanks and to replace discharged cargo. Also closer supervision of structural stresses during discharging and ballasting operations and efficient measures to monitor and counteract corrosion in tanks were called for.

3.4.2 RONAN POINT, LONDON, 1968

Ronan Point was a 22-story block of flats, with 5 flats per floor, on a concrete podium containing garages and a car deck. It was built using the Larsen Nielsen system, an industrialized method chosen to save using skilled labor. At about 05.45 on 16 May, 1968 there was an explosion in Flat 90, at one corner of the eighteenth floor (see Figure 3.18). There was the same layout in the flats above and below Flat 90. A full account of the incident is given in the Ministry of Housing and Local Government (1968).

The explosion blew out the non–load-bearing face walls of the kitchen and living room and also the external load-bearing flank wall of the living room and bedroom of the flat, thus removing the corner floor slabs of the floor above, which then collapsed (see Figure 3.19). The flank walls and floor above this collapsed in turn, and the weight and impact of the wall and floor slabs falling on the floors below caused progressive collapse of the floor and wall panels right down to the level of the podium (see Figure 3.20).

The first Fire Brigade officer to enter the remains of Flat 90 found that a fire had good hold on the contents of the kitchen and bathroom and part of the entrance hall, being fed by ignited town

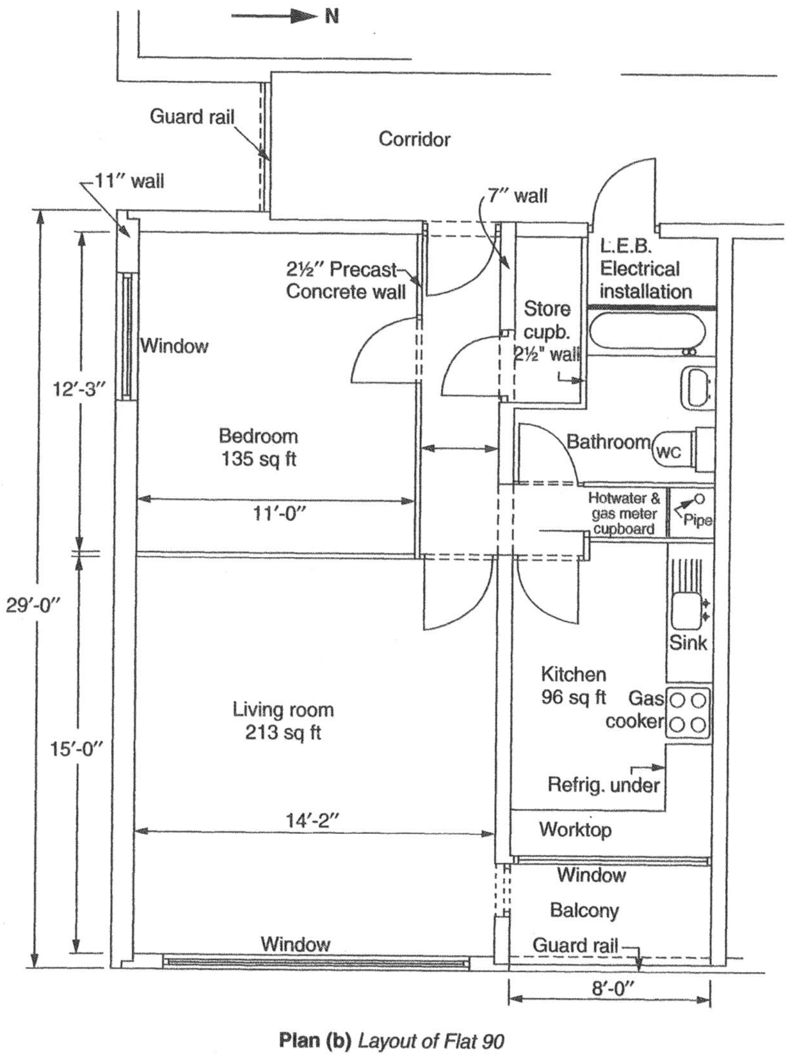

Figure 3.18. Plan of Flat 90, Ronan Point, before the explosion

gas that was escaping from a supply pipe in the kitchen. One jet of water quickly brought the fire under control, and when the main gas cock was turned off, the burning gas was immediately extinguished.

There were four deaths due to multiple crushing injuries, two being on the 19th and two on the 17th floor. Seventeen people were injured, one dying later from an unrelated cause and the others suffering no permanent injury or disfigurement. The person in whose flat the explosion occurred found herself lying on the floor when she recovered consciousness, made her own way out and suffered from minor shock and burns.

All the living rooms at the corner were almost completely destroyed on each floor, as were the bedrooms on the upper floors, down to the sixteenth floor. Below this, the bedroom floors held and damage was not extensive (see Figure 3.20). There was a main load-bearing cross wall

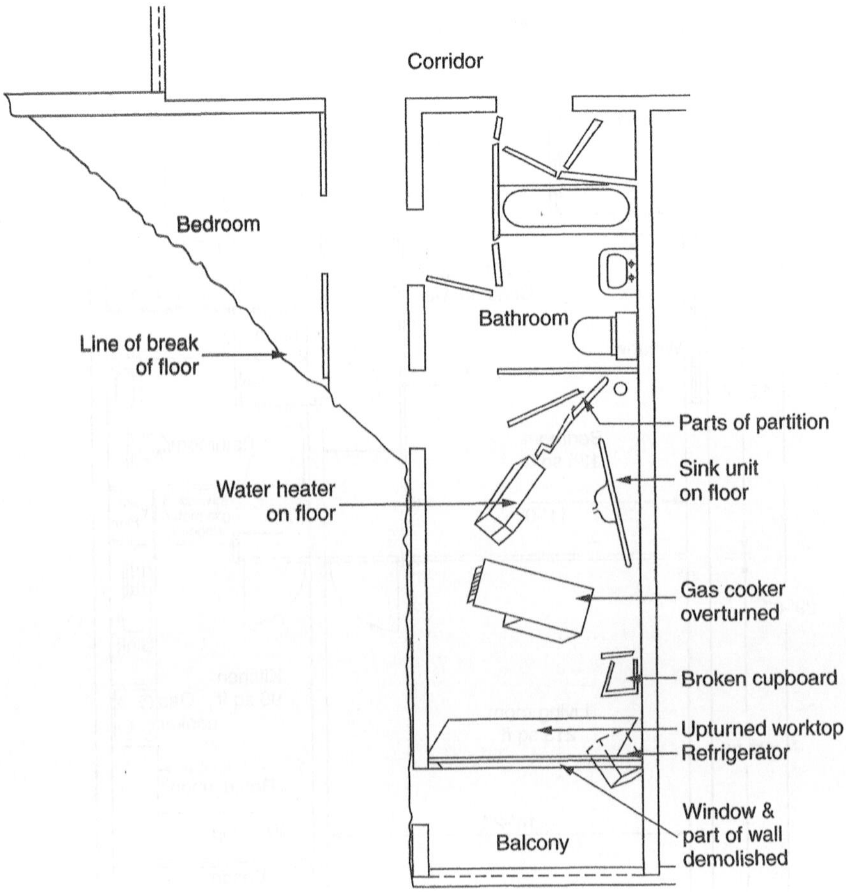

Plan (c) *The extent of the damage to Flat 90*

Figure 3.19. Plan of Flat 90, Ronan Point, after the explosion

between the living room and the kitchen and it did not fail. Except for damage to the kitchen of flat 90 caused by the explosion, the kitchens were relatively unaffected. Apart from the one corner, the block was very little affected either by the explosion or the subsequent collapse. The flat immediately opposite Flat 90 suffered from blast, as did the fire doors and lift doors on that floor. There was virtually no blast damage elsewhere and no visible damage.

The Inquiry by the Ministry of Housing and Local Government (1968) stated that the cause of the disaster was a town gas explosion in Flat 90. The gas had escaped into the flat due to the failure of a substandard brass nut joining the flexible connection from the gas cooker to the gas supply pipe, and the explosion occurred when the tenant struck a match to light her cooker. It was concluded that the explosion was not of exceptional violence, being approximately 83 kPa (12 lb/in^2) in the hall and probably not more than 70 kPa (10 lb/in^2) in the kitchen. This was within the "normal" range of domestic gas explosions. It was estimated from fire brigade reports of 1966 that the frequency of gas explosions involving town gas in premises supplied with gas was approximately 8 per million dwellings, of which 3.5 per million were of sufficient violence

Figure 3.20. Aerial view showing Ronan Point and Merritt Point

to cause structural damage. In the light of these figures, the Inquiry accepted that town gas was a safe and acceptable domestic fuel, but the nature of the risk per dwelling in high blocks is transformed if progressive collapse can occur as the result of an explosion in one flat. It was noted that when structural damage was involved, the cause was far more likely to be faulty equipment than a fault on the part of the user.

There were three possible ways of dealing with the hazard: preventing an explosion by disconnecting gas supplies from tall blocks; preventing the load-bearing walls being pushed out by using ventilation, explosion relief and strengthening; preventing progressive collapse by providing alternative load paths.

The Inquiry rejected the first method, on the grounds that gas was justifiably regarded as a safe and acceptable fuel in domestic premises and was likely to become even more popular when supplies from the North Sea came on line. Furthermore, there remained the possibility of explosions from other substances than town gas, as well as other forms of accidental damage. However, a statutory requirement for inspection of gas installations was recommended. The Inquiry concluded that if the windows had been open in Flat 90 then the explosion would not have occurred, and recommended that consideration should be given to improving ventilation in flats in high blocks. Finally, the Inquiry recommended new provisions in the Building Regulations to deal with progressive collapse, taking into account wind, fire, and explosion.

3.4.3 PIPER ALPHA PLATFORM, NORTH SEA, 1988

The Piper Alpha Platform came on stream in 1976 (Drysdale and Sylvester-Evans, 1998). At the time of the disaster it was producing some 125,000 barrels of oil per day, one third of its total capacity. At about 2200 on 6 July, 1988, an explosion occurred at the Production Level, followed swiftly by a major fire. The platform was destroyed and evidence from the survivors was thus critical at the Cullen Inquiry (1990). The east elevation of the Platform before the explosion is shown in Figure 3.21.

The Production Level was divided into four modules

A – the well heads
B – separation of oil
C – gas compression with gas condensate collected in a drum beneath this module
D – control room, workshops, switch gear, electrical power generator, diesel power generator

The accommodation modules were located above Module D, including the east replacement quarters (ERQ), which was a four-story building used as the main muster area. Firewalls separated the modules – 4.5 h rated between A/B and B/C, and 6 h rated between C/D – but they were not rated for explosion overpressure.

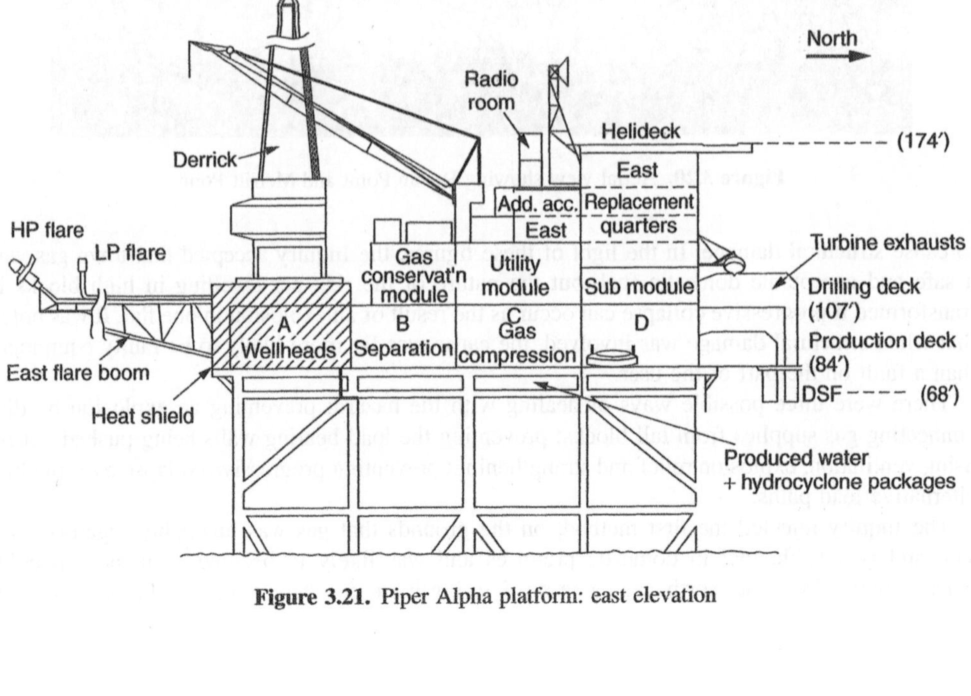

Figure 3.21. Piper Alpha platform: east elevation

The Cullen Inquiry concluded that the initial explosion was in module C, fueled by a release of condensate from a blind flange that had not been fitted securely. The ignition source was not established. The overpressure from the explosion caused failure and breakup of B/C and C/D firewalls. Ejection of wall panels caused a rupture of pipe work in module B – thus releasing crude oil – and also failure of a condensate pipeline passing from C to B, which created a fireball some 28 m in diameter. Unburned fuel-rich gases from the fire in module B, and residual fuel from module C flowed towards module D to produce a fire beneath the ERQ. A crude-oil fire developed in module B and continued until at least 2250. A riser failed catastrophically at 2220 resulting in a massive fireball and leaving a torch flame to develop beneath the whole platform, at which time the atmosphere in the ERQ had already deteriorated badly. A second riser failed at 2250 and a major fireball developed. Between 2300 and 2330, the two other risers failed, the derrick collapsed and the center of the platform sagged. Shortly thereafter, the ERQ toppled into the sea.

There were 61 survivors out of 226. After the explosion occurred, the smoke plume prevented access to either modules B or C, but personnel were able to move through module A and then make their escape at a lower level. Of the 135 bodies recovered (79 being from the ERQ), only 4 indicated death from burning. Most died from inhalation of smoke and gas. Smoke prevented access to lifeboats and helicopter access to the helideck was not possible. Areas of refuge were not effectively sealed from smoke.

As a result of the disaster, much effort has been expended worldwide in the mitigation of effects from explosions and fires. It is possible to estimate overpressures using various empirical relationships. Blast walls and relief panels have been installed accordingly. Pipeline inventories have been isolated from fixed installations. Emergency shut down valves have been increased in number and installed in protected locations. The importance of preventing hydrocarbon leaks has been recognized. Nevertheless, over a two-year period up to October 1994 some 523 releases were reported to the UK Health and Safety Executive, 70% of which involved more than 10 kg of hydrocarbon. Fortunately, the frequency of ignition was only 3% (HSE, 1996). Efforts to mitigate the effects of smoke include pressurization of secure areas, air locks, barriers, water spray, sealing of wall penetrations.

The above measures, together with new lifetime management systems, have made considerable improvements, leading Drysdale and Sylvester-Evans (1998) to suggest that the risk of a major disaster offshore may have been reduced by about 90%.

References

Best, R L (1978). Tragedy in Kentucky. *Fire Journal*, **72**(1), 18–35.

Bright, R G (1977). An analysis of the development and spread of fire from the room of origin (Zebra Room) to the Cabaret Room. Appendix C of Best, R L. *Reconstruction of a Tragedy: The Beverly Hills Supper Club Fire*, National Fire Protection Association, Quincy, MA.

Clark, H M, de Cabrera, F M, Droz, L, Gurman, J L, Kaye, S, Levin, B, Landron, F, Rechani, P R, Rodriguez, J R and Yoklavich, M F (1990). Analysis of carboxyhemoglobin and cyanide in blood of victims of the Dupont plaza fire in Puerto Rico. *Journal of Forensic Sciences*, **35**(1), 151–168.

Costello, J D (1980). *Report on the Disaster at Whiddy Island*, Bantry, Co. Cork on 8 January, 1979, The Stationery Office, Dublin.

Cullen, L (1990). *The Public Inquiry into the Piper Alpha Disaster*, Department of Energy, HMSO, London.

DHSS (1975). *Report of the Committee of Inquiry into the Fire at Fairfield Home*, Edwalton, Nottinghamshire on 15 December, 1974, HMSO, London.

Drysdale, D D and Sylvester-Evans, R (1998). The explosion and fire on the piper alpha platform, 6 July 1988. A case study. *Philosophical Transactions of Royal Society of London, Series A*, **356**, 2929–2951.

Emmons, H W (1983). The analysis of a tragedy. *Fire Technology*, **19**(2), 115–124.

Fennell, D (1988). *Investigation into the Kings Cross Underground Fire*, HMSO, London.

Government Office (1974). *Report of the Summerland Fire Commission*, Government Office, Isle of Man.

HSE (1996). *Offshore Accident and Incident Statistics Report 1995*, Offshore Technology Report O.T.O./96/950, HMSO, London.

Keane, J R (1982). *Report of the Tribunal of Inquiry on the Fire at the Stardust*, Artane, Dublin on the 14th February, 1981, The Stationery Office, Dublin.

Klem, T J (1987). Ninety-seven die in arson fire at Dupont plaza hotel. *Fire Journal*, **81**(3), 74–77.

Klem, T J (1989). Los Angeles high-rise bank fire. *Fire Journal*, **83**(3), 72–74.

Lawson, R G (1984). *Beverly Hills: The Anatomy of a Nightclub fire*, Ohio University Press, Athens, OH.

Lund, D P (Ed.) (1995). Defining fire and smoke spread dynamics in the Dupont plaza fire of 31 December 1986 (Panel). *Proceedings, International Conference on Fire Research and Engineering*, Society of Fire Protection Engineers, Boston.

Ministry of Housing and Local Government (1968). *Report of the Inquiry into the Collapse of Flats at Ronan Point*, HMSO, Canning Town, London.

National Fire Protection Association (1990). Guide for automatic fire detector spacing. Appendix C in *Standard on Automatic Fire Detectors*, NFPA 72E, Quincy, MA.

Nelson, H E (1987). *Engineering Analysis of the Early Stages of Fire Development – The Fire at the Dupont Plaza Hotel and Casino*, December 31, 1986, NBSIR 87–3560, National Bureau of Standards, Gaithersburg, MD.

Nelson, H E (1989). *Engineering View of the Fire* of May 4, 1988 in the First Interstate Bank Building, Los Angeles, CA, NISTIR 89–4061, National Institute of Standards and Technology, Gaithersburg, MD.

Nelson, H E (1990). *FPETOOL: Fire Protection Engineering Tools for Hazard Estimation*, NISTIR 4380, National Institute of Standards and Technology, Gaithersburg, MD.

Popplewell (1985). *Committee of Inquiry into Crowd Safety and Control at Sports Grounds, Interim Report*, Home Office, HMSO, London.

Rasbash, D J (1991). Major fire disasters involving flashover. *Fire Safety Journal*, **17**, 85–93.

Routley, J G. (1989). *Interstate Bank Building Fire*, Los Angeles, CA, May 4, 1988, USFA Fire Investigation Technical Report Series, Report 022, TriData Corp, Arlington, VA.

US Fire Administration (USFA) and Polaris Research Development (1987). *The Dupont Plaza Fire: Lessons in Hotel Safety*, USFA, Emmitsburg, MD.

4 REQUIREMENTS FROM PUBLIC AND PRIVATE AUTHORITIES FOR FIRE SAFETY

4.1 Introduction

Fire safety has been a concern of mankind for centuries, even millennia. It has been expressed in edicts usually following major fire or explosion disasters and guided by such understanding of fire and explosion as was available at the time. In the wealthier countries of the world, the present generation has inherited a large swath of legal and other requirements and associated advice. The legal requirements emanate from many government departments since they tend to gravitate to the department concerned with the risk or hazard rather than the phenomena of fire and explosion as such. In addition, local authorities, insurance companies, standards bodies, and professional organizations are concerned with setting requirements for fire safety. There is also a major international element concerned with transport fire hazards and the requirements of international trade. Most of the requirements of the relevant legal instruments are prescriptive in nature. However in recent years, particularly as more is understood of the nature of fire hazard, a functional and more open approach to fire safety legislation has been developing.

The first task for the engineer designing fire safety for a specific hazard area is to become acquainted with the legal requirements concerning the safety for this area and the relevant advice that exists on how to fulfill these requirements. In this chapter, a summary is given of legal and other requirements for fire safety in the United States (Sections 4.2 to 4.6), Canada (Section 4.7), and the United Kingdom (Section 4.8), which may be regarded as examples of countries where extensive fire safety requirements exist.

Legal requirements for fire safety change frequently and do not endure in the same way as scientific relationships. In the narrow sense, much of this chapter is already out of date. However, many of the past requirements embody ideas that were considered to be a good practice at the time and they rightly command respect. Knowledge of these requirements is an aid to judgment, which is a vital part of the evaluation of fire safety. Fire safety engineers need to know what the law currently requires of their project, but they should also understand how these requirements have been developed.

4.2 US regulatory environment

Over the years, most US jurisdictions have developed building regulations designed to provide for public safety. Building codes and similar instruments are adopted sets of rules designed to

Evaluation of Fire Safety D. Rasbash, G. Ramachandran, B. Kandola, J. Watts and M. Law
 ISBN: 0-471-49382-1

ensure that the buildings are structurally sound, fire resistant, and generally safe for the building's occupants. In the course of time, the rules have become more sophisticated as new materials, new fire protection techniques, and methods for protecting the public have been developed. In addition, the codes have become the vehicles for enactment of regulations designed to promote certain societal goals, such as providing accessibility for disabled persons and energy conservation.

Compared to other developed nations, the system of legal requirements in the United States is highly fragmented. There are many sources of regulations and many routes leading to the promulgation of standards. The fifty states that make up the United States all have their own differing constitutions and in many states the regulation of fire safety is left to the local community.

With a large population that usually does not feel very close to its federal government, there are misgivings about intervention by a higher authority. Allegiance to the idea of "self-regulation" persists. Thus, most regulation of the public health, safety, and welfare is relegated to the State Governments and in turn to municipalities and other local or regional jurisdictions. This autonomy of local governments has led to a wide variety of regulations creating some difficulty for design professionals who practice across jurisdictional boundaries.

4.2.1 CODES AND STANDARDS

In the regulatory system, the terms *code* and *standard* are often used interchangeably. In addition, the phrase "model code" occurs frequently in the literature. This interchanging in practice is acceptable as long as the content makes clear what is intended. There are some specific differences in the concepts associated with these terms and they should be noted.

A code is a legal document that sets down minimum requirements to protect the public health, safety, and general welfare. A jurisdiction may choose to write its own code. In 1968, for example, 1.6 million dollars was spent in writing the New York City building code. However, it is more common for a jurisdiction to incorporate a model code into its legal requirements. Various organizations write model codes that are designed to be suitable for adoption by a variety of legislative authorities.

A standard is a rule for an orderly approach to a specific activity. Standards are narrower in scope than codes. They address a particular topic such as a method of test or a component specification. Like model codes they are written for general use and are not mandatory until written into a law or adopted by a local regulatory authority. Codes usually incorporate or reference standards. The Building Code of the City of New York references over 300 standards. The United States has more than 620 private sector groups that issue standards (SP 806, 1996). See Boring *et al.* (1981) and Cote and Grant (1997) for more detailed information on US codes and standards.

Bodies that issue model codes and standards in the United States are described in Sections 4.3 to 4.6.

4.2.2 CONSENSUS CODES AND STANDARDS

Model codes aim to reach consensus in order to achieve a wide acceptance. They are written to express, insofar as practical, a common agreement among interested parties (manufacturers, designers, users, government, etc.) on what constitutes the appropriate design and implementation of specific fire safety elements. Consensus standards generally meet certain criteria to ensure due process and avoid restraint of trade.

Consensus is usually something more than a majority agreement but less than unanimous agreement. Most often it involves compromises of various positions. Consensus standards may have various degrees of consensus depending on the parties represented in the standardization process.

Full consensus implies participation of and agreement among all interests. These might, for example, be categorized as producers, users, consumers, and general interest. Full consensus standards are considered the most desirable where public weal is at issue.

American National Standards Institute (ANSI) is a voluntary federation of over 400 US organizations that write standards. Its function is to coordinate national consensus standards, and only standards that meet the ANSI criteria for full consensus are accepted. ANSI is not itself a standards-writing organization. Every standard with an ANSI designation was written by another organization and submitted to ANSI for adoption as an American National Standard. ANSI is the single official representative of the United States to the International Standards Organisation (ISO). To become an international standard, therefore, a US standard must first be accepted by ANSI.

Limited consensus codes and standards, in which a nonrepresentative group such as building officials or members of a particular industry makes the decisions, are also prevalent. The subject of the US model building codes has become a contentious issue because members of the fire service and other interests are excluded from the voting body.

The character of standards prepared by the US government has also been troublesome when small groups of people – who may or may not be experts in the field – autocratically decide what the standards shall be. The threat of a mandatory national building code haunts many in the building industry. Consequently, it is probably correct to observe that fear of federal intervention has been a salient factor in motivating the bodies that write US codes and standards.

4.2.3 LIMITATIONS OF EXISTING CODES

A recognized limitation of many US codes is that they are designed for new buildings only. Another limitation is their prescriptive nature, which is not sympathetic to innovative building design or indeed to the upgrading or change of use of existing buildings. For these, a performance approach is often more appropriate.

Existing and historic buildings

Building codes and standards have generally been written to prescribe methods, materials, or criteria for new construction. They assume that an architect, owner, or builder is starting from scratch. When applied to existing buildings, they can result in a major modification of those buildings. Often, such modifications do not lend themselves to historic buildings, and can result in elimination of some of the historic structure's most important features, for example, ceiling heights, decorative wood detailing, or open monumental stairways (Watts, 1999a).

The earliest approaches to rectifying these conflicts were simple provisions that allowed wide discretion to the code enforcement officers to determine the appropriate level of safety. Often, this discretion was only exercised when the project had no change of use or occupancy. More specific guidance was developed in the rehabilitation codes and documents adopted in the late 1970s. However, these standards also failed to provide adequate guidance for the projects where the retention of historic elements and character were significant factors. Most recently, several states have adopted specific codes for historic buildings (Kaplan and Watts, 2000).

Performance initiative

The United States is following the lead of other countries such as the United Kingdom, Australia, and New Zealand in pursuing a performance approach to fire safety. The application of component performance evaluation has always been available through equivalency clauses in the codes. Now, however, the *Life Safety Code* (NFPA 101, 2000) has a systemic performance option and the

Society of Fire Protection Engineers has published the *SFPE Engineering Guide to Performance-Based Fire Protection Analysis and Design of Buildings* (SFPE, 2000).

4.2.4 CODE ADOPTION AND ENFORCEMENT

In the United States, there are at least 14,000 regulatory building codes; some estimates are as high as 20,000 (Toth, 1984). While the three principal model building code groups have brought some order and uniformity within their areas, building and fire codes are the responsibility of local governments at city, county, or state level. Local governments are free to develop their own codes or they may adopt a model code. Most states and local jurisdictions use the model codes only as a starting point on which to append amendments and additions.

Fire safety regulations are enforced through the power of building permits, licensing authority, and inspections required for occupancy of buildings. In most jurisdictions, the building department administers the building codes and the fire department or fire marshal is responsible for compliance with the fire code.

Typically, the building department reviews plans to ensure that a new building meets the requirements of the building code and inspects the completed structure for compliance before permitting occupancy. Once an occupancy permit has been issued, the building department has no further responsibility (except for major changes in construction or occupancy type). The fire codes then become applicable and are in force for the life of the building. Thus there is a traditional distinction between building codes and fire codes in terms of content, enforcing authority, and period of application.

A fire marshal is the chief fire prevention officer of a state or municipal jurisdiction. Where a statewide fire code has been adopted, there is typically an office of the fire marshal in the state's governmental organization. A fire marshal is often also charged with the responsibility for fire investigation. In most jurisdictions, a fire marshal or equivalent has the responsibility to maintain fire safety in existing buildings occupied by the public.

The National Association of State Fire Marshals (NASFM) was formed in 1989 to represent its members, the most senior fire officials of each state and their staffs. It was believed that through such an organization, common, uniform communications could be disseminated throughout each state. A common forum for problem solving, research, and development could be established through this organization. It also gave the group the ability to speak as a body in a national arena with one voice, representing the senior decision makers of each state.

4.3 National Fire Protection Association

The National Fire Protection Association (NFPA) is a nonprofit organization with approximately 68,000 members. Membership represents a broad range of interests including practically anyone who has an interest in fire safety. About 10% of the membership is from outside the United States, representing more than 70 countries. Basic technical activity of NFPA involves development, publication, and dissemination of current consensus standards. There are more than 291 NFPA technical documents developed by 211 technical committees made up of over 5500 individuals. NFPA is the publisher of the *Fire Protection Handbook* (Cote, 1997) and the *National Fire Codes*. In addition to the *Life Safety Code* described below, some of the more widely used NFPA codes and standards that make up the *National Fire Codes* include the following:

NFPA 1, Fire Prevention Code
NFPA 13, Installation of Sprinkler Systems
NFPA 30, Flammable and Combustible Liquids Code

NFPA 70, National Electrical Code
NFPA 72, National Fire Alarm Code
NFPA 550, Fire Safety Concepts Tree
NFPA 909, Code for the Protection of Cultural Resources
NFPA 914, Code for Fire Protection of Historic Structures
NFPA 1600, Disaster Emergency Management.

4.3.1 NFPA CODES AND STANDARD-MAKING PROCESS

The process for promulgating and changing NFPA technical documents is summarized in Figure 4.1. Any interested individual may submit a proposal for a new standard or a change to an existing standard. The appropriate technical committee will discuss, develop, and revise the proposal and issue a "Report on Proposals" that is made available to any interested party. Every comment received along with the corresponding action by the committee is then published as "Report on Comments," also available to any interested party.

Submitters of comments then have the right to present their comment to the association members at either the spring (annual) or fall meeting, where the body of assembled members will vote on the committee's report. Usually, the members will endorse the action of the technical committee, for it is understood that the technical committee should have both the thoroughness and expertise to deal with that particular subject. Yet, many times the body assembled will favorably receive a motion based on a comment from the floor and reverse a technical committee action. A 13-member Standards Council, which reviews the procedural actions of the committees and reports to the Board of Directors of the NFPA, makes the final determination before a new fire standard or a revision of an existing one is issued.

4.3.2 PERFORMANCE OPTION

In 1993, NFPA established an in-house task group to study the implications of performance-based design and the role of the NFPA in the development of performance codes and standards. In consequence of this study, NFPA is pursuing a dual-track approach for its codes and standards. Many

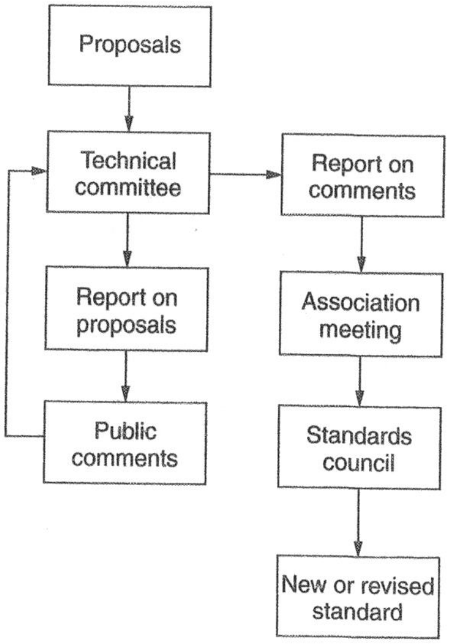

Figure 4.1. NFPA Process for making codes and standards

future NFPA documents will include both performance-based and prescriptive-based options. Maintaining both the prescriptive and performance options within a single document is intended to formalize the options, keep both the approaches on par, and encourage mutual improvements in the codes and standards (Puchovsky, 1997).

In future, the NFPA documents will include sections on fire safety goals, objectives, assumptions, fire scenarios, and evaluation. While incorporation of these elements is prompted by the development of the performance-based option, many of these aspects will also apply to the prescriptive option and their consideration will help the prescriptive requirements to become more scientifically based. The *Life Safety Code*, described below, and several other NFPA standards have incorporated a performance option.

4.3.3 LIFE SAFETY CODE

In the United States, the NFPA *Life Safety Code*® (NFPA 101) is the most widely used guide to life safety from fire in buildings. The Life Safety Code has its origins in the 1918 *Factory Exits Code* that evolved as a result of a disastrous fire that killed 146 factory workers in 1911. Subsequently, a *Department Store Exits Code* was published and this was shortly followed by the publication of a *School Exits Code*.

The requirements for these and other occupancies were combined with specifications for building construction and automatic fire protection into the *Buildings Exit Code*, first proposed in 1922. The *Building Exits Code* was finally adopted and published in 1927. During the next 37 years, there were 18 published revisions of this code, greatly expanding its content. In 1963, the document was reorganized and renamed the *Code for Safety to Life from Fire in Buildings and Structures*, or simply, the *Life Safety Code*. Since then, there have been 11 new editions that bring us to the year 2000 edition in effect today.

Unlike the model building codes, the *Life Safety Code* deals explicitly with existing buildings. It is divided into chapters for both the new and existing buildings. It is recognized that a significant majority of buildings that will be occupied in the future are already here, and that it is not always economically feasible or physically possible to meet the standards for new construction when a building is rehabilitated. The US Federal Government mandates compliance with the *Life Safety Code* for health care facilities treating Medicaid or Medicare patients, because of the *Code*'s coverage of existing buildings.

The 2000 edition of the *Life Safety Code* introduces a performance-based option. In this approach, life safety goals and objectives are translated into performance criteria. Fire models and other calculation methods are then to be used in combination with the building design specifications, specified fire scenarios, and explicit assumptions, to calculate whether the performance criteria are met. If the criteria are met, then compliance with the Code under the performance-based design option has been achieved (Watts, 1999b).

4.3.4 NFPA BUILDING CODE

A new effort undertaken by NFPA in the year 2000 is the production of a building code that will complement other NFPA codes and standards. It will be based on the current *EPCOT Building Code*, which was promulgated for the Reedy Creek Improvement District in Florida. The intent is to issue the *NFPA Building Code* in 2002.

4.4 Model building and fire codes

Building codes are directed at preventing structural collapse and limiting the extension of fire. Fire codes are usually directed at processes, materials, and equipment within a building for the purposes of fire prevention and protection of people and property.

From 1800 to 1900, fires destroyed 11 major US cities, killing an untold number of people and destroying hundreds of millions of dollars worth of property. As a direct result of these fires, large cities began to develop and enforce building codes. Chicago, for example, developed a building code in 1875 as a direct result of the National Board of Fire Underwriters (NBFU) threatening to discontinue the insurance business in the city after the Great Fire of 1871. By the turn of the century, most major cities had their own building codes.

Extensive losses by the fire insurance companies in the 1904 Baltimore, Maryland conflagration, prompted NBFU to publish in 1905 the *Recommended National Building Code* to guide the municipalities concerned with reducing fire hazards in and about buildings. This was the only nationally recognized "model" building code until 1927, when the *Uniform Building Code* was published by the forerunner to the International Conference of Building Officials (ICBO). The Southern Building Code Congress International (SBCCI) published in 1945 what is now the *Standard Building Code*, in reaction to the unique problems affecting construction in the South. The Building Officials and Code Administrators International (BOCA) first published its *Basic Building Code* in 1950. In 1985, BOCA began using the title *National Building Code*, assumed from the now defunct original NBFU model code.

These model code groups were initially established to enable building officials and their respective jurisdictions to seek solutions to common problems and avoid inconsistencies in code development and enforcement. Since codes are usually adopted at the local level, each group covers a specific regional area in part to account for the geographic or climatic differences.

Each group writes, maintains, revises, and distributes a series of model codes including a building code, fire code, mechanical code, and plumbing code and other codes and documents helpful to jurisdictions in administering codes. These model codes are published every three years and updated annually. They are often modified by the city, county, and state jurisdictions producing thousands of variations.

4.4.1 MODEL CODE WRITING ORGANIZATIONS

There are now three principal building code organizations, all nonprofitable:

International Conference of Building Officials (ICBO)
This organization is owned and controlled by its member cities, counties, and states, which tend to be heavily, but not exclusively, concentrated in the western states. The ICBO (1999a,b) publishes the *Uniform Building Code* and the *Uniform Fire Code*.

Southern Building Code Congress International (SBCCI)
This organization is also a professional society of BOCA. One of its major aims is to develop basic code provisions appropriate for the climate and building influences of the southeastern states. Its membership tends to be largely, but not exclusively, concentrated in the southern states. The SBCCI (1997) publishes the *Standard Building Code* and the *Standard Fire Code*.

Building Officials and Code Administrators International, Inc. (1996a,b)
This is a service organization for professional code administration and enforcement. Active membership is open to governmental units, departments, or bureaux that administer, formulate, or enforce laws, ordinances, rules, or regulations relating to construction, fire safety, property maintenance, development, or land use. Other categories of membership are open to the private sector. Its membership tends to be most heavily concentrated in the northeastern and midwestern states. The BOCA publishes the *National Building Code* and the *National Fire Code*.

4.4.2 INTERNATIONAL CODE COUNCIL (ICC)

The ICC was established in 1994 by the three code groups described in Section 4.4.1. It is a nonprofit organization dedicated to developing a single set of comprehensive and coordinated national codes. Up to now, technical disparities among the three model codes have made it difficult for the building industry professionals to operate in more than one region. The ICC intends to offer a single, complete set of construction codes without regional variations. The ICC published the International Building Code (IBC) and the International Fire Code in the year 2000, in completion of this goal.

Any interested individual or group may submit a code-change proposal to the International Codes for consideration by the ICC code development committees. Upon receipt of the code-change proposals, they are checked for completeness, and accepted or corrected proposals are published and made available to the membership.

The appropriate code-changes committee conducts announced public hearing on the proposals, in order to obtain as much factual information as possible about each proposed change and to guide the code-changes committee in making recommendations to the membership on disposition of the proposals. The recommendations of the committee are approval as submitted, approval as modified, or denial. The committee's recommendations are published along with the reasons and substantiation for their actions. Only eligible ICC voting members may ratify the committee decisions. The results of votes are published in annual reports of the ICC code development hearings.

All the interested parties are invited to file a challenge to the committee recommendations. Challenges are also published prior to consideration. The challenge will afford the interested party an opportunity to testify at the hearings held during the annual conferences. The final action on all the challenged code-change proposals is based on voting at the annual conferences of BOCA, ICBO, and SBCCI. The eligible voting members adopt the proposed changes as part of the code or reject them. These actions are then incorporated into the next supplement to the pertinent code or the next edition.

4.5 Other nonprofit organizations

The many other types of fire safety related organizations in the United States include standards organizations, engineering societies, and insurance interests.

4.5.1 PRODUCT STANDARDS AND TESTING

The American Society for Testing and Materials (1999) is a developer and publisher of technical information designed to ensure the quality of commodities and services and the safety of products. It has approximately 26,000 members organized into committees, about half being for the purpose of developing tests of given phenomena, and the other half for developing standards for given classes of products. These committees maintain over 5000 standards. The committee membership is balanced by requiring that the number of industrial members must always be less than the number of general interest and consumer members. The ASTM committee that is of greatest interest to the fire protection field is ASTM E-5, Committee on Fire Test Standards, which maintains product testing standards in the areas of combustibility, fire resistance, smoke and toxicity, and fire hazard assessment.

A proposal for a new standard may be originated by a member of ASTM E-5 or by someone outside the organization. The prescribed process of adopting a new ASTM fire standard is as follows. First, the proposal is brought to the attention of the chairman of the appropriate subcommittee. The subcommittee chairman, having determined the reasonableness and validity of the proposal will then appoint a task group. The proposer and other interested parties (both pro and con) will form the group, charged to prepare a need statement and one or more drafts of the proposed fire test method.

When the task group has reached some agreement on the content of the proposal it may conduct a series of comparative tests. Such tests, if properly structured, will establish both repeatability and reproducibility of the proposed fire test method. Repeatability is a measure of the degree of repetition of the test by a single laboratory. On the other hand, reproducibility is a measure of the variation in a test result when the test is conducted on the same product at different laboratories.

At this point, there will be the first formal vote by the subcommittee on a draft of the proposed fire test method. All information will be available for consideration, including the calculated measures of repeatability and reproducibility. Each negative vote must include the comments explaining the negative.

It is an important part of the ASTM procedure for due process that the subcommittee must resolve every negative. Each negative must be discussed with the voter and within the task group, and a determination made as to whether it is persuasive. If a negative vote is found to be nonpersuasive, the subcommittee's decision must be explained. There are also requirements on the minimum number of affirmative votes necessary to approve the standard.

This process is repeated for the main committee and the society as a whole. In each case negative votes must contain reasons, and these votes must be resolved, and there are required minima on the votes returned and the number of affirmative votes to carry the proposal forward. Finally, a committee on technical operations referees the entire development and voting procedure.

Committee E-5 has collected all standards related to fire safety in one publication, *Fire Test Standards*. This volume includes all the standards promulgated by Committee E-5 as well as those of other ASTM committees that involve fire characteristics of assemblies, products, or materials. The standards are also available individually.

4.5.2 UNDERWRITERS LABORATORIES

Underwriters' Laboratories (UL) is an independent organization dedicated to testing products for public safety. It was founded in 1894, after the Chicago Columbian Exposition of 1893 revealed both the awesome power of electricity and its propensity for starting fires. The UL was originally sponsored by the insurance industry. It soon became an authority on safety of electrical products and on fire prevention, later becoming independent and expanding into other safety areas. It develops and revises the standards used by UL, and other third-party safety testing and certification organizations, to evaluate the safety of consumer products.

4.5.3 ENGINEERING SOCIETIES

The Society of Fire Protection Engineers (SFPE) was organized in 1950 for the professionals engaged in fire protection or fire safety engineering. Its purpose includes advancing the science of fire safety, maintaining high ethics among members, and fostering fire safety engineering education. SFPE has chapters in Australia, Canada, Europe, and the United States. The Society sponsors the production of the *Handbook of Fire Protection Engineering* (DiNenno, *et al.*, 2001), a definitive collection of applied technical aspects of fire safety. It has recently started the development of guides of practice in fire safety engineering, the first of which is the *SFPE Engineering Guide to Performance-Based Fire Protection Analysis and Design of Buildings* (SFPE, 2000).

Other US engineering societies also produce standards that pertain to fire safety evaluation. These include construction standards from the American Society of Civil Engineers (ASCE), pressure vessel and elevator standards from the American Society of Mechanical Engineers (ASME), and process safety standards from the American Institute of Chemical Engineers (AIChE).

4.5.4 INSURANCE ORGANIZATIONS

As discussed in Section 4.4, much of the fire safety regulatory milieu has its roots in the insurance industry. The NBFU developed the first US model building code, and originated many of the fire safety standards that are now within the purview of NFPA. The industry, comprised of hundreds of individual companies, associations, and service organizations maintains a strong presence in the area of requirements for property protection. Service organizations of note include Factory Mutual Research and the Insurance Services Office.

Factory Mutual Research (FMR) is a scientific research and testing organization managed by FM Global, the merged insurance companies of the former Factory Mutual System. The FMR focuses on fire loss phenomena and fire loss control, particularly the areas of protection, materials, structures, and risk engineering methodologies. Research into industrial hazards, test data, loss statistics, and field experience is used to develop protection guidelines covering a comprehensive list of property-loss-prevention topics. These property conservation standards are accepted by a variety of jurisdictions, industries, and insurance companies. The FMR also maintains a third-party product certification service. Loss prevention products and materials are tested and approved for listing by FMR.

Insurance in the United States is regulated by individual states. The Insurance Services Office (ISO – not to be confused with the International Standards Organisation) functions according to the state laws as an insurance rating organization, an insurance actuarial service or advisory organization, and a statistical agent. The ISO files insurance rating schedules with the state governments. These are presently known as SCOPES (*Specific Commercial Property Evaluation Schedule*, ISO Commercial Risk Services, Inc. (1990)). They also administer the *Fire Suppression Rating Schedule* (ISO Commercial Risk Services, Inc. (1980)). This document is the basis by which ISO evaluates and classifies over 24,000 municipal fire departments in the United States.

4.6 US federal agencies

The US federal government had a minimal role in fire safety prior to the 1970s. From the perspective of the general public, it was decisively nonregulatory, only providing for the protection of federal property and its employees and the general public while on federal property. Developments in various technologies, growing public awareness of safety, and increasing social conscience of the legislature has intensified federal activities in fire safety.

Today, federal involvement with fire safety is multifaceted, complex, and regulatory. Many individual federal organizations promulgate, adopt, and enforce fire safety standards. There is a degree of fire safety function within each of the 12 executive branch departments and 10 independent agencies of the US federal government. The Department of Housing and Urban Development (HUD) is responsible for federal housing programmes. It sets standards for publicly funded buildings and construction, including combustibility and other fire safety features. The Nuclear Regulatory Commission is concerned with fire hazards in nuclear plants. The Department of Health and Human Services and the Veterans Administration are concerned with safety in hospitals. The General Services Administration is concerned with fire safety in federal buildings. The list goes on, since the US government does not have a central agency dealing with fire safety, nor does it have a single central enforcement agency. In addition, the US government is exempt from state and local codes. Therefore, it must provide its own criteria for fire safety. Although most now recognize local requirements, each agency has individual responsibility for protecting its property from fire loss.

The most significant fire safety activities of the US Government include regulation of the workplace, consumer products, and transportation.

4.6.1 WORKPLACE FIRE SAFETY

The Department of Labor, Occupational Safety and Health Administration (OSHA) has the major task of developing and enforcing standards to protect health and safety in places of work. To a large extent, OSHA has adopted NFPA standards to provide fire safety in the workplace. However, OSHA has recently issued a standard for the management of process hazards of highly hazardous chemicals. It requires a comprehensive programme to prevent or reduce the risk of major industrial incidents that might expose the employees to the hazards of toxicity, fire, or explosion. It specifies a holistic approach that integrates technologies, procedures, and management practises.

Fire hazards in mines are the province of the Mine Health and Safety Administration.

4.6.2 CONSUMER PRODUCTS

The Consumer Product Safety Commission (CPSC) is an independent Federal regulator agency established in 1972. The purpose of the commission is to protect the public against unreasonable risks of injury from consumer products. To this end, it assists the consumers to evaluate the comparative safety of the products. It develops uniform safety standards for the products. It aims to minimize conflicting state and local regulations. It promotes research and investigation into the causes and prevention of product-related deaths, illnesses, and injuries. It has the primary responsibility for establishing mandatory product safety standards where appropriate. In addition, it has authority to ban hazardous consumer products. It conducts extensive research on consumer product standards, engages in broad consumer and industry information and education programmes, and operates a comprehensive injury information-clearing house. It has authority for the enforcement of the Flammable Fabrics Act and similar fire safety legislation.

4.6.3 TRANSPORT

The Department of Transportation (DOT) establishes the national transportation policy. Within the Department are several subagencies involved with the regulatory actions relating to fire safety. The National Highway Traffic Safety Administration (NHTSA) carries out programmes related to the safe performance of motor vehicles and similar equipment. They issue standards that prescribe safety features and levels of safety-related performance for automobiles and trucks including flammability regulations. They also undertake extensive investigations of major fires involving motor vehicles. The Federal Aviation Administration (FAA) is charged with the responsibility for air safety and has regulations for the control of flammability of combustible aircraft materials. It is also concerned with crashworthiness and the prevention and control of an ensuing fire. Fire hazards on vessels are within the purview of the Coast Guard, an agency of the Department of Commerce.

4.6.4 NONREGULATORY ACTIVITIES

The Federal Emergency Management Agency (FEMA) is an independent agency created in 1979 to provide a single point of accountability for all federal, state, and local levels in preparing for and responding to the full range of emergencies – natural, man-made, and nuclear. The focus is on hazard mitigation, preparedness planning, relief operations, and recovery assistance. FEMA's National Emergency Training Center library has an extensive collection of references on emergency management. The only two federal units uniquely concerned with fire safety are within FEMA. The United States Fire Administration (USFA) conducts and supports training, planning, and educational efforts for the fire service – federal, state, and local – and the public at large. The National Fire Academy (NFA) in Emittsburg, Maryland conducts training courses for the fire service personnel on-site and in outreach programmes. It also develops and disseminates training and instructional materials.

The National Institute for Standards and Technology (NIST) does not develop standards in the area of fire safety. However, it contributes indirectly to the technical aspect of the development of fire standards. The Building and Fire Research Laboratory, a component of NIST, is the focal point for fire research in the United States. It has a multidisciplinary technical staff supported by extensive testing laboratory and research library facilities.

4.7 Canadian regulations

The National Research Council of Canada publishes a number of national model codes: The National Building Code (NBC) provides minimum requirements for health, life safety, and structural sufficiency in new buildings. The National Fire Code (NFC) provides minimum fire safety requirements for buildings, structures, and areas where hazardous materials are used, and ensures an acceptable level of fire protection and fire prevention in the ongoing operation of buildings.

These Codes are model documents only and must be adopted by an authority having jurisdiction in order to come into effect. The national model codes are either adopted unchanged as the regulations of a province, territory or municipality or, in some cases, altered to suit local needs.

Building codes in Canada are generally concerned with fire safety, structural sufficiency, and health. They apply to the construction of new buildings and to the demolition or relocation of existing ones. They also apply when the use of a building changes or when it is significantly renovated or altered. The scope of building codes in Canada is generally restricted to health, safety, and accessibility; however, some provincial building codes also address energy conservation.

Fire codes generally apply to buildings that are already in use and they regulate the activities that cause fire hazards. They require the maintenance of fire safety equipment and egress facilities, and they control the combustibility of furnishings. They also provide for the safe use of combustible materials and dangerous goods in both new and existing buildings, structures, and areas. They require fire safety plans in anticipation of emergencies. Fire codes reduce the likelihood of fires, particularly those that may present a hazard to the community, and limit the damage if fire occurs. Unlike building codes, fire codes may contain retroactive requirements, which apply to all buildings, regardless of when they were built. The enforcing authority must exercise judgment in the application of such requirements.

4.7.1 NATIONAL RESEARCH COUNCIL

Under the British North America Act and its successor the Constitution Act, the responsibility for building regulation in Canada rests with the provinces and territories. In the past, this responsibility was generally delegated to the municipalities. Not surprisingly, a multiplicity of regulations developed as each municipality tried to deal with its own needs. These variations from one municipality to the next made it very difficult for designers, product manufacturers, and contractors to conduct business in more than one region. In 1937, the Department of Finance asked the National Research Council (NRC) to develop a model building regulation that could be adopted by all the municipalities in Canada. The result of that initiative was the publication of the first edition of the National Building Code of Canada (NBC) in 1941.

The postwar construction boom resulted in the demand for a revised NBC. In 1948, NRC created the Associate Committee on NBC to update and maintain the document and to provide for a broader input. The Associate Committee revised the Code in 1953 and has published a new version about every five years since. The NBC 1995 is the 11th edition.

In 1956, NRC created the Associate Committee on NFC that produced the first edition of NFC in 1963. The two Associate Committees were disbanded in October, 1991 and replaced by the Canadian Commission on Building and Fire Codes (CCBFC).

4.7.2 CANADIAN COMMISSION ON BUILDING AND FIRE CODES

The National Building Code of Canada (NBC) and the National Fire Code of Canada (NFC), are prepared and maintained by the CCBFC and are published by NRC (1995a,b). They are the recommended model codes that may be adopted by an appropriate authority. The Institute for Research in Construction (IRC) provides secretarial and technical support for the CCBFC and its related committee operations. These services are coordinated within IRC by the Canadian Codes Centre (CCC). The CCC staff also provides the committees with a communication link to the specialist research. They provide information to Code users on the scope, application and intents of the Codes, and on the code development process.

The National Model Code documents are developed and maintained using a broad-based consensus process. Individuals in all segments of the construction industry have the opportunity to influence the changes in the Codes, either directly, through committee membership, or indirectly, by suggesting changes or commenting on proposed changes.

The CCBFC is aided in its work by standing committees that are responsible for the various technical areas in the Codes. Those areas of expertise are given below.

Fire safety and occupancy
Building services
Structural design
Houses
Environmental separation
Hazardous materials and activities
Energy conservation in buildings

In turn, Standing Committees rely on Topic Groups and Task Groups for advice on areas of special interest within the committee's jurisdiction. Topic Groups are ongoing as they relate to the need of the special interests, while Task Groups have short-term objectives. Expertise from outside the Standing Committees can be used on both of these groups.

Members of these committees and groups are drawn from all segments of the construction industry: regulators, fire services, architects and engineers, manufacturers and product suppliers, building owners and developers, and building users. They are appointed as individuals and not as delegates from a specific association or company. They are also selected in a way that provides representation from all the geographic regions of the country. In all, over 200 members work on about 25 committees, topic groups, and task groups.

The standing committees are open to suggestions from any source. Suggestions should be supported by valid technical arguments in order to be considered by the committees, which are unlikely to be influenced by statements of opinion or nontechnical arguments related to such considerations as market share and international trade.

An important feature of the Code development and maintenance process is the extent of public involvement. During every five-year Code cycle, all technical changes agreed to by standing committees are circulated for a three-month public comment period. This allows for a feedback from those most affected by a proposed change, and increases the range of expertise available on any subject. The appropriate technical committee reviews each comment.

Following consideration of the comments received, the standing committees submit final sets of changes to the CCBFC for approval. A period of about 20 months is required from the time the standing committees decide on the final changes they are going to recommend until the Code documents are published. This means that the standing committees must receive proposals for changes to the current codes at least two years before the end of the cycle.

The consultation process that is used to develop the National Code Documents has been altered for the next revision cycle. Provinces and territories will be more directly involved throughout by coordinating local reviews to occur simultaneously with the national code-change process. Proposals to the CCC will automatically be distributed to all the provinces and territories and when the standing-code committees come to decisions, these will be similarly distributed. CCC will flag issues that are problematic for local jurisdictions, for further consideration by CCBFC and its committees before the proposed changes go out to public review.

4.7.3 OBJECTIVE-BASED CODES

In the 1995 Strategic Plan, the CCBFC included a call for efforts to make the National Code Documents clearer and easier to use. The mechanism chosen was to convert the Codes to objective-based format. To undertake the formative work on this project, the Commission appointed a Task Group on Planning for Objective-Based Codes. When the Commission had accepted the plan that was developed, a Task Group on Implementation of Objective-Based Codes was formed, with broader membership reporting jointly to the Commission and to the Provincial/Territorial Committee on Building Standards.

A limitation of the current prescriptive requirements is that they typically evolved at a time when alternatives that are now available did not exist. The present codes are good at indicating to users what they have to do to conform but they are not so at indicating why they must do it. Many newly developed products or designs that may achieve the same – or even better – results are not covered. That makes it difficult for new, innovative, and possibly more cost-effective products and designs to gain acceptance in the marketplace. This problem is greatly reduced with performance requirements. However, there are two reasons why the CCBFC has not fully adopted the performance approach:

1. Not all the code users want to deal with performance requirements.
2. Not enough information to develop full performance-based Codes is available; that is, quantitative performance criteria cannot be defined for the majority of the Codes' current prescriptive requirements.

The CCBFC has therefore developed the concept of "objective-based codes." There are five major points in this concept:

1. The fundamental objectives the Codes seek to address (e.g. health, safety) will be stated up front.
2. From the objectives will be derived a number of more specific functional requirements that products, materials, procedures, and systems must satisfy. These will be stated in qualitative terms.

3. The present requirements of the Codes, whether prescriptive or performance-oriented will remain available in the Codes as acceptable solutions.
4. Where the necessary information is available, quantitative performance criteria will be provided for the evaluation of alternatives to the acceptable solutions.
5. The intent behind every Code requirement will be stated, as will its relationship to the Code's objectives.

The Task Group on Implementation of Objective-Based Codes is guiding the work of converting the Codes to objective-based form. This is a massive undertaking that is absorbing a major portion of the efforts of the Commission's standing committees and CCC staff. One part of this effort is the analysis of every single requirement in each of the two Codes (NBC, NFC) to identify its intent and the objective(s) to which it is related. (It is estimated there are more than 5000 requirements in the two codes.)

The new codes will be published in two parts: Division A will contain the statements of the codes' objectives and functional requirements. This division will be revised only when some fundamental change is deemed necessary. It is expected that Division A will have a "tree-like" structure of increasingly specific objectives, subobjectives, and functional requirements.

Division B will be the part for application, setting out the quantitative performance criteria with which the solutions must comply and also providing "deemed-to-comply" solutions drawn from the current editions of the codes. This division will be revised on a regular schedule, like the present codes.

4.7.4 INSTITUTE FOR RESEARCH IN CONSTRUCTION

During the postwar construction boom, NRC established the Division of Building Research to respond to the needs of an industry that was rapidly expanding. The Division's name was changed to the Institute for Research in Construction (IRC) in 1986. One of its original mandates was to provide research support for the National Building Code of Canada. The IRC is now involved in every aspect of the development of the National Codes.

The essential link between the Code committees and the IRC research staff is provided through the CCC. The committees receive a continuous stream of suggestions for changes in the Codes from all segments of the construction industry. IRC advisors evaluate these proposals, from both the technical and enforcement points of view, and suggest an appropriate course of action. However, the committees make final decisions on the technical content of the Codes, not the IRC staff.

When the committees need more information to make informed decisions, studies are performed to provide the missing data. These studies are performed not only by IRC but also by provinces, manufacturing groups, and various consortia having similar interests.

4.7.5 CANADIAN STANDARDS

Standards are publications that establish accepted practices, technical requirements, and terminology. The Canadian National Codes reference more than 200 documents directly and many more indirectly. Generally, these are standards prepared by the standards-writing organizations in Canada that are accredited by the Standards Council of Canada (SCC) such as Canadian Gas Association (CGA), Canadian General Standards Board (CGSB), Canadian Standards Association (CSA), Underwriters Laboratories of Canada (ULC) and Bureau de normalisation du Québec (BNQ). Standards from American organizations such as the ASTM and NFPA are also referenced in Canadian codes.

Evaluation

Conformance of building products, materials, or systems is evaluated by a number of organizations, such as the CSA, CGA, and ULC, which provide full third-party certification for safety-related products or systems. The NBC does not require such certification, only that the product or system meets certain minimum requirements. Code enforcement officials, however, often rely on certification as a guarantee that such is the case. To provide the construction industry with a national evaluation service for innovative materials, products, and systems, NRC created the Canadian Construction Materials Centre. This service includes the evaluation of new and innovative products for which no standards exist, and of products for which the standards exist but no third-party certification programme has been established. Most provinces and territories use the Centre's evaluation reports as a basis for accepting new products.

The SCC is a federal Crown corporation with the mandate to promote efficient and effective standardization. The SCC accredits testing laboratories and standards-writing organizations. The SCC represents Canada at the international level through membership of the International Electrotechnical Commission (IEC) and the International Organisation for Standardisation (ISO), both of which are charged with promoting the development of voluntary international standards as a means to facilitate international trade. The SCC coordinates Canadian participation in the technical committees (TCs) of ISO and IEC in conjunction with SCC-accredited standards organizations.

4.8 United Kingdom

4.8.1 BUILDINGS

Read and Morris (1993) give an extensive survey of the requirements for buildings in the United Kingdom. Broadly, legislation to control the building fire hazard, as distinct from the content hazard, exists in two main areas – the design of a new building and the provision of fire precautions once it is occupied. Requirements are drawn up nationally, and the local authority carries out the enforcement through its building control and fire departments. An existing building that undergoes alteration is treated as a new building.

Regulations cover the requirements for means of escape, internal fire spread (linings), structural fire protection, external fire spread and facilities for the fire service. At a national level, three separate sets of Building Regulations exist: England and Wales together, Scotland and Northern Ireland. Within Inner London, there are certain powers not covered by these regulations and some previous Acts remain. All the Regulations can be considered as functional but in Scotland and Ireland, unlike England and Wales, the Technical Standards supporting them are mandatory.

Many statutory provisions exist for occupied premises; in particular, those introduced under the Fire Precautions Act 1971. Hotels and boarding houses (1972), factories, shops, and railway premises (1976) have been designated under this Act and the Home Office has issued guides to each of them. The precautions relate mainly to the maintenance of means of escape, raising the alarm, and first aid fire fighting. There are other types of premises, not designated, which have fire safety requirements under other Acts, including cinemas, theaters, schools, houses in multiple occupation, clubs and licensed premises (license to sell liquor), and hospitals. However, since most of these premises contain parts that can be described as offices, encroachment under the Fire Precautions Act has sometimes caused confusion.

The enforcement of Building Regulations is generally within the Building Control department of the local authority and the Fire Precautions Act within the local fire authority.

Insurance requirements within the United Kingdom generally take into account the rules published by the Loss Prevention Council (1986). This body incorporates the Fire Offices' Committee, which has been active in the field for over a century. The requirements are contained within the Rules for Construction of Buildings Grade 1 and 2. Grade 1 construction could generally qualify for a reduction in the premium and Grade 2 would not normally incur an additional rate.

Numerous Codes of Practice and Standards support the statutory requirements for fire safety in buildings in the United Kingdom. They deal with fire precautions in different types of buildings and the performance standards for materials and fire safety equipment. The main source of standards is the British Standards Institution, although the International Standards Organisation (ISO) and the Conseil Europeenne de Normalisation (CEN) have produced standards that have been agreed internationally. Various Codes of Practice and Guides have been produced by government departments, particularly by the Home Office, the Department of Health and Social Security (hospitals) and the Department of Education and Science (schools). A large number have been produced by the British Standards Institution. They cover fire precautions for major risk areas – flats and maisonettes, office buildings, shops, for example – and major fire protection measures – such as the fire design of structures, automatic detection and alarm systems, and fire fighting lifts. Read and Morris (1993) give an extensive list of these various Codes and Standards. Malhotra (1992) has given an account of Standards and Codes produced by CEN.

4.8.2 INDUSTRIAL AND PROCESS HAZARDS

Industrial and process hazards come under the aegis of the Health and Safety Commission whose executive arm is the Health and Safety Executive (HSE), part of the Department of Employment. The main statutory instrument is the Health and Safety at Work etc. Act 1974 (HMSO, 1974), an act that was at least hastened into existence by the Flixborough disaster in the previous year. This act, which covers all the safety and health matters in places of employment, has sponsored a number of sets of Regulations. Of special importance as far as fire safety is concerned are the Fire Certificates (Special Premises) Regulations (1976) and the Control of Industrial Major Accident Hazard Regulations (CIMAH, 1984). The first is concerned with the identification of industrial areas of special risk either because of the presence of a particularly dangerous quantity of flammable or toxic material or because an inherently dangerous process is carried out. Because of their special knowledge of the processes and the materials concerned, the enforcing authority for fire safety is the HSE rather than the fire authority (HSE, 1985). The second is concerned with safety requirements and management for processes that can be identified as providing a major accident hazard. The requirements in these regulations tend to be functional rather than prescriptive. A major requirement is that a safety case is produced in which the firm concerned assesses the hazards and puts forward its own way of dealing with them. Following the Piper Alpha disaster in the North Sea in 1987, the Health and Safety Commission also took over the responsibility of regulation of safety of offshore gas and oil installations. This, of course, has a high content of fire and explosion hazard. Major new regulations in this field require that, as with CIMAH, every offshore installation must produce a safety case for acceptance by the HSE. The HSE is also the body in the United Kingdom concerned with the enforcement of European Commission directives on health and safety in industry.

The HSE has provided an extensive background of codes and guidance notes. There has also been much activity in this field by professional organizations, particularly the Institution of Chemical Engineers (1992) in the provision of training and videos. These tend to focus on the quantification of hazard and risk. The Engineering Council (1992, 1993) has also produced documents of Risk Assessment for general use by professional engineers.

4.8.3 CONSUMER FIRE SAFETY

It is in the domain of consumer items that the general public, probably unaware, are likely to encounter day-to-day requirements for fire safety. These items are mainly of two kinds. First, there are those that contain combustible materials which, when ignited, can either impose an immediate threat to individuals or cause rapid spread of fire. The second are heat and power sources as used by individuals that may be the sources of ignition and spread of fire. The requirements have developed over a lengthy period and have been promulgated by different government authorities. Currently, the consumer protection section of the Department of Trade and Industry is the main department concerned with these matters. They published a consultative document on consumer safety (DTI, 1976) that gave a survey on consumer safety at the time and what was needed to improve it. This document indicates the extensive reliance on British Standards in support of fire safety requirements that have been laid down by legislation.

Fabrics and furniture are the two major types of combustible items that have come under fire safety control. Flammable clothing is a major cause of burn accidents. British Standard BS 5438 (1976) describes the way in which the materials are tested, using a vertical sample 62 cm high ×17 cm wide, and BS 5722 (1984) defines flammability criteria for sleepwear and dressing gowns on the basis of the results of the BS 5438 test. Curtains and drapes may be easily ignited and spread fire because they are free hanging. Accordingly, when they are in public places it is a normal requirement by the fire authority that they are treated, so that they are resistant to a small source of ignition. Certification according to one or more of a number of relevant British Standards may be required as evidence, for example BS 5438 as above, BS 5867 (1980) and, for PVC drapes, BS 2782 (1987, 1988).

Furniture was not considered a major fire safety hazard until the 1960s when it became clear from the fire service reports that modern furniture in domestic premises was causing fires to develop very rapidly. The main reason for this was the introduction of polyurethane foam as the major filling in upholstery and mattresses. Research programmes were put in hand that resulted in an array of test methods and culminated with the Furniture and Furnishings Fire Safety Regulations 1988. The test methods used in these regulations, BS 5852 (1990) and BS 6807 (1990) have been summarized by Paul (1989). Carpets are not usually a cause of rapid spread of fire when used as floor coverings that is, when they are disposed horizontally facing upwards. This may be subject to revision as far as stairs are concerned following the evidence at the King's Cross fire disaster concerning fire spread up an escalator (Chapter 3). The ignitability of carpets is usually tested by the hot metal nut test BS 4790 (1987).

Heaters, particularly domestic heaters, have been responsible for starting many fires. A fireguard is the major form of protection for an appliance with an exposed heating element. The open coal fire used to be the major ignition source in the United Kingdom, although its importance declined with the onset of central heating. They are protected by fireguards specified in BS 6539 (1991) and spark guards specified in BS 3248 (1986). Fireguards are required for all other open heaters, including gas and electric covered by BS 1945 (1991) and BS 6778 (1991) respectively. Portable kerosene heaters are covered by the legislation that followed a fire incident in 1959 in which five children were killed. A radiant drip feed heater was in the hall and the mother, who had to go out for a few minutes, left the front door open exposing the heater to a draught. This unbalanced the flame and caused the appliance to flare and involve the fuel reservoir. The standard, BS 3300 (1974) requires inter alia that all portable kerosene heaters should operate safely in a draught up to 8 m/s. Catalytic heaters use a gaseous fuel, usually from an LPG cylinder. They need to be tested according to BS 5258 (1983a) to ensure that the catalyst is operating uniformly and not discharging undue amounts of unburned vapors (BS 5258, 1983b). Central heating systems are normally much safer than open fires or portable space heaters. The main precaution when a gas or oil fuel is used is to have the presence of the pilot reliably monitored and to interlock between the existence of a pilot and the main fuel supply (BS 5258, 1983a). Electrical power supplies are covered by Regulations of the Institution of Electrical Engineers (1981).

4.8.4 TRANSPORT

Fire safety in road, rail, marine, and aviation transport is included under this heading. In the United Kingdom the government department ultimately concerned is the Department of Transport, having individual sections dealing with all of these. However, the HSE has a major interest particularly in the transport of dangerous goods. Moreover, many of the requirements are dominated by international agreements because of the highly international nature of transport, particularly for marine and aviation hazards.

Special requirements for road transport fall into two categories, those for vehicles licensed to carry passengers and those for vehicles licensed to carry dangerous goods. The Department of Transport certificates of fitness requirements cover buses and coaches. Freight vehicles are covered by the Health and Safety at Work etc. Act, particularly under the HSE 1981 Regulations. As far as railway rolling stock is concerned, regulations for materials of construction and installation of fire extinguishers are covered by BS 6853 (1987). In addition, British Rail has in-house regulations and procedures.

For shipping, the major source of legislative requirements is the International Maritime Organisation (IMO), which until recently was the Inter-governmental Maritime Consultative Organisation (IMCO). Its input has been mainly through the Safety of Life at Sea Conventions (SOLAS) held in London 1913, 1929, 1948, 1960 and 1974. These Conventions gave rise to a number of acts covering Merchant Shipping and the Carriage of Goods by sea. Rushbrook (1979) gives a detailed account of these acts and their applications in his book Fire Aboard. Another major source of fire safety requirements for shipping is the International Chamber of Shipping. Its Oil Companies International Marine Forum has published guides on Oil Tanker and Terminal Safety (1974) and Ship to Ship Transfer of Liquefied Gases (1980).

The authority having jurisdiction over aviation standards for fire safety in flight and also on the ground at airports and refueling zones in the United Kingdom is the Civil Aviation Authority (CAA). This organization works very closely with its sister organization in the United States, the FAA and the relevant international body, the International Civil Aviation Organisation (ICAO) to develop what are mainly international standards. Following the disaster at Manchester Airport in August 1985 when 55 passengers died, new rules have been issued (1987) requiring improvement of fire-retardant qualities of the seats, wall, and ceiling linings of aircraft.

4.8.5 FIRE SAFETY AUDITS AND CHECKLISTS

In a given hazard area, certain precautions may have been required by legislation, imposed by one of the sources described above, or directly by the management to bring about sufficient fire safety. The basic method of assessing fire safety is by continued inspection and maintenance to see

(a) that the risk has not changed,

(b) that the fire precautions called for are in place and can be expected to operate effectively.

This assessment is normally accomplished in a disciplined way by checklists and safety audits. The organization of these is an integral part of fire safety management. The management should be responsible for formulating the audits and seeing that they are followed. This discipline is necessary whatever the manner in which the precautions have been developed and defined. A number of model checklists have been produced by official, semiofficial, and professional organizations. Examples are the Home Office publications on danger of fire in the home, publications of the Fire Protection Association (undated) on managing fire safety in a wide range of specific industries and a guide to safety audits in the Chemical Industry produced by the Chemical Industries Association (1977). These checklists draw attention to the potential hazards that may exist in typical premises and safety measures that may be used to counteract them. Special attention was paid to managing fire safety and fire safety auditing in the Fennell Report (1988) of the King's Cross Underground fire (Section 3.2.3.) and the Symposium of the Institution of Mechanical Engineers (1989) that followed.

References

American Society for Testing and Materials (1999). *Fire Test Standards*, 5th Edition, American Society for Testing and Materials, West Conshohocken, PA.

Boring, D F, Spence, J C and Wells, P E (1981). *Fire Protection through Modern Building Codes*, 4th Edition, American Iron and Steel Institute, Washington, DC.

BS 1945 (1991). Specifications for Fire Guards for Heating Appliances (gas, electric and oil burning).

BS 2782 (1987, 1988). Methods of Testing Plastics, Part 1, Tests for Flammability, Method 140D (1987), 140E (1988).

BS 3248 (1986). Specifications for Spark Guards for use with Solid Fuel Appliances.

BS 3300 (1974). Specification for Kerosene (paraffin) Unflued Space Heaters, Cooking and Boiling Appliances for Domestic Use.

BS 4790 (1987). Method for Determination of the Effects of a Small Source of Ignition on Textile Floor Coverings (hot metal nut method).

BS 5258 (1983). Safety of Domestic Appliances: (a) Part I. Specification for Central Heating Boilers and Circulators (b) Part II. Flueless Catalytic Combustion Heaters.

BS 5438 (1976). Methods for Tests for Flammability of Vertically Oriented Fabrics and Fabric Assemblies Subjected to a Small Igniting Flame.

BS 5722 (1984). Specification for Flammability of Fabrics and Fabric Assemblies Used as Sleepwear and Dressing Gowns.

BS 5867 (1980). Specification for Fabrics for Curtains and Drapes, Part 2, 1980, Flammability Requirements (when tested by BS 5438).

BS 5852 (1990). Method of Test for Assessment of Ignitability of Upholstered Seating by Smouldering and Flaming Ignition Sources.

BS 6539 (1991). Specifications for Fireguards for Use with Solid Fuel Fire Appliances.

BS 6807 (1990). Methods of Test for Assessment of Ignitability of Mattresses, Divans, and Bed Bases with Primary and Secondary Sources of Ignition.

BS 6778 (1991). Specifications for Fireguards for Use with Portable Free Standing or Wall Mounted Heating Appliances.

BS 6853 (1987). Code of Practice for Fire Precautions in the Design and Construction of Railway Passenger Rolling Stock.

Building Officials and Code Administrators International (1996). National Fire Code, Country Club Hills, IL.

Building Officials and Code Administrators International (1996). *National Building Code*, Country Club Hills, IL.

Chemical Industries Association (1977). *Safety Audits - A Guide for the Chemical Industry*.

Cote, A E (Ed.) (1997). *Fire Protection Handbook*, 18th Edition, National Fire Protection Association, Quincy, MA.

Cote, A E and Grant, C C (1997). Building and fire codes and standards. *Fire Protection Handbook*, A E Cote (Ed.), 18th Edition, National Fire Protection Association, Quincy, MA, Section 1, Chapter 4.

DiNenno, P J *et al.* (Eds.) (2001). *SFPE Handbook of Fire Protection Engineering*, 3rd Edition, National Fire Protection Association, Quincy, MA.

DTI (1976). *Consumer Safety: A Consultative Document*, Department of Trade and Industry, HMSO, London.
Engineering Council (1992). *Engineers and Risk Issues*, Code of Professional Practice, London.
Engineering Council (1993) *Guidelines on Risk Issues*, London.
Fennell, D (1988). *Investigation into the King's Cross Underground Fire*, HMSO, London.
Fire Protection Association (undated). *Planning Programme for the Prevention and Control of Fire*, London.
HMSO (1974). Health and Safety at Work Act, London.
HSE (1985). *Control of Industrial Major Accident Hazards*, HMSO, London.
Institution of Chemical Engineers (1992). Safety Training Packages, I. Chem. E., Institution of Chemical Engineers, Rugby, UK.
(IEE) (1981). *IEE Wiring Regulations*. Institution of Electrical Engineers, 15th Edition, London.
Institution of Mechanical Engineers (1989). *The King's Cross Underground Fire: Fire Dynamics and the Organisation of Safety*, M.E.P., London.
International Conference of Building Officials (1999). *Uniform Building Code*, Whittier, CA.
International Conference of Building Officials (1999). *Uniform Fire Code*, Whittier, CA.
ISO Commercial Risk Services, Inc (1980). *Fire Suppression Rating Schedule*, Parsippany, NJ.
ISO Commercial Risk Services, Inc (1990). *Specific Commercial Property Evaluation Schedule*, Parsippany, NJ.
Kaplan, M E and Watts, J M Jr (2000). *Draft Code for Historic Buildings*, Association for Preservation Technology, Williamsburg, VA.
Loss Prevention Council (1986). Rules for Construction of Buildings, Grades 1 and 2.
Malhotra, H L (1992). Recent developments in the European fire scene. *Fire Surveyor*, **21**, 4.
National Research Council (1995). *National Building Code of Canada*, Ottawa.
National Research Council (1995). *National Fire Code of Canada*, Ottawa.
Oil Companies International Marine Forum (1974). *International Tanker and Terminal Safety Guide*, 2nd Edition, Applied Science Publishers Ltd, London.
Oil Companies International Marine Forum (1980). *Ship to Ship Transfer Guide (Liquefied Gases)*, Witherby & Co. Ltd, London.
Paul, K T (1989). Furniture and furnishing regulations. *Fire Surveyor*, **18**(1), 5–13.
Puchovsky, M (1997). Performance-based fire codes and standards. *Fire Protection Handbook*, A E Cote (Ed.), 18th Edition, National Fire Protection Association, Quincy, MA, Section 10, Chapter 13.
Rushbrook, F (1979). *Fire Aboard*, Brown, Son and Ferguson.
Read, R E H and Morris, W A (1993). *Aspects of Fire Precautions in Buildings*, Building Research Establishment, UK.
SFPE (2000). *SFPE Engineering Guide to Performance-Based Fire Protection Analysis and Design of Buildings*, Society of Fire Protection Engineers, Bethesda, MD.
Southern Building Code Congress International (1997). *Standard Fire Code*, Birmingham, AL.
SP 806 (1996) *Standards Activities of Organizations in the United States*, U.S. Government Printing Office, Washington, DC.
Toth, R B (Ed.) (1984). *Standards Activities of Organizations in the United States*, NBS Special Publication 681, National Bureau of Standards, Gaithersburg, MD.
Watts, J M Jr (1999a). Rehabilitating existing buildings, fire protection engineering. *Fire Protection Engineering*, **1**(2), 6–15.
Watts, J M Jr (1999b). Y2K performance-based life safety code. *Proceedings, Third International Conference on Fire Research and Engineering, Society of Fire Protection Engineers*, Bethesda, MD, pp. 146–151.

PART II QUANTIFYING FIRE SAFETY

5 PHYSICAL DATA

5.1 Introduction

In the last 20 or 30 years, a surge of experimental and other investigations relevant to fire safety has taken place. A great deal of the results of this work has entered the public domain in the form of abstracts of the fire research, papers in journals, journals dedicated to fire safety, text books, and in recent years, volumes of symposia held by the International Association of Fire Safety Science. Table 5.1 lists some of the major publications in this field. This work is forming an ever-increasing basis for the quantitative approach to fire safety.

The areas of major input of this work, particularly those yielding physical data for evaluating fire behavior and fire safety, will be outlined in this chapter. Physical data have particular relevance for the Subsystems (i) to (v) in Table 2.4. Most models using physical data up to the present time have been within Subsystems (ii) and (v), the Fire Development and Direct Detriment subsystems. It is impossible to describe the whole field in detail in a single chapter. However, the intention is to survey the main areas in which available data can provide input into quantitative modeling of fire safety and to indicate major gaps and contradictions where these exist.

5.2 Burning and ignition

5.2.1 MECHANISMS OF BURNING

Step 3 in Table 2.3 is concerned with the identification of fuels, that is, materials that can burn. Fires manifest two major mechanisms of burning, gas-phase flaming combustion, and solid-phase smoldering or glowing combustion. In both cases, the air diffuses to the fuel. Gas-phase flaming combustion is the main mechanism in practice, the gases and vapors being produced from liquid or solid fuels by heat feedback from the flames of the fire itself. Explosions are usually associated with propagation of flame through premixed fuel–air mixtures at concentrations between the flammability limits (see Section 5.14).

The most important property of a fuel is the heat of combustion (H). This is measured in standard apparatus and extensive information is available in the literature for gaseous, liquid, and solid fuels. In general, liquid fuels vaporize entirely when they burn and the only form of combustion is flaming combustion. However, the majority of solid fuels, particularly cellulosic

Evaluation of Fire Safety D. Rasbash, G. Ramachandran, B. Kandola, J. Watts and M. Law
 ISBN: 0-471-49382-1

Table 5.1. Published sources of quantitative information in fire safety engineering

Refereed journals:
Fire and materials – Wiley and Sons
Fire safety journal – Elsevier
Fire technology – National Fire Protection Association
Report of the Fire Research Institute of Japan
Journal of Fire Protection Engineering – Society of Fire Protection Engineers
Combustion and Flame – Combustion Institute
Fire Science and Technology – University of Tokyo.
Major symposia.
Fire safety of combustible materials – Edinburgh University
1st, 2nd, International Symposium on Fire Safety Science – Hemisphere
3rd and 4th International Symposium on Fire Safety Science – Elsevier
5th and 6th International symposium on Fire Safety Science – International Association of Fire Safety Science
1st, 2nd, and 3rd International Conference on Fire Research and Engineering – Society of Fire Protection Engineers.
Compendium.
S.F.P.E. Handbook of fire protection engineering – Society of Fire Protection Engineers, National Fire Protection Association (1995)
Other series.
Fire research abstracts and reviews 1958–1977. National Academy of Sciences, National Research Council, USA
Fire research. Annual reports of the Fire Research Station, 1947–1973, London, H.M.S.O.

fuels, only partly vaporize or decompose to produce flammable vapors known as *pyrolysate*. These vapors move away from the fuel surface and at least in the initial stages usually have relatively free access to necessary air for combustion, particularly diffusion to the flaming zone directly above the fuel. The remaining fraction is the char, which is usually substantially less than half. It burns at the solid interface with air and combustion is limited by oxygen access to the surface. As a result there is glowing, which can continue for a substantial time after the fuel has been denuded of its volatile fraction. However, some solid fuels, particularly certain polymers, do not produce char.

The major hazard condition of fire is governed by the flaming combustion regime. The ease of production of the fuel vapors, particularly the heat required to produce them (L), is a major factor in assessing the potential hazard of the fuel in a fire. For liquid fuels, this can normally be estimated as the heat required to heat the fuel to the temperature at which it vaporizes plus the latent heat of vaporization. For solid fuels, however, the process of forming the pyrolysate usually involves chemical decomposition as well as vaporization and it is necessary to measure it directly. This can be done by exposing the fuel to a known increment of heat transfer rate, usually radiation, under conditions following the onset of pyrolysis and measuring the rate at which the fuel weight loss increases.

5.2.2 PROPERTIES OF FLAMING COMBUSTION OF LIQUIDS AND SOLIDS

The ratio H/L is the major parameter that influences the way a liquid or a solid fuel can contribute to a fire. For a continuing fire, the ratio needs to be at least unity unless there is an independent

Table 5.2. H/L values for fuels (Tewarson, 1980)

Fuel[a]	H/L[b]
Red oak (solid)	2.96
Rigid PI foam (43)	5.14
Polyoxymethylene(granular)	6.37
Rigid PU foam (37)	6.54
Flexible PU foam (1-A)	6.63
PVC (granular)	6.66
Polyethylene 48% Cl (granular)	6.72
Rigid PU foam (29)	8.37
Flexible PU foam (27)	12.26
Nylon (granular)	13.10
Flexible PU foam (21)	13.34
Epoxy/FR/glass fiber (solid)	13.38
PMMA (granular)	15.46
Methanol (liquid)	16.50
Flexible PU foam (25)	20.03
Rigid polystyrene foam (47)	20.51
Polypropylene (granular)	21.37
Polystyrene (granular)	23.04
Polyethylene (granular)	24.84
Rigid polyethylene foam (4)	27.23
Rigid polystyrene foam (53)	30.02
Styrene (liquid)	63.30
Heptane (liquid)	92.83

[a]Numbers in parenthesis are PRC sample numbers.
[b]H measured in an oxygen bomb calorimeter and corrected for water as a vapor for fuels for which data are not available; L is obtained by measuring the mass loss rate of fuel in pyrolysis in N_2 environment as a function of external heat flux for fuels for which data are not available.

supply of heat to the fire. Table 5.2 gives values of H/L for a range of common fuels and indicates values varying between 3 and 90. For luminous flames, the heat transfer from a flame of a given size back to the fuel surface will not depend greatly on the nature of the fuel. However, the amount of fuel vapor fed into the flame will depend on the ratio of H/L, which will be the main determinant of the amount of flame produced, particularly flame height and width. For fuels with a given heat of combustion, those with high values of H/L will tend to burn out quickly and produce large flames. They will thus have a major influence on the rapidity of fire spread. Those with a low value of H/L will tend to burn for prolonged periods and have a more deep-seated effect on structures exposed to them.

Diffusion flames from fuel sources of small dimensions and low fuel flow rates tend to be laminar (Drysdale, 1985a). Oxygen reaches the fuel vapor flow by molecular diffusion. The flame appears as a smooth surface and the combustion takes place in a thin reaction zone at this surface. Flame heights are long compared with the dimension of the fuel source and at the flame tip the burnt air associated with the flame is approximately the stoichiometric amount. However, as the dimension and the flow rate from the fuel source increases, vortexes, and bulges appear some distance above the fuel source. These may be due to a lateral inflow of air above the fuel source and as a result more air is entrained into the flame.

In practise, upward moving flames from a fuel source of dimension greater than 100 mm are mostly turbulent, although there may be a laminar region near the bottom of the flame. This

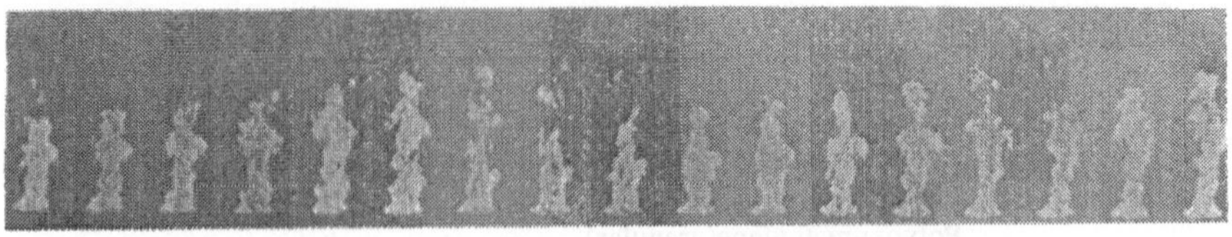

Figure 5.1. Cine record of flames of petrol fire, showing upward movement of flame (plate)

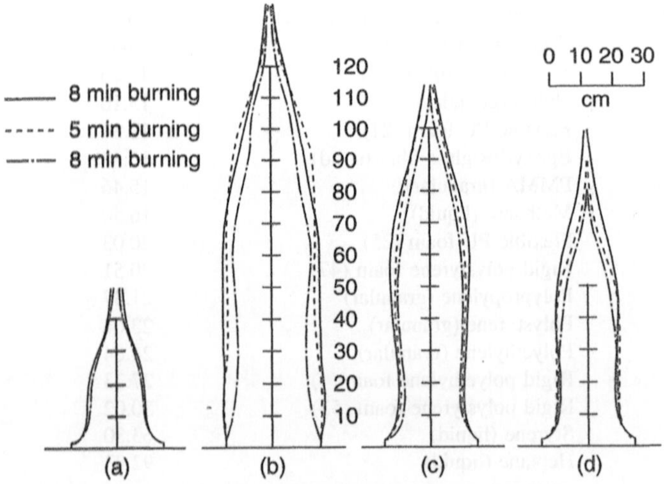

Figure 5.2. Flame profiles

turbulence often manifests itself by bulges of flame forming, moving upward with the flame and breaking off with a frequency that depends on the flame dimension. This gives rise to an intermittency of the flame at heights greater than about half the vertical height. This is exemplified in Figure 5.1, which shows a flame sequence for a petrol fire burning in a 0.3-m-diameter vessel with a 20-mm ullage, at a rate of 1.65 g/s (Rasbash *et al.*, 1956). Assuming the heat of combustion to be that of octane, the theoretical heat output of the fire would be 1040 kW/m^2 of fuel surface. The natural buoyancy of the flame causes the turbulence. Air is entrained into the body of the flame, and except for a region near the fuel surface, the combustion reaction takes place in the volume of the flame. At the flame tip, the temperature will be about 500 °C and about 12 times the stoichiometric airflow needed is associated with the buoyant column (Heskestad, 1986). Pools of flammable liquid and most articles that burn in building fires burn this way. Figure 5.2 (Rasbash *et al.*, 1956) shows mean shapes of continuous parts of the flame for ethanol, kerosene, petrol, and benzene for the 30-cm diameter vessel. The theoretical heat outputs cover a range of 21 to 163 kW (300–2300 kW/m^2 of fuel surface) and the influence of heat output on flame height and width is clearly indicated. The height and mean flame diameter of the part of the flame, continuous for 90% of the time, varied as the 0.61 power and the 0.30 power of the theoretical heat output of the fire respectively. McCaffrey (1979) studied in detail methane flames of a similar kind emanating from a 0.3-m square porous burner at different flows corresponding to heat outputs between 14.4 and 57.5 kW (160–639 kW/m^2). Mean relationships for the temperatures, velocities, and mass flow rates within the continuous flame and the plume above the flame have been used extensively in deterministic fire modeling (Chapter 12). Figure 5.3 shows turbulent flames above a developing fire in a burning wood crib 0.91 × 0.91 × 1.07 m high with stick section 2.5 cm^2 (O'Dogherty

(a) 704 kW (b) 1407 kW (c) 2813 kW

Figure 5.3. Sizes of a wood crib fire at different rates of theoretical heat output (plate)

et al., 1967), which suggests that the tendency to bulge formation may lessen if there is no immediate narrowing of the flame above the fire source. The flames above the crib in this fire were probably already turbulent and in (b) and (c) where they had spread across the whole crib cross section were burning with a theoretical heat output of about 1500 and 3000 kW/m^2 of crib cross-sectional area. Markstein (1978) has presented mean flame shapes for a number of plastics burning on a 0.31-m square base with theoretical heat outputs varying between 250 and 550 kW/m^2 (Figure 5.4). The shape of the PMMA and polypropylene flame are similar to that of alcohol in Figure 5.2, with same heat output per unit area (250–350 kW/m^2). However, the polystyrene flame shape (540 kW/m^2) differs from the kerosene flame shape of similar heat output.

Flames that burn from a fuel surface at an angle to the horizontal tend to attach themselves to the surface when the angle exceeds about 15°. The flames burn along the surface and remain turbulent as the angle is increased to vertical and beyond. However, as the angle approaches downward facing, the flames become laminar and then cellular (de Ris and Orloff, 1974). If the fuel is entering the flame as a high-velocity jet, then substantial air is introduced immediately into the flame and does not need to be engendered by natural buoyancy. This shortens the flame and increases its temperature and intensity of combustion. Flames of this kind are more common in fires in the process industries in which substantial quantities of gaseous and liquid fuels may be handled under pressure.

5.2.3 IGNITION

Step 4 in Table 2.3 is concerned with sources of ignition. In order to produce an ignition leading to a flame, it is necessary first that there exist a gas (or vapor pyrolysate) – air mixture capable of producing a flame, and secondly, either an ignition source capable of producing pilot ignition or temperature conditions of the mixture that could lead to spontaneous or auto ignition. The range of flammable mixtures in air between lower and upper limits of fuel gases and vapors have been extensively documented, for example, (Bond, 1991, Drysdale, 1985a,b, Kanury, 1977). These tabulations also provide information on the energies of ignition sources required to ignite such mixtures. For a given fuel, this is usually a minimum near the stoichiometric mixture where

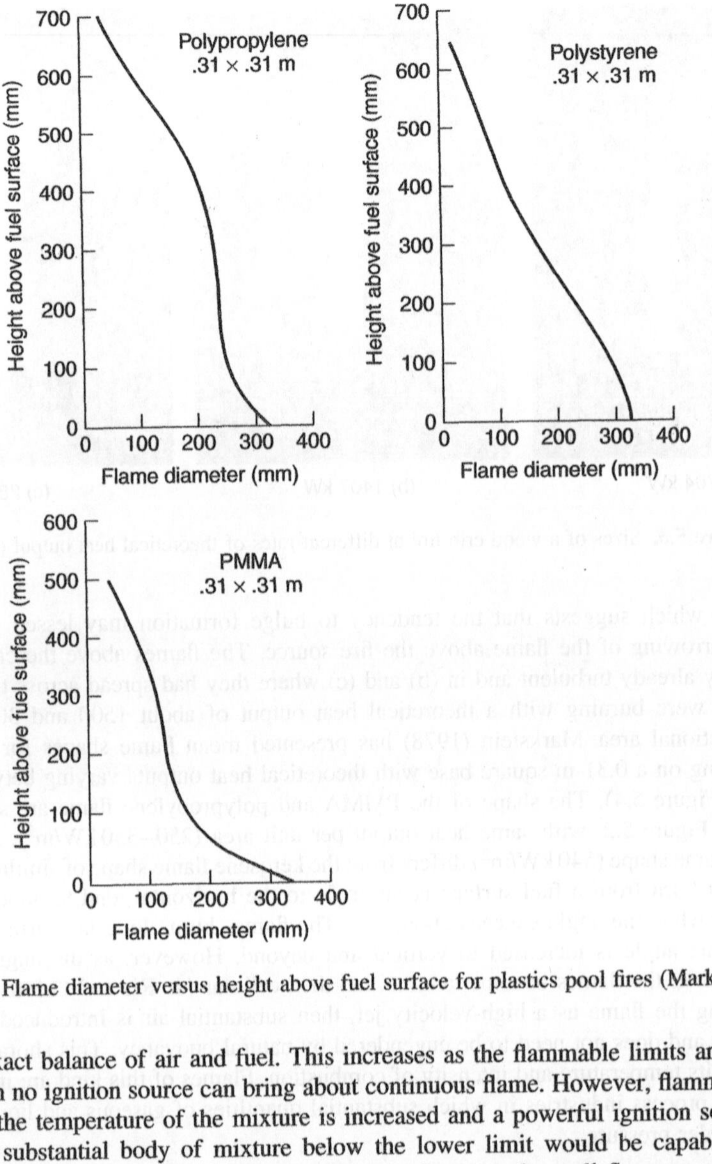

Figure 5.4. Flame diameter versus height above fuel surface for plastics pool fires (Markstein, 1978)

there is an exact balance of air and fuel. This increases as the flammable limits are approached beyond which no ignition source can bring about continuous flame. However, flammability limits do widen as the temperature of the mixture is increased and a powerful ignition source capable of heating a substantial body of mixture below the lower limit would be capable of producing significant flame propagation within the mixture. In general, small flames and electrical or mechanical sparks are needed to ignite flammable mixtures, the minimum energy of sparks being about 0.3 MJ for a wide range of flammable vapors and gases with air and one-tenth of this amount for a hydrogen–air mixture.

To ignite a liquid or solid fuel, it is usually necessary that the fuel be heated so that a flammable vapor–air mixture at or above the lower limit exists near the surface of the fuel. For flammable liquids the temperature at which this occurs is known as the flash point, which again is extensively documented for different fuels (Bond, 1991). Volatile liquids such as hexane, alcohol, or gasoline have flash points below room temperature and can be ignited by a small spark or flame. However, less volatile liquids and almost all common combustible solids such as cellulosic and plastic fuels require a substantial heat input into the fuel before sufficient flammable vapor is produced at the surface of the fuel. In general, the heat input required is orders of magnitude different from what is required to ignite the gas mixture itself and very much dependent on the physical conditions and the geometric configuration of the fuel.

For fire to continue with solid and liquid fuels, it is not sufficient for the fuel to be heated to the point at which a flammable vapor–air mixture is produced. It is also necessary for the flame to feed back sufficient heat to the fuel surface to allow the flame to continue. The temperature when this occurs is the fire point, which is usually a few degrees higher than the temperature at which flashing ignition takes place. The fire point condition is also associated with a critical flow rate of volatiles, which depends on the nature of the fuel, the geometry, and the oxygen concentration (Rasbash, 1975a, 1976). If the heat input from the ignition source to the fuel is maintained, then the extra heat from the established flame at the fire point results in extra heat being imparted to the fuel surface that increases the temperature of the fuel surface and hence the rate of production of fuel vapors and the rate of burning. However, if the heat of the ignition source is removed at or even following ignition the flame may be extinguished if the heat transfer from the flames is not sufficient to produce the necessary volatiles and compensate for heat losses associated with heating the fuel to the fire point temperature (Drysdale, 1985c).

For flammable liquids, the flash point and fire point temperatures are measured in standard forms of apparatus in which a small quantity of liquid in a vessel is heated slowly (at 5 to 6 °C/min). Stirring the liquid or circulation currents helps keep the liquid at a uniform temperature during this heating process. However, ignition phenomena for solid fuels are usually measured by exposing the solid surface to radiant heat. As the temperature of the surface increases, it is possible to distinguish a flashing condition followed by a fire condition when a small igniting source is placed in the flammable vapors that are emitted. The time taken for this to occur depends on the ignition temperature – which is the surface temperature needed to produce the fire condition – the thickness, and thermal properties of the fuel as well as the level of radiant heat flux.

At high rates of radiant heat transfer, the time needed for the surface of an exposed solid to reach a given temperature will depend on whether the solid may be regarded as thick or thin. Thin materials are heated uniformly across their thickness and a direct heat balance gives the time, t_s:

$$q''t_s = (T_s - T_0)\rho c\tau \qquad [5.1]$$

q'' = heat flux absorbed by exposed face
t_s = time to reach fire point temperature T_s
T_0 = initial temperature of solid.
ρ = density of solid
c = specific heat of solid
τ = thickness of solid

For thick materials there will be a temperature gradient behind the exposed surface and:

$$q''t_s^{1/2} = (T_s - T_0)\left(\frac{\pi}{4}k\rho c\right)^{1/2} \qquad [5.2]$$

k = thermal conductivity of solid
$k\rho c$ = thermal inertia of solid

In general, a slab may be regarded as thick if the thickness

$$\tau > 2\left(\frac{kt_s}{\rho c}\right) \qquad [5.3]$$

Equations [5.1] and [5.2] rely on the assumption that heat losses are negligible compared to heat absorbed and that the material being heated is inert and unchanging. These assumptions may not be justified when materials are being heated to their fire point. Fire points are generally in the range of 300 to 400 °C and significant heat losses both by convection and radiation do occur as the fire point is approached. Indeed, these heat losses give rise to a critical heat transfer rate for bringing about ignition. Further complications occur when materials char before they ignite, so that their absorptivity to radiation increases, char builds up following long exposure time, and char oxidation takes place. In addition, materials can melt or distort and composite materials can delaminate and thus have their character changed from thick to thin. Also, since critical thickness of slabs (equation [5.3]) increases as the exposure time t_S increases, a slab that is thick for high rates of heating may become thin for rates of heating near the critical value.

Whilst being mindful of the above difficulties, it is possible to interpret experiments on ignition by radiation to obtain data useful for predicting ignition. Thus, a plot of a power between 0.5 and 1 of the reciprocal of ignition time against the radiant flux will, when extrapolated to zero, give the critical rate of radiation for ignition.

In potential fire situations, it is desirable to know what are the critical heat transfer conditions needed to heat a material to the fire point and the time it would take to bring about ignition for heat transfer rates in excess of the critical rate. To calculate these, it is broadly necessary to know the fire point of the fuel and the thermal and other properties of the fuel that govern the heating and heat loss processes. For solid fuels, it is very difficult experimentally to measure the fire point temperature, although some measurements have been carried out by Thomson and Drysdale (1989) and are given in Table 5.3. Fire points are usually deduced from critical heat transfer rates that are themselves deduced from experimental measurements of time to ignition at different radiant heat transfer rates. These in their turn are then related to the heat loss condition; usually convective and radiative heat loss under the experimental conditions. The recent advent of the cone calorimeter (Babrauskas and Grayson, 1992) and standardized pilot ignition apparatus have facilitated such measurements. Mikkola (1992) plots radiant heat against $t_{ig}^{-1/2}$ for thick materials and t_{ig}^{-1} for thin materials as required by equations [5.1] and [5.2]. More recently, Delichatsios *et al.* (1991) have put forward a method that relies on using experimental values based on thick fuels and high heat transfer rates and plotting $(1/t_{ig})^{1/2}$ against the radiant heat and correcting for the varying heat loss from the surface prior to ignition. Janssens (1992b) has put forward the following equation relating ignition time, radiant heat flux (q''), and critical heat flux (q''_{cr}) for thick fuels:

$$q'' = q''_{cr}[1 + 0.73\, k\rho c/(h_{ig}t_{ig})^{0.547}] \qquad [5.4]$$

where h_{ig} is a total heat transfer coefficient from the surface.

[a]Table 5.3

Fire points of polymers	°C
Polymethyl methacrylate (PX)	310
Polymethyl methacrylate (FINN)	309
Polyoxymethylene	281
Polyethylene	303
Polypropylene	334
Polystyrene	366

[a]Taken from Thomson and Drysdale (1989).

5.2.4 IDENTIFICATION AND POWER OF IGNITION SOURCES

Given a fuel with a certain fire point, whether a source of ignition will heat the fuel to the fire point will depend upon the heat output and the heat transfer conditions from ignition source to fuel and the heat loss conditions from fuel to environment. Fire point measurement of solid fuels is based on experiments with flat plates, on the order of 10-cm dimension. It is likely that items with much smaller dimensions, particularly in a draught, may have higher values of the fire point because a higher critical flow rate of fuel vapor is needed (Rasbash, 1976). However, there may also be a higher convective heat transfer rate as well as a smaller amount of heat required to heat the material because of the smaller dimension.

An ignition source may be a source of heat or flaming combustion that can vary from a lighted match, a flame from a cooker, a burning waste paper basket, substantial rubbish fires, or sources of fire deliberately and maliciously introduced. Each of these can be characterized by a total heat output and a power output according to the amount of fuel and its rate of burning (Babrauskas and Walton, 1986). Oxygen consumption calorimeters may provide a ready method of measuring such outputs, even for quite large fire sources. Various standardized ignition sources exist for tests on furniture (BS 5852, 1990). However, the important parameter governing fire spread is not so much the heat output but the heat transfer to combustible surfaces to raise the temperature to the relevant fire point.

At the fire point, ignition will depend on whether there is a small pilot source of ignition capable of igniting the vapors produced. If the heating is occurring by direct contact with a flame, then the flame will also act as an ignition source. A spark, ember, or flamelet from a more distant fire can also act as an ignition source. In the absence of these, it is necessary to heat the fuel to a temperature at which the fuel vapors evolved will oxidize in air and spontaneously ignite. This temperature will depend on the chemical composition of the fuel. For cellulosic fuels, it is about 200 °C higher than the pilot ignition temperature of 300 to 350 °C. For liquid fuels, spontaneous ignition of vapor–air mixtures may be measured in vessels held at different temperatures (Mullins and Penner, 1959). The spontaneous ignition temperature decreases as the size of the vessel increases. Bond (1991) extensively documents spontaneous ignition temperatures, now generally called *auto ignition temperatures*.

It is unlikely that a lighted cigarette will produce ignition of a flammable gas/air mixture. However, this source of ignition is capable of bringing about smoldering in solid materials that burn, particularly if they are finely divided. The temperature of the smoldering zone depends on the air speed and flaming may follow, particularly if thin fuel is encountered under suitable air speed conditions (Drysdale, 1985d). Smoldering ignition may also occur spontaneously within the bulk of a porous solid due to self-heating oxidation processes within the solid, possibly preceded by microbiological processes. For a given fuel, occurrence of ignition depends on the dimensions of the fuel bulk and the heating and heat loss environment of the fuel (Bowes, 1984).

5.3 Spread of fire

The phenomena associated with the spread of fire are major inputs into both Subsystem (i) fire occurrence, and Subsystem (ii) fire development (Table 2.4). Inclusion in Subsystem (i) is very much dependent on how a fire is defined. If fire is associated with a certain minimum detriment, then this in its turn will be associated with a certain minimum fire size. Thus, a lighted match thrown down will not lead to a fire unless it falls on thin material that is easily ignitable and even in the presence of such a material, whether it leads to a fire for which the fire brigade is called or with a significant detriment, will depend on the nearness and disposition of other flammable fuels. A piece of paper burning on a wooden floor is unlikely to set the floor alight and even if it does, there is a high chance of self-extinction when the paper burns out. A vertical surface is more likely to ignite. Ignition and spread of fire is even more likely to occur in a vertical channel where heat loss is restricted. As far as fire development in buildings is concerned, there are two main regions of hot gases, usually moving, that can transfer convective and radiative heat to combustible surfaces. First, the flames and plumes that rise from burning articles, and second, the hot gas layers that form under ceilings and may extend to fill a room. The complex structure of these bodies of hot gas gives rise to substantial complexities in estimation of heat transfer rates and there is a degree of incoherence in data available for this. Conductive transfer in the fuel being heated plays a major part in the time for the surface to reach the fire point temperature. Chapters on radiative, convective, and conductive heat transfer are given in the S.F.P.E. Handbook (1995, 1988), which cover some of the necessary ground.

5.3.1 RADIATION FROM FLAMES

Radiative transfer depends on the emissivity, the temperature, and dimension of the flames. There are two components – radiation from gases and radiation from soot. In general, for luminous flames the radiation is dominated by the soot luminosity and is controlled by the soot concentration in the flame, which varies according to the material burning (Delichatsios *et al.*, 1992). Table 5.4 gives absorption coefficients and temperatures of flames from a number of common fuels with luminous flames as well as ethanol that has nonluminous flames. The flames listed are all well within the turbulent regime. There was evidence that the absorption coefficient increased as the flame thickness increased. In particular, the coefficient increased from 0.7 to 1.4 for wood as flame thickness increased from 15 to 200 cm. There are data in the literature for gaseous laminar flames. These have not been included since the reaction zone is thin, absorption coefficients are much higher and cannot be regarded as representing the total thickness of the flame. Orloff and de Ris (1982) used values of absorption coefficient of 0.6 and 1.3 m^{-1} for turbulent methane and propane flames respectively together with a temperature of 1200 K to obtain estimates of flame dimensions and volume from radiation flame mapping. These values are in line with the data in Table 5.4. An absorption coefficient of 0.85 m^{-1} has been given for heptane (Ndubizu *et al.*, 1983) and a similar value for propane is implied in data by Delichatsios (1993). This is substantially less than the value for petrol and kerosene given in Table 5.4 but the figures for these fuels were probably influenced by the higher hydrocarbons in the wide boiling point range of the fuels. The emissivity ε and the radiation q''_{rad} from the flames are given by equations [5.5] and [5.6] below

$$\varepsilon = 1 - e^{-\alpha L} \quad [5.5]$$

$$q''_{\text{rad}} = \varepsilon\sigma T^4 \quad [5.6]$$

α = absorption coefficient(m^{-1})
L = flame thickness (m)
T = temperature of flame (K)
σ = Stephan Boltzmann constant (5.67×10^{-8} W/m^2 K^4)

The data in Table 5.4 indicate approximate similarity in absorption coefficient of flames of fuels of similar chemical composition. Benzene and polystyrene have the empirical formula CH, they produce much soot and consequently the flames have a high absorption coefficient (3.1–5.3/ m), and emissivity approaches unity as flame thickness increases to 1 m. Kerosene, petrol, and polypropylene, empirical formula ~CH_2, give absorption coefficients 1.6 to 2.6 m^{-1}, and 1-m thick flames will have an emissivity of about 0.9. The presence of combined oxygen and nitrogen reduces the absorption coefficient. Thus, polymethyl methracrylate ($C_5H_8O_2$) gives a value of 1.3 to 1.5. Wood ($C_6H_{10}O_5$) gives a value of 0.8 with a comparable flame thickness; a 2-m

Table 5.4. Flame radiation properties of burning fuels

Fuel (B.P. range °C)	Surface shape & dimension	Height above surface (mm)	Flame thickness (mm)	Absorption coefficient $\alpha(m^{-1})$	Flame temperature (K)	Rate of burning (gm/s)	Reference
Ethanol (77–79)	Circular	150	180	0.37	1491	13.2	Rasbash *et al.*, (1956)
Petrol (30–200)	0.30 m	300	220	2.0	1299	23	
Kerosine (155–277)	diameter	300	180	2.6	1263	14	
Benzene		300	220	3.9	1194	27	
		300	290	4.1	1194	43	
		300	300	4.2	1194	60	
Polyoxy methylene	0.3 m square	5.1	c.60	c.0.3	1380	6.4	
PMMA		150	150	1.3	1380	10.0	
Polypropylene		100	250	2.2	1310	8.4	Markstein (1978)
Polyurethane		51	162	1.3	1408	n.a.	
Polystyrene		50	310	5.3	1190	14.1	
		100	300	4.8	1180	14.1	
		200	230	3.1	1020	14.1	
		250	200	4.2	1000	14.1	
PMMA	Circular 0.73 m diameter	200	520	1.5	1350	20.0	Markstein (1978)
Wood	Cribs of	300	150–2000	0.7–1.4	1300	n.a.	Hagglund and Persson 1976
8% moisture	varying length and 1.2 m width	500	250–1600	0.5–1.15	1300	n.a.	

thick flame is required to produce an emissivity of 0.94. Ethanol (C_2H_6O) produces very little soot and has a coefficient of 0.37/m. Emissivity will tend to rise to a maximum of about 0.4 as flame thickness increases since radiation occurs in only part of the spectrum. The above observations suggest that there could be room for interpolation to obtain absorption coefficients for polymers with intermediate empirical formulas. Such interpolation could be guided by measurement of smoke point lengths of laminar diffusion flames burning the pyrolysate or the relevant vapor or gas (de Ris, 1988).

In fire development, one is usually concerned with heat transfer from flames greater than 0.1 m thickness and emissivity can be based on the absorption coefficients in Table 5.4. Usually, these flames are sufficiently turbulent for combustion to be regarded as taking place across the whole flame thickness.

To estimate the radiation it is also necessary to know the dimensions of the flame, particularly the flame height and width. The flame height has been related to the rate of burning and dimension of the burning fuel for freely burning fires and a number of working correlations exist (Heskestad, 1988, McCaffrey, 1988). There is, however, little systematic information on measured flame width, which is important as it controls flame emissivity in a lateral direction to nearby objects.

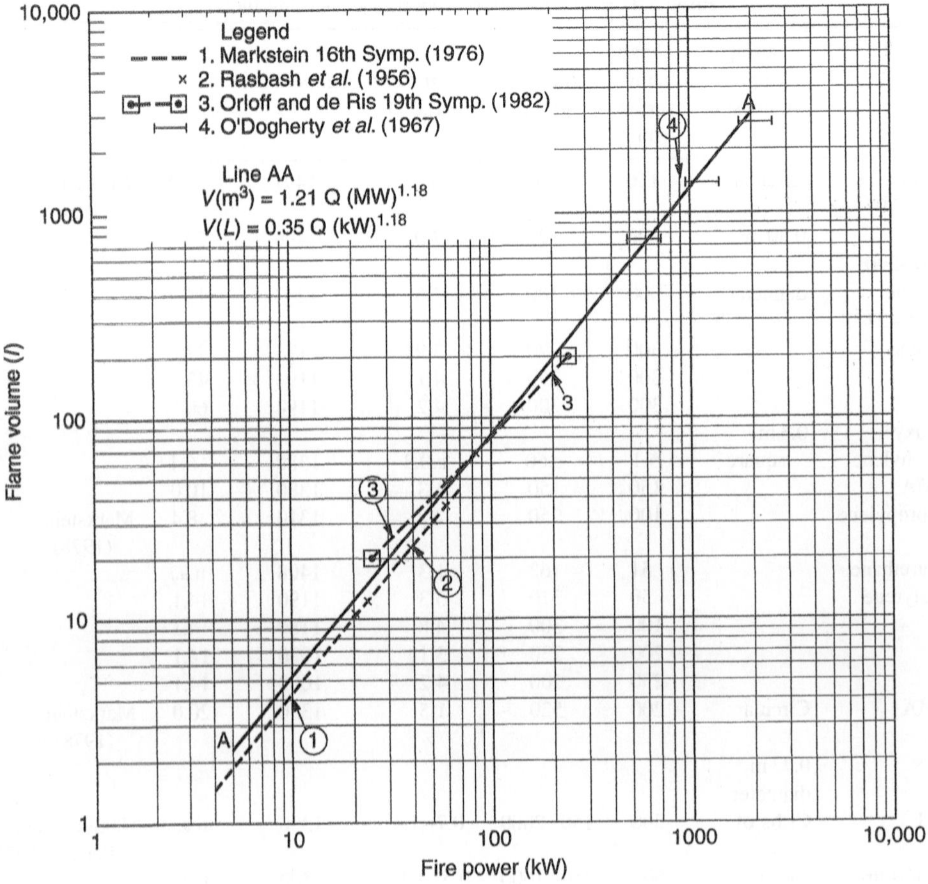

Figure 5.5. Dependence of flame volume on fire power

An approximate value of the flame width may be arrived at by obtaining a measure of flame volume from the heat output. Then by assuming a certain shape to the flame, flame diameter may be estimated as a function of height. Figure 5.5 plots available information on the volume of turbulent flames as a function of heat output. Four sets of information were used

1. Propane burning at flow rates of 44 to 412 cm^3 through a 12.7-mm-diameter nozzle (Markstein, 1976).
2. Flame volumes derived from Figure 5.2 (Rasbash *et al.*, 1956).
3. Propane, methane, and PMMA burning from 0.38- and 0.76-m-diameter surfaces (Orloff and de Ris, 1982).
4. Measured flame volumes from Figure 5.3 (O'Dogherty *et al.*, 1967).

The heat output of the hydrocarbon fires from Figure 5.2 was corrected for efficiency of combustion (petrol 0.92, kerosene 0.91, and benzole 0.69; obtained from Tewarson (1995)). The points relevant to Figure 5.3 were plotted for a range of "chemical" heat of combustion of 12.4 kJ/g given by Tewarson and a theoretical 18.6 kJ/g given in the reference. A mean relation between flame volume and heat output Q in the range 20 to 2800 kW is given by

$$V(\mathrm{m}^3) = 1.21\,Q(\mathrm{MW})^{1.18} \quad [5.7a]$$

or

$$V(\mathrm{L}) = 0.35\,Q(\mathrm{kW})^{1.18} \quad [5.7b]$$

The flame volumes in Figure 5.2 led initially to the conclusion that the intensity of flaming combustion in flames controlled by convective turbulent buoyancy was independent of flame power over the range studied (20–160 kW) and was approximately 1.9 MW/m^3, kW/L, or W/cm^3. Similar assumptions have also been utilized by other authors (de Ris, 1978, Back *et al.*, 1994). However, equation [5.7] indicates that over the wider range of heat outputs, the combustion intensity decreases somewhat as the power of the flame is increased, particularly if heat output is corrected for incomplete combustion.

Figures 5.2, 5.3, and 5.4 suggest that the main part of the flame has a cylindrical mean shape, topped in the upper portion by a cone. The height of the flame L_f may be obtained from an appropriate formula relating flame height with heat output and fuel dimension. If it is assumed that the flame is cylindrical throughout the height, the flame diameter may be estimated as $\sqrt{4V/\pi L_\mathrm{f}}$. If it is thought to be cylindrical through half or two-thirds of its height and conical above, then the diameter of the cylindrical portion would be respectively 1.22 and 1.13 times larger. It has become customary in zone modeling to assume that the shape of the flame is conical on a base of area equal to that of the fuel source. However, if this assumption is adopted, then the flame height would be given by $12V/\pi D_\mathrm{b}^2$ where D_b is the diameter of the flame base. More empirical information on mean flame shapes would be helpful.

Estimates of radiation made in this way should only be expected to apply to the lower half of the flames where the flame is continuous in time and where the bulk of the measurements of flame radiation properties in Table 5.4 were made. A more general estimate of radiation may be obtained from the observation that the total radiant heat output of buoyant flames is a constant fraction of the total heat output depending on the burning conditions and the soot-producing properties of the fuel, which in turn is related to the measured smoke point length of the laminar diffusion flame (Markstein, 1984, Delichatsios *et al.*, 1993). With turbulent jet flames of up to 40 kW produced by a 12.7-mm-diameter nozzle, Markstein obtained a radiative fraction varying from 0.18 for the least sooty methane flames to 0.429 for the most sooty 1,3 butadiene tested. Lower fractions, particularly for methane, ethane, and propane, were obtained for laminar flames of heat output 140 to 300 W flowing from a 4.4-mm-diameter nozzle. This radiation may be considered as emanating uniformly from a point in the center of the flame as far as distant objects are concerned.

Markstein and de Ris (1990) also obtained measurements of forward radiant emission from a slot burner 16 mm deep and 380 mm long, with a heat output of 10 to 60 kW under conditions of free burning and when against a cooled wall 2.2 m high. Comparison was also made with a free burning jet burner of 28-mm diameter of similar heat output range. The flames were buoyancy-controlled turbulent diffusion flames, although in all cases, there was a fully laminar region near the burner exit. The fuels used were methane, ethane, ethylene, and propylene and again the radiative fraction for both the free burning jet fire and the line fire varied from about 0.18 to 0.4 of the total heat release rate according to the soot-producing capacity of the fuel. However, the total radiative fraction was substantially reduced by 18 to 36% when the line fire was placed against the cooled wall. Also the peak radiance was reduced to about 40% of the value under free burning conditions, although the flame height was about 1.8 times as great. The heat transfer to the wall was not measured but one must assume that at least for thin slot flames the cooling caused by convective heat transfer to the wall considerably reduced both forward and backward radiant transfer. There is evidence (Kulayne, 1984) that with a flame thickness of 95 mm, the effect on total radiance is much less, but one would still need to allow for a reduction in peak radiance due to flame lengthening when adding convective heat transfer to free burning radiative heat transfer to obtain a total heat transfer from a flame to a wall with which it is in contact.

5.3.2 CONVECTION FROM FLAMES

To estimate the convective heat transfer, it is necessary to know the velocity of the moving stream as well as the temperature. Heskestad (1988) has also provided information on these parameters. This can then be related to convective heat transfer by

$$q''_{\mathrm{conv}} = h(T_{\mathrm{h}} - T_{\mathrm{b}}) \tag{5.8}$$

T_{h} = Temperature in flame or plume at point of contact.
T_{b} = Temperature of cold surface being heated.
h = Heat transfer coefficient.

The heat transfer coefficient h depends on the velocity of the flame or hot gas and the dimension and shape of the heated object and is given in standard textbooks of heat transfer. For objects that are small compared to the size of the flame, information on velocity and temperature of the flame can be used directly to estimate convective heat transfer. However, in situations in which the flame moves against wall or ceiling surfaces, information based on free upward-moving flames and plumes may not be readily applicable. Estimates based on direct measurements of total heat transfer may be more reliable than calculated values (Section 5.3.3).

Convective heat transfer to a surface that is producing flammable vapors is reduced by the vapor flow. With incombustible surfaces and with very low fuel flow rates corresponding to critical burning rates, ca. 1 to 4 g/m^2 s, the heat transfer may be as high as 20 to 25 kW/m^2. However, radiative heat transfer to the surface, particularly for fuels burning with luminous flames, increases the burning rate substantially and diminishes the contribution of convective heat transfer. The rate of burning per unit area of surface (m'') is given by the general relationship (Spalding, 1955, Rasbash *et al.*, 1956):

$$m'' = (h/c)\ln(1 + B) \tag{5.9}$$

h is the relevant convective heat transfer coefficient, c is the specific heat of the gas (usually taken as air at room temperature), B is the transfer number.

The value of h/c for turbulent natural convection is, at a flat vertical or upward-facing surface, 10 g/m^2 s. The transfer number B is approximately the heat of combustion of air (~3000 J/g) divided by the convective heat transfer to the surface H_{f}, associated with the production of 1 g of fuel volatiles that burn at the surface. H_{f} can be greatly in excess of the heat of pyrolysis of the pyrolysate that burns at the surface. It also covers heat losses from the fuel surfaces and heat of pyrolysis of fuels that do not burn that may be evolved from the rear of the sample. If H_{f} is large, then B is small and the convective heat transfer $m''H_{\mathrm{f}}$ will tend towards $\frac{h}{c}BH_{\mathrm{f}}$, that is, $3000\frac{h}{c}$. If H_{f} is small, particularly as a result of a large contribution from radiation, B will become large and the convective heat transfer will fall in proportion to the ratio $\ln(1 + B)/B$.

There is no coherent approach to estimating how heat transfer is shared between radiation and convection because there is a dearth of experimental information on heat transfer to vaporizing fuel surfaces from flames. Sometimes, it is assumed that radiative heat transfer can be neglected because of the small dimension of the fire or nonluminosity of the flame. Radiation to the surface may be estimated and convective heat transfer obtained by difference to account for observed vaporization and heat loss from the surface. Since absorption of radiant heat by the vapor flow is often ignored, this tends to overestimate the contribution of radiant heat. Of the four fires featured in Figure 5.2, the alcohol flames were nonluminous and radiation contributed only 17% of heat toward the vaporization. The convective heat transfer allowed an estimated burning rate based on equation [5.9] of 13.4 g/m^2 s, which compared very well with the measured value of 13.2 g/m^2 s. However, for the three other fires with luminous flames, the estimated radiant heat transfer uncorrected for vapor flame absorption more than covered the amount needed for vaporization (Rasbash *et al.*, 1956), making it impossible to use equation [5.9] to estimate burning rate as B becomes infinity when convective heat transfer is zero. An analysis of heat transfer to the fuel surface from a flame burning vertically up a PMMA slab 0.406 m wide (Orloff *et al.*, 1974), with a 7% correction for radiative absorption by the vapor, showed the radiative heat transfer increasing from 13,530 to 21,510 W/m^2 as the height increased from 38.1 to 152.4 cm above the fire base. This was due to increasing flame thickness. Convection (estimated by difference) decreased from 6480 to 5540 W/m^2. Equation [5.9], in this case, predicted observed burning rates with a lower value of h/c of about 6 g/m^2 s. More experimental information is needed to help quantify h/c. In the interim it may be taken as 10 g/m^2 s when convection dominates, and 6 g/m^2 s when radiation dominates. Instead of the parameter h/c, Delichatsios (1986) uses a parameter $0.088\rho_\infty(\nu_\infty g \Delta T_m / T_\infty)^{1/3}$, where ρ_∞, T_∞, ν_∞ are density, temperature, and kinematic viscosity of ambient air, g is the acceleration due to gravity, and ΔT_m is the maximum temperature rise in the flame. In calculating B, he also reduces the heat of combustion of air by the fraction not used in providing convective heat output of the flame. In general, there is a tendency in modeling turbulent fires with luminous flames to assume that the convective heat transfer will cover the reradiant heat loss from the surface and to assume that the rate of burning is obtained by dividing the radiant heat transfer by the heat required to produce the fuel vapors (L).

5.3.3 MEASUREMENTS OF HEAT TRANSFER FROM FLAMES

The data in Table 5.4 apply to upward-moving flames from a horizontal base. Here, the intensity of combustion in the flames is governed by free buoyant air entrainment. Underneath a ceiling, the air entrainment will be more restricted, and with a combustible ceiling, a thick zone of combustible products is likely to buildup between the ceiling and flaming zone. Hinkley *et al.* (1968, 1984) experimented with town gas and wood cribs, burning with heat inputs varying between 140 to 600 kW, at different distances below a corridor-shaped ceiling. He obtained heat transfer rates to the ceiling, increasing up to a sharp peak of 170 kW/m^2 at the point of impingement of an air rich flame. The peak value did not depend on the heat input and, according to criteria of McCaffrey (1979), occurred at a point within the continuous flame zone. There was also an exponential decrease of heat transfer with horizontal distance from a virtual origin some distance behind the point of impingement. You and Faeth (1979), working with heat inputs of up to 3.5 kW give heat transfer rates in the area of flame impingement of up to approximately 40 kW/m^2 and a power decrease at distances beyond this. The peak heat transfer rate corresponded to conditions when about the upper 40% of flame height impinged on the ceiling. It is likely that a substantial proportion of the heat transfer in the above two cases was convective. Kokkala (1991) working with natural gas flames of 2.9 to 10.5 kW obtained heat transfer rates of 60 kW/m^2 when flame height L_f was 1.5 to 3.5 H_r, H_r being the ceiling height above the burner surface. The radiative fraction was between 40 to 60%. You and Faeth incorporate the heat transfer rate to a ceiling q'' in a dimensionless number, $q'' H_r^2 / Q$, where Q is the heat output of the fire. The constant product of this number and the one-sixth power of a plume Rayleigh number that also contains Q and H suggests that the peak heat transfer rate should decrease with $Q^{0.15}$ at the point of impingement within the flame. However, Alpert and Ward (1984) state that the peak convective flux increases with $Q^{0.2}$ to a value not greater than 100 kW/m^2, and it occurs when the top of the flames are impinging on the ceiling. However, Kokkala found that the product referred to increased by a factor of about 4 in the flame impingement region $0.7 < L_f/H_r < 2$. It is difficult to reconcile these observations.

Heat transfer from comparatively thin methane line burner flames alongside a vertical surface of 20 to 30 kW/m^2 (Hasemi, 1986, Saito *et al.*, 1986) have been given and 60 kW/m^2 from 150 kW propane flame from a burner of dimension 0.3 m, placed next to the vertical surface (Williamson *et al.*, 1991). Back *et al.* (1994) measured peak heat transfer rates q''_p of 40 to 120 kW/m^2 for square propane fuel sand burners of dimension 0.280 to 0.7 m and heat release rate Q of 50 to 500 kW/m^2 on a gap in a wall adjacent to the burners. Back correlated heat fluxes obtained by equation [5.10]

$$q''_p = E[1 - \exp(-k_a Q^{1/3})] \qquad [5.10]$$

where E is the blackbody emissive power of the flame and was given the value of 200 kW/m^2 and k_a, a variant of the absorption coefficient with the value of 0.09 kW$^{-1/3}$. The relationship depends on the assumption that flame volume is directly proportional to heat output Q and the linear dimension for flame thickness can be taken as proportional to $Q^{1/3}$. q''_p was stated as being independent of flame aspect ratio L_f/D_b. This approach differs from that which accompanies equation [5.7], which aims at calculating relevant flame width. In developing equation [5.10], convective heat transfer that can be a substantial part of the lower heat transfer rates was neglected and the blackbody radiation of 200 kW/m^2 employed is about twice as great as might be expected from measured flame temperature (900 °C). A burning chair, 460 mm deep, with a polyurethane foam back and a seat, 50 mm thick, with PVC covers next to a wall gave a heat transfer of 115 kW/m^2 on the walls (Rogowski, 1984, Morris, 1984). Babrauskas (1982) measured the radiation from a number of items of furniture and recorded fluxes up to 80 kW/m^2 and 20 kW/m^2 at 0.05 and 0.88 m from the burning item. If the efficiency of mixing of the combustion air with fuel vapors is greater than that which normally occurs in free entrainment buoyant flames, then considerably higher flame temperatures and higher heat transfer rates may be achieved. Thus heat transfer from a jet flame may be as high as 600 to 700 kW/m^2 (Odgaard and Solberg, 1981). A particularly high heat transfer rate of 250 kW/m^2 and a flame temperature of 1350 °C were measured on a full-scale test of a rig representing the Dublin Stardust fire disaster (Morris, 1984). These high values may have been due to turbulence engendered in the entrained air feeding the fire. There is room for a great deal more systematic information on heat transfer rates between flames of various kinds and surfaces within and outside the flames, related to the parameters that control radiative and convective heat transfer.

5.3.4 RADIATION FROM HOT GAS LAYERS

Radiant heat transfer from a hot gas layer will again depend on the soot, carbon dioxide, and water vapor concentrations. The absorption coefficient of soot is given by $7 f_v/\lambda$, where f_v is the volume fraction of the soot (de Ris, 1978). It is thus inversely proportional to the wavelength λ, and since the wavelength corresponding to the maximum in black body radiation increases from 2.06 to 7.27×10^{-6} m as temperature decreases from 1400 to 400 K, there will be a corresponding decrease in absorptivity and emissivity. The gases absorb radiation only in parts of the spectrum, but here again absorption is a function of gas temperature. It is fairly constant for carbon dioxide but decreases with temperature for water vapor. The concentration of the above species is determined from data on efficiency of combustion, the chemical nature of the fuel and the extent of dilution by air entrainment. The soot concentration may also be obtained from measurements of obscuring power to light since as long as the fuel has been burning in a plentiful supply of air the obscuration is mainly due to soot and an obscuration of 1(b_n)/m will correspond to a soot concentration of 130 mg/m^3 (Section 5.10). Given a mean wavelength of light as 0.55×10^{-6} m, this indicates a soot density of 1.66 g/cm^3. Thus, if smoke concentration is measured in b_n/m, f_v will be equal to 78×10^{-9} for 1 unit of b_n/m.

Given all the necessary data, it is possible to work out a mean absorption coefficient for the smoke layer and thence radiation to neighboring surfaces. If temperature through the layer needs to be averaged, then it will be the mean fourth power of the absolute temperature that will need to be obtained (Orloff *et al.*, 1978). However, estimation of the radiation is still a relatively complex matter and it is usual to make simplifications. Thus, if one is concerned with a layer several meters thick of smoke and one is estimating heat transfer that might cause physical injury or ignition of a fuel, it may be assumed that the layer is radiating with unit emissivity to be on the safe side. If less than 2% of the fuel is converted into soot, the gaseous radiation will tend to dominate, and if the fuel conversion is more than 2%, soot radiation will dominate. Quintiere (1977) has provided a formula for emissivity ε of a smoke layer:

$$\varepsilon = 1 - \exp[-(0.33 + 0.47C_s)l] \qquad [5.11]$$

l being the thickness of the layer and C_s the smoke concentration in grams per meter cube or mg per liter.

It is assumed that the layer contains 12% carbon dioxide and 12% water vapor.

5.3.5 CONVECTION FROM HOT GAS LAYERS

The convective heat transfer from the upper layer of hot gas to the ceiling and walls with which it is in contact is a subject for which there is still a wide disparity in the available information in the literature. If the layer were static, then the heat transfer to the surfaces would be governed by turbulent natural convection. According to the S.F.P.E. Handbook, this is given by

$$\frac{hx}{k} = \text{Nu} = 0.16(\text{Gr Pr})^{0.33} = 0.16[(g\beta\Delta T x^3/\nu)(\nu/\alpha)]^{0.33} \qquad [5.12]$$

which, when allowance has been made for variation of k, ν, α with temperature, can be simplified to

$$h = 2.66\Delta T^{0.25} \qquad [5.13]$$

ΔT = temperature difference between gas and surface.
β = coefficient of expansion of the gas
x = linear dimension (this cancels in the relationship in equation [5.12])
k, ν, α = thermal conductivity, kinematic viscosity, thermal diffusivity of the gas
Nu, Gr, Pr = Nusselt No., Grashof No., and Prandtl No. respectively.

However, the plume of hot gas rising from the fire causes an outward flowing ceiling jet that is characterized by a stagnation point where the center of the plume hits the ceiling, an increase in velocity to a point on the order of the radius of the plume, and a decrease in velocity beyond. When reaching the walls the ceiling jet turns and forms a wall jet. A simple approach would be to apply a formula for turbulent forced flow over a plate (S.F.P.E. Handbook, 1988):

$$\text{Nu} = 0.0296\,\text{Re}^{0.8}\,\text{Pr}^{0.33} = 0.0296\left(\frac{Vx}{\nu}\right)^{0.8}\left(\frac{\nu}{\alpha}\right)^{0.33} \qquad [5.14]$$

Re is the Reynolds No., V is the gas velocity, and x is the distance from the stagnation point. It is not clear at what value of x the relationship would be expected to hold, since the boundary layer of the ceiling jet forms in a different way to the boundary layer of flow over a flat plate. Atkinson and Drysdale (1992) indicate that over a large part of the ceiling, the heat transfer follows turbulent natural convection, although the relation they use has a constant of 0.193 rather than 0.13 in equation [5.10]. Near the area of impact of the plume they use

$$h = 0.45\frac{k}{(v)^{1/2}}\frac{V}{x} \qquad [5.15]$$

The relation represents the heat transfer expected from heated laminar jets. It is stated that for turbulent jets it would be 1.4 to 2.3 times larger but doubt is also expressed that the coefficient in equation [5.13] may be much smaller.

Some experiments carried out by Zukoski (1987) suggested that when a comparatively thin flowing layer of hot gas moves under a ceiling representing a thermal gravity current, a higher heat transfer occurs given by

$$\mathrm{Nu} = 0.013\,\mathrm{Re} \qquad [5.16]$$

The enhanced heat transfer was attributed to the development of a rolling motion in the layer of hot gas.

Cooper (1982) and Cooper and Woodhouse (1986) have put forward relationships for estimating relevant heat transfer coefficients and temperature differences for convective heat transfer from ceiling jets up to distances from the stagnation point twice the height of the ceiling above the fire source. These relationships are based on the assumption that both momentum and heat transfer for a plume against a ceiling follow a similar relationship to a turbulent jet against a wall. This has been criticized in that the characteristic length scale of the turbulence produced by natural buoyancy, particularly if the flames are reaching a substantial height below the ceiling, are likely to be much larger than in a forced turbulent jet and this would tend to overestimate the heat transfer.

A comparison between the above approaches may be made by applying the different approaches to a specific case. Kung *et al.* (1988) gives information on plume axis and ceiling jet temperatures and velocities for two rack storage fuels for a range of convective heat outputs. A typical output of 1000 kW with the top surface of the storage 5 m below the ceiling will give an axial temperature rise of 200 K and upward velocity 7.5 m/s at the ceiling, where the plume has a nominal radius 0.72 m and a maximum temperature rise and outward velocity of 44 K and 1.64 m/s, 5 m from the stagnation point. The temperature and outward velocity correlations given also indicate that at the nominal plume radius, these will be the same as in the upward axial stream. Table 5.5 gives estimates of heat transfer coefficients and convective heat transfer for the different relationships given above using this information.

The wide range of heat transfer coefficients from 6.7 to 104 $\mathrm{W/m^2K}$ is demonstrated. There is a lack of measurement of these coefficients for appropriate-sized fires to allow discrimination between the values that should be used particularly in the ceiling area near the plume. The conditions under which Zukoski's correlation applies also need to be checked. Modeling approaches, see Mitler (1978) and Chapter 12, recognize the range of coefficients but state that the increase of coefficient from low to high takes place over a moderate increase in temperature of the ceiling layer and applies to the whole layer. At the very least, this will overestimate the heat transfer in areas remote from the plume. The incoherence of the information leaves a gap since heat transfer to the ceiling plays a large part in predicting both flashover and flameover (Section 5.5). To be on the safe side, one would need to use low values of coefficients for predicting flashover, since more heat is maintained within the hot layer, but high values for predicting flameover where combustible layers on a ceiling may catch fire. The rate of heat transfer from the hot gas layer will also influence the time taken for buildup of the layer since the volume of the gas will decrease as it loses heat.

Table 5.5. Convective heat transfer rates from hot gas layers

Position on ceiling → / Method of calculation ↓	At plume axis-stagnation point		At nominal plume radius (0.72 m)		At 5 m from stagnation point	
	Coefficient h (W/m^2K)	Heat transfer q''_{conv} (kW/m^2)	Coefficient h (W/m^2K)	Heat transfer q''_{conv} (kW/m^2)	Coefficient h (W/m^2K)	Heat transfer q''_{conv} (kW/m^2)
1. Turbulent natural convection (equation [5.12])	10.0	2.0	10.0	2.0	6.9	0.30
2. Turbulent forced convection	[a]	[a]	18.3	3.66	6.7	0.294
3. Atkinson (equation [5.15])	44	8.8	44	8.8	[c]	[c]
4. Zukoski (equation [5.16])	n.a.	n.a.	104	20.8	27.9	1.23
5. Cooper (equations in references)[b]	61	11.35	37.8	5.55	11.8	0.487

[a]Cannot be applied.
[b]On the basis of estimates of heat transfer coefficient and ΔT for 1000 kW convected heat from fire 5 m below the ceiling.
[c]As for turbulent natural convection, but ca. 1.5 times greater due to larger coefficient.

5.3.6 FIRE SPREAD ALONG SURFACES

In practice, given a heat transfer regime, it is frequently necessary to estimate the rate at which flame will spread along a surface. This will depend on whether the flame flow is along the surface and is in the same direction as fire spread or if it is not. The former condition, known as *cocurrent flame*, operates when flame spreads upward along a vertical surface or horizontally below ceilings and the combined convective and radiant flux from the flame operates on the surface. Also, for upward-facing slopes in excess of about 15°, the entrainment conditions may cause the flame to bend over and heat the surface in the same way (Markstein and de Ris, 1972, de Ris and Orloff, 1974). With fire spreading sideways along a vertical surface and spreading on a horizontal surface facing upward, the heat transfer from the flame is limited to conduction at the leading edge, supplemented by a limited amount of radiation onto fuel in advance of the flame from the flame itself. This is known as *countercurrent flame* since the airflow feeding the flame is opposed to the direction of flame spread. Heat transfer to the unburned fuel and the rate of flame spread are far less than for cocurrent flame. The rate at which the surface fuel heats to the fire point depends on whether the material is thick or thin as indicated in equations [5.1] and [5.2]. Parameters for thin or thick materials may be obtained by plotting $t_{ig}^{-1/2}$ or t_{ig}^{-1} respectively against radiant heat transfer and measuring the slopes of the lines produced. Quintiere and Harkleroad (1984) have analyzed conductive sideways fire spread for a wide range of materials based on results of the LIFT test standard spread of flame test and have given ignition temperatures and thermal inertia values based on these results. This information may be used for estimating countercurrent flame spread rates. Nevertheless, actual heat flux measurements given in the literature for countercurrent flame spread show a variation by a factor of 20 (Babrauskas and Witterland, 1995). These authors measured peak values at the flame edge for the LIFT test varying from 25 kW/m^2 for wood particle board to 92 kW/m^2 for rigid FR, PU foam. There has also been analysis of upward spread of flame on surfaces by flame established on the surface at heat transfer rates from the flames up to 30 kW/m^2 (Quintiere *et al.*, 1986). A review of flame spread along surfaces has been given by Janssens (1992b).

5.4 Circumstances favorable to rapid fire spread

In addition to cocurrent flame spread over surfaces, where both convective and radiant heat transfer act on the fuel at a distance well ahead of the flame, a number of other circumstances can help promote rapid fire spread. Thus, at a sharp corner the heat transfer from the flames can feed into two sides, and at a corner where two surfaces meet at an angle less than 180°, there is less heat loss from the surfaces. Fire at a corner of the latter kind has been made the base of a number of room fire tests. At cavities, there is also less heat loss than under open conditions, particularly if there is fuel on more than one side of the cavity. In a similar way, where two opposing surfaces are close, there can be less heat loss with mutual radiation if both sides are burning. Delamination, particularly of thick paint layers on a wall or ceiling, can cause the fuel condition to change from thick to thin and make it burn much more rapidly. Moreover, since the paint burns on both sides when away from a wall or ceiling, this may result in a thickening of the flame that increases the heat transfer to the fuel ahead. Very rapid spread of flame along corridors, where the fuel is a multilayer of paint, has been known to occur and is probably due to this mechanism (Meams, 1986). For fires burning on a slope, there is an increased configuration factor between the flame and the surface, particularly if the flame bends over toward the slope. In the presence of a wind, flames will also bend over toward the surface with increased heat transfer rate. Fire may also be spread by expulsion of sparks or brands, release of molten drops, by flowing of a burning liquid, or by flame propagation, particularly through dust clouds. In the latter case, the tendency of burning dust to be deposited on exposed surfaces will assist fire spread.

5.5 Sudden massive flaming in buildings

A major hazard of fire in buildings is the sudden increase in flaming combustion that might occur in less than a minute to involve the whole of a space after fire has been burning locally for a much longer time. Such catastrophic increases can be a major feature of disasters. In the past, this phenomenon has been given the general name of flashover, but it has recently become the practice to allocate different names to three major mechanisms, whereby massive flame increase is caused, namely flashover, flameover, and the phenomenon known as *flashback* in the United Kingdom and *backdraft* in the United States.

Flashover is reserved for the most common of these and can occur in any room in which there is the normal type distribution of individual items of fuel in the lower part of the room. It is ascribed to downward radiation from hot gases that accumulate under the ceiling. A hot gas layer of unit emissivity, at a temperature of 500 to 600° will radiate downward to an upward-facing combustible surface at about 20 kW/m^2, which is above the critical radiation intensity for most combustible materials. There will therefore be a tendency for all upward-facing surfaces to ignite, within a short period, across the whole room. The occurrence of a sufficiently hot layer can be related to the size of a local fire Q and the heat loss from the hot layer to the room surfaces. A number of correlations have been proposed (Walton and Thomas, 1995, 1988); perhaps the most popular being that due to McCaffrey *et al.* (1981):

$$Q = 610(h_K A_T A_o \sqrt{H_o}) \qquad [5.17]$$

Q = Power of flame (kW)
h_K = effective heat transfer coefficient (kW/m^{2}K)
A_T = total area of compartment surfaces (m^2)
A_o = area of opening (m^2)
H_o = height of opening (m)

Another mechanism that may contribute to flashover is the occurrence of flame propagation at the interface between fuel-rich gases in the lower layer (Beyler, 1984). Hinkley's experiments (1984) on downward heat transfer from flames moving under a combustible ceiling suggest a heat transfer of 40 kW/m^2 under these conditions. At the Dublin disco fire, the flames under the ceiling produced a downward rate of heat transfer of 60 kW/m^2 on the furniture below. This led to spontaneous ignition in a few seconds. The catastrophic spread of fire at the Interstate Bank Building at Los Angeles and to the furniture in the alcove and ballroom at the Stardust fire in Dublin (Section 3.2.5 and 3.2.2) are examples of flashover.

Flameover is a very rapid flame spread along an extensive continuous flammable surface that has either been preheated by a local fire or brought close to a high local heat transfer rate. In general, such surfaces involve walls or downward facing ceilings. An experiment with a small fire burning in a compartment at the end of a corridor lined with hardboard illustrates this effect (Malhotra *et al.*, 1971). Figure 5.6 shows the fire after 7 and 8 min, respectively, indicating the rapid fire development. Very rapid fire that spread up the perspex walls and ceilings at Summerland (Section 3.2.1) was a flameover. Because of the tendency of flames to bend over and attach themselves to an upward slope of greater than 15°, flameover can also be a feature of fire spreading up a stairway in a building and also of heather and forest fires in hilly country. The catastrophic spread of fire up the wooden escalator at the King's Cross fire (Section 3.2.3) can thus be described as flameover as can the subsequent spread of fire to the paint layers on the ceiling of the shaft (Moodie and Jagger, 1989). In general, for flameover to occur, the ignited fuel must burn for a sufficient time to heat the unburned fuel ahead of the pyrolysis zone until it reaches the fire point. The delamination of thickly painted surfaces, if it occurs at temperatures well below the fire point, will substantially reduce the time required to heat the paint to the fire point and thus help to promote flameover on walls and ceilings.

Because of fire disasters that have taken place in the past, the contribution which combustible walls and ceilings can make to rapid fire growth has long been appreciated, and over the last half century, a wide range of test methods for classifying materials used in this way have come

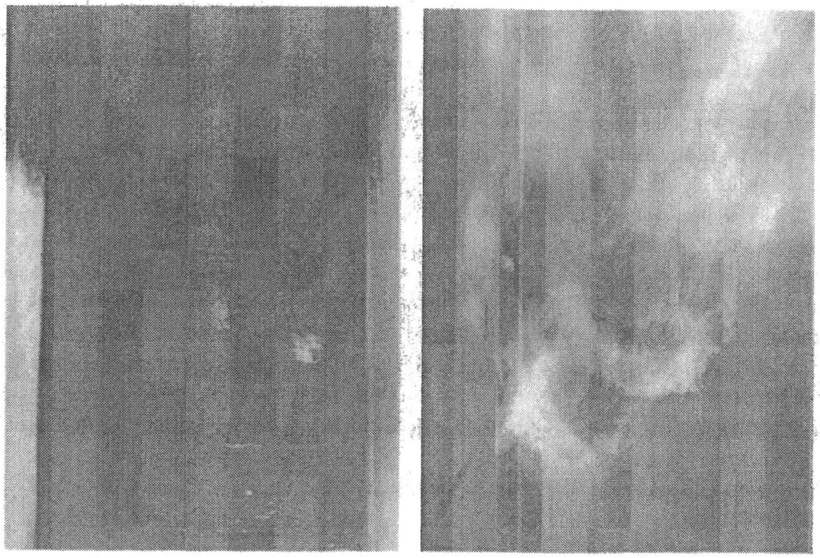

(a) 7 min (b) 8 min

Figure 5.6. Fire spread down a 13-m long corridor lined with hardboard into existence. These tests, which are given the general name of reaction to fire tests, have been developed in different countries, usually independently of what has gone on elsewhere. A great deal of regulatory requirements in the countries concerned for the use of these materials in places such as living accommodation, public assembly buildings, and means of escape are based on the test results. In general, these tests do not directly provide information that can lead to quantification of rapid fire spread. Moreover, comparative tests with different materials have shown poor agreement between the order in which the different materials are placed with regard to their contribution to fire hazard (Emmons, 1968). There are several reasons for this, including the wide range of heat transfer regimes used for exposure, the different geometrical disposition of the samples, and the tendency of different materials not only to respond differently to these factors but also to complicate response by any tendency to change shape by distortion, delamination, melting, or intumescence. This has led to difficulties in efforts made to harmonize the use of tests in different countries (Malhotra, 1992). One cannot expect these difficulties to be resolved until the behavior of materials in the tests themselves can be forecast from basic information on the fire properties of the materials. It is encouraging to note that useful steps have already been taken toward this end. As indicated earlier, a standard spread of flame test has been used by Quintiere and Harkleroad to estimate properties of fire point temperature and thermal inertia for a range of materials. The capacity of different materials to produce flameover in a room corner test has also been related to basic fire properties of the material as measured in a cone calorimeter and other such tests (Karlsson and Magnussen, 1991, Quintiere, 1993). A great deal more remains to be done.

Flashback or backdraft is a phenomenon associated with the buildup of flammable vapors in an air-starved fire in a room or cavity. Poor ventilation can cause a fuel-rich atmosphere to develop. This occurs in two ways. First, the vapors produced in a local flaming zone in which the air has limited access are incompletely burned and tend to move to other portions of the space, particularly the upper part in which combustion products are cooled and oxygen concentration is not sufficient to support flaming combustion; the higher the value of H/L, the greater will be the tendency for this to occur. Secondly, smoldering combustion will continue at concentrations of oxygen too low to support flaming combustion. Heat from this process can feed unburned vapors from within the fuel into the atmosphere. Air can then enter, for example, by a fanlight window breaking, a door opening, or a partition burning through or falling away. Flames possibly accompanied by a pressure pulse are then likely not only to involve the whole space but also extend greatly beyond the space as well. This is a source of injury to firemen on opening a door of a room in which an underventilated fire is burning (Bukowski, 1995). The ejection of flames from a cavity such as in the Summerland Disaster is also an example of this type of phenomenon (Section 3.2.1). It should be noted that one volume of flammable gas such as propane could react with 20 times its volume of air to produce more than 100 times its volume of flame. Thus, the sudden eruption of flame from a limited volume of unburned gases in a room through an opening into the rest of the building can, for a short period, fill much of the building with flame as well as leaving the room burning fiercely with continuing flames coming through the opening. Extensive flaming into the rest of the building following flashover in a room can occur if the opening to the rest of the building is high up in the room and is comparable to other openings. This phenomenon occurred at the Puerto Rico fire (Section 3.2.6).

5.6 Sudden massive flaming during fires in process industries

The process industry handles very large tonnage of flammable and liquefied flammable gases. These can be responsible for massive fire disasters if they become suddenly released. There are three major ways in which experience has shown that this can occur – the boiling liquid expanding vapor explosion (BLEVE), the open flammable cloud explosion, and the boilover.

The BLEVE tends to involve liquefied flammable gas stored in spherical or cylindrical tanks under pressure. When a fire heats the contents of a tank, the vessel may burst open violently and the contents will undergo explosive physical evaporation. On ignition, the vapor can produce a large, intense but short-lived fireball that, in the case of a large spherical tank, can produce dangerous radiation over an area of several hundred meters diameter. The violent rupture of the tank is one of a number of phenomena that may occur on heating and has been associated by Venart *et al.* (1992) to shock failure of the vessel following coherent bubble collapse of vapor nuclei. The bubble collapse is due to an increase in pressure that would follow the choking of the flow of vapor through a small hole that might develop in the shell due to heating. Venart has suggested that the phenomenon be renamed *boiling liquid compressed bubble explosion* (BLCBE). gives a model for predicting radiant heat and blast hazards from LPG bleves.

An open flammable cloud explosion can follow the massive leak of the order of tens of tons of flammable gas, vapor, or mist into the atmosphere. Flammable fuel/air mixtures ignited in the open, while capable of giving extensive flash fires, do not normally give rise to dangerous pressures unless ignited by a powerful detonating source. However, when a massive leak of fuel is dispersed into a space in which there are many obstacles and semienclosures, ignition even by a small source can give rise to significant pressure over a large area. Although excess pressures may not exceed 1 bar, the pressure rise is sufficient to destroy buildings and disrupt storage tanks and process plants. Such incidents therefore tend to be followed almost immediately by fires over large areas several hundred meters square. These explosions were also called *unconfined vapor cloud explosions* (Gugan, 1979). The word "unconfined" has tended to be dropped in recent years following general agreement that it is the partial confinement that gives rise to the pressure effects. The word "vapor" may also be misleading since, as indicated above, flammable gases and mists have been known to cause disasters of this kind. There is also the possibility of confusion with physical vapor explosions (see below).

A boilover may occur in a fire in a tank containing crude oil and certain fuel oils. The fire produces a hot zone at a temperature in excess of 200 °C. The fire burns steadily and the hot zone proceeds down the tank. Near the bottom of the tank, there is usually some water, either as a separate liquid or dispersed among the tank contents. Water may also be present at an intermediate portion in the tank if it lodges on a floating roof that has sunk because of fire-fighting activities, (Steinbrecher, 1987). When the hot zone reaches the water, there can be a comparatively sudden release of water vapor that pushes the hot contents of the tank out of the top, produces tall flames, and spreads the burning liquid over a large area. This can be fatal to firemen and others in the vicinity of the tank. A boilover following an explosion and fire in a fixed storage tank at Tacon, Venezuela in 1982 killed 150 people and ignited the contents of a second tank.

It is possible that the interaction between the hot zone and the water can produce an explosion known as a rapid phase transition (RPT) or a rapid vapor explosion (Fletcher, 1991) because of the sudden production and expansion of vapor. This would help to explain why the boilover phenomenon occurs so suddenly. Such explosions can also occur when water mixes with molten metal, (meltwater detonations) and when water mixes with certain liquefied gases (Hogan, 1982). In the latter case, it is the liquefied gas that produces the vapor. After a collision at sea, there may be an interaction between water and flammable liquid gas, for example, liquefied natural gas followed by an extensive flash fire, or even an open flammable cloud explosion if the vapors encroach upon a built-up area.

5.7 Production and movement of smoke and toxic gases

Smoke and toxic gases from fires, particularly carbon monoxide and hydrogen cyanide, are major agents leading to fire casualties (Kingman *et al.*, 1953). Recent years have seen much input into quantifying these products. Smoke output is usually measured by the opacity to light that may be produced by a given volume of gas, through combustion of a given amount of material. A number of ways of expressing opacity exist in the fire safety literature.

The bel and the decibel are units for comparing levels of intensity on a logarithmic basis and are therefore appropriate for the Beer Lambert Law when used to quantify the transmission of light through smoke. The opacity in bels per meter (b/m) and decibels per meter (db/m) is given by equations [5.18a] and [5.18b]:

$$\text{Opacity (b/m)} = \frac{1}{d}\log_{10}\frac{I_0}{I} \qquad [5.18a]$$

$$\text{Opacity (db/m)} = \frac{10}{d}\log_{10}\frac{I_0}{I} \qquad [5.18b]$$

I_0 = intensity of light at beginning of light path.
I = intensity of light at end of light path
d = length of light path (meters)

Natural logarithms are also widely used for expressing light attenuation in smoke. In order to reduce confusion, it was suggested (Rasbash, 1995a) that the unit ben (b_n) be used when natural logarithms are used as indicated in equation [5.18c]:

$$\text{Opacity (b}_n\text{/m)} = \frac{1}{d}\log_e\frac{I_0}{I} \qquad [5.18c]$$

The term "optical density" has been widely used in smoke measurement, but in different texts it represents either $\log_{10}(I_0/I)$ or the opacity as indicated by any of the equations [5.18a,b,c].

Opacity as expressed by equation [5.18b] is used in the European Standard for smoke detectors (Section 5.11). It was suggested (Rasbash and Philips, 1978, Rasbash and Pratt, 1979) that the unit db/m be called an *obscura* (ob). This would lead to a unit of smoke output of obm^3 and of smoke potential or specific smoke output (Rasbash, 1995b) of obm^3 per unit mass of fuel volatiles. When expressed in this way, the specific smoke output of freely burning organic materials tended to fall between values of about 0.2 obm^3/g for wood to 7 obm^3/g for polystyrene. The term extinction coefficient is also in common use to express opacity as in equation [5.18c] but has also been traditionally used in a similar manner in equation [5.18a]. Rasbash has used the word "smokiness" for opacity to light in the context of smoke production at fires. Thus, a unit of smokiness of 1 ob would be equal to 0.23 b_n/m and 0.1 b/m. Because of widespread usage, it would be helpful if the unit of smokiness of b_n/m were given a name.

Using information provided by Seader and Ou (1977), it is possible with reasonable accuracy to relate smokiness directly to particulate mass concentration. For smoke from flaming fires when soot is the main constituent, a smokiness of 1 db/m, 1 b_n/m, and 1 b/m correspond respectively to 30, 130, and 300 mg of smoke particles per meter square of smoke. For nonflaming fires when liquid droplets are the main constituent, the relevant values are higher, namely, 53, 227, and 530 mg/m^3, respectively. The latter results are, however, in poor agreement with Tewarson's data (see below), which indicate even higher figures for particulate concentration. A value of 340 mg/m^3 of particulate matter of a smokiness of 1 b/m was obtained for a range of conditions for smoke from domestic solid fuel heaters (Shaw *et al.*, 1952).

An increasing amount of information is becoming available for the output of smoke from well-ventilated fire tests such as the cone calorimeter and furniture calorimeter (Babrauskas and Grayson, 1992, Mulholland, 1988, Tewarson, 1995). While these give useful comparative information on the smoke-producing propensity of different materials and specific items, insufficient information is available as yet for scaling up to large fires or to postflashover conditions. There are indications that under the latter conditions, specific smoke output for wood may be considerably higher. However, for small freely burning fires, smoke output is not greatly dependent on air–fuel ratio, although there is a tendency for smoke to change from a basis of soot formation to a basis of decomposition products as the air–fuel ratio moves into insufficiency (Tewarson, 1995). This may lead to an increase in the obscuring capacity of the smoke for wood but a decrease for certain plastics. Output of carbon monoxide and other toxic gases is very dependent on the air–fuel ratio and is much higher for ratios to stoichiometric less than unity than for ratios greater than unity (Tewarson, 1995). Another approach to predicting smoke outputs from fires is through collation with smoke point laminar flame heights (Delichatsios, 1993).

Increasing information is also becoming available on the hazard levels associated with smoke and toxic gases. The main effect of smoke, particularly in the early stages of a fire and at comparatively low-smoke concentrations, is to reduce visibility and thence cause confusion and hinder escape. There is now substantial information on the effect of smoke opacity on visibility (Rasbash, 1967, Jin, 1971). In general, the visibility for given illumination conditions is approximately inversely proportional to the opacity. As a rule of thumb approach, a smokiness of 1 ob (or 1 db/m) corresponds to a visibility of about 10 m for non-self-illuminated objects under conditions of diffuse lighting. For a self-illuminated sign, the visibility can be three times as great. There is evidence that a general visibility of 10 m marks a borderline between what is acceptable for ease of escape. Thus, an analysis of the responses of people who had been involved in real fires (Wood, 1972, Rasbash, 1975b) indicated an increasing tendency for people to turn back from smoke as the visibility was reduced below 10 m (Figure 5.7). This has some bearing on the sensitivity of smoke detectors that are in general required to operate at a smokiness of less than 1 ob (Section 5.11). At smoke concentrations greater than 1 ob, lachrymation and other effects of the smoke may cause incapacity, particularly for nonsooty white smokes produced by smoldering. Thus, Jin found visibility to drop off at an extinction coefficient greater than 0.4 b_n/m (1.7 ob). In setting up criteria for smoke visibility in a model, a reasonable approach would be visibility to a safe point. Thus, within a room, visibility to the door would be a possible approach.

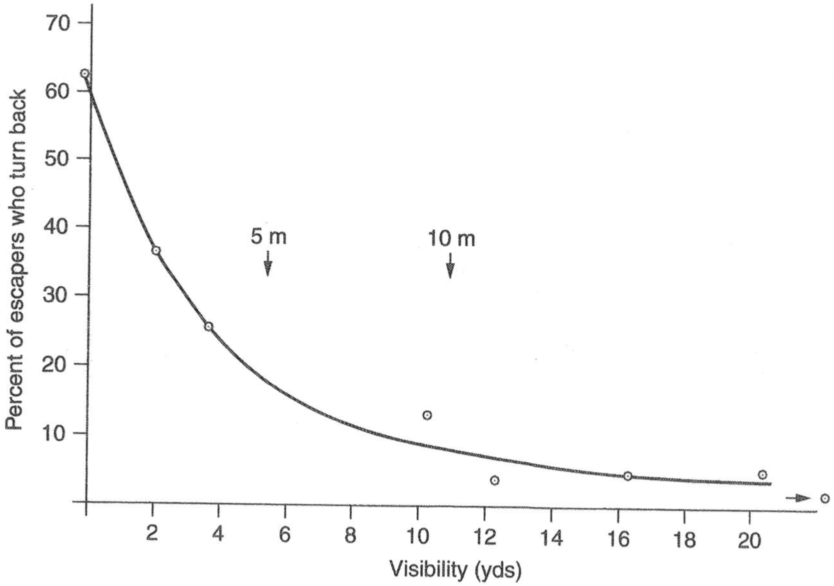

Figure 5.7. Effect of visibility on percentage of people who try to move through smoke and turn back

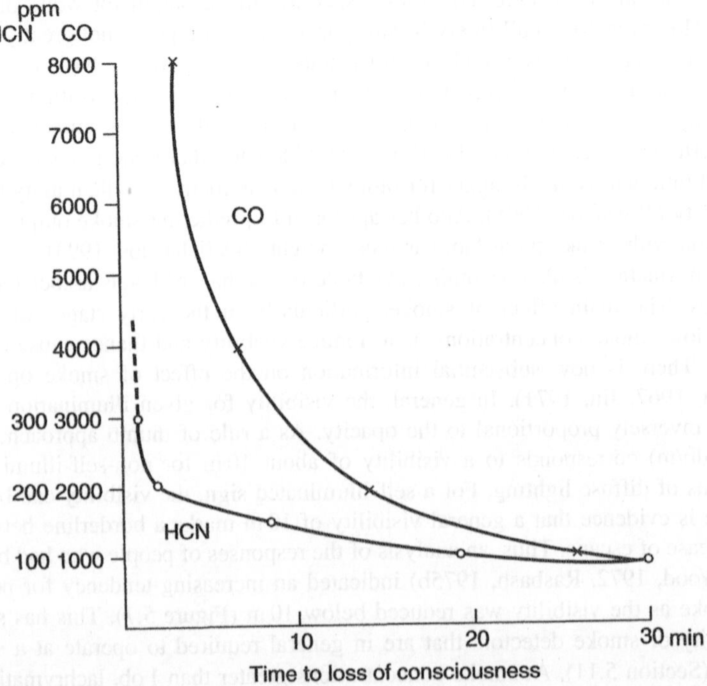

Figure 5.8. Comparison of the relationship between time to incapacitation and concentration for HCN and CO exposures in primates

For a door leading to a corridor with a smoke-stop door at each end, visibility to the end of the corridor would be reasonable. In this situation, an illuminated exit sign could be employed to enhance the visibility. A criterion S used with the EXITT models (Levin, 1989, Fahy, 1991) stated to follow Jin's work, is defined as follows:

$$S = 2 \cdot \sigma \cdot \frac{H_r}{D_s} \qquad [5.19]$$

σ = extinction coefficient (b_n/m), H_r = height of room, D_s = depth of smoke layer, a recommended value of S being 0.4. For a room completely filled with smoke, this would correspond to an extinction coefficient (smokiness) of 0.2 b_n/m or 0.9 ob.

At higher concentrations of smoke, the effect of lethal gases begins to dominate. The main concern is the propensity of these gases to produce incapacity, particularly by loss of consciousness (Purser, 1995). The time for this to occur decreases as the concentration of the gas increases but the precise relationship varies with different toxic gases as indicated in Figure 5.8. Carbon monoxide shows a smooth relationship between concentration and time for incapacity, whereas hydrogen cyanide tends to manifest a critical concentration of 200 ppm in which incapacity will occur within a few minutes and much less effect at lower concentrations.

5.8 Postflashover fires in buildings

Postflashover fires can be divided into three major regimes according to the fuel and ventilation conditions in the compartment concerned (Thomas *et al.*, 1967). When the ventilation openings are small compared with the floor area, air flowing into the compartment is driven by buoyancy head across the vent, and provided the mean temperature in the enclosure is in excess of 300 °C, is given by

$$M = 0.5A_0\sqrt{H_o} \quad [5.20]$$

M = flow of air (kg/s)
A_o = area of opening(m^2)
H_o = height of opening (m).

The rate at which fuel vapors are produced is generally proportional to the area of fuel surface, particularly that portion of the area that can see a substantial thickness of radiant flames within the enclosure. For certain fuels, particularly cellulosic fuels, combustion will also be taking place at the char surface following the removal of volatiles. If the air provided under the above conditions is more than sufficient to burn the fuel vapors produced as well as combustion at the char surface, then a fuel-controlled regime operates. The combustion that takes place within the enclosure is then controlled by the fuel, the rate increasing as the area of the fuel surface increases. If the airflow is insufficient for combustion at the fuel surface and to burn the volatiles produced, then a ventilation-controlled regime operates. Under these conditions, the amount of combustion taking place within the enclosure is independent of fire load. Volatiles that do not burn within the enclosure will burn when they pass through the opening and reach the outside atmosphere. The rate of production of fuel vapors will depend primarily on radiation to exposed surfaces from the flames, although in the combustion of thick cellulosic materials, conductive heat transfer from the glowing surface will drive off fuel vapors from within the depths of bulk fuel.

A third regime operates when there is a large vent present comparable to the floor area of the enclosure and combustion within the enclosure is supporting a flame that reaches the full height of the enclosure. Under these conditions, the air entering the enclosure is controlled by entrainment into the upward-moving flame and is given by

$$M = 0.13A_o\sqrt{H_o} \quad [5.21]$$

If the airflow and the heat being produced within the enclosure are known and it is assumed that the gases are well mixed, it is possible to set up a heat transfer balance. The heat produced is equated to the sum of the heat transferred to the inner surfaces, the heat radiated through ventilation openings, and the heat convected through openings by the flowing hot combustion products, including unburned air or volatiles. A major result of this calculation is the temperature history within the enclosure during the burnout period of the fuel (Walton and Thomas, 1995, 1988). This temperature may then be used to estimate the way in which exposed items within the enclosure, particularly elements of structure, will be heated. Early experiments on the burning of wood in enclosures, particularly in the form of wood cribs, showed that under ventilation-controlled conditions the rate of loss of weight of the fuel (R_f) was not only independent of the fuel load it was also approximately stoichiometrically related to the input of air:

$$R_f = 0.09A_o\sqrt{H_o} \quad [5.22]$$

This was probably due to the fact that combustion in the latter stage is dominated by the combustion of the char and that only a small fraction of the wood surface was exposed to radiant heat from the flames. However, the observations led to the general simplification, in the calculations for application when cellulosic fuels dominate, that no excess volatiles were produced and all the oxygen entering through the vent was burned. This would have the effect of increasing the temperature in the enclosure, as well as increasing the time over which combustion takes place. Fuels with high values of H/L, including many polymers and liquid fuels, tend not only to reach ventilation-controlled conditions more readily but will also produce more volatiles than can burn within the enclosure. The size of flame outside the vent in which excess vapors burn will depend on the amount of fuel vapors and the dimensions of the opening (Drysdale, 1985e).

5.9 Interaction between fire and structures

The heat from a fire can affect structures in a way that will prevent them from fulfilling their normal function. There are two major ways in which failure of a structural element can lead to major damage or increase of fire size. The first is to reduce the load-bearing capabilities of the element to the extent that it produces collapse of the structure. The second is to allow heat or flame to penetrate the structural element so that it leads to fire spreading on the remote side. The latter can apply particularly to the walls and floors of buildings that separate fire-resistant compartments and to items such as doors, ducts, and services that may pass through these items.

As soon as heat from a fire falls on any element of structure, the heating process that may lead to a failure will begin. However, for fires in buildings, particularly those with moderately sized compartments, it has become conventional to divide the fire development into two phases, a preflashover phase in which the effect of heat on elements of structure is generally ignored and the postflashover stage in which it is assumed that all the fuel in the compartment may burnout completely and during which time heat is transferred from the flames in the room to the interior surfaces of the room. For structures that support process plant, one is not normally concerned with flashover as in compartments and the direct effect of flames on elements supporting structural load dominates. On the basis of knowledge of the fire properties, the heat transfer from such fires may be estimated in the manner outlined earlier (Section 5.3). However, heat transfer rates for some fires, particularly those within flaming fuel jets may be very high, and there is a dearth of information as to their value. It is interesting to note that following the Piper Alpha disaster (Section 3.4.3) this is one of the areas in which more information is being sought for the purpose of quantifying fire safety (Renwick and Tolloczka, 1992). On the other hand, there is quite extensive information on radiation to distant objects from flare stacks and large pool fires (Mudan and Croce, 1995, 1988).

In testing an element of structure for fire resistance, it is normal to expose it in a furnace in which the gas temperature is increased according to a preset time–temperature curve. For tests for components used in buildings, there is an international standard time–temperature curve expressed by the relationship:

$$T = T_o + 345 \log_{10}(0.133t + 1) \qquad [5.23]$$

T_o and T are temperatures at time $t = 0$, $t = t$ (s) respectively.

For structures that may be exposed in the open to a petroleum fire, a relationship giving higher temperatures is used. Both relationships are plotted in Figure 5.9. Although the temperature of the gases in the furnace may be controlled, major differences in the apparent performance of structural items in different furnaces may arise because of differences in the heat transfer to the items of structure. These occur for the following reasons:

1. According to the method of measuring the temperature, an intermediate value between the gas and the wall temperature will be obtained.
2. The radiation from the gas flames may vary depending on the fuel used and the thickness of the flame (i.e. the geometry of the furnace).
3. The radiation from the surface walls may vary according to their insulating properties and emissivity.

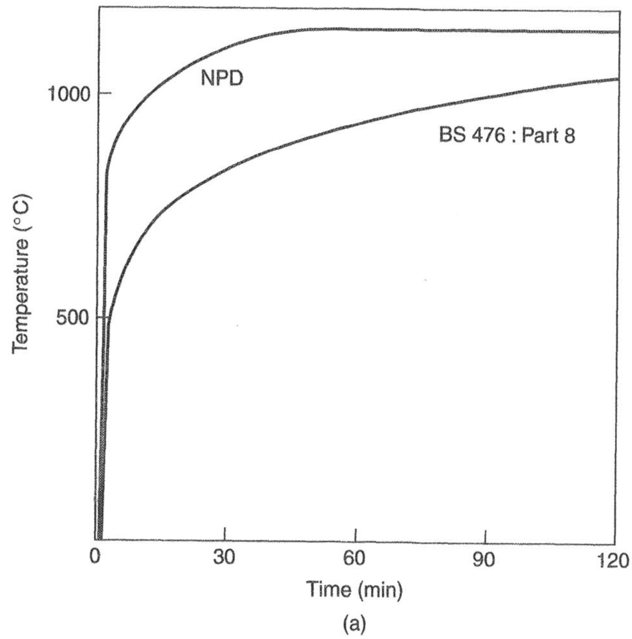

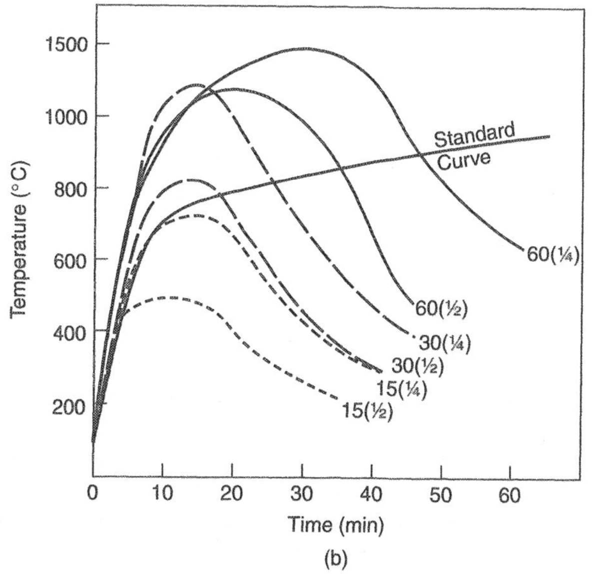

Figure 5.9. (a) Standard five curves and (b) compartment fire curves

A summary of the various methods of estimating structural fire safety in buildings has been given in the Workshop of C.I.B. No.14 (1983, 1986). According to the sophistication of the heat exposure model and the structural model used (Figure 5.10), nine different approaches are possible. The three modes of heat exposure H_1, H_2, H_3 correspond to exposure to (1) a furnace time–temperature curve, (2) a time in the furnace equivalent to the heat exposure under the conditions that would occur in practice, and (3) conditions that would actually occur in practice.

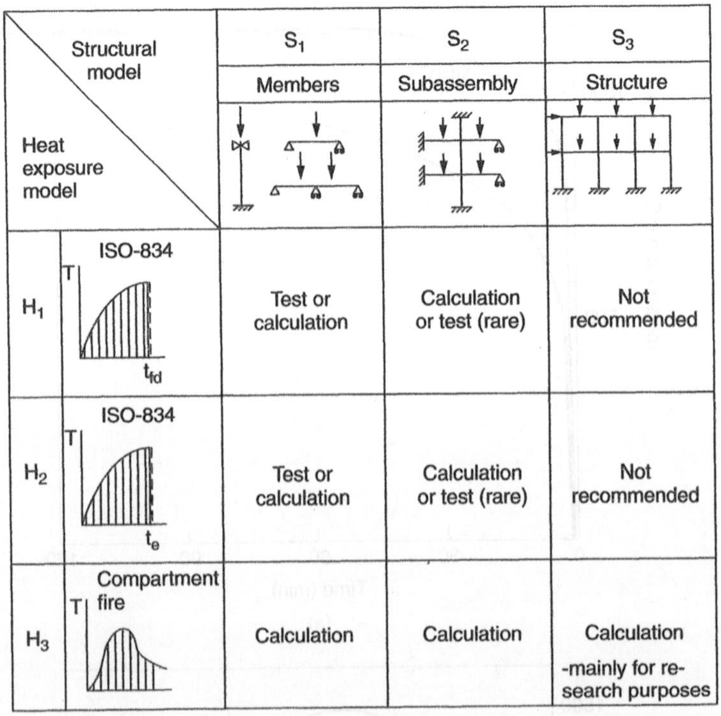

Figure 5.10. Heat exposure models and structural models

The three structural models (S_1, S_2, and S_3) correspond to simply supported single element of structure, elements combined into a subassembly and complete structures.

The simplest and most widely used approach (S_1H_1) is to obtain by direct test, a measure of the way a single element of structure will respond when exposed in a furnace in which temperature is varied according to the temperature time curve. The ability of the element to reach a required fire resistance time is determined. These tests are expensive and only single tests are usually carried out. One has to be aware that even for a given furnace there may be differences between test samples that give rise to different results. For this reason, a calculation of the response of an element of structure when exposed in the furnace may be more appropriate. However, such a calculation requires a sufficient depth of knowledge of the thermal properties of the element of structure. In many load-bearing structures, steel plays an integral part in the load-bearing capability of the structure and the calculation requires estimating the time when the load-bearing steel element reaches a certain critical temperature. This will depend on the thickness of the protective layer of insulating material that covers the steel element. The assumption that the outer layer of insulating material, that is, the surface exposed in the furnace, follows the temperature–time curve of the furnace gases would lead to an estimate of the fire resistance on the safe side.

The temperature–time curve in real fires is different from that specified in the furnace. The concept of equivalent fire exposure H_2 has been developed to assist in calculations of fire resistance requirements in postflashover fires in rooms leading to burnout. The exposure is influenced not only by the nature of the fuel and the fire load in the room but also by the potential ventilation and the heat loss to different surfaces, the latter being controlled by the area and thermal properties of the surfaces. A number of approaches to specifying equivalent exposure time in a furnace have been put forward dealing with one or more of these factors, notably Ingberg (1928), Law

(1971), Pettersson (1985), DIN 18230 (1986), Harmathy (1980). Ingberg's approach deals with fire loading only. The other approaches bring in most of the other factors to a varying extent. However, Harmathy's approach takes specific cognizance of the fact that the furnace with which comparison is being made may vary. The DIN method is the only one that takes account of the nature of the fuel, although the factor introduced will depend on the ventilation in an unspecified way.

Given a risk situation in which postflashover conditions may vary, for example, the nature and output of the fire load or the ventilation conditions (e.g. window alone – window plus door), a probabilistic approach to whether an element of structure will fail on burnout may be made by using one of the above approaches to equivalent fire exposure. Harmathy (1987) has made a comparison between the different methods of estimating equivalent fire exposure. Butcher (1991a) has indicated that when applied to premises with very high fire loads, estimates based on equivalent fire exposure can give values that are excessively high. In general, these equivalent exposure models rely on the assumption that it is the total amount of heat that enters the structural element, which governs its behavior rather than the rate at which it is being absorbed.

If the temperature–time curve of a fire can be prescribed, for example, by the heat balance calculations referred to in Section 5.8, it is possible to follow approach H3 and estimate the response of certain elements of structure directly. To do this it is necessary to specify a heat transfer associated with the specified temperature. A reasonably safe assumption is that the flame is radiating to the surface with unit emissivity and transferring heat to the surface also by natural convection. However, as indicated above, this may not be the case if a highly turbulent jet flame is present when forced or impingement convection may be more appropriate. Where significant insulation covers sensitive elements, then a safe assumption would be that the exposed surface temperature is equal to the gas temperature.

ECCS (1983, 1984, 1985, 1988) and the Institution of Structural Engineers (1978) have given detailed information on performance of steel and concrete structures in fires. Information is also available for timber members (White, 1988), concrete members (Fleischmann, 1988), and brick walls (Fisher, 1975). The calculations above have referred mainly to the failure of a load-carrying element. Elements of structures, particularly walls and ceilings, can also fail through conductive heat transfer through the element causing a dangerous temperature rise on the unexposed surface and by loss of integrity of the element, leading to flames and hot gases being ejected from the unexposed side. Knowing the thermal conductivity of the elements, the first of these can be calculated (see Drysdale, 1985f). However, there is as yet no reliable way of calculating loss of integrity. The tendency for this to occur can be assessed from a fire resistance test result.

5.10 Defenses against smoke

In view of the uncertainties noted in Section 5.7, in scaling up smoke and toxic gas output, particularly with regard to flashover and the influence of air–fuel ratio, models for handling smoke are frequently not developed to the point of estimating smoke or toxic gas effects in real hazard situations. Such models assume a defense mechanism is in place that would prevent the smoke as a whole gaining access to spaces in which people are at risk.

The most important of such defenses against smoke movement is the closed door, particularly a door that leads from a room that could be on fire to other spaces such as corridors or stairs, which may be used by escaping people. Smoke-stop doors are also used for subdivision of corridors, thus limiting the size of the portion that can be at risk, and for entry into enclosures intended to be smokefree, particularly staircase enclosures. The difficulty about fire and smoke doors is that they are only effective if they are closed. If they are required to be closed for normal use, then they can be made self-closing. However, if their function is superfluous apart from fire safety, there is a tendency for them to be permanently propped open and thus they would not provide protection should a fire occur. This can normally be overcome by installing door holders that release the doors automatically in response to detection of fire or local smoke.

The next defense against smoke is to exploit layering and there are a number of zone models of fires in a building that specifically do this. These will be dealt with in Chapters 11 and 12. Given a fire of known heat output, the buildup of smoke in the upper part of a room can be estimated or modeled, as can the way the smoke flows through a door or doors into neighboring spaces. In general, spaces are considered to be safe as long as the buildup of the major smoke concentrations remains above the heads of the people in the rooms and as long as the temperature of the smoke layer does not exceed values that will bring about dangerous radiation to people below. However, because of cooling of the smoke, there is a limit to which this approach can be applied. Also imperfections in the layering, for example, due to currents at the walls and entrainment of smoke into air entering through a door or vent into the smoke layer can bring about contamination of the desired clear layer at lower levels with smoke from the upper layer. To ensure that sufficient visibility is maintained, the input of smoke into the lower layers may need quantification and this will depend on the nature of the fuel burning. Models for quantification of smoke volume rely heavily on estimates of entrainment into moving streams of smoke and air. While there is reasonable agreement on air that is entrained into a rising plume above a fire (Zukoski, 1994), there are differences in approach for smoke that moves from within a compartment to another space such as an atrium (Morgan and Marshall, 1975, Law, 1986, McCaffrey plume model, see Chapter 11, Section 11.4.9). Full-scale tests and the greater use of field models based on computational fluid dynamics (Chapters 11 and 12) would help settle these differences.

The third major defense against smoke is to install a smoke control system. Systems are usually based on either smoke extraction or pressurization. The former usually relies on the existence of a smoke reservoir in the upper part of a room or building from which smoke can be extracted by natural buoyancy through openings in the roof or chimney (Thomas and Hinkley, 1964, Hinkley, 1995, Cole, 1989) or by mechanical extraction. Openings, usually in the lower part of the space, are required to allow air to flow in to replace the smoke that is being extracted. Pressurization relies on building up a pressure in spaces that need to be kept smoke free, particularly escape stairs and corridors, to counteract the tendency of smoke to flow into them (Butcher, 1991). It is necessary for air that enters the fire compartment to be able to escape to prevent buildup of pressure that would defeat the process of pressurization. This can be achieved by having openings to the outside atmosphere that come into play during the process or by combining pressurization with extraction from the fire compartment. An extensive literature now exists on smoke control systems (Klote, 1995, Klote and Milke, 1992).

5.11 Fire detection

Various methods are available for the detection of fires. The methods most widely used for buildings, particularly where there are life safety risks, are heat detectors and smoke detectors. Infrared and ultraviolet detectors are used more for industrial processes, particularly where there is an outdoor risk. There is a growing number of standards, both national and international, covering this equipment.

Heat detectors are designed to respond to the increased temperature that occurs in the environment of a fire, particularly the temperature in the stream of gases in the fire plume. The heat input into the detector is predominantly by convective transfer from this plume and the response may be due to a specific property within the sensitive element of the detector. Commonly used properties are expansion (e.g. sprinkler bulb), melting of a link, or a change in electrical properties. Fixed temperature detectors respond when the temperature of a sensitive element reaches a certain value, for example, a melting point. Rate-of-rise heat detectors respond when the temperature of the airstream flowing past increases beyond a certain minimum rate. The sensitivity of a heat detector depends on the time constant, which is the ratio between heat capacity (C) of the sensitive element and the product of the heat transfer coefficient (H) and the area of the element (A). Heat detectors tend to have a time constant of 20 to 40 s, but a standard sprinkler has a time constant of 2 min. The European standard (EN54, Part 7, 1984) for heat detectors specifies a sensitivity test in which the detector is exposed in a wind tunnel to an airstream of 0.8 m/s, the temperature of the air increasing at a number of constant rates varying between 5 and 30 °C/min. According to the sensitivity grade of the detector, there is a maximum time in which the detector may alarm depending on the rate of temperature rise. There is also a minimum time specified to avoid the occurrence of false alarms.

Smoke detectors rely on detection of particulate matter in the smoke. Light obscuration and light scatter detectors rely specifically on the capability of the particles to obscure and scatter light respectively and ionization chamber detectors rely on the capability of particles to reduce an ionization current. Obscuration and scatter detectors are not sensitive to particles of size significantly below that of the wavelength of light, that is, about 0.3 μ, whereas ionization detectors are sensitive down to 0.01 μ. However, none of these detectors will detect a nonsmoky fire such as that of alcohol or even whisky. Apart from this, ionization detectors are most responsive to a freely burning fire in the early stages although sensitivity and output tend to reduce as the smoke ages. Light scatter detectors are substantially less responsive to a highly sooty smoke than smoke obscuration detectors. However, there is usually sufficient nonsoot particulate matter from freely burning fire, that is, condensed water, to make scatter detectors sufficiently sensitive to small sooty fires. Light scatter detectors are very responsive to smoke produced by smoldering.

Smoke detector standards rely heavily on the capability of the detector to respond to smoke produced by different test fires that are monitored for opacity, ionization current, and temperature in the vicinity of detectors under test. The European standard for point smoke detectors (1984) requires that the detector should be able to detect smoke from four specified smoke-producing fires whose smoke opacity in the vicinity of the detectors varies up to 2 db/m. The most sensitive detectors need to operate before an opacity of 0.5 db/m is reached and the least sensitive need to operate before the opacity reaches 2 db/m. The British Standard on point smoke detectors containing an alarm, particularly for use in domestic premises (BS 5446, 1990), requires that detectors should respond within 10 s of smoke in the vicinity, reaching an opacity of 0.5 db/m for a slow burning or fast burning wood fire and 0.8 db/m for a liquid hydrocarbon or polyurethane foam fire. The Underwriters Laboratories (1970) standard, calls for a response at 0.6 db/m for a cellulose smoke and 1.4 db/m for a kerosene smoke. As a broad rule of thumb, therefore, smoke detectors may be considered as operating when the smoke in the vicinity reaches an obscuration of 1.0 db/m.

The response of both heat and smoke detectors may be related to the output of heat and smoke from the fire, using relationships for ceiling jet flow for a steady or growing fire (Evans, 1995, Schifiliti, 1995). Except for clean burning materials such as fiberboard, smoke detectors respond much earlier than heat detectors. Also, for freely burning fires, an obscuration of 1 db/m occurs well below the value at which the carbon monoxide or hydrogen cyanide content renders the gas toxic (Rasbash, 1975b).

5.12 Fire suppression

There are three major types of fire suppression:

(i) Early manual fire fighting, particularly use of extinguishers,

(ii) Automatic fire suppression, particularly sprinklers,

(iii) Activity of fire brigades.

It is necessary in fire modeling procedures to be mindful of the contribution that all three types may make to the control and final suppression of fire.

Manual fire fighting with an extinguisher can only be expected to be effective in the early stages of the fire while the fire size is small and generally of a dimension under 1 m. The capability of fire extinguishers is classified by test procedures using trained personnel (BS 5423, 1987), for either a wood crib fire of varying length or a liquid fire of varying area. Capabilities are governed by the extinguishing agent and by the weight, the upper limit of total weight being somewhat in excess of 20 kg. The training of the user is a major factor in the effectiveness of the use of an extinguisher and one needs to be mindful of the fact that a large proportion of people using an extinguisher will be using it for the first time. There is evidence to indicate that the time to extinction of a given fire using a water extinguisher is inversely proportional to the cube root of the number of attempts made (Rasbash, 1962, p.38). This would lead to an estimate that a trained operator would extinguish a fire about 3 to 4 times larger than an untrained operator would.

Sprinkler systems may be regarded as the major force of active defense against fire. Over a period exceeding a century in which they have been in use, detailed standards have been developed, particularly for the insurance industry, for their design and maintenance to cope with various fire risks (NFPA, 1980, Comite Europeen des Assurances, BS 5306 Pt.2, 1990). Until fairly recently, sprinkler systems were used primarily for property protection rather than for life safety. The heat detection element used with the sprinkler was comparatively insensitive, with a response time of about 2 min. As a result, fire in a room of normal height could reach a heat output of about 0.5 to 1 MW before a sprinkler would operate. This insensitivity probably developed as a response to the required robustness of sprinkler heads in the very wide ranging locations in which they are employed and as means of avoiding spurious operations by heat sources that were not fires. However, recently for specialized uses, sprinkler heads have been developed with much lower time constants so that they can come into action at fires at a much smaller size (Theobald, 1987). One use of these "fast response" sprinklers is for life safety since a fire of 0.5 to 1 MW at sprinkler response is too dangerous for life, particularly if the sprinkler also causes combustion products in the upper part of the room to mix with a safer atmosphere in the lower part. Another major use of fast response sprinklers is in high bay warehouses where it is essential that the fire be detected at a very early stage (Field, 1985). In this situation, a fire can grow with great rapidity as it travels upward through the vertical aisles. To counter this, it may be necessary not only to have sprinklers with a fast response but also with the capability of suppressing a rapidly growing upward-moving fire. Research at the Factory Mutual laboratories has met this requirement by the development of powerful downward momentum, large drop size sprinklers capable of overcoming the upward thrust of the flames (Yao and Marsh, 1984).

In spite of the fact that sprinkler systems have been in use for over a century, there is still a wide range of opinion in quantitative statements on their effectiveness and reliability. This is mainly due to a different criterion being used as to what constitutes effective action. Different regimes of maintenance have a major influence on reliability. There is also the necessity in many areas in which there is a danger of water freezing in the pipes for these pipes to be filled with air. This delays the onset of water flow to the fire, thereby having a substantial effect on sprinkler performance. According to UK Fire Statistics (Rasbash, 1975c) sprinklers are described by the fire brigade in attendance as having achieved control of the fire in only some 80% of incidents, whereas in Australia, sprinkler systems (Marryat, 1971) are described as being effective in more than 99% of incidents. Miller (1974) rated sprinkler effectiveness at 86%. In practice, it is difficult to separate the contribution of a sprinkler to suppressing a fire and that of the fire brigade, and figures of effectiveness of sprinklers usually embrace the combined effect of both. Thus, a fire on which a sprinkler is operating may still not be under control but may be substantially less advanced than if no sprinklers were operating, thus easing the task of the fire brigade. A detailed study based on area damaged by fire carried out by the UK Home Office (Rutstein and Gibert, 1978) indicates a success rate for sprinklers, that is, combined reliability plus effectiveness of 94 to 96%. Certainly, expected cost of fire damage for sprinklered premises, for those fires in which sprinklers would be expected to operate, is substantially smaller than for unsprinklered premises, this forming the basis for reduced premium rates for sprinklered premises (Chapter 10). In fire safety modeling, it is usually assumed that within the limit imposed by reliability, the convected heat output and the consequential smoke and toxic gas output will not increase following sprinkler operation.

Sprinkler and other water application systems are the major forms of automatic suppression, but other systems are in use for specialized purposes, particularly in areas in which there is a risk of gas or liquid fires and in which water may cause unacceptable damage. These systems include in particular dry powders, carbon dioxide, and halon. Available information indicates that they are less reliable than sprinkler systems (Miller, 1974). Because of environmental hazards, there is now a requirement to phase out halon systems. The main advantage of these systems, particularly those based on Halon 1301 when used for total flooding, is that they do not give rise to a toxic atmosphere and this particular advantage is difficult to replace. Nevertheless, introduction of a small amount of carbon dioxide into an inerting gas based on argon and nitrogen allows a reduction in tolerable oxygen concentration because of increase in breathing rate (Coxon, undated). A number of perfluorohydrocarbon compounds have also been found comparatively effective (Moore, 1996). One possibility is to use water sprays that are sufficiently fine to extinguish flaming combustion. However, these sprays would need to be much finer than those used at present in sprinkler systems. Buoyancy-controlled diffusion flames could be extinguished by sprays of mass median drop size between 0.3 to 0.6 mm (Rasbash, 1986a). However substantially smaller drop sizes would be needed to extinguish forced jet flames and premixed flames (Jones and Thomas, 1993), although under favorable conditions disintegration of drops could be brought about by the force of the jet or the explosion blast. There has been in recent years much activity in producing water mist systems much finer than sprinklers for use in suppressing diffusion flame combustion (Smith, 1995).

A major entity in the extinction of fire is the fire service. Given an acceptable standard of service and equipment, the most important factor governing the effectiveness of the fire service is the time of call following ignition and the time for arrival. The latter depends on the number and distribution of fire stations and a balance needs to be achieved between the extra effectiveness caused by shortening the response time and the cost in providing extra fire stations. This is discussed further in Chapter 10. A major contribution of the fire service to fire safety is in rescuing people from a fire. However, because of the delay caused by the response time the direct contribution that the fire service makes in this respect is rarely taken into account in quantitative fire safety design and models of fire safety usually rely on people making safe egress from a building independent of fire service action.

5.13 Interaction between fire and people

In addition to direct harm that fire can afflict on people, there are two further areas of interaction between fire and people of direct importance to fire safety design. The first is connected with the way people cause fires and the second is the way people react to fires and escape safely from them. People are main agents for bringing together the contributory elements of fire, particularly ignition source plus combustible material, plus situation for fire spread. Although there is statistical information on this aspect of fire occurrence, there has been little systematic analysis of this information. Moreover, this is an area in which it is very difficult to carry out direct observational experiments. However, there have been investigations where the effect of education and training of people have been observed by measuring their effect on fire occurrence. These are dealt with in Chapter 9.

Detailed codes, for example, NFPA life safety codes for the provision of escape, have been available in fire safety literature for a considerable time. However, it is not until fairly recently that any systematic investigations of the way people behave in fire have been carried out (Wood (1972), Bryan (1989)). Moreover, it is also in recent decades that the movement of people making their exit from buildings has been studied in detail (Predtechenskii and Milinskii, 1978, Fruin, 1971, Pauls, 1980, Kendik, 1986). It is now possible to integrate this information to predict quantitatively the time it will take for people to make their escape from a building (Nelson and McLennan, 1988). This has become an increasing feature of the quantitative modeling of life safety in fire, where it can be compared with complementary information on the time for a threat to develop on the escape route (Chapter 12). In estimating escape time, allowance needs to made for inefficiencies that may be engendered by variations in local circumstances and people's behavior.

5.14 Explosions

The prevention of explosions and protection from their potential effects form a major aspect of fire safety. The vast majority of harmful accidental explosions that take place in buildings or in plant occur as a result of the ignition of fuel–air mixtures. The fuel can be in the form of flammable gases, vapors, dusts, or mists. The combustion process takes place in seconds or in fractions of a second. Unlike fires, there is no lengthy period of development that allows time for escape, following inception of the explosion process. The major damage is caused by pressure and blast effects and to a lesser extent by burns caused by flames and hot gases passing over people. Unstable substances and physical effects resulting particularly from intermixing of incompatible liquids may also be the source of dangerous explosions.

The major defenses against explosions lie in preventing flammable fuel–air mixtures from forming sizable pockets, in explosion relief, and automatic extinction (Rasbash, 1986b, Zalosh, 1995). Explosion prevention relies heavily on the engineering of gas and vapor-handling systems to prevent leaks (King *et al.*, 1977). In situations in which leaks may occur, major defenses are the provision of ventilation to keep the volume of the flammable fuel/air pocket down to a nondangerous size (Harris, 1983), and the use of electrical and other apparatus designed not to ignite the pocket (BS 5501, 1977). Automatic flammable gas detection may also be an important defense, particularly for industrial plant handling flammable gases, as is also the provision of inert spaces within such plant. Explosion relief relies on the existence of weak panels on the side of a space through which expanding combustion gases produced by the explosion can be safely expelled. For fuel–air explosions in volumes of approximately cubical shape, there generally needs to be a substantial fraction of one side of the volume capable of acting as a vent. As far as explosions in buildings are concerned, such relief is often provided by windows and these play a major part in the prevention of the destruction of the building concerned. In a single-story building, a light roof can provide a similar effect. However, if an explosion takes place in a compartment like a basement in which there is little or no natural explosion relief, the explosion can result in the destruction of the whole building. Unfortunately, basements are often places that house flammable fuel–gas apparatus and also they may be places into which flammable gas may leak from cracked fuel pipes outside the building. If the timescale of the explosion is on the order of 1 s, then this is usually sufficient to enable the explosion to be detected and extinguishing material to be injected automatically into the path of the flame propagation, thus suppressing the explosion. If the timescale is reduced to considerably less than 1 s, particularly by the onset of highly turbulent combustion or detonation, then automatic extinction processes generally become impracticable. Explosion relief may also become very difficult. However, movement of gas and vapor explosions from one item of plant to another may still be stopped by the insertion of flame arresters (HSE, 1980). These can be designed even to suppress a detonation (Barton *et al.*, 1974).

Pressure pulses and even violent explosions can take place during fires for a number of reasons. In addition, to pressure effects that may become manifest, these might cause considerable spread of fire. As mentioned earlier (Section 5.5), a common phenomenon is the development of fuel-rich pockets in a fire enclosure due to lack of ventilation, which then become diluted with air at a later stage. In general, the pressure effects of flame propagation are not considerable as they are usually vented by the opening that supplies the diluting air. However, firemen have been known to be knocked over by the pressure pulse. It is also possible that such pressure effects could be sufficient to dislodge fire-resistant partitions, thus reducing their effectiveness. Perhaps situations more likely to give rise to dangerous pressure effects occur when heat flowing through a compartment wall causes destructive distillation of organic material on its other side that can explode when ignited. A similar situation arises when smoldering materials produce a flammable smoke that can accumulate in an enclosed volume and explode when ignited. This may occur when a slight change of conditions brings about a sufficient intensification of the smoldering source. There is also of course the danger of cylinders of compressed flammable gas or even compressed air or nitrogen becoming heated by the fire and exploding, as well as aerosol containers with flammable fuel that may be present. A building used to house or store flammable liquids, for example, a spirit bond, can present a considerable hazard of sudden intensification of a fire caused by explosion and may need to be specially designed to protect firemen (Home Office, 1973).

5.15 Fire scenarios in fire safety design

A major input into fire safety modeling is the introduction of relevant fire scenarios into the modeling procedure. As far as buildings are concerned, particularly those containing solid everyday organic fuels, these scenarios usually assume that a fire is already established and is producing a heat output that may be varying with time. The output and movement of heat, smoke, and toxic gases within the building is then estimated and used to test the effectiveness of fire safety systems that may be in place or to design a new or amended system that will give sufficient safety. The main concern is usually the safety of the people within the building and their ability to move or be moved to a safe place. The safety of the firemen and also of the property itself may also be addressed. As far as process industries are concerned, major fire damage usually follows the leakage of gaseous liquid fuel from enclosed systems. The total fire scenario will encompass the way in which such leakages can occur, when, where, and how the leakage can be ignited and the consequences of ignition, particularly if it may lead to a devastating fire or explosion. Apart from people at risk within the plant, people may be at risk beyond the boundaries of the plant and their safety is a major consideration as well.

The simplest scenario for a building fire is to stipulate a constant heat output. Such a procedure has been in use for some time as an aid to the design of smoke control in shopping centers (Gardner and Morgan, 1990). In this case, the assumption is made that the existence of a sprinkler system will limit the power of the fire to 5 MW, and the movement of smoke from a fire of this magnitude is followed and necessary smoke control procedures designed. The next simplest approach is to assume that the fire either involves a single major item whose output of heat, smoke, and toxic gases is available from published data or a test or that it follows a prescribed growth law. The growth stage of many specific fuel items is described by a square law:

$$Q_t = b(t - t_i)^2 \qquad [5.24]$$

Q_t = heat output at time t (kW)
t_i = time of incubation of fire (s)
b = growth factor

Data for a wide range of fuel items based on equation [5.24] have been given in the N.F.P.A. Code 204M (1991). The square law is the expected law of operation for fires growing horizontally across a fuel of uniform properties and also for crib fires (Heskestad and Delichatsios, 1978). For fires growing in rack storage, the convective heat release produced by the initial growth period has a third power dependence on time (Yu, 1990). An even simpler version of the square law may be used to express the growth of fire in a building according to the overall combustibility of the contents (NFPA92B, 1991):

$$Q_t = 1000(t/t_g)^2 \qquad [5.25]$$

Q_t = heat output at time t (kW)
t_g = growth time to reach 1000 kW

Fires with growth time t_g of 600, 300, 150, and 75 s are classified as slow, medium, fast, and ultrafast respectively. An estimate of a mean value of t_g may be made from individual items of data represented by equation [5.24] by putting t_i equal to zero and equating b to $1/t_g^2$.

With models based on a square law of fire growth, it would be possible to predict the time for fires to operate detectors and sprinklers, the increasing output of smoke and its flow into the hot layer, and to neighboring spaces. With a generic fire law such as in equation [5.25], it may be possible to assume that it also applies beyond the original item on fire, to cover the growth of fire to flashover, or full room involvement in the room of origin when all items in the space become involved.

An alternative approach would be to use exponential growth curves:

$$Q_t = Q_0 e^{at} \qquad [5.26]$$

Q_t = heat output at time t
Q_0 = heat output at zero time
a = growth factor

This law can be justified if a constant fraction of the heat output is used to heat the surrounding fuel to the fire point. Some data are available for certain fuel items (Friedman, 1978). It is possible also to use data estimated by Ramachandran from statistical information on final fire sizes and time of growth of fire (Chapter 7). These growth rates depend on the nature of the occupancy and the analysis provides both an initial fire size and growth rate as indicated in equation [5.26] with confidence limits. The data, being based on final fire size, will include the effects of fire spread, flashover, compartmentation, and fire control that occur in practice.

The more detailed forms of modeling involve the estimation of fire spread beyond the item first ignited and the consequential effect on heat and smoke output. As indicated in Section 5.3, this necessitates both knowledge of heat transfer, particularly from the flames, plume, and hot layer under the ceiling to combustible items that are not burning, in addition to the fire point and physical properties of these items. One approach is to assume the ignition of a major combustible item in the room of known heat and smoke output as a function of burning time (Babrauskas, 1995) and to estimate heat transfer rates to other items. To do this, it will be necessary to estimate appropriate dimensions to the flame as a function of heat output, possibly as indicated in Section 5.3.1 and to supplement the rate of burning because of additional radiative heat transfer from the hot gas layer. Whether a neighboring item can be ignited prior to the burnout of the original item can then be estimated. Another approach is to define an "established fire" and locate it in a number of places in a room and estimate the progress of the fire thereafter. Guidance on where to locate the established fire may be obtained from statistical information of the location of the ignition sources in the premises concerned or next to major combustible items in the space.

A convention that has developed is that an established fire is one in which capacity for radiant heat transfer exceeds that for convective heat transfer. Assuming that convective heat transfer from a flame to a neighboring surface not actually burning is 20 kW/m^2, the data in Table 5.4 suggest that an established wood or cellulosic fire would have a thickness of about 18 cm. Assuming a cylindrical flame of height twice the diameter would indicate a flame volume of about 9 L and hence a heat output of 16 to 18 kW (Figure 5.5), which is approximately the heat output of a waste paper basket fire. However, when placed next to a surface, the flame would lengthen (Section 5.3.1) and the combined convective plus radiative heat transfer would probably be nearer 30 rather than 40 kW/m^2 over an area of about 0.05 m^2 in the lower part of the flame in the period prior to ignition. The ignition of the item will depend on the burnout time of the established fire, which for cellulosic fires could be given by assuming a burning rate of 1 g/s. Following ignition of the item, the procedure indicated above for a major burning item could be followed. The development of fire within the space concerned as a function of time can then be estimated. The effect on this development of manual and automatic intervention at various times can also be estimated. Carrying out this procedure for a number of ignition locations and intervention scenarios can give a probability distribution for the fire development scenarios that would include the possibility of attaining flashover.

In the process industries, the fire and explosion scenarios following loss of confinement, particularly of flammable liquids and gases, depend to a large extent on the way confinement is lost and the way the fuel is ignited. Confinement may be lost through human error in operating the plant, structural failure of containing vessels and pipes, or failure of gaskets, flanges, valves, and so on. In addition, there may be missiles and local pressure rises caused by explosions in neighboring items of plant. The mechanism of containment loss will determine the size of opening associated with the loss that together with the pressure of containment will govern the rate and duration of the fuel flow. The range of fire scenarios that could follow containment loss could vary from a local fire at the leak, which could lead to escalation by heating the vessel contents, a pool fire, a running fire with the burning fuel flowing over elements of structure, a jet fire when the fuel is issuing from a high-pressure source that may also impinge on elements of structure, a large fireball or a bleve, or an open flash fire or flammable cloud explosion (Section 5.6). The latter two will tend to occur if ignition has been delayed. Whether a jet flame or a fireball will follow ignition of a leak depends on the size and duration of the leak (Makhviladze *et al.*, 1995). Scenarios need to incorporate all the elements that can be integrated into a fault tree or event tree. Detailed hazard and operability studies on the plant concerned can provide data for incorporation in the analysis (Chapter 17). Apart from hazards to people and plant caused by direct involvement in the flame, radiant heat particularly from a fireball or a large pool fire might affect people at a distance (Mudan and Croce, 1995, 1988, Shield, 1995) and pressure effects from an open flammable cloud explosion can be highly damaging to plant within the area covered by the explosion flame and to both buildings and people at substantial distances beyond the flame (Puttock, 1995, Barton, 1995).

Nomenclature

a	growth factor of fire (exponential)
A	area of element (detector)
A_0	area of aperture in compartment
A_T	Total area of compartment surfaces
b	Symbol for bel (common logarithms); growth factor of fire (square law)
b_n	Symbol for ben (natural logarithms)
B	Transfer number

c	Specific heat
C	Heat capacity of detector
C_s	Soot concentration g/m^3
d	Length of light path
db	decibel
D_b	Diameter of flame base
D_s	Depth of smoke layer
E	Black body emissive power of flame
f_v	Volume fraction of soot
g	Acceleration due to gravity
Gr	Grashof number
h	Heat transfer coefficient
h_{ig}	Heat transfer coefficient from surface to environment
h_K	Effective heat transfer coefficient
H	Heat of combustion (kj/g); heat transfer coefficient (detectors)
H_f	Convection heat transfer associated with production of 1 g fuel volatiles
H_o	Height of opening
H_r	Height of room
I	Intensity of light at end of light path
I_0	Intensity of light at beginning of light path
k	Thermal conductivity
k_a	Absorption coefficient
l	Thickness of smoke layer
L	Heat required to produce fuel vapors (kJ/g); flame thickness
L_f	Height of flame
m''	Rate of burning per unit surface area
M	Mass flow of air
Nu	Nusselt number
Pr	Prandtl number
q''	Heat flux absorbed by surface
q''_{conv}	Convection heat flux from flame
q''_{cr}	Critical heat flux for ignition
q''_p	Peak heat flux from flame
q''_{rad}	Radiation heat flux from flame
Q	heat output (power) of fire (kW)
Q_0	Heat output at zero time
Q_t	Heat output at time t
Re	Reynolds number
R_f	Fuel mass loss rate
S	Criterion for visibility in exit model
t	time
t_i	Time of incubation of fire
t_{ig}	Time to ignite a solid
t_g	Growth time of fire to 1000 kW
t_s	Time to reach temperature T_s
T	Absolute temperature
T_∞	Temperature of ambient air
T_b	Temperature of cold surface
T_h	Temperature of hot gas in flame or plume

T_0	Initial temperature of solid
T_s	Fire point temperature of solid fuel
ΔT	Temperature difference between gas and surface
ΔT_m	Maximum temperature rise in flame
V	Flame volume (m^3 or liters); velocity of gas
x	Linear dimension
α	Absorption coefficient (radiation); thermal diffusivity
β	Coefficient of expansion of gas
ε	emissivity
λ	Wavelength of radiation
ν	Kinematic viscosity
ν_∞	Kinematic viscosity of ambient air
ρ	Density of solid
ρ_∞	Density of ambient air
σ	Stefan Boltzmann constant; extinction coefficient (b_n/m eq [5.19])
τ	Thickness of solid

References

Alpert, R L and Ward, E J (1984). Evaluation of unsprinklered fire hazards. *Fire Safety Journal*, **7**, 127–143.

Atkinson, G T and Drysdale, D D (1992). Convective heat transfer from fire gases. *Fire Safety Journal*, **19**, 217–245.

Babrauskas, V (1982). Will the second item ignite. *Fire Safety Journal*, **4**, 281–292.

Babrauskas, V (1995). Burning rates. *SFPE Hand book of Fire Protection Engineering*, 2nd edition, Society of Fire Protection Engineers, Boston.

Babrauskas, V and Grayson, S J (1992). *Heat Release in Fire*, Elsevier Applied Science, London, New York.

Babrauskas, V and Walton, W O (1986). A simplified characterisation of upholstered furniture heat release rates. *Fire Safety Journal*, **11**, 181–192.

Babrauskas, V and Witterland, I (1995). The role of flame flux in opposed flow flame spread. *Fire and Materials*, **19**, 275–291.

Back, G, Beyler, C, di Nenno, P and Tatem, P (1994). *Wall incident heat flux resulting from an adjacent fire*. Proc. 4th. Intl. Symp. Fire Safety Science, pp. 241–251.

Barton, R R *et al* (1974). *The use of reticulated metal foam as flashback arrester elements*. Proc. of Symp. on Chemical Process Hazards with special Reference to Plant Design, pp. 223–233.

Barton, R F (1995). Fuel/gas explosion guidelines-practical applications. I. Chem. E, *Major Hazards Onshore and Offshore II*, pp. 285–296.

Beyler, C L (1984). Ignition and burning of a layer of incomplete combustion products. *Combustion Science and Technology*, **39**, 287–303.

Bond, J (1991). *Sources of Ignition, Flammability Characteristic of Chemicals and Products*, Butterworth, Heinemann, Oxford.

Bowes, P C (1984). *Self Heating Evaluating and Controlling the Hazards*, Building Research Establishment, Elsevier.

B.S. 5501 (1977). Electrical Apparatus for Potentially Explosive Atmospheres, Parts 1–9.

B.S. 5423 (1987). Specifications for Fire Extinguishers.

B.S. 5306 part 2. (1990). Sprinkler Systems.

B.S. 5446 (1990). Specification for Automatic Fire Alarm Systems for Domestic Premises. Part I. Specification for Self Contained Smoke Alarms and Point Type Smoke Detectors.

B.S. 5852 (1990). Methods for Test Assessment of the Ignitability of Upholstered Seating by Smouldering and Flaming Ignition Sources.

Bukowski, R W (1995). Modeling a backdraft: the fire at 62 Watts street. *Fire Journal*, **89**(6), 85–89.

Butcher, E G (1991a). Fire resistance and buildings. *Fire Surveyor*, **20**(6), 6–12.

Butcher, E G (1991b). Pressurisation systems as a fire safety measure. *Fire Surveyor*, **20**(3), 29.

Cole, J (1989). Smoke movement in single storey buildings. *Fire Surveyor*, **18**(1), 25–32.

Cooper, L Y (1982). *Convective heat transfer to ceilings above enclosed fires. 19th. Symposium (International) on Combustion*, pp. 933–939.

Cooper, L Y and Woodhouse, A (1986). The Buoyant plume – driven adiabatic ceiling temperature revised. *Journal of Heat Transfer*, **108**, 822–826.

Coxon, P (Undated). Inergen, Wormald Britannia.

Delichatsios, M A (1986). *A simple algebraic model for turbulent wall fires. 21st. Symp. (International) on Combustion*, pp. 53–63.

Delichatsios, M A (1993). Some smoke yields from buoyant turbulent jet flames. *Fire Safety Journal*, **20**, 299–312.

Delichatsios, M A, Panagiotou, Th. and Kiley, F (1991). The use of time to ignition data for characterising the thermal inertia and the minimal critical heat flux for ignition or pyrolysis. *Combustion and Flame*, **84**, 323.

Delichatsios, M A *et al*(1992). The effects of fuel sooting tendency and the fire flow on flame radiation in luminous jet flames. *Combustion Science Techniques*, **84**, 199.

de Ris, J (1978). *Fire radiation – a review. 17th. Symp. (International) on Combustion*, pp. 53–63.

de Ris, J (1988). *A scientific approach to flame radiation and material flammability. Proc. 2nd. Intl. Symp. on Fire Safety Science*, pp. 29–40.

de Ris, J and Orloff, L (1974). *The role of buoyancy direction and radiation in turbulent diffusion flames. 15th. Symp. (International) on Combustion*, pp. 175–182.

DIN 18230 Pts 1 and 2 (1986). *Structural Fire Protection in Industrial Buildings*, Beath Verlag GmbH, Berlin, p. 30.

Drysdale, D D (1985). *An Introduction to Fire Dynamics*, John Wiley & Sons.

(a) 78 – Limits of flammability.

(b) 186 – Ignition of flammable vapor/air/mixture.

(c) 215 – Ignition during discontinuous heat flux.

(d) 274 – Transition from smoldering to flaming combustion.

(e) 313 – Fully developed fire behavior.

(f) 339 – Methods of calculating fire resistance.

ECCS (1983). European recommendations for fire safety of steel structures: calculation of the fire resistance of load bearing elements and structural assemblies exposed to the standard fire. *European Convention for Constructional Steelwork*, Elsevier, Amsterdam-Oxford-New York.

ECCS (1984). Calculation of the fire resistance of composite concrete slabs with profiled steel sheet exposed to the standard fire. *European Convention for Constructional Steelwork*, Technical Note Publication No. 32, Brussels.

ECCS (1985). Design manual on the European recommendations for the fire safety of steel structures. *European Convention for Constructional Steelwork*, Technical Note Publication No. 35, Brussels.

ECCS (1988). Calculation of the Fire Resistance of Centrally Loaded Composite Columns Exposed to the Standard Fire. *European Convention for the Constructional Steelwork*, Technical Note Publication No. 55, Brussels.

Emmons, H W (1968). Fire research abroad. *Fire Research Abstracts and Reviews*, **10**, 133–143.

EN 54 (1977). Part 5, Components of Automatic Fire Detection Systems. Heat sensitive detectors – point detectors containing a source element. (1984) Part 7, Specification for point type smoke detectors using scattered light, transmitted light or ionisation.

Evans, D D (1995). Ceiling jet flows. *SFPE Handbook of Fire Protection Engineering, pp. 2–32.*

Fahy, R F (1991). *EXIT 89*. An evacuation model for a high rise building. *Proc. 3rd. Intl. Symp. of Fire Safety Science*, Elsevier, pp. 815–823.

Field, P (1985). Effective sprinklered protection for high racked storage. *Fire Surveyor*, **14**(5), 9–24.

Fisher, K (1975). The performance of brickwork in fire resistance tests. *Jubilee Conference, Structural Design for Fire Resistance*, Midlands Branch, Institution of Structural Engineers, Eng.

Fleischman, C (1988). Analytical Methods for Determining Fire Resistance of Concrete Members, *S.F.P.E. Handbook of Fire Protection Engineering*, pp. 3–113.

Fletcher, D F (1991). An improved mathematical model of melt water detonations. Parts I and II, *International Journal of Heat and Mass Transfer*, **34**, 2435–2460.

Friedman, R (1978). Quantification of threat from rapidly growing fire in terms of relative material properties. *Fire and Materials*, **2**, 27–33.

Fruin, R F (1971). *Pedestrian Planning Design*, Metropolitan Association of Urban Design and Environment Planners Inc, New York.

Gardner, J P and Morgan, H P (1990). *Design Principles for Smoke Ventilation in Enclosed Shopping Centres*, Building Research Establishment, Department of Environment.
Gugan, K (1979). Unconfined Vapour Cloud Explosions, I. Chem. E.
Hagglund, B and Persson, L E (1976). An experimental study of the radiation from wood flames. FoU-Brand, 1, 2–6.
Harmathy, T Z (1980). The possibility of characterising the severity of fires by a single parameter. *Fire and Materials*, **4**, 71.
Harmathy, T Z (1987). On the equivalent fire exposure. *Fire and Materials*, **11**, 95–104.
Harris, R J (1983). *The Investigation and Control of Gas Explosions in Buildings and Heating Plant*, British Gas, E. & F. N. Spon, London.
Hasemi, Y (1986). Thermal modelling of upward flame spread. *1st International Symposium Fire Safety Science*, pp. 87–96.
Health and Safety Executive (1980). *Guide to the use of Flame Arresters and Explosion Reliefs*, H.S.G. 11, HMSO, London.
Heskestad, G (1986). Fire plume air entrainment according to two competing assumptions. 21st. Symp (International) on Combustion, pp. 111–120.
Heskestad, G (1988). Fire plumes. *SFPE Handbook of Fire Protection Engineering*, pp. 1–107.
Heskestad, G and Delichatsios, M (1978). Reference not identified.
Hinkley, P L (1995). Smoke and heat venting. *SFPE Handbook of Fire Protection Engineering, pp. 3–160.*
Hinkley, P L *et al*(1968). *The Contribution of Flames under Ceilings to Fire Spread in Compartments. Part 1, Incombustible Ceilings*, Fire Research Note 712, Fire Research Station, Borehamwood.
Hinkley, P L *et al*(1984). Contribution of flame under ceilings to fire spread in compartments. *Fire Safety Journal*, **7**, 227–242.
Hogan, W J (1982). *The Liquefied Gaseous Fuels Spill Effects Programme*, A Status Report, Fuel, Air Explosions. University of Waterloo Press, pp. 949–967.
Home Office (1973). *Fire Precautions in New Single Storey Spirit Storage and Associated Buildings*, HMSO.
Ingberg, S H (1928). Tests of severity of building fires. *NFPA Quarterly*.
Institution of Structural Engineers (1978). *Design and Detailing of Concrete Structures for Fire Resistance*, London.
Janssens, M (1992a). Room fire models. *Heat Release in Fire*, Vol. 113, V Babrauskas and S J Grayson (Eds.), Elsevier Applied Science, London, New York.
Janssens, M (1992b). Determining flame speed properties from cone calorimeter measurements. *Heat Release in Fire*, Vol. 265, V Babrauskas and S J Grayson (Eds.), Elsevier Applied Science, London, New York.
Jin, T (1971). Visibility through smoke. Part III, *Bulletin of Japanese Association of Fire Science and Engineering*, **22**(1, 2), 11–15.
Jones A and Thomas G O (1993). The action of water sprays on fires and explosions – a review of experimental work. *Trans. I. Chem. E*, **71**, Part B, 41–49.
Kanury, A M (1977). *An Introduction to Combustion Phenomena*, Vol. 130, Gordon & Breach.
Karlsson, B and Magnusson, S E (1991). Combustible wall lining materials. Numerical simulation of room fire growth and the outline of a reliability based classification procedure. *Proceedings of the 3rd. Int. Symp. on Fire Safety Science*, pp. 667–678.
Kendik, E (1986). Methods of design for means of egress towards a quantitative comparison of national code requirements. *Proceedings of the 1st. Int. Symp. Fire Safety Science*, pp. 497–678.
King, P J *et al*(1977). *Report of the Inquiry into Serious Gas Explosions*, Department of Energy, London.
Kingman, F E T *et al*(1953). The products of combustion in burning buildings. *Journal of Applied Chemistry*, **3**, 463–468.
Klote, J H (1995). Smoke control. *SFPE Handbook of Fire Protection Engineering*, pp. 4–230.
Klote, J H and Milke, J (1992). *Design of Smoke Control for Buildings*, N.B.S. Handbook 141, ASHRAE and SFPE.
Kokkala, M A (1991). Experimental study of heat transfer to ceiling from impinging diffusion flame. *3rd. Intl. Symp. Fire Safety Science*, Elsevier.
Kung, *et al*(1988). Ceiling Flow of Growing Rack Storage Fires. *21st. Symp. (International) on Combustion*, pp. 121–128.
Law, M (1971). *A Relationship between Fire Grading and Building Design and Contents*, Fire Research Station, Fire Research Note No. 877.
Law, M (1986). A note on smoke plumes from fires in multi-level shopping malls. *Fire Safety Journal*, **10**, 197.
Levin, B M (1989). *A simulated model of occupant decisions and actions in residential fires. 2nd. Int. Symp., Fire Safety Science*, pp. 561–570.

Makhviladze, G *et al*(1995). Criterion for the formation of a 'cloud like' release upon depressurisation of a gas vessel. I. Chem. E, *Major Hazards Onshore and Offshore, II*, pp. 97–112.

Malhotra, H L (1992). Recent developments in the European fire scene. *Fire Surveyor*, **21**(4), 4–8.

Malhotra, H L *et al.* (1971). *The Fire Propagation Test as a Measure of Fire Spread and Correlation with full Scale Fires in Corridors*, Fire Note 876, Fire Research Station, Borehamwood.

Markstein, G H (1976). Scaling of radiative characteristics of turbulent diffusion flame. *16th. Symp. (International) of Combustion*, pp. 1407–1419.

Markstein, G H (1978). Radiative properties of plastic fires. *17th. Symp. (International) on Combustion*, pp. 1053–1062.

Markstein, G H (1984). Relation between smoke point and radiant emission from buoyant turbulent and laminar diffusion flames. *20th. Symp. (International) on Combustion*, pp. 1055–1061.

Markstein, G H and de Ris, J (1972). Upward fire spread over textiles. *14th Symp. (International) on Combustion*, The Combustion Institute, Pittsburgh, pp. 1085–1097.

Markstein, G H and de Ris, J (1990). *Wall-fire Radiant Emissions, Part I.: Slot-burner Flames, Comparison with Jet Flames*, Technical Report FMRC J.I. OQON5.BU, Factory Mutual Research Corporation, Norwood, MA.

Marryat, H W (1971). *Fire Automatic Sprinkler Performance in Australia, New Zealand 1886–1968*, Australian Fire Protection Association, Melbourne.

Meams, R (1986). A case for flame retardant paints. *Fire Surveyor*, **15**(2), 4.

McCaffrey, B J (1979). *Purely Buoyant Diffusion Flames: Some Experimental Results*, National Bureau of Standards, NBSIR, pp. 79–1910.

McCaffrey, B J *et al*(1981). Estimating room temperatures and the likelihood of flashover using fire test data correlations. *Fire Technology*, **17**, 98–119, **18**, 122.

McCaffrey, B J (1988). Flame height. *SFPE Handbook of Fire Protection Engineering*, pp. 1–298.

Mikkola, E (1992). Ignitability of solid materials. *Heat Release in Fire*, Vol. 225, V Babrauskas and S J Grayson (Eds.), Elsevier Applied Science, London, New York.

Miller, M J (1974). Reliability of fire protection systems. *Chem. Eng. Progress*, **70**, 62–67.

Mitler, (1978). *The Physical Basis for the Harvard Computer Fire Code*, Division of Applied Science, Harvard University, Home Fire Project Technical Report.

Moodie, K and Jagger, S E (1989). *Results and Analysis from the Scale Model Tests The King's Cross Underground Fire: Fire Dynamics and the Organisation of Society*, Institution of Mechanical Engineers, pp. 27–40.

Moore, P E (1996). Fluorocarbon halon alternatives. *Fire Safety Engineering*, **3**(6), 20.

Morgan, H P and Marshall, N R (1975). *Smoke Hazards in Covered, Multi-level Shopping Malls: An Experimentally Based Theory for Smoke Production*, Building Research Establishment, CP48/75, Borehamwood.

Morris, W A (1984). Stardust disco investigation – some observations on the full scale fire tests. *Fire Safety Journal*, **7**, 255.

Mudan, K S and Croce, P A (yr1995, 1988). Fire hazard calculations for large open hydrocarbon fires. *SFPE Handbook of Fire Protection Engineering*.

Mulhollland, G. W (1988). Smoke production and properties. *SFPE Handbook of Fire Protection Engineering*, pp. 1–369.

Mullins and Penner (1959). *Explosions, Detonations, Flammability and Ignition*, Pergamon Press, London.

National Fire Protection Association (1980). National Fire Code 13, Standard for Installation of Sprinkler Systems.

National Fire Protection Association (1991). Guide for Smoke and Heat Venting, Code 204M.

National Fire Protection Association (1991). *Smoke Management in Malls, Atria and Large Areas*, NFPA 92B.

Nelson, H E and McLennan, H A (1988). Emergency movement. *SFPE Handbook of Fire Protection Engineering*, pp. 2–106.

Ndubidzu *et al*(1983). A model of freely burning fuel fires. *Science and Technology*, **31**, 233.

Odgaard, E and Solberg, D M (1981). *Heat Loads from Accidental Hydrocarbon Fires*, A.P.S. Seminar, Fire Protection Offshore, No. 81, p. 75, Det Norske Veritas.

O'Dogherty, M J, Nash, P and Young, R A (1967). *A Study of the Performance of Sprinkler Systems*, Fire Research Technical Paper, No. 17, HMSO, London.

Orloff, L and de Ris, J (1982). Froude modelling of pool fires. *19th. Symp. (International) on Combustion*, pp. 885–895.

Orloff, L *et al.* (1974). Upward turbulent fire spread and burning of fuel surface. *15th. Symp. (International) of Combustion*, pp. 183–192.

Orloff, L *et al*(1978). Radiation from smoke layers. *17th. Symp. (International) on Combustion*, pp. 1029–10938.

Pauls, J L (1980). Effective width Model for Evacuation Flow in Buildings, *Prac. Engineering Applications Workshop*, SFPE, Boston.

Pettersson, O (1985). Structural fire behaviour. *Proc. of 1st. Int. Symp. Fire Safety Science*, Vol. 229, Hemisphere.

Predtechenskii, V M and Milinskii, A I (1978). *Planning for Foot Traffic Flow in Buildings*, American Publishing Co. Inc, New Delhi.

Purser, D A (1995). Toxicity assessment of combustion products. *SFPE Handbook of Fire Protection Engineering*, pp. 2–85.

Puttock, J S (1995). Fuel Gas Explosion Guidelines – The Congestion Assessment Method, I. Chem. E. *Major Hazards Onshore and Offshore II*, pp. 285–296.

Quintiere, J G (1977). *Growth of Fire in Building Compartments*, Fire Standards and Safety, ASTM STP 614, Philadelphia, pp. 131–167.

Quintiere, J G *et al.* (1986). Wall flame and implications for upward flame spread, *Combustion Science and Technology*, **48**.

Quintiere, J G (1993). A simulation model for fire growth in materials subject to a room corner test. *Fire Safety Journal*, **20**, 313–340.

Quintiere, J G and Harkleroad, M (1984). New concepts for measuring flame spread properties. *Symposium, Application of Fire Science to Fire Engineering*, ASTM, SFPE.

Rasbash, D J (1962). The extinction of fires by water sprays. *Fire Research Abstracts and Reviews*, **4**(1–2), 28–53.

Rasbash, D J (1967). Smoke and toxic gases produced at fires. *Trans. J. Plast. Inst. Conference*, Supplement No. 2, pp. 55–62.

Rasbash, D J (1975a). The relevance of fire point theory to the evaluation of the fire properties of combustible materials. *Proc. Int. Symp. on the Fire Properties of Combustible materials*, Edinburgh University, pp. 169–178.

Rasbash, D J (1975b). Sensitivity criteria for detectors used to protect life. *7th International Seminar on Problems of Automatic Fire Detection*, Tech Hoch, Aachen, pp. 137–154.

Rasbash, D J (1975c). Interpretation of statistics on the performance of sprinkler systems. *Control*, **2**(2), 63–73.

Rasbash, D J (1976). Theory in the evaluation of fire properties of combustible materials. *Proc. 5th Int. Fire Protection Seminar*, Vol. **1**, Karlsruhe, V.F.D.B., p. 113.

Rasbash, D J (1986a). The extinction of fire with plain water. *Proc. 1st Int. Symp. on Fire Safety Science*, pp. 1145–1163.

Rasbash, D J (1986b). Quantification of explosion parameters for combustible fuel-air mixtures. *Fire Safety Journal*, **11**, 113–125.

Rasbash, D J (1995a). Units for expressing light attenuation in smoke. *Fire and Materials*, **19**, 51–52.

Rasbash, D J (1995b). Measurement of smoke. *Fire and Materials*, **19**, 241.

Rasbash, D J and Philips, R P (1978). Quantification of smoke produced by fires. *Fire and Materials*, **2**, 102–109.

Rasbash, D J and Pratt, B T (1979). Estimation of smoke produced in fire. *Fire Safety Journal*, **2**, 23–37.

Rasbash, D J *et al*(1956). Properties of fires of liquids. *Fuel*, **35**, 94–106.

Renwick, P and Tolloczka, J (1992). Offshore safety. *The Chemical Engineer*, **525**, 17–18.

Rogowski, B (1984). A critique of fire test methods used to assess individual products involved in the Artane Fire. *Fire Safety Journal*, **7**, 213.

Rutstein, R and Gibert (1978). *The Performance of Sprinkler Systems*, Memorandum 9/78, Home Office Scientific Advisory Board.

Saito, K *et al.* (1986). Upward turbulent flame spread. *1st Int Symp. Fire Safety Science*, pp. 75–86.

Seader, J D and Ou, S S (1977). Correlation of the smoking tendency of materials. *Fire Research*, **1**, 3–9.

Schifiliti, R P (1995). Design of detection systems. *SFPE Handbook of Fire Protection Engineering*, pp. 4–1.

S.F.P.E. Handbook (1995, 1988). *Fire Protection Engineering*, Society of Fire Protection Engineers, National Fire Protection Association.

Shield, S R (1995). The modelling of BLEVE fireball transients, I. Chem. E. *Major hazards Onshore and Offshore II*, pp. 227–236.

Smith, D P (1995). Water mist: fire suppression systems. *Fire Safety Engineering*, **2**(2), 10–15.

Spalding, D B (1955). *Some Fundamentals of Combustion*, Butterworths, London.

Steinbrecher, L (1987). Analysis of tank fire 'Classic'. *Fire Engineers Journal*, **47**, 15–17.

Tewarson, A S (1980). *Fire and Materials*, **4**, 185–191.

Tewarson, A S (1995). Generation of heat and chemical compounds in fires. *SFPE Handbook of Fire Protection Engineering*, pp. 3–534.

Theobald, C R (1987). Heated wind tunnel for testing fast acting sprinklers. *Fire Surveyor*, **16**(20), 9.

Thomas, P H and Hinkley, P L (1964). *Design of Roof Venting Systems for Single Storey Buildings*, Fire Research Technical Paper, No. 10, HMSO, London.

Thomas, P H *et al.* (1967). *Fully Developed Compartment Fires*, Joint Fire Research Organisation, HMSO, London.

Thomson, H E and Drysdale, D D (1989). Critical mass flow rate of the fire point of plastics. *Proc. of 2nd. Int. Symp. Fire Safety Science*, pp. 67.

Underwriters Laboratories (1970). Standards for Single and Multiple Station Smoke Detectors, UL 217, Northbrook.

Venart, J E S *et al*(1992). To BLEVE or not to BLEVE: anatomy of a boiling liquid expanding. *AIChE Meeting on Loss Prevention*, American Institute of Chemical Engineers, New Orleans, LA, pp. 1–21.

Walton, W D and Thomas, P H (1995, 1988). Estimating temperatures in compartment fires. *SFPE Handbook of Fire Protection Engineering*, pp. 3–134.

White, R H (1988). Analytical methods or determining fire resistance of timber members. *SFPE Handbook of Fire Protection Engineering*, pp. 3–130.

Williamson, R B *et al.* (1991). Ignition sources in room fire tests. *3rd Int. Symp. Fire Safety Science*, pp. 657.

Wood, P G (1972). *The Behaviour of People in Fires*, Fire Research Station, FR Note No. 953.

Workshop, I. B. W14 (1983). A conceptual approach towards a probability based design guide in structural fire safety. *Fire Safety Journal*, **6**, 1–79.

Workshop, C. I. B. W14 (1986). Design guide structural fire safety. *Fire Safety Journal*, **10**, 77–136.

You, H Z and Faeth, G M (1979). Ceiling heat transfer during fire plume and fire impingement. *Fire and Materials*, **3**, 140.

Yu, H Z (1990). Transient plume influence in measurement of convection heat release rates of fast growing fires using a large scale fire products collector. *ASME Journal of Heat Transfer*, **112**, 186–191.

Zalosh, R G (1995). Explosion protection. *SFPE Handbook Fire Protection Engineering*, pp. 3–312.

Zukoski, E E (1987). *Heat transfer in unwanted fires. ASME – JSME Thermal Engineering Conference*, Honolulu, pp. 311–317.

Zukoski, E E (1994). *Mass flux in fire plumes. Proc. 4th. Intl. Symp. in F ire Safety Science*, pp. 137–147.

6 SOURCES OF STATISTICAL DATA

6.1 Introduction

It is generally accepted that there is a need for statistical information on fire and this should provide a background against which decisions can be made on economically justifiable levels of effort and expenditure on activities concerned with fire prevention, protection, and insurance. Fire statistics enable a detailed evaluation of fire risks insofar as they have existed in the past, thus providing the basic data from which programmes for fire precautions dealing with risks in different sections or areas of a community may be established. In this way, they support the appraisal of a wide variety of risk situations to guide policies of administrative activity and framing of fire safety regulation, legislation, standards, and codes.

In recent years, fire scientists and engineers have used information from fire statistics to increase knowledge about the technical nature of fire in real situations. Such data, which augment results of experiments carried out under controlled conditions, can provide improved tools for developing quantitative models of fire safety. Such models coupled with information on the behavior of people in fire situations would enable management and code makers to adopt a more flexible approach in identifying appropriate measures that would meet a quantitative level of fire safety acceptable to the community.

Fire statistics should, therefore, be available and useful to different groups and individuals, all of whom, while working toward the same ultimate goal of reducing the loss of life and property by fires, may have different immediate objectives. It may appear that the administrator and research worker require statistics for quite different purposes. This is, however, rarely true since research work involving the use of statistics is an essential step in the process of arriving at administrative decisions.

From time to time, it is likely that specific problems will arise, which will require the collection of very detailed information on a particular sector of the fire field. It is impossible to foresee all eventualities of this type and devise a statistical system that would cover all of them. Moreover, such a system would be too cumbersome to handle. There is, however, a solid block of information which, when collected on a continuing basis, can be of great value to all users of fire statistics and can also serve as the starting point for the collection of additional statistics through special surveys. The continuing statistics should be sufficient to reveal important trends

Evaluation of Fire Safety D. Rasbash, G. Ramachandran, B. Kandola, J. Watts and M. Law
 ISBN: 0-471-49382-1

and changes in the frequency of fires, the occupancies in which they occur, their causes, their sizes, and their costs in terms of damage to life and property. They should also provide necessary data for forecasting the extent of future demands on fire-fighting resources and for evaluating the effectiveness of fire safety measures.

Regular analysis of the continuing statistics can be of great value in itself, but for some purposes they have to be related to other statistical information obtained from a variety of different sources. These ancillary statistics include population data, housing and construction statistics, industrial output, fire brigade expenditure, transport statistics, socioeconomic data, and meteorological data. Information provided by fire statistics should also be related to those collected by insurance organizations and to the results of fire tests and experiments. Fire statistics cannot, therefore, be regarded as a completely self-contained collection of data suitable for solving all fire problems.

The object of this chapter is to briefly review the sources of statistical data on fires and fire losses available in different countries, particularly the United Kingdom, the United States, and Japan. In addition to major and special databases, the review covers other minor sources, special studies or surveys, and ancillary statistics. The ways in which these statistics can be used in fire safety evaluation are broadly outlined, with attention drawn to detailed discussion in other chapters. Topics also discussed in this chapter include gaps in current statistics, limitations in their use, and international comparisons.

6.2 Fire departments and brigades

6.2.1 UNITED KINGDOM

Good national fire statistics emanate mainly from a well-organized public fire service. In the United Kingdom, for example, the formation of the National Fire Service during World War II gave the opportunity of organizing statistics on a national scale. Fire statistics have been collected since that war by the Home Office, through the fire brigades of the local government authorities. The brigades furnish information on fires on a voluntary basis and not under any statute of the national government.

Until 1977, the brigades in the United Kingdom submitted a standard report form, K433, which was completed for every fire with the exception of chimney fires confined to (did not spread beyond) chimneys. The background to the development of this form and the way the analyses of fire incidents were carried out was described by Wallace (1948). However, to simplify the tasks of fire brigades, a second form, K433H, was introduced in 1960 in which rather limited information was given for fires that were confined to grassland, heathland, or railway embankments. In 1970, K433H was revised and its use extended to cover a much larger group of minor fires. Replacing K433, a new form, FDR1, was introduced in January 1978, taking into consideration the needs of various government departments and other organizations for fire statistics. FDR1 was revised in 1994.

The main questions on K433 and FDR1 for building fires include the following details:

(a) *Time*: The day and the month of the fire, time of discovery, the time the call was received by the brigade, the time of the arrival of the brigade at the scene of fire, the time the fire was brought under control, and the time the last appliance returned to the fire station.

(b) *Place*: The address of the fire, the name of the occupier, and the trade or business carried out.

(c) *Location of fire*: Type of property where fire started, descriptions of the building, floor of fire origin, place (e.g. room) where fire started and use of room, for example, production, storage, office, assembly area, kitchen, or bedroom.

(d) *Construction of building*: Approximate year of construction, number of basements, number of floors, approximate dimensions of premises and room of fire origin, and materials of construction and linings and so on, directly affected by fire.

(e) *Extinction of fire*: Number of heads activated and performance of sprinklers and drenchers (if installed), methods of fighting the fire before arrival of the fire brigade, for example, portable extinguishers, methods used by the brigade, for example, hose reel, jet, number of jets used, and other appliances employed.

(f) *Spread*: Whether the fire spread beyond the material or object ignited first and, if so, whether it spread beyond the room or floor of origin or beyond the building of origin.

(g) *Cause of the fire*: Most likely cause or source of ignition, material or item ignited first, defect, act, or omission giving rise to ignition and material or item mainly responsible for the development of the fire.

(h) *Life risk*: Approximate number of persons at the time of discovery of fire who left the affected property because of the fire and of those who escaped by unusual routes; name, age, and other details of people sustaining fatal or nonfatal injuries and of those who were rescued.

(j) *Explosions and dangerous substances*: Whether explosion caused the fire or the fire caused the explosion, materials involved, and details about dangerous substances affecting fire fighting or development of fire.

In FDR1 (since 1978), the fire brigade is required to estimate the time interval from ignition to discovery according to the following four classifications:

(i) Discovered at ignition
(ii) Discovered under 5 min after ignition
(iii) Discovered between 5 and 30 min after ignition
(iv) Discovered more than 30 min after ignition.

In the revised form that came into existence in 1994, the fourth classification relates to discovery of a fire between 30 min and 2 h with a further classification (fifth) for discovery taking place more than 2 h after ignition.

Information on the construction of building (item (d)) is only required to be given in FDR1 if there was any evidence of heat damage to the structure. This form also contains information on total horizontal area damaged by direct burning and total area damaged, including smoke and water damage.

Until 1973, the fire brigade reports were collated and the statistics processed by the Fire Research Station (FRS), Borehamwood, Herts, United Kingdom. This work is still being carried out at Borehamwood but, since 1974, the Home Office has been exercising direct control of the Fire Statistics Section. Much of the information in the fire reports has been coded and put on magnetic discs (instead of tapes) during recent years. On the basis of computer tabulations, some of the information is rearranged into statistical tables that are published annually, but there is generally a time lag of about two years for the publication of statistics for any year. This booklet is currently known as *Fire Statistics, United Kingdom* and can be purchased from the Home Office.

The current published information represents only a fraction of the data stored on computer discs. The information stored can be identified from the code lists and can usually be obtained from the Home Office, which charges a small amount for each computer run (or tabulation) to cover mostly the cost toward computer time. Further information may be obtained from the original reports (FDR1) that are not usually open to outside inspection. Occasionally, copies of the original reports may be obtained on request directly from the fire brigades concerned. Using published and unpublished data, several statistical analyses and reviews have been carried out, particularly by the FRS, in order to evaluate fire risks in different types of occupancies and due to various causes and other factors (see, for example, Fry (1969), Chandler (1978), and North (1973)). Statistics provided by fire brigades in the United Kingdom have also been used by several research workers for developing and validating statistical models dealing with fire protection problems. (These models are described in the other Chapters of Part II).

An important development during the last five years has been the establishment of FINDS, Fire Information National Data Service, an information system for fire brigades throughout the United Kingdom. This system is operated by Chief and Assistant Chief Fire Officers Association (CACFOA Research Ltd), a charity organization set up by a fire brigade association. Among other information of interest to the fire service, FINDS is intended to provide a national directory of research and development, fire investigation reporting and collection system, national emergency equipment and manpower coordinating service, interbrigade communication and message service, management information service, and a national data reference and retrieval service. In due course, it is expected to include information contained in FDR1. FINDS can be accessed via an international data network (X25) by users in the United Kingdom and other countries through their existing computer network or personal computer. The user has to pay certain annual subscription charges and a one-off joining fee. FINDS includes electronic mail and data transfer facilities.

6.2.2 UNITED STATES OF AMERICA

In the United States, the development of the National Fire Incident Reporting System (NFIRS) in the late 1970s made detailed representative national fire statistics possible for the first time. This system is operated by the Federal Emergency Management Agency's (FEMA's) United States Fire Administration (USFA). NFIRS provides annual computerized data based on fire incidents with data classified according to a format specified in National Fire Protection Association (NFPA) 901 standard for fire incident reporting.

During the early years, only a few states were involved in NFIRS. Now, roughly three-fourths of all states have NFIRS coordinators who receive fire incident data from participating fire departments and combine the data into a state database. These data are then transmitted to FEMA/USFA. Participation by the states and by local fire departments within participating states is voluntary. NFIRS captures about one-third of all US fires each year. More than one-third of all US fire departments are listed as participants in NFIRS, although not all of these departments provide data every year.

One of the strengths of NFIRS is that it provides more detailed incident information than any US National database not limited to large fires. NFIRS is the only database capable of addressing national patterns for fires of all sizes by specific property use and specific fire cause. It also captures information on the construction type of the involved building, the avenues and extent of flame spread and smoke spread, and the performance of detectors and sprinklers. Analysis based on NFIRS has been widely carried out since 1982, when the second edition of FEMA/USFA's "Fire in the United States" was published – the first study based primarily on NFIRS.

One weakness of NFIRS is that its voluntary character produces annual samples of shifting composition since it does not cover all fires reported to US fire departments. Participation rates of fire departments in NFIRS are not necessarily uniform across regions and sizes of community, both of which are factors correlated with frequency and severity of fires. This means that NFIRS may be susceptible to systematic biases arising from the fact that it is not based on a random sample of fires, fire departments, or populations.

For the reasons mentioned above, some analysts combine NFIRS-based percentages with NFPA-survey-based totals (next paragraph) to produce national estimates of numbers of fires, deaths, injuries, and dollar loss for subparts of the fire problem. However, the calculation rules used to produce these estimates have varied among users. Hence, Hall and Harwood (1989) have presented a detailed consensus procedure for such calculations and the supporting rationale. This could be called a *multiple calibration approach* and it makes use of the annual NFPA survey. In principle, it would be possible to use the NFPA survey to calibrate NFIRS separately for each of the 17 different property categories for which the survey directly provides national fire experience totals. In the basic approach described by Hall and Harwood (1989), national estimates of scaling ratios are calculated for four major property classes (residential structures, nonresidential structures, vehicles, and others) and for each measure of fire severity (fire incidents, civilian deaths, civilian injuries, and direct property damage).

The NFPA annual survey is statistically designed and is based on a stratified random sample of roughly 3000 US fire departments, which represents 10% of nearly 30,000 US fire departments; stratification is according to the size of the population. The survey collects information on the total number of fire incidents, civilian deaths and injuries, and the total estimated property damage (in dollars) for each of the major property use classes defined by NFPA 901. Similar data are collected specifically for incendiary and suspicious fires separated only into structure versus vehicle; cause-related information is obtained only for these fires (not for all fires in the sample). The results for these fires and the totals for the first category mentioned above are analyzed and reported in NFPA's annual study: Fire Loss in the United States. The survey also collects data on the number of injuries to firefighters on duty, by type of duty, and nature of injury or illness. These statistics are analyzed and published in NFPA's annual report – US Fire Fighter Injuries. In the NFPA survey, information is also obtained on the type of community protected (e.g. county versus township versus city) and the size of population protected. These data are used in the statistical formula for projecting national estimates from sample results and leads on multiple death and large-loss fires and firefighter fatalities. The NFPA survey provides a valid basis for measuring national trends in fire incidents, civilian deaths and injuries, and direct property loss, as well as for determining patterns and trends by community size and major region.

The NFPA Fire Incident Data Organisation (FIDO) system provides detailed information on fires that are deemed to be of high technical interest. It covers virtually all incidents reported to fire departments involving three or more civilian deaths, one or more firefighter deaths, or large dollar loss (redefined periodically to reflect the effects of inflation and defined since 1980 as one million dollars or more in direct property damage). FIDO also captures a selection of smaller incidents of technical interests in certain types of properties – high-rise buildings, presence of hazardous materials, and performance of detectors or sprinklers. Candidates for FIDO are selected from several sources such as newspapers, insurers' reports, NFIRS, and respondents to the NFPA annual survey. Once notified of a candidate fire, the NFPA seeks standardized incident information from the responsible fire department and solicits copies of other reports prepared by concerned parties.

The strength of FIDO is its depth of detail on individual incidents. Information captured by FIDO, but not by NFIRS, includes types and performance of systems for detection, suppression, and smoke and flame control; detailed information on factors contributing to flame and smoke spread; estimates of time between ignition and detection and between detection and alarm; indirect loss and detailed breakdowns of direct loss; and escapes, rescues, and number of occupants. FIDO supports three annual NFPA reports – US Fire Fighter Deaths, Multiple Death Fires in the United States, and Large-Loss Fires in the United States. One weakness of FIDO is that it mostly covers larger incidents and does not permit comparisons of characteristics of large and small fires.

6.2.3 JAPAN

Municipal fire departments in Japan prepare a report on every fire incident and every fire death according to a unified format suggested in the Handbook for Reporting System of Fire Incident and Fire Death. This system has been designed by the Fire Defense Agency of Ministry of Home Affairs, Japan. The Agency collects all the fire reports, processes the information contained in the reports, analyzes the information and publishes the results in White Book on Fire Service in Japan.

The reports contain information on the following main items:

(a) *Occupancy type*: Buildings, forests, vehicles, vessels, aircrafts, and others.

(b) *Physical damage*: (For houses and households) According to three categories – totally burnt, half burnt, or partly burnt.

(c) Area Burnt (m^2)

(d) *Casualties*: Injuries and fatalities including those who died within 48 h after they were injured in fires.

(e) *Financial damage*: (In yen)

(f) *Methods of acknowledgment (reporting fires)*: Such as fire alarm, fire alarm service telephone (Dial 119), subscriber's telephone, and police telephone.

(g) *Equipment used for initial control of fires*: Simple fire-fighting equipment, fire extinguisher, fixed fire-fighting apparatus, and others (and no initial fire fighting).

(h) *Causes of fires (or sources of ignition)*: Accidental (several categories), incendiary and suspected incendiary, spontaneous combustion and reignition, and natural disasters.

Special features of Japanese fire statistics relate to analyses of fatalities by building structures (wooden, fire-protective, simple fireproof, fireproof, and others) and by floors (including basement). Fatalities are also classified according to age and causes such as carbon monoxide poisoning, suffocation, burns, bruise, and bone fracture. Sekizawa (1991) has carried out a detailed analysis of the characteristics of fatalities in residential fires.

6.2.4 OTHER COUNTRIES

A number of other countries collect national statistics through fire brigades or departments, although, in general, they are rather less detailed and do not go back as far as the UK Fire Service Statistics. In Canada, national fire statistics are compiled by the Association of Canadian Fire marshals and Commissioners and Fire Commissioner of Canada, in the Netherlands by the Inspectorate of Fire Services (Ministry of Interior), in New Zealand by the New Zealand Fire Commission, in Denmark by the Danish State Fire Inspectorate, in Finland by the Ministry of Interior, and in Norway by the Directorate for Fire and Explosion Prevention. National French statistics provide detailed information on breakdown of fires in different risks according to the month of the year and statistics of fires in Paris according to the occupancy. Because Germany has a decentralized Federal Administration, as in the United States, its national fire statistics are only available from recent years, but they provide useful information on the spread of fire and the agencies for causing fires.

In Australia, although various state fire services have been collecting data for a considerable time, it was only in 1983 that a national standard was published for the collection, processing, and analysis of fire statistics. This standard describing the Australian Fire Incident Reporting System (AFIRS) has been revised recently taking into account the requirements of the fire services and the community. Despite the standard, some nonuniformity existed in the adoption of AFIRS by different states. Implementation of AFIRS on a uniform basis is currently being attempted by a task force comprising representatives from the Commonwealth Scientific Industrial and Research Organisation (CSIRO) and the state fire services.

In 1990, CSIRO and the Australian Assembly of Fire Authorities (AAFA) agreed to collaborate on the collection and analysis of national fire statistics. Fire incident data was to be collected by each AAFA member brigade and entered into a database at CSIRO. The first collection of data, requested in 1991 for the year 1989–1990, represented 81% of the majority of calls that fire brigades attended in all states except Queensland, Northern Territory, and Australian Capital Territory. An analysis of these data is contained in the first Australian National Fire Incident Statistics, 1989–1990, published by CSIRO in October 1992. This publication includes several interesting tables useful for fire risk analysis. An example worthy of special mention is the relationship between fire brigade attendance time and the extent of flame damage, separately for fires in which sprinklers operated and for fires in which sprinklers did not operate. Similar tables have also been produced for dollar value of fire damage.

6.3 Insurance organizations and fire protection associations

6.3.1 UNITED KINGDOM

Fire protection associations (FPAs) and insurance organizations in many countries constitute another major source of statistical data, particularly on fires with large financial losses greater than a threshold level, which occur mainly in industrial and commercial properties. In the United Kingdom, for example, information on such fires has been published by the British Insurance Association (BIA), now known as the Association of British Insurers (ABI), for several years. These are preliminary estimates of losses made by the loss adjusters soon after the occurrences of fires; they are not final claim amounts settled by the insurance companies concerned. The losses relate to direct material (property) damage. Since 1965, information on these large fires, split into building and contents categories, has been made available to the Fire Research Station by the BIA through the Fire Protection Association (FPA). The threshold level of loss for these fires, which was £10,000 until 1973, has been gradually increased to £50,000 over the years, taking into account inflation and the need to keep the number of large fires to be reported at a manageable size.

Since 1965, the large-loss data furnished by the FPA have been merged with the fire brigade statistics by locating the fire reports (K433 or FDR1, since 1978) for these incidents. United Kingdom Fire and Loss Statistics published by the FRS for the years 1968 to 1973 included a separate section containing some tables based on large-loss fires. Mainly because of economic reasons, the publication of these tables was discontinued by the Home Office when it assumed the responsibility for fire statistics in 1974. However, the practice of merging the large-loss data with fire brigade statistics is still being continued by the Home Office. The matched data are returned to the FPA, which then publishes an analysis of the data in its journal *Fire Prevention*.

The primary source for the statistics on large fire losses compiled by the FPA is a Loss Report Form completed by insurance loss adjusters and sent at present to the Loss Prevention Council (LPC). This form has been in existence for more than 20 years, although it was revised a few years ago and split into two parts. Loss Report Form A is to be completed by the loss adjusters in the event of fire, explosion, and sprinkler leakage losses involving material damage estimated at £50,000 or more and/or a fatality. Form A is not required for household losses except where fatalities are involved. Loss Report Form B is to be completed by the loss adjusters and insurers in the event of fire, explosion, or sprinkler leakage in sprinklered premises where a loss of any value has occurred. The original version of the Loss Report Form covered also incidents in premises equipped with automatic fire alarm installation or with fixed carbon dioxide or fixed dry powder extinguishing systems. Information from Form B is reported quarterly to the ABI.

In addition to an estimate of total direct (material) loss, split into building and contents, the Loss Report Form includes information on the dimensions of the room in which the incident started and the approximate area damaged by fire and water. The form also contains other useful items of information about the sprinkler installation and its performance during the time of occurrence of the incident. An estimate of the consequential loss (if known) was required in the original form but this item has not been included in the revised form apparently because of the fact that an evaluation of this (indirect) loss is a complex statistical problem (see Ramachandran (1995)).

The statistics provided by the Loss Report Form are regarded as confidential to the LPC and the participating insurance companies. Both the forms, A and B, are analyzed annually by the LPC and collated with similar reports from Comite Europeen des Assurances (CEA) member countries. These analyses provide the only source for European statistics on sprinkler installations.

6.3.2 UNITED STATES OF AMERICA

As mentioned in Section 6.2.2, the NFPA collects detailed information on fires with large losses (one million dollars or more) in direct property damage through its FIDO system. Figures for financial losses in these fires and other data obtained from the insurers' reports are then collated with the information contained in the fire departments' reports as done by the FPA in the United Kingdom. FIDO appears to be the only source easily accessible by research scientists, statisticians, and others interested in the analyses of large fires.

The Insurance Services Office (ISO) in New York collects, analyzes, and distributes insurance performance statistics for approximately 80% of all insurance companies throughout the United States. A principal reason for doing this is to help establish an actuarially sound rate structure for the myriad of insurance coverages. A product of this effort is the collection of loss data for major industrial occupancy types. Fire loss data is compiled for four major loss areas defined by the ISO as buildings, contents, business interruption, and other time elements.

In ISO statistics, a major classification relates to the fire-resistance characteristics of a building – whether fire-resistive or non-fire-resistive. A fire-resistive building is one in which noncombustible roof decks are supported by a concrete or steel frame. Non-fire-resistive buildings have combustible floors and roof supported by a wooden or masonry frame. Other information contained in ISO statistics relates to various building characteristics such as type of structure and its conditions, size, height, safety devices, hazards, and so on. These statistics necessarily include information about the installation of sprinklers, automatic fire alarms, and portable fire extinguishers, and their performance in fires.

The results of statistical analyses of fire losses are incorporated in insurance-rating schedules such as the ISO Commercial Fire Rating Schedule, which is widely used in the United States. In general, tabulated values and conversion factors are based on actuarial analyses of fire losses (claims) paid by insurers and reported to the ISO. The ISO schedule is now the property of a subsidiary corporation, ISO Commercial Risk Services Inc.

Perhaps, for purposes of loss control and fire protection engineering, the most comprehensive database is the one maintained by the Factory Mutual Research Corporation (FMRC). This organization is part of Factory Mutual Engineering and Research, a world leader in property loss control, particularly in the industrial and commercial sectors. Apart from loss data, the FM Engineering Risk Data Base includes information on building characteristics (construction type, number of stories, and total floor area), sprinkler system (coverage, type, and water supply), type of detector system, and special items such as exposure to special hazards (flammable liquids, gases, explosive dusts, and radioactive material). There are other items related to insurance information.

FMRC's Loss and Operational Analysis Department collects, stores, retrieves, and analyzes loss data gathered worldwide. The department prepares individual studies in addition to reports on overall loss statistics, burning costs, loss ratios, and loss trends that frequently reveal patterns of change. These data explain where, how, and why losses occur. The raw data on individual incidents are confidential to FMRC as with insurance companies and ABI in the United Kingdom.

Along with test data, the loss statistics collected by FMRC provide inputs to Factory Mutual Engineering and Research Loss Prevention Data Sheets written by experts from various engineering disciplines. The data sheets are incorporated into Loss Prevention Data books that currently have 10 volumes containing key sections on automatic sprinklers, construction, heating and mechanical equipment, electrical hazards, chemical processes, storage, extinguishing equipment, boilers, pressure vessels, and much more. The data books provide industry with a constant flow of loss control information as equipment, technology, and patterns of usage change, and as new hazards are discovered.

6.3.3 OTHER COUNTRIES

As in the United Kingdom and the United States, fire/loss protection/prevention organizations in several countries collect statistics and other data to support their publicity campaigns in making people (children and adults) aware of the dangerous effects of fires and educational programmes and training in various aspects of fire safety. These statistics are also useful in the evaluation of fire risk and the effectiveness of fire safety measures. Data on financial losses are, however, generally limited to large fires since collecting those figures for all the fires is a costly and time-consuming exercise.

Among insurance organizations in countries other than the United Kingdom and the United States, some valuable statistics are produced in Switzerland by Amelioration des Establissements Cantonaux d' Assurance Contre I' Incendie. These data provide information on fire costs as well as the extent of insurance cover for various types of risks. Some useful statistics are collected by the Insurance Organisation Technical Committee (SKAFOR) in Denmark, Insurance Companies Association in Sweden and Federation of Finnish Insurance Companies in Finland. A comprehensive database of all Scandinavian fire losses is maintained by an insurance Statistical Bureau located in Stockholm. Among organizations in other countries, the Insurance Rating Bureau in Japan collects interesting statistics that are useful for determining insurance tariff structures as well as for carrying out risk analyses.

6.4 Special databases

6.4.1 UNITED KINGDOM

In many countries, some government and private institutions collect national data on specific areas in the fire field, which are of particular interest to them in determining appropriate safety measures and maintaining safety levels required by regulations and so on. In the United Kingdom, the Railway Inspectorate (Health and Safety Executive) collects some statistics on fires in trains, railway stations, and so on, in addition to information on accidents involving passenger trains as well as trains carrying particularly dangerous goods such as LPG. Likewise, some data on fires in underground trains and stations and fires in buses and bus depots are collected by London Transport, on fires in road vehicles by the Central Government Department of Transport, on fires in mines by the Safety in Mines Research Establishment, and on fires in aircrafts and airports by the Civil Aviation Authority. For many years, the National Health Service (NHS) has been collecting statistics on fires in hospitals and other health care premises in England. In

1994, NHS Estates, an Executive Agency of the Department of Health, introduced a standard fire incidence report form. The statistics collected through this form for 1994/1995 were analyzed and published in 1996 for the first time. However, the databases mentioned above contain only some basic information on numbers and causes of fires and so on, and practically no information on items such as time factors and damage, which are needed for mathematical models used in fire risk evaluation.

A report is completed for every fire occurring in a property belonging to the government (Crown). The report has items relating to cause of fire, watching service (patrol etc.), discovery (by whom and at what time), method of extinguishment, casualties, damage, and cost of damage. Information is also recorded about the delay in summoning the local authority fire brigade, if called, and in the arrival of the brigade. The reports on fires in Crown properties are collated by a government department for determining the fire safety requirements of these properties. The analyses based on these reports are not published. Military establishments also compile reports on fires in properties belonging to them.

A major database deserving a special mention is that developed by the Institution of Chemical Engineers, which compiles yearly data on fires occurring worldwide in chemical plants, oil refineries, and premises involved in the manufacture of chemical products. These data are pooled on an international scale to aid research workers in several countries engaged in the analyses of risks associated with chemical processes. The research studies in the chemical industry have led to the development of specialized evaluation techniques such as Fault Trees (Chapter 14) and HAZOP (Chapter 17) applied in hazard analyses of processes that might lead to detonation and in producing the likelihood of certain types of nuclear accidents. The Royal Society for the Prevention of Accidents (ROSPA) maintains a database of fires and other accidents in homes, which are analyzed and published periodically.

6.4.2 UNITED STATES OF AMERICA

The US Consumer Product Safety Commission (CPSC) maintains a computerized database on fires drawn from a sample of hospital emergency room cases. The National Electronic Injury Surveillance System (NEISS), dating back to 1972, focuses on product-related injuries in the home. Information on fire casualties related to consumer products is similar to that collected for casualties under NFPA 901 but is much more detailed regarding the type of product involved. NEISS is particularly useful for analysis of serious injuries due to electrical shock or burns not caused by fire. Severe burns receiving specialized care are addressed to some extent in annual surveys by the American Burn Association. Admissions and some other factors are tabulated for patients passing through the nation's specialized burn care units.

The National Transportation Safety Board (NTSB) compiles accident reports on aircraft and railway accidents and on highway accidents involving hazardous materials. The information collected tends to emphasize the circumstances of the accident with little discussion of the ensuing fire. Some of these reports, however, have supplements addressing issues such as human factors in escape and often contain much more fire-related information. The National Highway Traffic Safety Administration (NHTSA) established a computerized file on fatal motor vehicle highway accidents, beginning in 1975. The US Coast Guard collects reports on accidents involving recreational boats and commercial vessels. As with other special databases on vehicle accidents, there is little coded standardized information on the cause and development of these fires.

The US Forest Service issues annual reports titled *Wildlife Statistics and National Forest Fire Report*, which cover fires occurring on national, state, and private forests and include information on numbers, estimates of the extent of damage (acres burned), and profiles of causes. The US Department of the Interior's Bureau of Land Management issues an annual report of Public Lands statistics that contains statistics on fires on the lands it owns and administers. The information addressed includes cause, extent of damage, rate of spread, and method of suppression. The various branches of the military have historically compiled their own databases for fires on military installations. Beginning in 1985, the Naval Safety Center in Norfolk, VA is the receiving point for fire reporting for all the US military services. Incident records from the center are also submitted to NFIRS. The International Association of Electrical Inspectors and Underwriters Laboratories maintain databases generated from clipping files, covering not only electrical fires but also electrical shocks.

6.4.3 INTERNATIONAL

In addition to national databases, international bases exist for information relating to fires and other accidents involving aircraft. Narrative summaries about these incidents appear in the World Airline Accident Summary published by the Air Registry Board in England. Standardized narrative reports extracted from original accident reports prepared by the national air safety organizations are published by the International Civil Aviation Organisation, headquartered in Canada, in its Airline Accident Digest series. The World Health Organization's Statistical Annual includes information on fire death rates by country. It can be difficult, however, to obtain data for many countries for the same year.

Lloyds Register (LR) of Shipping, with its headquarters in London, collects detailed information on all serious casualties, including total losses, to all merchant ships of 100 gross tonnage and above and on all incidents (serious and nonserious) to tankers, including combination carriers and gas carriers/tankers. This worldwide database includes only incidents in which fire and/or explosion was the first event reported (except where first event was a hull/machinery failure leading to fire/explosion). Following this definition, casualties involving fires and/or explosions after collisions, stranding, and so on are categorized under "collision" and "stranding;" scavenge fires and crankcase explosions are included in this category. LR has been collecting the casualty statistics, from its agents and surveyors in over 130 countries, since 1978 for ships and since 1975 for tankers. Some of these statistics are contained in Casualty Return, an annual publication of LR; these include information on voyage (to and fro), cargo, circumstances, and place.

6.5 Other data sources

6.5.1 MINOR DATABASES

There are also other databases on fires that are not maintained at a national or industry level. An important example in this respect relates to fires in premises belonging to leading motorcar manufacturers such as Ford and Vauxhall. Detailed reports on these fires are compiled by the industrial or works fire brigades of the manufacturers, which have the responsibility to organize the initial attack on the fires and prevent them from developing into large sizes. The local (or state) authority fire brigade (department) is called to a fire only if it cannot be put out by the works brigade. Major department stores such as Sainsbury's, Marks and Spencers, Woolworth's, and Sears compile detailed reports on fires in their premises.

Manufacturers of fire protection systems such as sprinklers, detectors, and alarms compile reports on fires in premises protected by their systems. Fire reports of Mather & Platts, for example, contain some information on fires in premises protected by their sprinkler systems. The reports, which include basic information on the type of the system, also cover small fires extinguished by the system and not attended by or reported to the local authority fire brigade. Similar statistics on sprinkler performance are collected by Grinnell Corporation.

Statistics are available for fires occurring in Switzerland in premises monitored by Cerberus automatic fire alarm systems. Cerberus carries out periodical analyses (surveys) of these fires in order to establish the effectiveness of their systems in reducing the financial loss by comparing the average loss in premises with and without their systems. The financial losses are based on insurance claims. "Fire" in the sense of these studies is an incident that has been detected by the fire alarm system and has led, or almost certainly would have led, to damage claimable from the insurance company. Cerberus studies including booklets containing detailed information on each fire incident are published and hence not confidential, although the names of the Cerberus system owners are withheld.

The Fire Extinguishing Trades Association in the United Kingdom collects statistics on the use of portable fire extinguishers. Information is obtained on the type of property where the extinguisher was used, type of extinguisher (carbon dioxide, powder, water, etc.), and whether the local authority fire brigade was called or not. These statistics include small fires put out by extinguishers and not reported to the fire brigade. The data are analyzed and published in *Fire Prevention*, the journal of the FPA.

6.5.2 RESEARCH STUDIES

Research studies carried out in some countries contain some statistical data particularly on fires of special interest. In the Fire Research Station, UK, for example, in-depth studies were carried out on the characteristics of multiple fatality fires, fires in hotels, hospitals, laundrettes, flats, retail premises, shopping complexes, and large-loss fires. Other studies of FRS included fires due to electrical causes, which were reported on a special form containing items such as apparatus primarily involved, make and age of apparatus, thermostatic controls, power taken by apparatus, fuse in circuit-feeding apparatus, allocation of fault and cause of fire (overheating, overloading, short circuit, etc.).

In the United Kingdom, before 1965, large fires were defined as those fought with five or more jets and reported on a special form (K433A). The form contained additional items of information – on approximate floor area of each story of the building involved in fire, area of floor damaged by fire on each story, time periods relating to fire development including buildup of jets, performance of automatic detectors and fire stop doors and use of water. A special survey was carried out by FRS in collaboration with the fire brigades to estimate the frequency of false alarms (as a percentage of genuine fires) in buildings equipped with automatic fire detection systems; the report form contained several items relating to the system (type of the system, type of wiring, type of connection to the brigade and suspected reason for false alarm).

The Home Office in the United Kingdom also carried out special studies using Operational Research Forms, SAF1 and SAF2, which were completed for all fires for which the regular form (K433) was completed except for certain categories of minor fires (derelict buildings, buildings under demolition etc.). Details required on these forms included those relating to time intervals between ignition and discovery and between call and "control," particulars of "construction" involved, total financial value of "constructions" and contents, time history of fire spread, extent of fire in terms of number of compartments and floors ignited, and of area damage and financial loss. The FPA has also carried out surveys of large fires and case studies of fires of particular interest.

In the United States, some special studies have produced databases or statistics that provide continuing value for fire analysts. In 1985, the CPSC completed a survey of unreported home fires and their characteristics; this database will be of use for many years. The CPSC has also conducted a number of special projects including in-depth investigation of samples of home fires involving such equipment as electrical systems or alternative heating systems. A similar in-depth study of a sample of mobile home fires was conducted in the late 1970s by the Department of Housing and Urban Development (HUD).

As of 1984, the NFPA (USA) fire investigation programme, supported by FEMA and the National Bureau of Standards (now National Institute of Standards and Technology) had generated several in-depth incident reports. No one-property use accounted for enough incidents to form a statistically significant database. However, some issues, such as those involving patterns in major fires in buildings with large life exposures, might be able to draw on this database for statistical significance. These incident reports contain detailed particulars on fire development, smoke spread, human factors in escape, and the performance of fire protection systems and features.

Some data on upholstered furniture fire losses were generated by a study carried out by the National Bureau of Standards (NBS) to evaluate three alternatives for reducing these losses. The results of this investigation were published in a report produced by Helzer *et al.* (1979). Another useful study of NBS related to the evaluation of three alternative residential fire loss reduction strategies (see Gomberg *et al.* (1982)). The two studies mentioned above provided a framework for systematically assessing the costs and losses occurring under different intervention strategies including smoke detectors, standard under consideration by the CPSC (in the first study), and residential sprinkler systems (in the second study).

Ontario Housing Corporation in Canada collects statistics on the performance of smoke detectors in fires in dwelling units owned by the Corporation. Data collected since 1975, the year of smoke detector installation, include information on place of origin of fire, cause of fire, fatalities, and property damage. An analysis of these statistics is contained in an annual report produced by the Corporation.

6.6 Ancillary statistics

6.6.1 POPULATION DATA

As mentioned in the introduction, a regular analysis of fire statistics cannot be of great value in itself in appraising various risk situations unless they are related to other relevant statistical information. Consider, for example, the regional variation in the distribution of dwelling fires and fire casualties (fatal and nonfatal) occurring in any year; there may also be year-to-year variations. These variations will be partly due to variations between regions in the population size and number of households, information about which can be obtained from census data collected and maintained by the central/federal government. Such data are available, for example, with the Office of Population Censuses and Surveys in the United Kingdom and the Bureau of Census in the United States. Summary statistics on the size of population (and working population) and number of households are given in the Annual Abstract of Statistics published by the Central Statistical Office, UK and in the Statistical Abstract published annually by the Bureau of Census, USA. Comparisons between regions of a country or between countries should be made on the basis of rates of fire incidents and fire casualties per, say, million population or households.

Composition of the population also affects the frequency of fires and fire casualties. For example, the proportion of households experiencing fires would vary according to the number of children at home. Age distribution of population is another factor. Studies in many countries have generally shown that children (below 5 years in age) and elderly people (over 65 years in age) contribute to a major proportion of fire deaths. This is apparently due to proportions of the young and elderly in the population. The occurrence of fires and fire casualties is also related, though this does not necessarily imply causality, to socioeconomic factors such as home-ownership (owner or tenant), poor and substandard housing, overcrowding, social class or household income, race, and lack of family stability. The significance of these factors has been discussed in various studies (see Section 7.8).

6.6.2 BUILDING STOCK

The frequency of fires in any risk category of buildings or probability of fire occurrence in any building (Chapter 7) depends on the total number of buildings at risk, involved and not involved in fires. The occurrence probability is also related to the size of a building in terms of, say, footprint (ground floor) area and number of floors or total floor area. The probable damage in a fire also depends on the size of a building. Some information on the number of residential buildings at risk and its distribution according to size is generally available in most countries.

It may be necessary to conduct special surveys to obtain the statistics mentioned above, for industrial and commercial buildings at risk (see, for example, Rutstein (1979)). In its Annual Abstract of Statistics, the Central Statistics Office, UK, publishes some summary statistics relating to the number of industrial establishments employing more than 10 persons. A detailed breakdown of these data according to different industrial groups, geographical areas, and employment size groups is available in Business Monitor published by the Business Statistics Office frequently, though not regularly. In the United Kingdom, some data on floor space in industrial buildings, warehouses, and shops are collected by the Department of the Environment and in distribution and other services by the Department of Trade. In the United States, similar statistics are compiled through a Census of Manufacturers and Industry Services.

The age of buildings can be used as a surrogate factor to assess the effectiveness of building (fire) regulations (or codes) that, in England and Wales for example, came into operation in 1965 replacing Model By-laws introduced in 1953. Hence, in general terms, post-1953 buildings, particularly those built after 1965, are expected to have better fire protection than those constructed before 1953. To test this hypothesis, it is necessary to assess the variation in the rate of fire incidence or fatality according to the age of a building (see Chapter 10). For this purpose, for each age group or period of construction, statistics are required for the number of all buildings of a particular type at risk in addition to a number of fires and fire deaths in buildings of that type. The age distribution of buildings at risk is likely to be available only for the stock of dwellings; special surveys may be necessary for industrial properties.

6.6.3 ECONOMIC DATA

Inflation is one of the economic factors contributing to increasing financial losses due to fires over a period of time. These losses can be corrected to some extent for the decreasing value of money with the aid of retail (consumer) price indices, which are available for most of the countries in publications such as Annual Abstract of Statistics in the United Kingdom. Such annual publications also contain estimates of gross national product (GNP), which is the total output, in monetary terms, of goods produced and services rendered in a given year. GNP is, therefore, an indicator of the economic strength of a nation. Fire loss for any year expressed as a percentage of GNP provides an inflation-free measure of the national economic effort wasted in fires (see next section).

In some countries such as Switzerland, GNP is composed of a relatively high proportion of the value of services rendered, for example, banking and tourism. Moreover, GNP is not a measure of the total amount of burnable value at risk in buildings and their contents (ex stock and stock). For the reasons mentioned above, GNP is a somewhat unsatisfactory denominator for expressing fire loss as a percentage. Gross capital stock is a realistic indicator of the total value-at-risk estimates; it may be available for some countries, for example, United Kingdom, separately for building structures and for plant and machinery that are defined as fixed assets. It is very difficult to estimate the value of consumer durables and other contents that can, perhaps, be assumed to be of the same order as the value of structures. Under this assumption and using the gross capital stock figures for structures and plant and machinery, Ramachandran (1970a) calculated, for various groups of industrial buildings, the loss in large fires per £100 of value at risk. Fixed assets valuation for any industrial building can be used as input for estimating all losses including consequential losses considered as output losses (see Ramachandran (1995)).

An estimate of the capital formation in a year is a measure of the increase in value at risk. Information on gross fixed capital formation (GFCF) is contained in National Income and Expenditure (Blue Book) published by the Central Statistical Office, United Kingdom, and in corresponding publications of other countries. Loss per GFCF may be a more realistic yardstick than loss per GNP for international comparisons (see Ramachandran (1970b) and Section 6.7.4).

6.6.4 OTHER ANCILLARY STATISTICS

A major cause for the occurrence of fire relates to faults in and misuse of appliances or equipment using energy sources (fuels) such as gas, electricity, oil, and solid fuel. The actual (intrinsic) safety of these items may not be entirely responsible for any change (increase or decrease) in fire risk over a period of years. Part of the change in fire risk may be due to changes in the amounts of energy consumed in addition to possible changes in the nature of usage of the fuels. Statistics on energy consumption compiled in most countries can provide an indication of the relative increase or decrease in fire risk that can be attributed to an increase or decrease in the amount of fuel used (see Section 7.8). Similarly, statistics on sales of consumer products such as radios, television sets, and cigarettes can be utilized to assess the increase or decrease in fire risk arising from the use of these items. Fire risk is also affected by severe weather conditions (see Section 7.8).

6.6.5 INTERNATIONAL SOURCES

The Organization for Economic Co-operation and Development (OECD) compiles national statistics relating to its member countries on items such as area, population, labor force, gross domestic product (GDP), gross fixed capital formation, percentage increase in consumer prices, energy consumption (per capita), and number of television sets (per 1000 inhabitants). These statistics are published by OECD and are available for 1963 and later years.

The statistical office of the United Nations Organisation (UNO) in New York compiles national statistical data on some of the ancillary items mentioned in this section. These items include population characteristics, consumer price indices, and gross national product. The statistics are published by UNO in Statistical Yearbook, Demographic Yearbook, Yearbook of Industrial Statistics, and National Accounts Statistics Yearbook.

6.7 Discussion

6.7.1 USE OF STATISTICS

Some of the general uses of fire statistics have been mentioned in the previous sections. These relate to various aspects of the fire problem constituting an overall picture of fire risk and their trends over a period of time. In the subsequent chapters, the uses of fire statistics specifically in a quantitative evaluation of fire risk have been discussed.

In equation [1.1], fire risk has been defined as the product of two components – fire frequency and the probability of harmful effects. Using probabilistic terminology, the first component may be defined as the probability of fire starting (Chapter 7) and the second as the probable consequences or damage in the event of a fire occurring. Fires cause damage to life (Chapter 8) in terms of fatal or nonfatal injuries or hurt and damage to property (Chapter 9). In addition to direct damage to life or property, some fires can cause indirect or consequential losses. Chapter 10 is concerned with the performance of fire prevention and protection measures. The performance of any fire safety measure can be assessed in terms of reduction (saving) in damage. In order to achieve these savings or benefits, property owners, society, and the nation at large have to incur additional expenditure (costs) toward fire safety measures (see Section 1.4). For economic justification, the probable benefits should exceed the costs.

6.7.2 LIMITATIONS IN THE USE OF NATIONAL STATISTICS

Inaccuracies in reports on fire incidents may cause some concern, but their significance is likely to decrease with increasing sizes of the samples of data that are analyzed. Quality control for a database is a difficult problem requiring considerable effort and time in trying to make sure that each incident report is as complete and accurate as possible. It is not easy to maintain a reasonable balance or trade-off between data quality and data quantity. However, an analyst needs to be aware of the completeness and limitations of data sources before conducting an analysis.

Two examples have been discussed in the Sixteenth Edition (page 2.30) of the NFPA Fire Protection Handbook (1986) to explain how differences in fire databases and assumptions can produce different results. The first example is concerned with the estimates of the US Department of Justice on the size of the nation's arson problem through its Uniform Crime Reports (UCR) based on reports from law enforcement agencies. NFPA, through its annual survey of fire departments, estimates the size of the nation's fire problem due to incendiary or suspicious causes. The UCR arson estimates and the NFPA incendiary fire estimate differ, but not significantly. This is because the UCR definition of arson is similar to the NFPA's definition of incendiary. NFPA and other fire organizations, however, traditionally regard the combination of incendiary and suspicious fires as the best indicator of the nation's problem with intentionally started fires.

The second example involves the 1981 estimates made by FEMA and NFPA of total US civilian fire fatalities. The NFPA estimate was based on a survey and the FEMA estimate on a multipart procedure that started with death certificate information reported to the National Center for Health Statistics. The FEMA estimate was just outside the upper limit of the range for the true value estimated in NFPA survey with due allowances for random (statistical) variations. Five significant differences in methods accounted for the discrepancy in the estimates obtained by NFPA and FEMA.

A similar discrepancy has been encountered in the United Kingdom in regard to the number of deaths given in the annual issues of Fire Statistics United Kingdom published by the Home Office (see page 111 of the issue relating to 1988 fires). These figures, based on fire brigade reports, differ from those in the Mortality Statistics published by the Office of Population Censuses and Surveys (OPCS) as deaths from "accidents due to fire and flames." This discrepancy is mainly due to the following reasons. Fire deaths resulting from deliberate actions appear elsewhere in the OPCS statistics. A death occurring when a preexisting disease in a person is exacerbated by the effects of fire attributed by OPCS to the disease but may be recorded in fire brigade reports as due to fire. Similarly, deaths from fires in motor vehicles are recorded by OPCS as "motor vehicle accidents" but are recorded as fatal fire casualties in fire brigade reports. Some deaths occur from fires to which no attendance is made by the fire service; these may be recorded in OPCS statistics.

The following examples relate to limitations on the usefulness of information contained in a database. A "room" as recorded in fire service reports is not necessarily a "fire compartment" with fire resistance as specified in fire regulations or codes. The figures for the number of fires that spread beyond the room of origin include fires that spread by the destruction of barrier elements (wall, ceiling, etc.) as well as those that spread by convection through a door or window left open or through some other opening. It cannot be assumed that in all the fires the spread beyond the room was due to the collapse of the barrier. Estimates for area damage given in fire service reports tend to be somewhat inaccurate. Information given in these reports is generally insufficient or even inaccurate to some extent for classifying buildings involved in fires according to fire-resistant properties of structural elements, although it may be possible to identify broad categories – high, medium, and low fire resistance. Figures for financial losses compiled by the FPA in the United Kingdom are preliminary (first) estimates from insurance loss adjusters that can differ considerably from final claim amounts settled or paid several months after the occurrence of the fires.

Another type of limitation is due to the fact that it is practically impossible to code all the items in fire incidence reports for processing and tabulation by computers. Items not included in computer discs can be extracted from original reports but this will be a tedious and time-consuming task. Also, a coding system may be such that it may not be possible to identify, uniquely or fully, fires in certain types of properties. In regard to the fire brigade data processed by the Home Office in the United Kingdom, for example, fires in the following types cannot be identified uniquely – railway stations (underground or above ground), railway tunnels, trains, and power stations. Fires in tunnels would include all outdoor tunnels, be it road, rail, or others. However, all fires starting in road vehicles in tunnels can be separately identified.

6.7.3 GAPS IN NATIONAL FIRE STATISTICS

A major gap in national fire databases is in fires not attended by or reported to a public fire department/brigade. These are generally fires controlled quickly by industrial fire brigades, sprinklers, portable fire extinguishers, and other first-aid methods, for example, buckets of water or sand. Allowances for these unreported "success stories" should be made in evaluating the effectiveness of suppression methods.

National databases, generally, do not collect information on number and sizes of rooms involved in fires and constructional features of buildings – age, condition, dimensions, material of construction of walls, ceilings, and floors. Fire brigades in the United Kingdom only provide information on these items for fires involving the structure. National bases also lack information on financial losses in fires smaller than a threshold level and financial value of buildings (and their contents) involved in fires. Most of the major data sources do not have statistics on the number of persons normally occupying fire-hit buildings; this information is required for evaluating life risk particularly in large buildings.

National fire statistics do not provide sufficient information on the extent of fire spread in a building. Among fires that have spread beyond the material first ignited, it is difficult to identify worst cases in which all the materials in the rooms of fire origin were ignited and the structural barriers of the rooms were seriously affected. It is also not possible to identify fires that spread to adjoining buildings. This information, if available, can be used to quantify the effectiveness of fire-resistant barriers in occupancies such as semidetached dwellings separated by party walls with a prescribed standard of fire resistance. Information on fire spread to adjoining buildings is available in UK fire statistics for the years 1962 to 1977 but not specifically for later years. Penetration of party walls was required to be reported in the fire report form introduced in 1978 but this information has not been coded.

There are also some gaps in national databases due to the nonavailability of other relevant statistics. A major item in this respect relates to buildings at risk, particularly in the industrial and commercial sectors. For all buildings, involved and not involved in fires, information is not available on the number and distribution of buildings according to size, age, financial value, use (production, storage, etc.), and presence or absence of fire protection devices such as sprinklers, detectors, and smoke control systems. Reliable statistics are also not available on installation and maintenance costs of fire protection devices including structural protection. Information on indirect/consequential losses is almost completely lacking apparently because it is very difficult to estimate these losses.

6.7.4 INTERNATIONAL COMPARISONS

Comparative estimates of the fire loss experience in various developed nations have been published for a number of years by the NFPA in the United States, FPA in the United Kingdom and a limited number of other organizations. Important measures of fire loss estimated are number of fires, monetary losses, and casualties. Perhaps, the most comprehensive up-to-date statistics on national fire costs currently available are those compiled by World Fire Statistics Center under the direction of Wilmot. The Center, which is sponsored by the Geneva Association for the study of Insurance Economics, produces periodical reports that also include indirect fire losses and costs of fire-fighting organizations, fire insurance administration, and fire protection to buildings. Certain adjustments are applied to figures received through a questionnaire sent to participating countries. For comparison purposes, direct and indirect fire losses and other costs are expressed as percentages of GDP and fire deaths per 100,000 persons. Wilmot (1996) has analyzed the latest figures compiled by this Center for 1991 to 1993. A summary of these figures has been produced in Table 1.1 of (Ramachandran (1998)). In most of the countries for which complete data are available, the total cost of fire exceeds three times the direct loss.

GNP (or GDP) is widely used as the basis for international fire loss comparisons, but it is not a satisfactory measure of the total amount of burnable value at risk as pointed out in Section 6.6.3. Gross Fixed Capital Stock (GFCS) or Gross Fixed Capital Formation (GFCF) appears to be a better denominator for expressing losses and costs as percentages although this measure only includes fixed assets and not consumer durables. The trend in this percentage for 1963 to 1968 is shown by (Ramachandran (1970b) and for 1984 to 1993 in Table 1.2 of (Ramachandran (1998)). In most of the countries, there was no significant increase in the percentage over the years.

Ramachandran (1970b) corrected the direct losses in different countries for inflation expressing them at 1955 prices in sterling (£) equivalents and also calculated corrected losses per fire. This analysis revealed that inflation and increasing frequency of fires were major factors contributing to the increase in fire losses over the period 1955 to 1968. The average loss per fire corrected for inflation did not register any significant increase in most of the countries studied. Per capita loss (per head of population) corrected for inflation was increasing in most of the countries and varied from country to country perhaps due to differences in living standards. Some allowance for these differences can be made by dividing the loss per head by the average hourly earnings in manufacture (see Fry (1964)).

Ramachandran (1970b) commented that figures for different countries are not strictly comparable because of differences in methods of collecting and classifying fire loss data. For instance, major databases in most countries record only those fires attended by public fire departments and exclude small fires extinguished by industrial fire brigades, sprinklers, and portable fire extinguishers – see Section 6.7.3. Fractions of fires reported to authorities can vary significantly from country to country. Some countries exclude chimney, brush, rubbish, or forest fires, while others include them. Some countries include all losses except those occurring in government properties. There are also wide differences in the values of the properties involved in fires. Methods of estimation too are likely to vary from country to country. Fluctuations in exchange rates do not help matters.

Some of the points mentioned above were considered by Rardin and Mitzner (1977) in a very detailed investigation supported by the US National Fire Prevention and Control Administration,

National Fire Data Center. The authors systematically reviewed the various hypotheses and theories that have been advanced to explain fire loss differences among nations. The additional factors enumerated by them fall into three broad categories. Firstly, there are differences between countries in human factors – economic and technological development and social and cultural patterns. Secondly, there are physical differences relating to building construction, contents and utility systems of buildings, weather, and so on. A third major class of theories centers on variations in the organization and functioning of the professional fire communities in different countries. There are other minor factors that include the severity with which fire safety codes are enforced and the influence of fire insurance in fire protection planning.

Acronyms

AAFA	Australian Assembly of Fire Authorities
ABI	Association of British Insurers
AFIRS	Australian Fire Incident Reporting System
BIA	British Insurance Association (now ABI)
CACFOA	Chief and Assistant Chief Fire Officers Association (UK)
CEA	Comite Europeen des Assurances
CPSC	Consumer Product Safety Commission
CSIRO	Commonwealth Scientific Industrial Research Organization (Australia)
FEMA	Federal Emergency Management Agency (USA)
FIDO	Fire Incident Data Organization (USA)
FINDS	Fire Information National Data Service (UK)
FMRC	Factory Mutual Research Corporation (USA)
FPA	Fire Protection Association (UK)
FRS	Fire Research Station (UK)
GDP	Gross Domestic Product
GFCF	Gross Fixed Capital Formation
GFCS	Gross Fixed Capital Stock
GNP	Gross National Product
ISO	Insurance Services Office (USA)
LPC	Loss Prevention Council (UK)
LR	Lloyds Register (UK)
NBS	National Bureau of Standards now NIST (USA)
NEISS	National Electronic Injury Surveillance System (USA)
NFIRS	National Fire Incident Reporting System (USA)
NFPA	National Fire Protection Association (USA)
NHS	National Health Service (UK)
NHTSA	National Highway Traffic Safety Administration (USA)
NIST	National Institute of Standards and Technology (USA)
NTSB	National Transportation Safety Board (USA)
OECD	Organization for Economic Co-operation and Development
OPCS	Office for Population Censuses and Surveys (UK)
ROSPA	Royal Society for the Prevention of Accidents (UK)
SKAFOR	Insurance Organisation Technical Committee (Denmark)
UCR	Uniform Crime Reports (USA)
UNO	United Nations Organization
USFA	United States Fire Administration

References

Chandler, S E (1978). *Fires in Hospitals*, Building Research Establishment, Fire Research Station, Current Paper CP 67/78.

Fry, J F (1964). The Cost of Fire. *Fire International*, **1**(5), 36–45.

Fry, J F (1969). *An Estimate of the Risk of Death by Fire when Staying in a Hotel*, Fire Research Station, Fire Research Note No 797.

Gomberg, A, Buchbinder, B and Offensend, F J (1982). *Evaluating Alternative Strategies for Reducing Residential Fire Loss – The Fire Loss Model*, National Bureau of Standards Center for Fire Research Report NBSIR 82–2551.

Hall, J R Harwood, B (1989). The National estimates approach to US fire statistics. *Fire Technology*, **25**(2), 99–113.

Helzer, S G, Buchbinder, B and Offensend, F J (1979). *Decision Analysis of Strategies for Reducing Upholstered Furniture Fire Losses*, National Bureau of Standards, Center for Fire Research. Technical Note 1101.

NFPA (1986). *Fire Protection Handbook*. 16th edition, National Fire Protection Association, Quincy, MA.

North, M A (1973). *The Estimated Fire Risk of Various Occupancies*, Fire Research Station, Fire Research Note No 989.

Ramachandran, G (1970a). *Fire Loss Indexes*. Fire Research Station, Fire Research Note No 839.

Ramachandran, G (1970b). *Fire Losses in Different Countries*, Fire Research Station, Fire Research Note No 844.

Ramachandran, G (1995). Consequential/Indirect loss. *SFPE Handbook of Fire Protection Engineering*, 2nd Edition, National Fire Protection Association, Quincy, MA, USA, Section 5, Chapter 7, pp. 63–68.

Ramachandran, G (1998). *The Economics of Fire Protection*, E. & F. N. Spon, London.

Rardin, R L and Mitzner, M (1977). *Determinants of International Differences in Reported Fire Loss*, Georgia Institute of Technology, Sponsored by National Fire Prevention and Control Administration, National Fire Data Center.

Rutstein, R (1979). The estimation of the fire hazard in different occupancies. *Fire Surveyor*, **8**, 21–25.

Sekizawa, A (1991).Statistical analyses on fatalities characteristics of residential fires. *Fire Safety Science – Proceedings of the Third International Symposium*, G Cox and B Langford (Eds), Elsevier Science Publishing Co, New York, pp. 475–484.

Wallace, J (1948). The fire census and technical field intelligence work of the Joint Fire Research Organisation. *Journal of the Royal Statistical Society, Series A (General)*, **CX1**, Part 11.

Wilmot, R T D (1996). *United Nations Fire Statistics Study*, World Fire Statistics Centre Bulletin No 12, The Geneva Association.

7 OCCURRENCE AND GROWTH OF FIRE

7.1 Introduction

Following the definitions adopted for Fire Statistics United Kingdom, fire locations can be grouped under two main headings: fires in occupied buildings and outdoor fires. Fires in derelict buildings may be included with outdoor fires with which they have more common characteristics than with fires in occupied buildings. An occupied building is one that is in use (not derelict); it need not necessarily have any people in it at the time of the fire. Buildings under construction are regarded as occupied and those under demolition as derelict.

Outdoor fires are of two types: secondary and nonsecondary. The first type of fires are those involving only single derelict buildings, single buildings under demolition or such outdoor locations as grassland, railway embankments, refuse or derelict vehicles, except such fires that involve casualties, rescues or escapes, spread beyond the location of origin, or attended by five or more appliances, all of which arrived at the fire ground and were used in fighting the fire. The second type of fires mainly includes those occurring in outdoor storage, outdoor machinery and equipment, road vehicles, caravans, ships and boats, and railway rolling stock.

A third category comprises fires confined to chimneys, which form a distinct type of fire, large in number but causing very little damage. It would be misleading to include chimney fires under either of the two main headings.

Most of the losses to life and property occur in occupied buildings to which the methods of fire-risk evaluation discussed in this book are particularly applicable. For this category of buildings, causes or sources of ignition can be classified into two broad groups according to their nature – human and nonhuman. The first group mainly consists of causes such as children playing with fire, for example, matches, arson (malicious or intentional ignition), and careless disposal of matches and smokers' materials. The second group includes electricity, gas, and other fuel sources, which may be further subdivided according to cooking, space heating, central heating, and other appliances; it also includes causes such as mechanical heat or sparks in industrial buildings, spontaneous combustion, and natural occurrences, for example, lightning.

Most of the fires start at a single point, although some fires such as those caused by arsonists can have multiple points of ignition. A fire in any room of a building usually starts with the ignition of one of the objects. The spread of fire within a room depends on the burning characteristics

Evaluation of Fire Safety D. Rasbash, G. Ramachandran, B. Kandola, J. Watts and M. Law
 ISBN: 0-471-49382-1

of the objects, particularly the rate of heat release, and the quantity of combustible materials. The materials used to cover the floor and line the walls and ceiling may promote fire spread. The spread of fire beyond a room can be restricted by limiting the size of a room and providing its structural boundaries with adequate fire resistance, thereby making it a fire compartment. An appropriate level of fire resistance can be determined by using engineering formulas that express the maximum potential severity expected in a fire as a function of fire load density, dimensions of the compartment, thermal inertia of compartment boundaries, and area of ventilation. Fire-resistant compartmentation of a building has long been the core of fire safety measures.

The spread of fire in a room or compartment is also influenced by the arrangement of objects contributing to the total fire load. Distances between objects and hence the degree of overcrowding is an important factor affecting the chance of flashover or full development of a fire, which would lead to the attainment of maximum severity. The physical and chemical processes evolved by various burning materials have multiple interactions at different times, which are also affected by imponderables such as wind velocity and direction, humidity, and temperature prevalent in a room at the time of occurrence of a fire. Because of the uncertainties caused by the factors mentioned above, the spread of fire in a particular building is a stochastic (not deterministic) phenomenon involving probabilities as discussed in Chapter 15. There is also a probability attached to the occurrence of an accidental (not arson) fire. Also, as defined in the first chapter, fires in the context of this book are those that spread beyond the points of ignition.

7.2 Probabilistic approach

For any group or type of buildings with similar fire risks, the probabilistic approach discussed in this chapter would provide overall or global values for quantitative measures of fire risk, which would be sufficient for most practical purposes. In this approach, the probable damage during a period, say, a year is expressed as the product of

(a) probability (F) of fire starting during a year,

(b) probable damage to life and property in the event of a fire occurring.

Life loss can be expressed in terms of number of fatal and nonfatal casualties (Chapter 8) and property damage in terms of floor area damaged (D), extent of spatial spread or financial loss (L). Some aspects of probable property damage are considered in this chapter, followed by a detailed discussion in Chapter 9.

Fire prevention measures (publicity campaigns, fire safety education, safety audits, etc.) can reduce the first component (F) of fire risk, while the second component can be reduced by adopting fire protection measures such as sprinklers, automatic fire alarms and detectors, structural fire resistance, smoke control systems, and means of escape facilities. The adverse effect of some "residual" risk, which is unavoidable, can be mitigated by fire insurance. It is practically impossible to eliminate fire risk completely, but the risk can be reduced to a small level acceptable to a property owner and the society at large. There is no such thing as absolute safety.

7.3 Probability of fire starting

In fire safety codes, the main objective of provisions against external fire spread is to ensure that the possibility of a conflagration due to external fire exposure hazard is reduced and fire spread from one building to another is prevented. The building where a fire starts is termed an *exposing building* while a nearby building to which a fire spreads is the *exposed building*. There is a historical basis for classifying fire risk according to these two broad categories, since in many countries there have been examples of serious fire incidents leading to conflagrations destroying several buildings. In such incidents, fires were able to spread horizontally with comparative ease from one building to another. External fire spread can occur across the space separating two neighboring buildings if the distance between the buildings is less than a critical distance depending upon the size, use, and external walls of the buildings.

According to UK fire statistics, the frequency of fires spreading beyond buildings is small (about 1%) for all fires starting in occupied buildings. This frequency varies from one type of building to another. These statistics on fire spread indicate to some extent the success of the current technical provisions for this purpose as well as the successful intervention by the brigade. Loss of life due to an externally spreading fire is rare. For the reasons mentioned above, the risk evaluation methods discussed in this book generally relate to an "exposing building."

The probability of fire starting (in an "exposing" building) depends on the nature and number of ignition sources present, which would vary from one type of building to another. Even within a building the nature and number of ignition sources would vary from one part of a building to another. In industrial buildings, for example, production area, storage area, and other areas would experience different frequencies of fires due to different sources of ignition and their number. In occupancies such as shops and department stores, the places where people assemble would differ from storage and other areas in regard to fire frequency. In dwellings, kitchen, living room, dining room, bedroom, and bathroom are broad categories of places of fire origin.

The main causes of ignition in different parts of a building can be identified using a two-way table such as Table 7.1, as suggested by Ramachandran (1979/80). For this example, "mechanical heat or sparks" are the main sources in the production and maintenance area, followed by "industrial appliances." In the storage areas, smokers' materials rank first followed by children playing with fire and malicious ignition (including doubtful cases). These are all human sources. Fires from human sources also predominate in miscellaneous areas other than storage.

The probability of fire starting due to a particular ignition source in a particular area of a building depends on the number of such sources present in that area and the duration for which the building is exposed to that risk. For example, the probability of fire due to mechanical heat or sparks is related to the number of machines in the production area and the duration for which the machines are operated. Likewise, the probability of fire due to an electrical appliance depends on the number of such appliances in any area and the duration for which they are used. Taking a human source as the third example, the probability of fire due to careless disposal of smoking materials depends on the number of smokers and the number of cigarettes, cigars, and so on, smoked.

In order to estimate the absolute value of the probability of fire starting due to a particular source, one has to relate the number of fires due to this source to the total number of such sources in the population of buildings and the duration for which the buildings are exposed to this risk. For this purpose, it is necessary to carry out a survey of buildings of the type considered, which is a costly and a time-consuming exercise. An approximate value of the probability may be estimated by adopting an indirect method discussed in the following paragraphs.

Statistical studies reviewed by Ramachandran (1970, 1979/80, 1988) show that the probability of fire starting in a building is given by

$$F(A) = KA^{\alpha} \qquad [7.1]$$

where A is the total floor area of the building and K and α are constants for a particular type of building. $F(A)$ is usually expressed on an annual basis.

The parameter K in equation [7.1] includes the ratio n/N where n is the number of fires in the risk category considered and N is the number of buildings at risk in this category. The parameter

Table 7.1. Spinning and doubling industry – places of origin of fires and sources of ignition

Sources of ignition	Production and maintenance		Assembly	Storage areas			Miscellaneous areas	Total
	Dust extractor (not cyclone)	Other areas		Store/stock room	Loading bay packing dept	Other areas		
A Industrial appliances								
(i) Dust extractor Electrical	14	3	–	–	–	–	–	17
Other fuels	12	–	–	–	–	–	–	12
(ii) Other appliances Electrical	6	111	–	–	–	–	–	117
Other fuels	–	22	–	1	–	–	2	25
B Welding and cutting equipment	–	10	–	6	–	–	7	23
C Motor (not part of other appliances)	–	7	–	–	–	–	–	7
D Wire and cable	1	12	–	–	–	–	2	15
E Mechanical heat or sparks Electrical	27	194	–	–	–	–	–	221
Others	52	387	–	2	–	–	–	441
F Malicious or intentional ignition	–	9	–	3	–	–	3	15
Doubtful	–	13	–	7	–	–	–	20
G Smoking materials	2	29	1	15	1	–	7	55
H Children with fire e.g. matches	3	4	–	12	2	4	5	30
J Others	4	29	2	3	2	–	12	52
K Unknown	11	78	–	14	–		9	112
Total	132	908	3	63	5	4	47	1162

α accounts for the increase in the value of the probability for an increase in building size. A value of unity for α would indicate that the probability of fire starting is directly proportional to the size of the building; this would also imply that all the parts of a building have the same risk of fire breaking out. This is not true since different parts have different types and number of ignition sources. Also, ignition sources are mostly associated with the walls of a building where electrical and heating appliances are usually present and the ratio of wall length to surface area would be expected to decrease as the area increases. For the reasons mentioned above, the probability of fire starting is not likely to increase in direct proportion to building size so that α would be less than unity. If two buildings are considered, one twice the size of the other, the probability for the larger building will be less than two times the probability for the smaller building. These theoretical arguments are confirmed by actuarial studies on frequency of insurance claims as a function of the financial value (size) of the risk insured (see, Ramachandran (1970), Benktander (1973)).

Rutstein (1979) has estimated the values of K and α for major groups of buildings in the United Kingdom (see Table 7.2). These are based on correlation between the frequency distribution of buildings involved in fires according to their size (total floor area) and the corresponding distribution of buildings at risk (involved and not involved in fires). Fire statistics provided the former distribution while a special sample survey was carried out for obtaining the latter distribution. According to Rutstein, with A in square meters, $K = 0.0017$ and $\alpha = 0.53$ for all manufacturing industries in the United Kingdom. Actuarial studies (Benktander, 1973) in some European countries confirm that the value of α is about 0.5 for industrial buildings.

It may be possible to obtain the distribution of buildings at risk according to size, by using some other statistics on the sizes of firms (business concerns) (see Ramachandran (1979/80)). In

Table 7.2. The fire risk in different occupancies – parameters of equations

Occupancy	Probability of fire per year[a]		Average damage in a fire (m^2)[b]	
	K	α	C	β
Industrial buildings				
Food, drink, and tobacco	0.0011	0.60	2.7	0.45
Chemical and allied	0.0069	0.46	11.8	0.12
Mechanical engineering and other metal goods	0.00086	0.56	1.5	0.43
Electrical engineering	0.0061	0.59	18.5	0.17
Vehicles	0.00012	0.86	0.80	0.58
Textiles	0.0075	0.35	2.6	0.39
Timber, furniture	0.00037	0.77	24.2	0.21
Paper, printing, and publishing	0.000069	0.91	6.7	0.36
Other manufacturing	0.0084	0.41	8.7	0.38
All manufacturing industry	0.0017	0.53	2.25	0.45
Other occupancies				
Storage	0.00067	0.5	3.5	0.52
Shops	0.000066	1.0	0.95	0.50
Offices	0.000059	0.9	15.0	0.00
Hotels etc.	0.00008	1.0	5.4	0.22
Hospitals	0.0007	0.75	5.0	0.00
Schools	0.0002	0.75	2.8	0.37

[a]Equation [7.1].
[b]Equation [7.4].

the United Kingdom, for example, the distribution of manufacturing units by employment size for different industries is given in Business Monitor, published periodically by the Business Statistics Office. A unit can have more than one building. An estimate may be available for average gross floor area per person (see Ramachandran (1970)).

Equation [7.1] gives the probability due to any cause and is the sum of probabilities of a fire starting in different parts of a building due to various causes. What is required is the probability due to a particular cause in a particular part. An indirect estimate of this probability is given by the product of the value given by equation [7.1] and the conditional probability that the fire is due to the particular cause and part given that the building is involved in fire. The conditional probabilities reflect the relative or comparative risks due to causes and parts and can be estimated from group statistics such as those in Table 7.1. The conditional probability due to, say, smoking materials in the store/stock room is 0.0129 (=15/1162). For a textile industry building of total floor area 2500 m^2, with $K = 0.0075$ and $\alpha = 0.35$, equation [7.1] gives a value of 0.116 for the overall probability of a fire starting during a year. Then, for this factory, an estimate of annual probability due to smoking materials in the store/stock room is 0.0015 (=0.0129×0.116). As explained below, this procedure enables the reevaluation of the probability of fire starting in the light of fire prevention methods adopted for any particular building.

For a particular building in any type or risk category, an estimate of the conditional probability (given fire) for the ith cause in the jth part of the building is given by

$$I_{ij} P_{ij} \qquad [7.2]$$

where P_{ij} is the probability for this cause, and the part is revealed by figures in a table such as Table 7.1. The parameter I_{ij} will be assigned the value zero if the ith cause is totally absent in the jth part of the building considered for risk evaluation. If the cause is present, I_{ij} should be given a positive value depending on the extent to which this cause can be responsible for starting a fire in the jth part; this value can be greater than unity. A value equal to unity can be assigned if the building is similar to the "average building" in this respect. Ramachandran (1979/80, 1988) has illustrated the application of this method with the aid of an example. As mentioned by him, the assignment of a value to the parameter I_{ij} has to be somewhat subjective with its accuracy depending on the extent and accuracy of relevant information used in the calculations.

Each possible cause or source of ignition in each part of the building considered should be identified and its I_{ij} value should be estimated. The aggregate probability of fire starting for the building is then

$$F(A) \sum_i \sum_j I_{ij} P_{ij} \qquad [7.3]$$

where $F(A)$ is the "global" value given by equation [7.1]. The value given by the part excluding $F(A)$ in equation [7.3] can be greater or less than unity. It will be equal to unity only if the building considered is identical to the average characteristics of the underlying population in regard to causes or ignition sources. The aggregate probability [equation 7.3] can be greater or less than $F(A)$. This allocation approach has also been used in fire-risk assessments of nuclear power plants (see Apostolakis (1982)).

7.4 Probable damage in a fire

The probable area damage is given by

$$D(A) = CA^{\beta} \qquad [7.4]$$

where, as in equation [7.1], A is the total floor area (size) of a building, and C and β are constants for a particular risk category or type of building. A fire in a large building is more likely than one in a small building to be discovered and extinguished before involving the whole building. The proportion destroyed in a large building would, therefore, be expected to be smaller than the proportion destroyed in a small building. These arguments suggest that the damage rate $[D(A)/A]$ would decrease with increasing values of A; in other words the value of β would be less than unity. This result is supported by actuarial studies (Benktander, 1973) and statistical investigations (Ramachandran, 1970, 1979/80, 1988). On the basis of a special survey, Rutstein (1979) has estimated the values of C and β for major groups of buildings in the United Kingdom (see Table 7.2). These figures relate to buildings with the minimum level of fire protection (without sprinklers). The product of equations [7.1] and [7.4] is an estimate of fire risk.

Provision of fire-fighting equipment in a building would reduce the damage rate and the value of β. Considering sprinklers, for example, with A in square meters and $C = 2.25$, Rutstein estimated a value of 0.45 for β for industrial buildings without sprinklers. He estimated an average damage of $D(A)$ equal to 16 m^2 for an industrial building of total floor area of 1500 m^2 equipped with sprinklers. These figures inserted in equation [7.4] would yield a value of $\beta = 0.27$ for an industrial building with sprinklers. In deriving this result, Ramachandran (1988) assumed that the parameter C associated with initial conditions will have the same value of 2.25 obtained for industrial buildings without sprinklers.

In another study, Ramachandran (1990) used the data for the textile industry (Section 7.5) to show that, with $C = 4.43$, β is about 0.42 for a building without sprinklers and 0.22 for a sprinklered building for fires in which the heat produced is sufficient to activate the system. These estimates based on average damage in a "reference building" of 8000 m^2 gave unrealistic (very high) values for A_s, the size of a sprinklered building equivalent in damage to the size A_u of a nonsprinklered building. Ramachandran, therefore, considered figures for spread beyond room (maximum damage) instead of overall damage and estimated that β has the value of 0.68 for buildings without sprinklers and 0.60 for buildings with sprinklers. The relationship between damage and building size is depicted in Figure 7.1. The figure is applicable to buildings larger than 105 m^2. In this case, if sprinklers operate in a fire, $A_s = 33{,}000$ m^2 would be equivalent in damage to $A_u = 10{,}000$ m^2. If a reliability investigation suggests a probability of 0.1 for the nonoperation of sprinklers, calculations would show that a sprinklered building of 28,000 m^2 would correspond to a nonsprinklered building of 10,000 m^2. Looking at Figure 7.1 from a different angle, the damage expected in a building of 10,000 m^2 would be 1200 m^2 if sprinklered compared to 2300 m^2 if not sprinklered. Such results can be used for determining rebates in fire insurance premium for buildings equipped with sprinklers.

Ramachandran (1990) applied the "power" relationship in equation [7.4] for estimating the damage likely to occur within a room as a function of the room size. In this case, he estimated that β has the value of 0.57 for a nonsprinklered room and 0.42 for a room with sprinklers. These results are based on a "reference room" of size 800 m^2 and $C = 4.43$. Figure 7.2 shows the relationship between the size of a room and damage expected within such an enclosure in the event of a fire. The figure is applicable to rooms larger than 32 m^2. According to this figure, if sprinklers operate, a sprinklered room of 4000 m^2 would be equivalent in damage to a nonsprinklered room of 500 m^2. In this case, the size of a sprinklered room will reduce to 3000 m^2 if a value of 0.1 is assigned to the probability of nonoperation of sprinklers.

Results such as those in Figure 7.2 would provide a statistical justification for increasing the maximum compartment size permitted in fire safety codes when buildings are equipped with sprinklers. The results can also be used for determining the maximum size of a basic (nonsprinklered) compartment according to an acceptable level of maximum property damage. In the figures quoted above, a maximum damage of 153 m^2 has been regarded as acceptable. Life

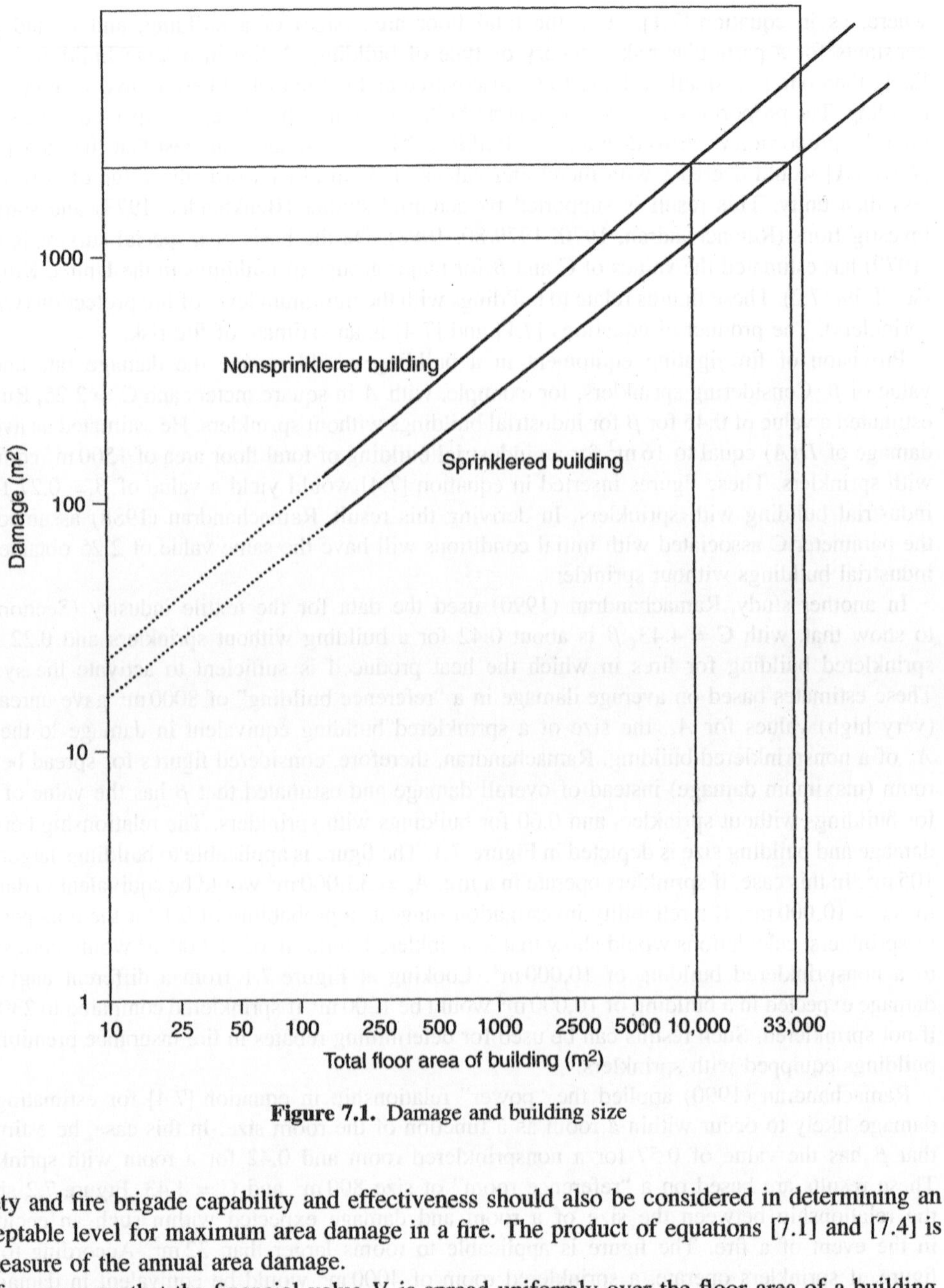

Figure 7.1. Damage and building size

safety and fire brigade capability and effectiveness should also be considered in determining an acceptable level for maximum area damage in a fire. The product of equations [7.1] and [7.4] is a measure of the annual area damage.

If we assume that the financial value V is spread uniformly over the floor area of a building, from equation [7.4], the loss $D(V)$ in financial terms expected in a fire is given by

$$D(V) = C'V^{\beta} \tag{7.5}$$

where

$$v = V/A$$

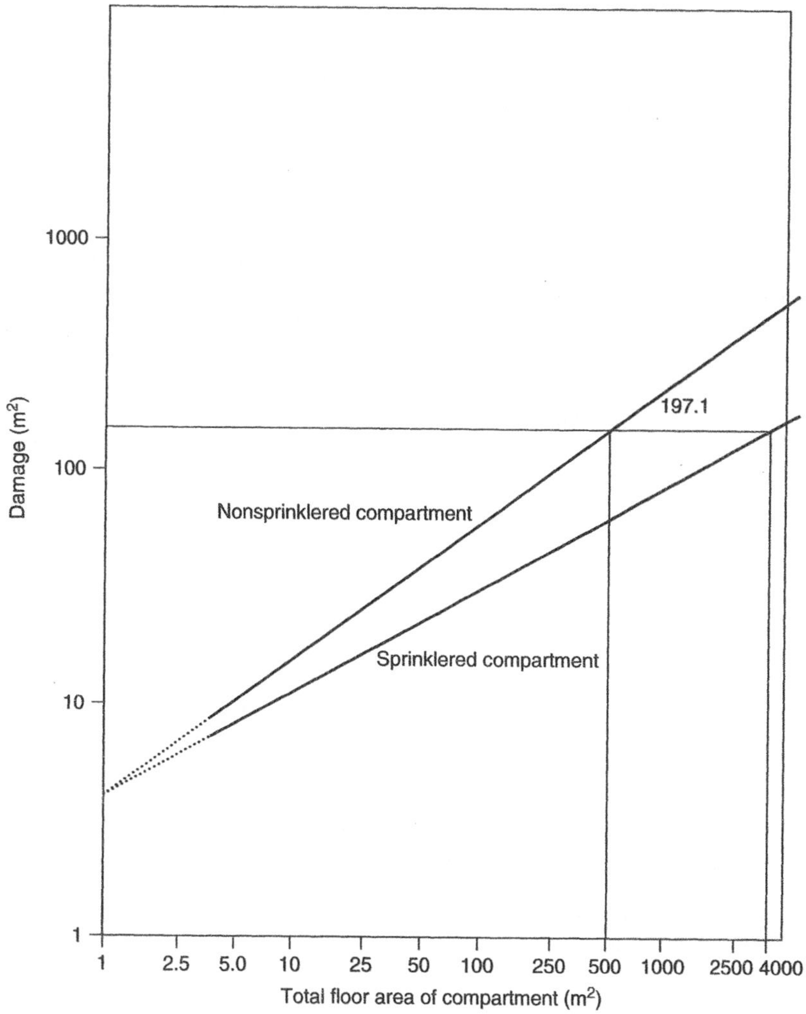

Figure 7.2. Damage and compartment size

is the value density per, say sq. meter of floor area and

$$C' = C.\upsilon^{-\beta}$$

Likewise, equation [7.1] may be transformed to

$$F(V) = K'V^{\alpha} \qquad [7.6]$$

where

$$K' = K\upsilon^{-\alpha}$$

and υ is the value density as defined earlier. $F(V)$ is the probability of fire starting during a year in a building with value V at risk. Equations [7.5] and [7.6] and their product are used in actuarial problems for determining approximately risk premiums for fire insurance (see Benktander (1973)).

Expected area damage in a fire can also be converted to financial loss by using in equation [7.4] an approximate value for loss per square meter of fire damage. At 1978 prices, this rate of loss was £140 for all manufacturing industries in the United Kingdom (see Rutstein (1979)). A better estimate of the expected value of $D(A)$ or $D(V)$ can be obtained through an appropriate probability distribution (Chapter 9).

7.5 Extent of spread

The probable area damaged in a fire can also be estimated by considering different categories of fire spread and the probabilities associated with these cases. Fire statistics produced in the United Kingdom enable the extent of spread to be classified as follows:

1. Confined to item first ignited.
2. Spread beyond item but confined to room of origin:
   (i) contents only
   (ii) structure involved.
3. Spread beyond room but confined to floor of fire origin.
4. Spread beyond floor but confined to the building of fire origin.
5. Spread beyond building of fire origin.

A fire starting in a room can spread upward to the next floor without involving the entire floor of origin. It is not possible to estimate the number of such cases. Hence, in the example shown in Table 7.3, the third and fourth categories have been combined to denote the event of fire spreading beyond the room of origin but confined to the building of origin. Fires spreading beyond the building of origin have also not been included in this example. For each category of spread, the damage shown in the table is the average for the category with the relative frequency indicated by the percentage figure. In the case of a sprinklered building, the percentage figures include one-third of fires in these buildings, which were estimated to be extinguished by the system but not reported to the local authority fire brigades (Rogers, 1977). These small fires were assumed to be confined to the item first ignited.

Table 7.3. Textile industry, UK. Extent of fire spread and average area damaged

Extent of spread	Sprinklered[a]			Nonsprinklered		
	Average area damaged (m^2)	Percentage of fires[b]	Time (min)	Average area damaged (m^2)	Percentage of fires	Time (min)
Confined to item first ignited	4.43	72	0	4.43	49	0
Spread beyond item but confined to room of fire origin						
(i) Contents only	11.82	19	8.4	15.04	23	6.2
(ii) Structure involved	75.07	7	24.2	197.41	21	19.4
Spread beyond room,	1000.00	2		2000.00	7	
Average	30.69	100		187.08	100	

[a]System operated.
[b]Includes fires not reported to the fire brigade.

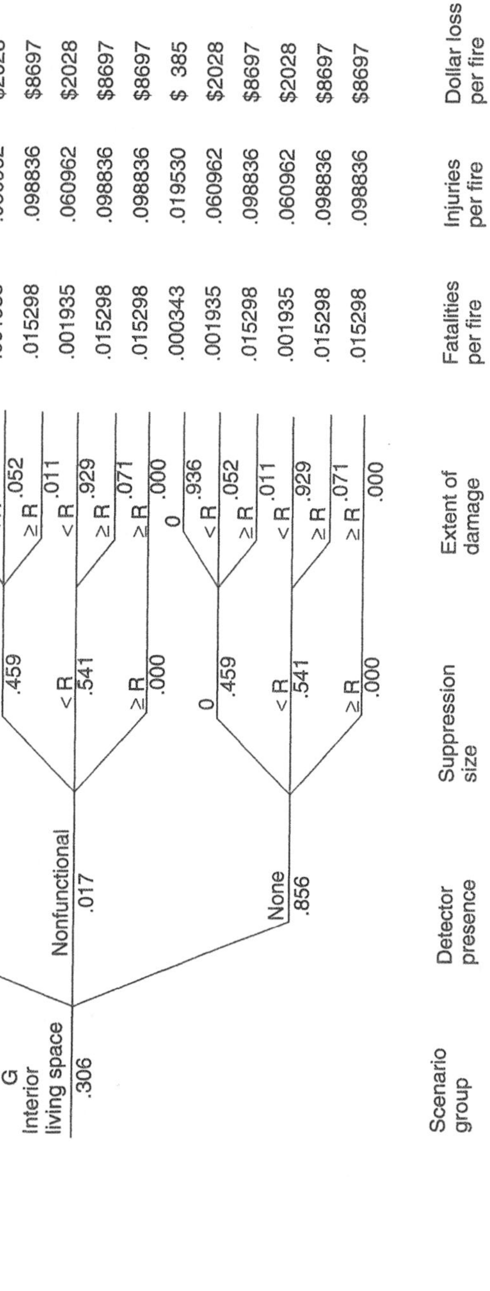

Figure 7.3. Sprinkler alternative probability tree

Ignition of an item in a room does not depend on the presence or absence of sprinklers. Hence, an overall average of 4.43 m^2 has been used for area damaged when a fire is confined to the item first ignited. Sprinklers reduce the area damaged in other categories of spread and the reduction is considerable when fire spreads beyond the room of origin; it is assumed that the entire building is provided with sprinklers. The percentage figures in Table 7.3 also indicate that, if they operate, sprinklers would increase the probability of a fire being confined to the item first ignited and reduce the probability of spreading beyond the room of origin. There is a probability of about 0.6 that sprinklers may not operate, mainly due to the fact that the heat generated may be insufficient for activating the sprinkler heads. Mechanical defect and the system having been turned off are reasons for the nonoperation of sprinklers in some cases. The results mentioned above are applicable for a "typical" textile industry building with rooms having an average size of 800 m^2.

Fire statistics collected by the US Fire Administration provide figures for probabilities and dollar losses for different categories of fire spread. In a study concerned with residential fire loss, Gomberg *et al.* (1982) estimated dollar losses for different spread categories, which were the same for both sprinklered and nonsprinklered buildings. This assumption appears to be somewhat unrealistic since, as revealed by Table 7.3, sprinklers would reduce the loss expected for each category of spread beyond the item first ignited. Gomberg *et al.* differentiated the probabilities of extinction to reflect the effectiveness of sprinklers. Their study also included the effectiveness of smoke detectors and life loss (fatalities and injuries).

Gomberg *et al.* (1982) used Probability Trees to assess the final extent of flame spread and the consequences in terms of dollar loss and life loss. Three possible levels of spread were considered – confined to the object of origin (O), spread beyond this object but confined to part of the room of origin (<R), and spread beyond room (>R). Figure 7.3 is an example reproduced from this study. The "suppression size" in this figure denotes the fire size at the start of a suppression activity. As with UK fire statistics, the US database does not provide probabilities for suppression size since only the final size after a fire was extinguished is recorded in fire reports. Hence, expert judgment was used for assessing the suppression size. The Probability Tree provided by the figures in Table 7.3 is discussed in Chapter 15 with reference to a stochastic model of fire spread.

7.6 Fire growth rate

A central parameter in the design of buildings and provision of fire protection measures is the rate at which a fire grows in the room of origin and subsequently spreads to other parts of a building. This rate depends primarily on the heat output from the materials ignited apart from other factors such as room dimensions and ventilation. Deterministic models (Chapter 11) and associated computer packages have been developed to estimate this rate and have been validated in the light of experimental data. However, experimental values for heat output are available only for a limited number of material assemblies. Therefore, it is difficult to use experimental data to estimate the growth rate for the development of a fire in a room or compartment containing several materials or objects arranged in a certain manner. Moreover, the performance of a material assembly in a real fire is likely to be different from its performance under experimental (controlled) conditions.

There is, therefore, a need to adopt a statistical approach for determining the growth rate for the spread of fire. In this approach, the (deterministic) growth of fire over a period of time is described by an exponential model (Ramachandran, 1986), according to which area damaged in T minutes is given by

$$A(T) = A(O)\exp(\theta T) \tag{7.7}$$

where

$$A(O) = \text{ area initially ignited}$$

$$\theta = \text{ fire growth parameter}$$

Equation [7.7] follows the suggestion of Thomas (1974) and Friedman (1978), according to whom the heat output from a fire increases exponentially with time. Area damaged in a fire is approximately proportional to heat output. Experimental results support the exponential model (see Labes (1966)), for example. In some cases, fire size may increase according to a square (parabolic) or some other power of T (see Friedman (1978) and Heskestad (1982)). According to Butcher (1987), there is very little difference between exponential and parabolic fire growth curves. It should be emphasized that $A(T)$ in equation [7.7] is the final (cumulative) size of a fire in terms of area damaged at the time (T) of its extinguishment. $A(T)$ is not the fire size at intermediate time T. Fire statistics do not and cannot provide information on the size of a fire at any specific time, say, when the fire brigade arrives at the scene of a fire.

Conceptually, $A(T) = 0$ for $T = 0$, but this condition is not satisfied by equation [7.7]. However, modifying this equation to force or bend the exponential curve to pass through the origin does not appear to be a sound engineering practice. Moreover, as pointed out by Butcher (1987), the initial stage of a fire, although small in size, can be very variable in length of time; it can last for hours (smoldering) or it can be over in minutes. Hence, the end of the first (early) stage may be taken as "zero time" and equation [7.7] adopted, which has been found to be reasonable for all practical purposes. If a fire survives the first ("infant mortality") stage, "established burning" would occur and the fire would grow steadily with heat output and area (or volume) destroyed increasing exponentially with time.

Fire statistics available in the United Kingdom provide, for each fire, information on $A(T)$ and the duration of burning T, as the sum of the following four periods:

T_1 – ignition to detection or discovery of fire
T_2 – detection to calling of fire brigade
T_3 – call to arrival of the brigade at the scene of the fire
T_4 – arrival to the time when the fire was brought under control by the brigade.

An estimate of T_1 is given according to the following classification:

(i) discovered at ignition ($T_1 = 0$)
(ii) discovered under 5 min after ignition
(iii) discovered between 5 and 30 min after ignition
(iv) discovered more than 30 min after ignition.

For estimating the total duration T, average values of 2, 17, and 45 min can be adopted for the second, third, and fourth classes of T_1 mentioned above. The growth of fire will be practically negligible during the fifth period of T from control to extinction of a fire.

Following the method described above, Ramachandran (1988) estimated the parameters of the exponential fire growth model for the raw data used in the preparation of Table 7.3. For fire spread beyond the initial stage (item) taken as zero time and the commencement of established burning, the overall growth rate θ was found to be 0.083 if not sprinklered and 0.031 if sprinklered. These values, estimated by the regression based on equation [7.7] were averages for fire spread throughout a building with a maximum duration of 250 min.

Apart from materials ignited, the rate of fire growth would also be affected by the structural barriers of a room and their fire resistance. Hence, the rate for fire spread within a room would be different from the overall rate for a building. For the data considered, fire growth within a room is described in Figure 7.4 for which average times for the second and third stages, since established burning, are as given in Table 7.3. Figure 7.4 is based on the following values in equation [7.7].

$$A(O) = 4.43\,\mathrm{m}^2$$

$$\theta = 0.117 \text{ for a sprinklered room}$$

$$\theta = 0.196 \text{ for a nonsprinklered room.}$$

The results for the example discussed above indicate that the fire growth rate is reduced by sprinklers and structural fire resistance. Early detection of a fire through, say, automatic fire detection systems will reduce T_1, which in turn will reduce the control time T_4. This is due to the fact that a fire detected soon after ignition will be in its early stage of growth when fire brigade arrives and hence can be controlled quickly. Consequently, the total duration of burning, T, and the damage $A(T)$ will be reduced considerably. This problem is discussed in Chapter 10 with reference to the economic value of fire detectors. The damage expected in fires can also be minimized by optimizing the siting of fire stations which will reduce the average of attendance time T_3 and the fire-fighting strategies, which will reduce the average of control time T_4 (see Chapter 10).

The exponential model in equation [7.7] also provides an estimate of "doubling time"

$$d = (1/\theta)\log_e 2$$

$$= (1/\theta)0.6931 \qquad [7.8]$$

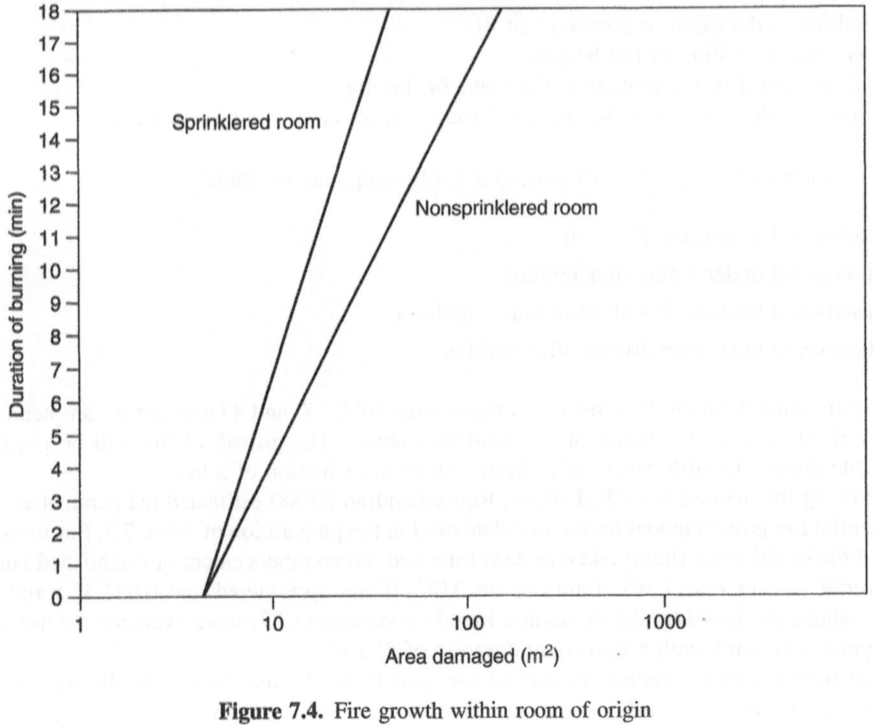

Figure 7.4. Fire growth within room of origin

which is the parameter generally used for characterizing and comparing rates of fire growth of different materials or objects. This is the time taken by a fire to double in size and is a constant for the exponential model. For example, if it takes 5 min for the area damaged to increase from 20 to 40 m^2, it will also take only 5 min for the damage to increase from 30 to 60 m^2, 40 to 80 m^2, 50 to 100 m^2, 80 to 160 m^2, and so on.

For the example in Table 7.3 and Figure 7.4, the doubling times are 5.9 min for a sprinklered room and 3.5 min for a room without sprinklers. With appropriate assumptions about the ratio of vertical rate of fire spread to horizontal rate, doubling times (and growth rates) as discussed above in terms of area damage (horizontal spread) can be converted to doubling times in terms of volume destroyed (see Ramachandran (1986)). As one might expect, doubling time in terms of volume involved is shorter than doubling time in terms of area. For some data quoted by Thomas (1981), the doubling time, apparently in terms of volume destroyed, ranged from 1.4 min to 13.9 min. At the Factory Mutual Research Laboratories (Friedman, 1978), the growth rates of a series of spreading fires involving various materials indicated doubling times ranging from 21 s to 4 min.

A regression analysis based on equation [7.7] would provide an estimate of the "expected" (average) value, $\overline{\theta}$, of the growth rate whose standard deviation according to statistical theory is

$$\overline{\sigma}_\theta = \sigma/\sqrt{n}.\sigma_T \qquad [7.9]$$

where σ is the "residual" standard error, n the number of observations (fires) used in the analysis and σ_T the standard deviation of T. With the aid of $\overline{\sigma}_\theta$ assuming a normal distribution, confidence limits can be obtained to denote the range within which the real value of $\overline{\theta}$ would fall. For example, if the lower and upper confidence limits are

$$\overline{\theta} - 1.96\overline{\sigma}_\theta, \qquad \overline{\theta} + 1.96\overline{\sigma}_\theta$$

the probability of average growth rate falling short of the lower or exceeding the upper is 0.025. Such confidence limits were obtained by Ramachandran (1986) for some average growth rates of the materials first ignited in certain industrial buildings, shown separately for three areas of fire origin – production, storage, and other areas.

Each fire provides a value of the growth rate θ whose average is $\overline{\theta}$. The standard deviation of such individual values of θ is given by

$$\sigma_\theta = \sigma/(\sigma_T^2 + \overline{T}^2)^{\frac{1}{2}} \qquad [7.10]$$

where the new term $\overline{T}$ is the average of the variable T. On the basis of σ_θ, the maximum of individual values of θ can be estimated according to any desired probability level. For example, the probability of an individual growth rate falling short of $(\overline{\theta} - 1.96\,\sigma_\theta)$ or exceeding $(\overline{\theta} + 1.96\,\sigma_\theta)$ is 0.025. The distinction between these limits of θ and those of $\overline{\theta}$ discussed earlier is explained in Figure 7.5 and Table 7.4 based on a research project carried out on behalf of the National Board of Fire Research in Sweden (see Bengtson and Ramachandran (1994)).

The individual growth rate θ and the average rate $\overline{\theta}$ both have the same expected value $\overline{\theta}$ but different standard deviations. The standard deviation $\overline{\sigma}_\theta$ of $\overline{\theta}$ [equation 7.9], is generally smaller than the standard deviation σ_θ of θ [equation 7.10]. Hence, the maximum of θ, which denotes the "worst-case" scenario, is greater than the maximum of $\overline{\theta}$.

Some indication of the faster rate at which smoke would spread can be obtained by applying the exponential model in equation [7.7] to total area damaged including smoke. An estimate of the total area damaged in each fire is available in the UK fire statistics but it also includes water damage. The rate of growth of smoke can also be derived from the rate for fire since the quantity of smoke produced is correlated with heat output.

Table 7.4. Average growth rate in all fires and growth rate in an individual fire

Building type	Average growth rate in all fires (θ)			Growth rate in an individual fire (θ)		
	Expected value (θ)	Standard deviation (σ_θ)	Maximum rate	Expected value (θ)	Standard deviation (σ_θ)	Maximum rate
Railway properties	0.0376	0.0021	0.0417	0.0376	0.0352	0.1066
Public car parks	0.0362	0.0025	0.0411	0.0362	0.0318	0.0985
Road tunnels and subways	0.0220	0.0024	0.0267	0.0220	0.0176	0.0565
Power stations	0.0208	0.0029	0.0265	0.0208	0.0210	0.0620

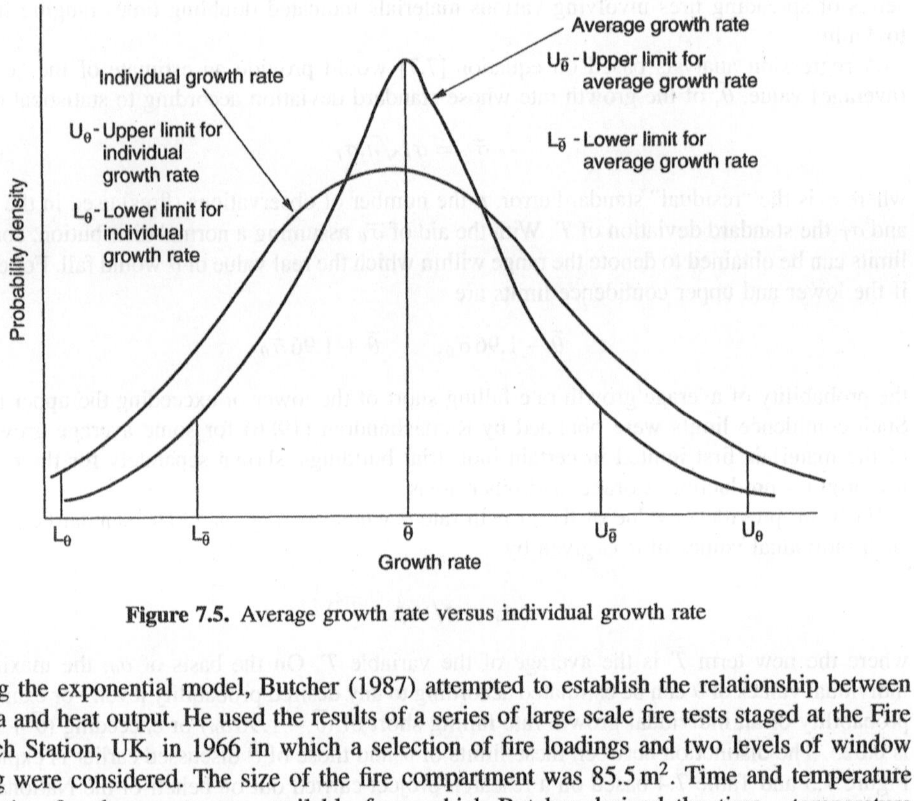

Figure 7.5. Average growth rate versus individual growth rate

Using the exponential model, Butcher (1987) attempted to establish the relationship between fire area and heat output. He used the results of a series of large scale fire tests staged at the Fire Research Station, UK, in 1966 in which a selection of fire loadings and two levels of window opening were considered. The size of the fire compartment was 85.5 m^2. Time and temperature information for these tests was available from which Butcher derived the time – temperature curve for a compartment with the largest fire load density of 60 kg/m^2. The value of heat output estimated from this curve was combined with the progressive area increases obtained by using the results on fire growth produced by Ramachandran (1986). The heat output thus obtained for each fire area, at the appropriate time, was integrated to provide a value for the total heat output for the growing and spreading fire for any time value in the fire's history. Ramachandran (1995) has developed a method for coupling deterministic rates for heat output and mass loss of fuel with statistical fire growth rate based on area damage and fire duration.

Bengtson and Laufke (1979/80) have used the exponential model and a combination of quadratic and exponential models to estimate the fire area and time when sprinklers operate in different hazard categories. The authors have also discussed the estimation of the time to flashover at different room volumes with and without installed fire ventilation system. Other topics discussed by them include operation time of smoke detectors, fire brigade efforts on extinguishing a fire, and effects on evacuation of people. Bengtson and Hagglund (1986) have described the application of an exponential fire growth curve in fire engineering problems.

7.7 Fire severity

Deterministic formulas have been derived (Harmathy, 1987) for the maximum potential fire severity, S_b, expected in a compartment fire, as a function of fire load (quantity of combustible materials), dimensions of the compartment, thermal inertia of compartment boundaries, and area of ventilation. On the basis of one such formula, Baldwin (1975) has estimated that for burnout times for rooms in office buildings, the probability, p_{S_b} of severity exceeding S_b is given by

$$p_{S_b} = \exp(-0.04 S_b) \qquad [7.11]$$

according to an exponential distribution.

The spread of fire in a compartment depends also on the distribution (arrangement) of the objects in a compartment contributing to the total fire load. This factor is generally not taken into account in fire load surveys, which only consider the variation of total fire load or fire load density from compartment to compartment. Such surveys would be more time-consuming and expensive particularly if details about the location of objects within compartments are also to be collected as carried out by Al-Keliddar (1982). Moreover, the potential fire severity estimated by a fire load survey needs to be considered alongside severity likely to be actually attained in a real fire. The potential severity, which relates to the complete burnout of a compartment, may only be attained with a very low probability in a real fire due to the stochastic nature of fire growth and spread (see Chapter 15).

According to Ramachandran (1990), severity in a real fire S_f is given approximately by

$$S_f = \mathrm{k}.\log_e \mathrm{d} \qquad [7.12]$$

where d is floor area damaged. Assuming an exponential distribution as in equation [7.11] and using equation [7.12], the probability of fire severity exceeding S_f is given by

$$p_{S_f} = d^{-\lambda \mathrm{k}} \qquad [7.13]$$

According to figures in Table 7.3, for example, for an unsprinklered "reference compartment" of 800 m^2 floor area, the probability of fire spreading beyond contents and involving structure (structural damage) is 0.28 with an average damage of 197.41 m^2. These figures suggest a value of 0.24 for the product λ.k.

According to Figure 7.2, the damage likely in an unsprinklered compartment of 1600 m^2 is 300 m^2 such that, with $\lambda.k = 0.24$ in equation [7.13], the probability of structural damage would be 0.25. If the compartment size is 2400 m^2, with $d = 374$ m^2, the probability of structural damage is 0.24. For a sprinklered compartment, similar calculations would show that $\lambda.k = 0.54$ and the damage likely in compartments of sizes 1600 m^2 and 2400 m^2 would increase to 98.21 m^2 and 116.44 m^2 from 75.07 m^2 for the "reference compartment" of 800 m^2. The probability of structural damage for these two larger compartments would decrease to 0.084 and 0.077 from 0.097. It may be observed that sprinklers, if they operate, would reduce the probability of structural damage in a compartment of any size.

The analysis described above shows that the probability of structural damage would decrease with the increasing size of a compartment, whether sprinklered or not. This hypothesis is supported by a result obtained by Harmathy *et al.* (1989) in regard to the phenomenon "flashover" defined in Section 5.5. According to Harmathy, the probability of flashover would decrease with increasing compartment size. In a bigger compartment, it would take a longer time for a fire to involve all the objects, and the extra time thus available would increase the chance of extinguishment or burning out and decrease the chance of structural damage. A larger room generally has a greater nonuniformity in the distribution of fire load and lesser degree of overcrowding of objects.

The severity expected in a fire would increase with an increase in damage according to equation [7.12]. For the example considered (Table 7.3, Figure 7.2), an increase in the size of a nonsprinklered compartment to $1600\,m^2$ from $800\,m^2$ would increase the maximum fire severity by 7.6% as given by (log300/log200). If the size is trebled to $2400\,m^2$, maximum fire severity would increase by 11.8% as given by (log374/log200). These results with similar percentage increases for a sprinklered compartment obtained by Ramachandran (1990) agree with deterministic calculations (Malhotra, 1987) provided the area of ventilation openings is maintained at a constant percentage of the surface area of external walls.

Because of an increase in fire severity, a larger compartment should in principle have a higher level of fire resistance but this is counteracted to some extent by a decrease in the probability of flashover or structural damage as defined in this section. Fire resistance required for a sprinklered or nonsprinklered compartment should, therefore, be determined on the basis of an acceptable value for the product of probability of flashover and probability of compartment failure (see Ramachandran (1990) and Chapter 10). The product denotes the probability of fire spreading beyond the compartment. It ought to be pointed out that structural damage can occur without flashover. Structural damage does not directly imply structural failure. But, during the postflashover stage, prolonged exposure to excessive heat would cause a serious damage to structure leading to structural failure. The approach proposed above would provide a sound basis for lowering the fire resistance requirement for a sprinklered compartment (see Section 10.7.3).

7.8 Special factors

An estimate of fire risk can be obtained by combining information on number of fires, amount of fire loss, and the number of fatal and nonfatal casualties from fire together with information on number of occupancies or people at risk. North (1973) attempted this problem for occupancies in the United Kingdom using data for the years 1968 to 1970. Some examples of North's estimates were as follows:

1. For all manufacturing industries, the risk of having a fire per annum per establishment was 0.092.
2. At 1968 to 1970 prices, the annual expected fire loss per establishment in all manufacturing industries was £610 with the greatest mean loss (£1600) occurring in the chemical industry.
3. Risk of death was greatest in hotels where it was 3.6 per person per 100 million exposure hours. This was almost 20 times greater than the mean risk in houses (0.19), about 10 times greater than the mean risk in hospitals (0.35) and 30 times greater than the mean risk in all manufacturing industries (0.12).

North's estimates gave to some extent an unfair picture of the fire risk in different industries since they were evaluated on the basis of "per establishment." Establishments vary in their size, in the number of buildings, in the amount of capital, and in the number of people they employ, and these will in turn vary with different industries. In order to assess the influence of all these factors on fire frequency, Hogg and Fry (1966) applied a rather complicated statistical technique called "Principal Component Analysis." Six main components were included in their analysis, which could be regarded as broadly descriptive of size, competitiveness, productivity, value of stock in relation to size, proportion of expenditure on administration, and sensitivity to external economic conditions.

Hogg and Fry found that the frequency of fire in an industry was dependent on the component measuring the "size" of the industry and on no other components. The "size component" was derived from the following factors:

Number of establishments
Purchases of materials, fuels etc.
Products on hand for sale
Stocks of materials and fuel
Payments for transport
Net output minus wages and salaries
Wages and salaries
Average number employed
New building work
Plant and machinery (acquisitions minus disposals)
Vehicles (acquisitions minus disposals).

The authors also ranked industries according to their relative fire frequency by calculating for each the number of fires that would have occurred if the industry had been of average size. Two lists were produced showing industries in descending order of the relative likelihood of fire in the production and storage areas. Miscellaneous wood and cork manufacture, furniture and upholstery, wooden containers, and baskets appeared high on both lists, while contractors' plant and quarrying machinery, industrial engines, engineers' small tools, and gauges were at the low end. The lists were subject to some uncertainty arising from chance variation and the order might have undergone changes in time, due to changes in an industry in either materials handled or production methods used.

By comparing the number of fires in buildings attributed to various types of fuel with the total amount of fuel used, it is possible to obtain correlations and predict trends. The number of fires for a given fuel may be plotted against the number of units of fuel used. This procedure followed by Chandler (1968) gave an approximately straight-line correlation for electrical fires occurring in the United Kingdom between the years 1956 and 1966. When this was extrapolated, it was estimated that 25,500 fires would occur in 1970 associated with an output of (210×10^9) kWh of electricity. The number of fires, which actually did occur in 1970 agreed with this very well.

For gas, the number of fires during 1957 to 1966 did not vary linearly with the amount of town gas sold. In fact, the trend showed there was a reduction in the frequency per 10^9 therms of gas sold. However, extrapolating on the trend did predict 7000 fires in 1970 and the number that occurred was 7100. Fires due to solid fuel showed a reduction in number because of a reduction in the amounts of solid fuel sold. With oil, the fire frequency per million tons of oil delivered dropped through the period 1955 to 1966; it was thought that this was due to the advent of central heating, which is much safer than portable oil heating.

Various studies carried out in the United States have demonstrated that fire incidence (with its consequences in terms of deaths and injuries) is related, though this does not necessarily imply causality, to a combination of factors (see, for example, Bertrand and McKenzie (1976), Munson and Oates (1977), and Gunther (1975)). These factors reflected poor and substandard housing, overcrowding, social class, race, lack of family stability, and proportions of the young or elderly in the population.

The only detailed analysis on socioeconomic aspects of fires in Britain was by Chandler (1979) in relation to fires in London. This study conclusively demonstrated the existence of correlations between fire incidence and housing and social factors. Of the housing variables examined, tenancy (owner occupied, private rented, etc.) and the lack of amenities were the most strongly correlated with fire incidence. People who own their homes might be expected to be more careful than those who live in rented properties. The social indicator most strongly correlated with fire incidence was the proportion of children-in-care, which was thought to reflect family instability. Although fire frequency appeared to be independent of the age distribution of the population, the incidence of casualties was generally highest among the young and elderly. Strong correlations were observed between malicious fires and serious offenses, and between serious crime rate and fires due to both smokers' materials and children. Updated results of London analysis were included in a later study by Chandler *et al.* (1984) with reference to Birmingham and Newcastle-upon-Tyne. A summary of the results obtained in the UK studies mentioned above along with other human aspects of fires in buildings is contained in a paper by Ramachandran (1985).

Severe weather conditions during winter give rise to an increase in the number of fires and fire casualties. Chandler (1982), for example, analyzed data for fires in dwellings in the United Kingdom during the severe winter of 1978 and 1979. He tabulated these data on a weekly basis and according to meteorological office regions. Data were obtained particularly for major sources of ignition such as space heating and electric blankets and for life risk fires involving casualties rescues and escapes. For each region, data were obtained from one weather station representing average weather conditions prevailing in the region. These data related to weekly averages of minimum daily temperature, rainfall, hours of sunshine, wind speed, vapor pressure, and relative humidity. Using these factors as independent variables, a multiple regression analysis was performed with each of a number of fire incidence variables as dependent variable. Following preliminary examination of the data, logarithmic transformations were used on data relating to fire incidence but not life risk fires.

Chandler found that in all regions, temperature and vapor pressure were significantly correlated with total fires and fires due to space heating, electric blanket, and wire and cable including leads. The same was true of life risk fires due to space heating. Fires due to cooking appliances, the major cause of domestic fires, were not generally influenced by severe weather conditions. The analysis suggested that in the temperature range 0 °C down to −5 °C there were an extra 30 fires per week in England and Wales for every degree drop in temperature. This result was in general agreement with the assessment based on fire frequencies for the 1962 to 1963 winter (see Gaunt and Aitken (1964)). The most vulnerable age group during the cold spell in the United Kingdom in early 1979 was those aged 65 and over, especially females, who were usually alone in the room of fire origin.

Symbols

A	total floor area (size) of a building
A_s	size of sprinklered building
A_u	size of unsprinklered building
$A(O)$	area originally ignited
$A(T)$	area damaged in T minutes
C	a constant in the equation for $D(A)$
C'	$= C\nu^{-\beta}$
D	floor area damaged
$D(A)$	expected area of damage in a building of size A
$D(V)$	expected loss in a building of financial value V
d	doubling time; floor area damaged
F	probability of fire starting during a year

$F(A)$	probability of fire starting during a year in a building of size A
$F(V)$	probability of a fire starting during a year in a building of financial value V
I_{ij}	probability of ith cause being present in the jth part of a building
K	a constant in the equation for $F(A)$
K'	$= Kv^{-\alpha}$
k	a constant in the equation for S_f
L	financial loss
N	number of buildings at risk in a given risk category
n	number of fires in a given risk category
O	object of fire origin
P_{ij}	probability of fire due to ith cause in the jth part of the building
p_{S_b}	probability of fire severity exceeding S_b
R	room of fire origin
S_b	potential fire severity
S_f	severity of real fire
T	duration of burning
V	financial value of a building
α	a constant in the equation for $F(A)$
β	a constant in the equation for $D(A)$
λ	a constant in the equation for S_f
θ	fire growth parameter
σ	residual standard error
σ_T	standard deviation of T
σ_θ	standard deviation of θ
v	$= V/A$

References

Al-Keliddar, A (1982). *Dwelling Design and Fire Hazard*, Ph.D. Thesis, School of Architecture, University of Manchester, UK.

Apostolakis, G (1982). *Nuclear Engineering and Design*, **71**, 375–381.

Baldwin, R (1975). *Economics of Structural Fire Protection*, Building Research Establishment, Fire Research Station, UK, Current Paper CP 45/75.

Benktander, G (1973). Claims frequency and risk premium rate as a function of the size of the risk. *ASTIN Bulletin*, **7**, 119–136.

Bengtson, S and Hagglund, B (1986). The use of a zone model in fire engineering application. *Fire Safety Science: Proceedings of the First International Symposium*, Hemisphere Publishing Corporation, New York, pp. 667–675.

Bengtson, S and Laufke, H (1979/80). Methods of estimation of fire frequencies, personal safety and fire damage. *Fire Safety Journal*, **2**, 167–180.

Bengtson, S and Ramachandran, G (1994). Fire growth rates in underground facilities. *Fire Safety Science: Proceedings of the Fourth International Symposium*, Ottawa, T Kashiwagi (Ed.), National Institute of Standards and Technology, Gaithersburg, MD, USA, pp. 1089–1099.

Bertrand, A L and McKenzie, L S (1976). *The Human Factor in High Risk Urban Areas: A Pilot Study in New Orleans, Louisiana, US*, Department of Commerce, Washington, DC.

Butcher, G (1987). The nature of fire size, fire spread and fire growth. *Fire Engineers Journal*, **47**(144), 11–14.

Chandler, S E (1968). Fire Research Station, UK, Fire Research Note No 716.

Chandler, S E (1979). *The Incidence of Residential Fires in London – The Effect of Housing and Other Social Factors*, Building Research Establishment, Fire Research Station, UK, Information Paper IP 20/79.

Chandler, S E (1982). The effects of severe weather conditions on the incidence of fires in dwellings. *Fire Safety Journal*, **5**, 21–27.

Chandler, S E Chapman, A and Hollington, S J (1984). Fire incidence, housing and social conditions the urban situation in Britain. *Fire Prevention*, **172**, 15–20.

Friedman, R (1978). Quantification of threat from a rapidly growing fire in terms of relative material properties. *Fire and Materials*, **2**(1), 27–33.

Gaunt, J E and Aitken, I S (1964). *Causes of Fires in Dwellings in London, Birmingham and Manchester and Their Relationship with the Climatic Conditions During the First Quarter of 1963*, Fire Research Station, UK, Fire Research Note No 538.

Gomberg, A, Buchbinder, B and Offensend, F J (1982). *Evaluating Alternative Strategies for Reducing Residential fire Loss – The Fire Loss Model*, National Bureau of Standards Center for Fire Research Report NBSIR 82–2551.

Gunther, P (1975). *Fire Journal*, **3**, 52–58.

Harmathy, T Z (1987). On the equivalent fire exposure. *Fire and Materials*, **11**, 95–104.

Harmathy, T Z *et al.* (1989). A Decision Logic for Trading Between Fire Safety Measures. *Fire and Materials*, **14**, 1–10.

Heskestad, G. (1982). *Engineering Relations for Fire Plumes*, Technology Report 82–8, Society of Fire Protection Engineers, Boston.

Hogg, J M and Fry, J F (1966). *The Relative Fire Frequency of Different Industries*, Ministry of Technology and Fire Offices' Committee Joint Fire Research Organisation, UK, Fire Note No 7.

Labes, W G (1966). The Ellis parkway and Gary dwelling burns. *Fire Technology*, **2**(4), 287–297.

Malhotra, H (1987). *Fire Safety in Buildings*, Report, Building Research Establishment, UK.

Munson, M J and Oates, W E (1977). *Community Characteristics and the Incidence of Fire: an Empirical Analysis*, Princeton University, Princeton, NJ.

North, M A (1973). *The Estimated Fire Risk of Various Occupancies*, Fire Research Station, UK, Fire Research Note No 989.

Ramachandran, G (1970). *Fire Loss Indexes*, Fire Research Station, UK, Fire Research Note No 839.

Ramachandran, G (1979/80). Statistical methods in risk evaluation. *Fire Safety Journal*, **2**, 125–145.

Ramachandran, G (1985). *The Human Aspects of Fires in buildings – A Review of Research in the United Kingdom Fire Safety: Science and Engineering*, ASTM STP 882, American Society for Testing and Materials, Philadelphia, pp. 386–422.

Ramachandran, G (1986). *Exponential Model of Fire Growth Fire Safety Science: Proceedings of the First International Symposium*, Hemisphere Publishing Corporation, New York, pp. 657–666.

Ramachandran, G (1988). Probabilistic approach to fire risk evaluation. *Fire Technology*, **24**, 204–226.

Ramachandran, G (1990). Probability based fire safety code. *Journal of Fire Protection Engineering*, **3**(2), 75–91.

Ramachandran, G (1995). Heat output and fire area. *Proceedings of the International Conference on Fire Research and Engineering*, Orlando, FL, Society of Fire Protection Engineers, Boston, MA, USA, pp. 481–486.

Rutstein, R (1979). The estimation of fire hazard in different occupancies. *Fire Surveyor*, **8**(2), 21–25.

Thomas, P H (1974). *Fires in Model Rooms: CIB Research Programmes*, Building Research Establishment, Fire Research Station, UK, Current Paper CP 32/74.

Thomas, P H (1981). Fire modelling and fire behaviour in rooms. *Eighteenth Symposium (international) on Combustion*, The Combustion Institute, pp. 503–518.

Rogers, F E (1977). Fire Losses was the Effect of Spinhler Protection of Buildings is a Vanity of Industries and Trades, Building Research Establishment, Current Paper CP9/77.

8 LIFE LOSS

8.1 Introduction

It is necessary to understand the characteristics of fire casualties, factors affecting life risk, and the behavior of building occupants at the time of a fire in order to incorporate appropriate life-safety measures into fire regulations, codes, and standards. A fair deal of information on these aspects can be obtained by analyzing the fire statistics that are available in many countries, particularly the United Kingdom, United States, and Japan. These statistics reveal some general trends and patterns as described briefly in this chapter.

Fire statistics show that loss of life due to the structural instability or collapse of a building involved in fire is a rare occurrence. Risk to life in a fire is mainly due to exposure to heat, smoke, and toxic gases produced by burning materials or products such as upholstered furniture. Smoke particles when dispersed in the air reduce visibility and can also lead to sensory irritation. Inhalation of heated air, whilst not instantly fatal, can lead to death in a short time if the temperature exceeds a certain level. Exposure of a person to a large dose of a toxic gas can cause incapacitation and eventual death.

The rates at which fatal and nonfatal casualties occur in fires are quantitative indicators of life risk consistent with the form in which statistics are available at present. Fatality rate per fire is a particularly useful measurement, especially if it can be related to the time factor and its components involved in the evacuation of a building. This problem is discussed with the aid of an exponential distribution applied to data available from UK fire statistics. This distribution arises from a Poisson probability model that can also provide an estimate of the multiple-fatality rate as a function of evacuation time. This rate provides an appropriate yardstick for measuring the life risk posed by fire, particularly in a building occupied by a large number of people.

Fatal Accident Frequency Rate (FAFR) is another measurement of life risk, which is generally taken into account for industrial buildings and plants. The FAFR due to fire in different risk areas is compared with other sources of death. The rate of death per person, per fire, or per annum is another useful indicator if data are available on the number of persons at risk in different types of buildings.

The last section of this chapter is concerned with a fire risk assessment method that was recently developed in the United States. This method is useful for analyzing the impact of changes made

Evaluation of Fire Safety D. Rasbash, G. Ramachandran, B. Kandola, J. Watts and M. Law
 ISBN: 0-471-49382-1

to a combustible product on its life-safety risk. The basic features of this method are discussed together with the results obtained for upholstered furniture, which was the product investigated in one of the case studies.

8.2 Characteristics of fire casualties

A proper perspective of the risk to life in fires can be obtained by analyzing the statistical data on fire casualties. Information on fire casualties, particularly deaths in the United Kingdom, was analyzed by Chandler during the 1960s and 1970s in a series of interesting research notes and reports (see, for example, Chandler (1971, 1972)). According to these studies, smokers' materials have been the main known cause of fire fatalities, especially among the elderly. In the United States, the National Fire Protection Association (NFPA) publishes every year in its journal an analysis of fire deaths including incidents involving multiple deaths. Sekizawa (1988, 1991) and Hall (1990) have investigated the reasons for the differences in fire death rates between United States and Japan. They have drawn some useful conclusions, a summary of which is as follows.

Japan's overall fire death rates per million population are far lower than those in the United States; fire death rates in the United States are two and a half times the corresponding rates in Japan (excluding incendiary suicides). However, the gap has closed to some extent in the past decade due to the fact that US fire death rates declined substantially prior to 1982 while rates in Japan declined very little. Deaths due to incendiary suicides have been increasing in Japan, partly offsetting the decline in fire deaths due to all other causes.

Sekizawa (1991) has identified two typical fire death patterns in residences in Japan – "Disaster-Vulnerable People and Daytime Fire" and "Nondisaster-Vulnerable People and Nighttime Fire." The former pattern can be described as the case in which a person who needs help to move encountered a fire and failed to escape when he/she was alone during daytime. The latter pattern involves a person capable of normal physical function getting killed in a fire mainly because of the delay in detection while asleep or in a drunken state during the nighttime. The majority of people in the disaster-vulnerable group are aged 65 and over, so that the fire death rate is much higher than that for the other group. Older adults are a high-risk group in the United Kingdom and United States as well. Fire death rates in Japan are nearly equal to the US rates for ages 71 to 80 and one-fourth higher for ages 81 and above.

For adults from 31 to 50 years of age, nearly two-thirds of all Japanese fire deaths are incendiary suicides. Without this uniquely Japanese problem, the US rates are more than triple the Japanese rate for this age group. In Japan, only children and older adults (age 61 and above) have most of their fatal fires from accidental causes. Similar fire death patterns in homes in the United States have been identified by Karter (1986).

In the United States, as in the United Kingdom, preschool children (ages 0–5) are a high-risk group, with a fire death rate nearly twice the average for all ages. In Japan, preschool children have a higher fire death rate than older children but about the same risk as the overall average if incendiary suicides are excluded. One reason for this difference is probably a higher incidence of single-parent families in the United States and a generally larger incidence of gaps in child supervision (Fahy, 1986). This factor together with proportion of children-in-care, perhaps, reflects family instability identified by Chandler (1979) as a social indicator strongly correlated with fire incidence. Preschool children in Japan also are more likely to sleep in their parents' rooms, which may give them an advantage in responding to fires. US preschoolers have a fire death rate four times the rate of their Japanese counterparts.

Most of the fire deaths occur in dwellings. In the United Kingdom and United States, the majority of deaths in accidental fires in dwellings were attributed to a relatively small number of specific causes such as careless disposal of smokers' materials, incidents with space heaters – mainly misuse or placing articles too close to them, ignition of matches mostly by children playing with them and misuse of cooking appliances. Electricity is the major fuel where deaths are caused by the misuse of space heaters or cooking appliances. As mentioned in Section 7.8, a contributory factor for an increase in fire deaths during severe winter is the use of portable heating appliances to supplement central heating.

In the United States, heating appliance deaths are mainly caused by portable heaters and space heaters, especially in poor and rural areas. Japan also has proportionally more trouble with heating equipments. The high risk of fire deaths among older adults in Japan may be partly due to the use of older types of portable heaters such as kerosene heaters that can cause serious fires when placed too close to combustibles in the small rooms typical of Japanese homes.

The pattern of nonfatal casualties from accidental dwelling fires in the United Kingdom is generally similar to that of fires caused by the misuse of cooking appliances, mostly electric or and the remainder gas. The next most common specific cause is the careless disposal of smokers' materials, followed by electric and gas space heating appliances, electrical wiring, electric blanket, and bed warmer. Television sets, washing machines, and dishwashers are other minor causes of nonfatal casualties. There is a falling trend in nonfatal casualties, where the source of ignition is an electric blanket or bed warmer. The characteristics of nonfatal casualties mentioned above are generally true for the United States and Japan.

8.3 Location of casualties

According to Table 8.1 based on the UK Fire Statistics for the period 1978 to 1991, half of the casualties in single-occupancy dwellings were found in the room of fire origin. Most of the remaining casualties were found elsewhere on the floor of origin or floors above the floor of origin. A comparatively smaller number of casualties was found in the floors below the floor of fire origin. Location of casualties in multiple-occupancy dwellings had a similar pattern except that an almost equal number of nonfatal casualties were found in the room and floor of fire origin.

It is understandable that occupants in the floors above the floor of fire origin have greater risk than those in the floors below. Fire, smoke, and toxic gases generally spread upwards and are more likely to be encountered by the people in upper floors whether they remain in their places of occupation or attempt to escape to safer places in or outside the building involved in fire. People in lower floors have a greater chance of avoiding combustible products and escaping safely.

It is apparent that while fire is a major threat to the occupants in its immediate vicinity, it is generally smoke and toxic gases that pose a greater threat than flame (heat) to the occupants who are remote from the fire. Smoke and fumes travel faster than fire to the occupied areas and escape routes. Even a small fire can generate considerable amounts of smoke and other combustible products and threaten a greater number of occupants outside the room of origin. Most building fires spread beyond the room by convection (advance of flame, smoke, and hot gases) rather than by the destruction of the structural boundaries (Harmathy and Mehaffey, 1985). A majority of fires in dwellings, where most of the casualties occur, are confined to the room of origin (see Table 8.2 based on the UK Fire Statistics).

Table 8.1. Location of casualties – single- and multiple-occupancy dwellings

Whereabouts of casualty and occupancy type	Number of persons	
	Fatal	Nonfatal
Single-occupancy dwellings		
Room of origin of fire	3539 (58.2)	26,259 (44.6)
Elsewhere on the floor of origin	1216 (20.0)	15,500 (26.3)
Floors above the floor of origin of fire	1267 (20.8)	14,835 (25.2)
Floors below the floor of origin of fire	54 (1.0)	2330 (3.9)
Total	6076 (100.0)	58,924 (100.0)
Multiple-occupancy dwellings		
Room of origin of fire	2347 (66.8)	15,353 (35.0)
Elsewhere on the floor of origin	823 (23.4)	18,245 (41.6)
Floors above the floor of origin of fire	330 (9.4)	9066 (20.6)
Floors below the floor of origin of fire	16 (0.4)	1233 (2.8)
Total	3516 (100.0)	43,897 (100.0)

Note: The numbers within brackets denote the percentage number of casualties.
Source: Fire Statistics, United Kingdom, 1978–1991.

Table 8.2. Spread of fire in dwellings (Fires starting in rooms or compartments within buildings)

Extent of spread	Number of fires	Percentage of fires
Confined to the item ignited first	308,844	40.2
Spread beyond the item but confined to the room contents only	184,020	24.0
Structure involved	215,464	28.1
Spread beyond the room but confined to the floor	25,540	3.3
Spread beyond the floor	33,527	4.4
Total	767,395	100.0

Note: Spread of fire only, not heat, smoke, and so on.
Source: Fire Statistics, United Kingdom 1978–1991.

8.4 Nature of injuries

Consistent with the location of casualties (Table 8.1), burns cause the highest percentage of fatalities in the room of fire origin. Elsewhere, gas or smoke is the major cause and accounts for more than 50% of the fatalities in the dwellings overall (see Table 8.3). Statistical studies and surveys (Bowes, 1974) carried out in the United Kingdom in the 1970s revealed that not only was

Table 8.3. Fatal casualties in dwellings by whereabouts of casualties and cause of death

Whereabouts of casualty and occupancy type	Number of persons		
	Cause of death		
	Overcome by gas or smoke[a]	Burns or scalds	Other or unknown causes
Single occupancy			
Room of origin of fire	1653	953	328
Floor of origin of fire	731	133	116
Elsewhere	868	130	118
Total	3252	1216	562
Multiple occupancy			
Room of origin of fire	1217	510	200
Floor of origin of fire	504	66	67
Elsewhere	217	37	40
Total	1938	613	307

[a]Including cases where "burns" and "overcome by gas or smoke" were joint causes of death.
Source: Fire Statistics, United Kingdom, 1978–1988.
A breakdown of figures for causes of death as in the table has not been published for the years 1989 to 1991.

a large proportion of fatal and nonfatal fire casualties being reported in the category "overcome by smoke and toxic gases" rather than heat and burns but also that there was a fourfold increase in the former category between 1955 and 1971. In the United States, in the absence of detailed national fire statistics during the 1970s, victims in a number of large fire disasters were reported, in newspapers and fire journals to have died from exposure to toxic smoke products.

It is an accepted fact that toxic products of combustion are the major causes of incapacitation and death in fires (Berl and Halpin, 1976, Harland and Woolley, 1979). In many fires, death or injury is not due to the immediate toxic effects of exposure to these products but results from the victim being prevented from escaping due to irritation and visual obscuration caused by dense smoke or to incapacitation caused by narcotic gases. Consequently, the victim remains in the fire and sustains fatal or nonfatal injury due to a high dose of toxic products inhaled during the prolonged exposure or due to burns. According to UK fire statistics, more than 50% of fatalities and more than 30% of nonfatal casualties are trapped by smoke or fire, because they either were unaware (asleep etc.) of fire or because of other reasons. Survivors from fires may also experience pulmonary complications and burn injuries that can lead to delayed death.

Increasing fire risk due to smoke and other combustion products led to the commencement of intensive research on combustion toxicology (Purser, 1988) during the 1970s. These studies have ranged from fundamental laboratory-based thermal decomposition experiments to large-scale fires with comprehensive gas analysis, bioassay, and detailed pathology of fire victims. Two types of models have been developed – the "mass loss" model and the "fractional effective dose" model. Both the models require as inputs the rates of generation of life-threatening combustion products and the estimate of the times when tenability limits are exceeded, resulting in incapacitation or death. For calculating these limits two major computer packages are available – ASKFRS (Chitty and Cox, 1988) developed by the Fire Research Station, UK and TENAB which is part of HAZARD developed by the Centre for Fire Research, National Institute of Standards and Technology, USA

8.5 Materials first ignited

Fires in the United Kingdom during the past few years when food fat was the material first ignited have risen steadily, but account for only a small number of deaths. The group of materials next most frequently first ignited was textiles, upholstery, and furnishings, accounting for more than 20% of dwelling fires and also for a large number of deaths. Within this group the major items were bedding, upholstery, or covers and clothing. Other individual items first ignited in a large number of fires in dwellings were electrical insulation and paper or cardboard.

The observations mentioned above were more or less true even 20 years ago. An analysis of UK fire statistics for 1970 showed that the chance of fatality in fire involving furniture and furnishings was twice that of other fires in houses (see Chandler and Baldwin (1976)). The majority of these fires involved upholstery and bedding and over 90% were caused by smokers' materials, electric blankets, space heating or the activities of children, and suspected arsonists. Nearly all fatalities were found in the fire room, overcome by smoke or toxic gases and the great majority were young or elderly (over 65). The main hazard appeared to be to people in armchairs and beds, using potential sources of ignition (smoking, space heating etc.), failing to respond to a fire in their vicinity through being asleep or otherwise incapacitated, and then being overcome by smoke or toxic gases. According to a study carried out by Clarke and Ottoson (1976), more than one-fourth of all US residential fire deaths resulted from upholstered furniture fires. Smouldering cigarettes inadvertently dropped on furniture were a common cause of these fires.

One possible reason for the increase in smoke-related casualties in homes during 1970s was the increased use of modern synthetic materials in furnishings and upholstered furniture. Cellulosic materials were replaced by thermoplastic fibers and urethane foam cushioning materials. The increase may not be directly related to modern materials but to the changes in living styles, which have led to more furnishings and upholstery material being used in homes. The introduction of synthetic upholstery fabrics has offered enormous scope to furniture designers where fashion appeal, color, durability, stain resistance, and so on are required and these fabrics have virtually replaced the traditional covers.

There is little doubt that the gradual change from natural to synthetic materials has brought certain benefits in fire performance. Natural materials tend to be prone to smouldering from small ignition sources, particularly when in contact with a lighted cigarette, whereas synthetic materials tend to be more resistant to this type of ignition. However, the synthetic fabrics are mainly thermoplastics and when subjected to a flame can burn rapidly with the fabric "melting" to expose the flammable infill fibers and foams. Natural fabrics (wool, cotton etc.) tend to form carbonaceous chars during flame exposure, which can act as an effective barrier to the penetration of fire. The results of a series of experiments (Woolley *et al.*, 1978) illustrated the ease of ignition of modern upholstery materials even with small ignition sources.

8.6 Casualty rate per fire

A simple yardstick for measuring life risk due to fires in any type of building is the ratio between number of fatal or nonfatal casualties and number of fires. Global trends in the risk can be observed by calculating the annual casualty rates per fire as shown in Tables 8.4 and 8.5. These figures have been extracted from Fire Statistics, United Kingdom published annually by the Home Office. (There is generally a delay of about two years for the publication of statistics for any year.) The figures for any other type of building can be obtained from the Home Office on payment of charges covering mostly the cost of computer time.

According to figures in Tables 8.4 and 8.5, both the fatal and nonfatal casualty rates do not vary significantly from year to year. In fact, there is some indication that the rates are gradually declining

Table 8.4. Fatality rate per fire

Year	Single-occupancy dwellings			Multiple-occupancy dwellings		
	Number of deaths	Number of fires	Fatality rate per fire	Number of deaths	Number of fires	Fatality rate per fire
1978	473	35,049	0.0135	251	15,830	0.0159
1979	575	38,629	0.0149	282	17,223	0.0164
1980	533	33,886	0.0157	268	15,683	0.0171
1981	496	35,230	0.0141	279	18,274	0.0153
1982	457	34,994	0.0131	257	18,826	0.0137
1983	432	34,667	0.0125	274	20,195	0.0136
1984	436	34,972	0.0125	250	21,020	0.0119
1985	438	36,905	0.0119	255	22,468	0.0113
1986	462	37,313	0.0124	283	22,389	0.0126
1987	441	36,669	0.0120	267	23,286	0.0115
1988	445	36,251	0.0123	269	24,331	0.0111
1989	394	34,947	0.0113	234	25,514	0.0092
1990	372	33,535	0.0111	246	25,328	0.0097
1991	382	33,876	0.0113	208	25,632	0.0081
Overall	6336	496,923	0.0128	3623	295,999	0.0122

Source: Fire Statistics, United Kingdom.

Table 8.5. Nonfatal casualty rate per fire

Year	Single-occupancy dwellings			Multiple-occupancy dwellings		
	Number of casualties	Number of fires	Casualty rate per fire	Number of casualties	Number of fires	Casualty rate per fire
1978	3503	35,049	0.0999	1835	15,830	0.1159
1979	3712	38,629	0.0961	2265	17,223	0.1315
1980	3463	33,886	0.1022	2017	15,683	0.1286
1981	3755	35,230	0.1066	2471	18,274	0.1352
1982	3966	34,994	0.1133	2605	18,826	0.1384
1983	4239	34,667	0.1223	2809	20,195	0.1391
1984	4546	34,972	0.1300	3112	21,020	0.1480
1985	4836	36,905	0.1310	3506	22,468	0.1560
1986	5459	37,313	0.1463	3736	22,389	0.1669
1987	5362	36,669	0.1462	3932	23,286	0.1689
1988	5590	36,251	0.1542	4399	24,331	0.1808
1989	5594	34,947	0.1601	4607	25,514	0.1806
1990	5233	33,535	0.1560	4647	25,328	0.1835
1991	5676	33,876	0.1676	4976	25,632	0.1941
Overall	64,934	496,923	0.1307	46,917	295,999	0.1585

Source: Fire Statistics, United Kingdom.

over the years. This trend is generally true for fires in the United States and Japan. An increase in the number of fires appears to be a major factor affecting an increase in the number of casualties. The results mentioned above, perhaps, indicate the fact that fire fighting and protection strategies including fire safety codes are performing effectively in the United Kingdom, United States, and Japan, while fire prevention activities aimed at reducing the frequency of fires need to be stepped up.

According to the figures quoted by Hall (1990), the overall civilian death rate per fire in 1987 was 0.0025 for the United States and 0.0316 for Japan. Excluding incendiary suicides the fatality rate for Japan was 0.0185. According to fire statistics for the years 1981 to 1991, the overall fatality rates for United Kingdom are 0.0026 for all fires and 0.0075 for occupied buildings. For United Kingdom, the overall rates for nonfatal casualties are 0.0346 for all fires and 0.1062 for occupied buildings compared to 0.0121 for United States and 0.1306 for Japan. The UK rates for both the fatal and nonfatal casualties are considerably higher than the corresponding rates for United States but lower than those for Japan. One reason for the differences in the rates for the three countries may be differences in the methods of classifying and estimating fire casualties data; there may be other reasons also. Any comparison of fire casualty rates between countries ought to be made with caution.

8.7 The time factor

In the event of a fire in a building, the casualty rate, fatal or nonfatal, would increase with the time (t) spent by the occupants under untenable conditions caused by combustion products. As a first approximation, the increase in the casualty rate per unit time, say, a minute, may be assumed to be a constant λ. The parameter λ can be estimated by relating the overall casualty rate discussed in the previous section to the average value of the variable t but sufficient data are not available at present for estimating this average time.

However, an approximate value for λ for any type of building can be obtained by relating casualty rates to delays in discovering fires with the aid of statistics published by the Home Office, UK. For example, based on these statistics for the 14-year period 1978 to 1991, total figures for the number of fires and fire deaths are given in Table 8.6 for two types of dwellings – single-occupancy and multiple-occupancy dwellings. The fatality rates estimated by these figures are also shown. Table 8.6 has been reproduced from a study on early detection of fire and life risk carried out by Ramachandran (1993a), which contains a similar analysis for nonfatal injuries.

The fatality rates for fires discovered at ignition are high apparently due to the fact that, as discussed in Section 8.3 a high percentage of fatal casualties were found in the room of fire origin. The figures in Table 8.6 for the other three categories of discovery time provide some indication of the reduction in fatality rate due to early discovery of fire. The discovery time for the last category, more than 30 min after ignition, can vary significantly within a wide range although an average time of 45 min has been suggested in Section 7.6.

For the reasons mentioned above, a reasonably good estimate of λ can be obtained by considering only the second and third categories – discovered under 5 min and discovered between 5 and 30 min. With average discovery times of 2 and 17 min, the increase in the fatality rate for these two categories divided by 15 provides an estimate of λ measuring the increase per minute as shown in Table 8.6. This method based only on discovery time is similar to "longitudinal analysis" adopted by Maclean (1979) for evaluating the relationship between fire brigade attendance time and fire loss.

Using the estimated value of λ, the linear relationship between fatality rate and discovery time of fires is depicted in Figure 8.1. For any discovery time D, the fatality rate per fire is given by $\lambda . D$ plus a constant K (Table 8.6) estimated by the intercept on the vertical axis. The parameter K, as discussed in the next section, denotes the overall and joint contribution to the fatality rate arising from other time periods involved in the evacuation process (Ramachandran, 1990). These periods are considerably short in relation to the total duration and hence K is only a small percentage of the overall fatality rate. The parameters λ and K are generally applicable to fires not discovered at ignition ($D > 0$). Figure 8.1 provides realistic estimates of the fatality rate for discovery times up to 40 min; extrapolation beyond 40 min would be somewhat unrealistic.

Table 8.6. Discovery time and fatal casualties

Discovery time and occupancy type	Number of deaths	Number of fires	Fatality rate (per fire)
Single-occupancy dwellings			
Discovered at ignition	445	76,243	0.005837
Discovered under 5 min after ignition	686	212,519	0.003228
Discovered between 5 and 30 min after ignition	2156	141,462	0.015241
Discovered more than 30 min after ignition	2766	53,677	0.051530
Total	6053	483,901	0.012509
Multiple-occupancy dwellings			
Discovered at ignition	204	27,805	0.007337
Discovered under 5 min after ignition	334	123,648	0.002701
Discovered between 5 and 30 min after ignition	1281	110,078	0.011637
Discovered more than 30 min after ignition	1703	28,125	0.060551
Total	3522	289,656	0.012,159

Note: Single-occupancy dwelling: $\lambda = 0.000801$
$K = 0.001{,}626$
Multiple-occupancy dwelling $\lambda = 0.000596$
$K = 0.001509$
Source: Fire Statistics, United Kingdom 1978–1991.

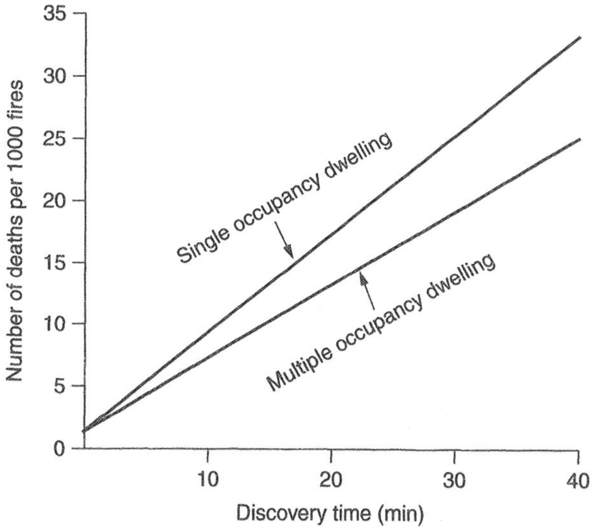

Figure 8.1. Discovery time and fatality rate (current risk level)

For fires that are not immediately discovered, automatic detection systems could reduce the fatality rate if it is assumed that these devices reduce the discovery time. The operation time of detectors depends on several factors such as their type (heat, smoke etc.), the location of the seat of the fire and the rates of development of heat and smoke. Assuming that the reduced discovery time is one minute on average, the reduced fatality rate would be $(\lambda + K)$, that is, 0.0024 for single-occupancy dwellings and 0.0021 for multiple-occupancy dwellings (see Ramachandran

(1993a)). It would be useful to mention in this connection that, (overall) fatality rates per fire of 0.0043 with detectors and 0.0085 without detectors have been estimated for single- and two-family dwellings in the United States (Bukowski *et al.*, 1987).

8.8 Evacuation model

Following the notation adopted by Ramachandran (1990), the discovery period, D, is a major component of the total time (H) taken by an occupant to evacuate a building in the event of a fire. This is the first among three main periods occurring sequentially in time since the start of the fire. The second period, B, is referred to as *recognition time* or *gathering phase* in human behavior studies (Canter, 1985). The third period, E, relates to the elapsed time since the commencement of the actual movement of an occupant until a safe place is reached inside a building, for example, entrance to a protected staircase or outside the building. Although commonly known as *evacuation time*, the period E really refers to an emergency or nonfire situation. Means of escape facilities such as maximum travel distance and number and widths of staircases should be designed according to the total evacuation time $H(= D + B + E)$ but only the subperiod E is generally considered explicitly in fire safety regulations, codes, and standards.

A combustion product such as smoke takes a time, F, to travel from the place of fire origin and produce untenable conditions on an escape route. If the total time H taken by an occupant to get through this route exceeds F, he/she is likely to sustain a fatal or nonfatal injury depending on the level of severity associated with F. For a safe evacuation, the condition $H \leq F$ should be satisfied, which is the objective of designing escape route facilities, smoke control systems, and emergency lighting. This model proposed by Marchant (1980) in terms of the ratio (H/F) has been modified by several authors to include additional periods in the total evacuation time H (see Sime (1986) for a review of these studies).

The evacuation time E and, hence, the total time H are affected by several factors governing the behavior of occupants at the time of a fire (see Canter (1985)). Depending on these factors some occupants may decide to evacuate while some may ignore the fire alarm and remain in their rooms. Some occupants who recognize the existence of the fire may help the group attempting to fight the fire by first-aid means such as portable fire extinguishers. Some occupants may be trapped in their rooms because of their physical conditions (bed ridden, etc.) or mental capability limitations or due to a temporary reduction in their ability because of sleep, drugs or alcohol. Some of the occupants, for example, patients in a hospital involved in evacuation may require assistance in preparing for escape and during escape.

Depending on their location relative to the place of fire origin and other factors mentioned above, some occupants of a building may "succeed" in fulfilling the condition $H \leq F$, while others may "fail" and consequently sustain fatal or nonfatal injuries. As discussed in the last section the casualty rate would increase with the duration of exposure to untenable conditions. This period denoted by t can be expressed as

$$t = H - F \quad [8.1]$$

An estimate of the increase in the fatality rate per minute of exposure for all the occupants in a building is provided by the parameter λ in Table 8.6 although it is only based on the discovery time. The value of λ is generally small since it is related to the occurrence of a rare event. It is, therefore, an approximation for the parameter p given by the exponential function

$$p = 1 - \exp(-\lambda) \quad [8.2]$$

The probability of no deaths during any period t since the onset of untenable conditions at time F is given by exp $(-\lambda t)$ and the probability of one or more deaths by

$$p_d(t) = 1 - \exp(-\lambda t) \quad [8.3]$$

Again, for values of t encountered in most of the fires λt would be small such that it is an approximation for $p(t)$. In terms of the discovery time D, from equations [8.1] and [8.3], the fatality rate per fire is given by

$$\begin{aligned} p_d(t) &= 1 - W\exp(-\lambda D) \\ &= 1 - W(1 - \lambda D) \\ &= K + W\lambda D \end{aligned} \quad [8.4]$$

approximately where

$$\begin{aligned} K &= 1 - W \\ W &= \exp[-\lambda(B + E - F)] \\ &= 1 - \lambda(B + E - F) \end{aligned} \quad [8.5]$$

The values of the parameter K for the two occupancy types considered are given in Table 8.6. In Figure 8.1, W has been amalgamated with λ since it is almost equal to unity.

A model similar to equations [8.4] and [8.5] was suggested by Ramachandran (1990) but with a different weight (not λ) for the time periods B, E, and F. It may be seen (Ramachandran, 1993a) that

$$K = \lambda(B + E - F) \quad [8.6]$$

Based on equation [8.6], it may be seen that $(B + E - F)$ on average, has the values of 2.0 min for single-occupancy dwellings, and 2.5 min for multiple-occupancy dwellings. The values of B, E, and F for any occupant would vary depending on the location of the occupant with reference to the place of fire origin, mobility of the occupant, and on the decision of the occupant to evacuate or not. From equations [8.4] and [8.6], with $W = 1$,

$$p_d(t) = \lambda(H - F) \quad [8.7]$$

approximately.

Considering multiple-occupancy dwellings as an example, in the absence of automatic detection systems, the overall fatality rate per fire is 0.0122 as shown in Table 8.4 or 8.6. This rate is consistent with an average discovery time of 18 min such that $t = 20.5$ min from equation [8.1]. In other words, the fatality rate of 0.0122 is the result of exposure to untenable conditions for a period of 20.5 min. Accordingly, $B + E - F$ is 2.5 min such that, from equation [8.6], $K = 0.0015$ as given in Table 8.6.

Installation of automatic detectors in multiple-occupancy dwellings would reduce the detection time to one minute and the fatality rate to 0.0021 as discussed earlier. In this case, the period of exposure to severe untenable conditions, $(H - F)$, is drastically reduced to 3.5 min. In the case of sprinklers, 3 min may be assumed for detection time. Sprinklers, if they operate satisfactorily, will reduce the fire severity and the rate of growth of fire and smoke; they also have a high probability of extinguishing the fire. This performance will increase the time (F) for the commencement of untenable conditions, by say, 4 min. Calculations would show that, if sprinklers are installed in multioccupancy dwellings, $B + E - F = -1.5$ and $H - F = 1.5$ such that the fatality rate is reduced to 0.0009 approximately (see Ramachandran (1993a)). A combination of detectors and sprinklers would reduce the fatality rate to almost zero. The results mentioned above may be adjusted by assigning appropriate probabilities for the nonoperation of detectors and sprinklers.

It may be seen that, for evacuating a multiple-occupancy dwelling, occupants will have, on average, an extra time of 17 min if only detectors are installed, 19 min if only sprinklers are installed and 21 min if the building is equipped with both the systems. In these three cases, there is a clear justification for allowing an increase in the design evacuation time E beyond the level for E specified for a building without these active fire protection devices. For office buildings in the United Kingdom, a design evacuation time of 2.5 min has been recommended in British Standard BS5588.

Depending on the reliability of active fire protection devices and factors such as physical or mental ability of escaping occupants, the increase in E can be up to a limit such that an acceptable level specified for fatality rate per fire is not exceeded. If this level is 0.005, for example, $t(= H - F)$ should not exceed 8.4 min. This level for t can only be achieved if the average discovery time, D, is less than 5.9 min in the absence of automatic fire detection systems; $(B + E - F) = 2.5$ min in this case as discussed earlier. With automatic detectors, $D = 1$ such that $(B + E - F)$ can be increased to 7.4 min. In this case, the design evacuation time, E, can be increased by 4.9 min if B and F are assumed as constants. Hence, E can be increased to 7.4 min for a building with detectors if it is 2.5 min for a building without detectors. Similar calculations would show that the design evacuation time should not exceed 9.4 min if only sprinklers are installed and 11.4 min if the building is equipped with both the systems. A relaxation (increase) in the design evacuation time (E) for a building equipped with fire protection systems would allow for an increase in the maximum travel distance specified in fire safety codes or standards.

The method described in this section is a simple technique based on a regression analysis for estimating approximately the correlation between the fatality rate and the time $t(= H - F)$ measuring the duration of exposure to untenable conditions. The method restricts the variable t to positive values and does not take into account the uncertainties (standard deviations) associated with the random variables H and F. If sufficient data are available for establishing the probability distributions of H and F, a more complex probabilistic method suggested by Ramachandran (1993b) can be applied to estimate the design evacuation time.

8.9 Multiple-Death fires

Fires can not only cause direct damage to human life in terms of fatal and nonfatal casualties but also indirect losses, for example, distress and financial loss to the families of the victims and to the society at large. The aggregate disutility or consequences due to fire deaths would be generally low for single-death fires and high for multiple-death incidents (Ramachandran, 1988). The disutility associated with a single fire with, say, ten deaths is greater than the total disutility caused by ten fires each with a single-death. Catastrophies have serious social and political consequences.

For the reasons mentioned above, it is necessary to analyze specially the characteristics of multiple-death fires and the trends in the rates of occurrence of these large incidents. Such studies are periodically carried out by the NFPA, United States, which publishes its findings in its journal. According to its recent report by Miller and Tremblay (1992), 52 catastrophic multiple-death fires in United States during 1991 claimed 342 lives. Just over half of these fires occurred in residential structures. The remaining 25 included 15 nonresidential structure fires, 9 vehicle fires and a wild fire. Nearly half of all deaths – 160 – occurred in residential structures, 92 in nonresidential buildings and 90 in fires outside of structures.

As revealed in the study mentioned above and earlier reports of the NFPA, the tragedies could have been prevented by adherence to NFPA codes and standards and basic fire safety principles.

Some of the recurring factors in multiple-death fires included the absence of operating smoke detectors and sprinklers, delays in fire department notification, and blocked or locked exits. Other major factors were inoperative or poorly functioning central heating systems, electrical code violations, and lack of escape planning, apart from socioeconomic factors such as poverty and homelessness.

In the United Kingdom, case studies of multiple-death fires have occasionally appeared in fire journals such as the one published by the Fire Protection Association. Only more than 20 years ago a detailed analysis of the characteristics of multiple-death fires was carried out by Chandler (1969). This report showed that during 1960 to 1966, about three-quarters of the multiple-death fires in buildings were in dwellings and most of these were in houses rather than flats. It is probable that most of these fires were in multiple occupancies, where more people were exposed to fires than in single occupancies. Among other occupancies, 10 people were killed in a department store fire, 19 in a fire in a club above an industrial building and 5 in living accommodation above a launderette. A third of the outdoor multiple-death fires were in vehicles and about a quarter in caravans. The winter months accounted for about two-thirds of the multiple-death fires.

According to Chandler (1969) the most prominent of the known sources of ignition in multiple-death fires were smoking materials, oil heaters, children playing with fire, and vehicle crashes. Furniture or furnishings were the leading materials first ignited among known cases. Compared with fatal fires as a whole, a much higher proportion of multiple-death fires involved young children, whereas the proportion of old people was lower. Only 21% of these large life loss fires were confined to the room of origin and 11% spread beyond the building of origin. Provision of self-closing doors at the entrances to stairways in multiple occupancies and improvement of housing conditions were suggested by Chandler as measures likely to reduce the frequency of multiple-death fires.

The annual booklet on fire statistics published by the Home Office, UK, includes a table showing the breakdown of the total number of deaths according to four categories – one death, 2 to 4 deaths, 5 to 9 deaths and 10 or more deaths. This information is given for each year for a 11 year period up to the year to which the publication relates; fires in the last category are listed. According to the list in 1991 statistics, there were 5 fires involving 10 or more deaths during the period 1981 to 1991. These included a 13-death fire in a house in Deptford (1981), a 16-death fire at Abbeystead Waterworks (1984), a 55-death fire at Manchester airport (1985), a 56-death fire at Bradford City football ground (1985), and a 31-death fire at Kings Cross underground railway station (1987).

The annual publication of the Home Office, UK also contains a table on fire fatalities showing the breakdown of the number of fires according to six categories – no death, one death, two deaths, three deaths, four deaths, and five or more deaths. The total figures based on these data for the period 1978 to 1991 are given in Table 8.7 for single- and multiple-occupancy dwellings. A discrete (discontinuous) distribution applicable to a random variable with integer values may be fitted to these data for estimating the probability of occurrence of a given number of deaths in a fire. Poisson is one such distribution that has been widely used in the statistical literature for modeling the occurrences of rare events. Adopting an extended form of this distribution (Beard *et al.*, 1969)

$$p(x, t) = \exp(-\lambda t)(\lambda t)x/x! \qquad [8.8]$$

where

$$x! = x(x-1)(x-2)\cdots\cdots\cdots 2\cdot 1$$

and $p(x, t)$ is the probability of exactly x deaths occurring in a fire due to exposure to untenable conditions caused by combustion products for a period of t minutes.

Table 8.7. Frequency distribution of number of deaths

Number of deaths	Single-occupancy dwellings		Multiple-occupancy dwellings	
	Number of fires	Percentage of fires	Number of fires	Percentage of fires
0	491,532	98.9151	292,747	98.9014
1	4794	0.9648	3002	1.0142
2	421	0.0847	194	0.0655
3	110	0.0221	40	0.0135
4	45	0.0091	10	0.0034
5 or more	21	0.0042	6	0.0020
Total	496,923	100.0000	295,999	100.0000

Note: Single-occupancy dwellings.
Average number of deaths per fire ($\lambda\bar{t}$) = 0.012705
$\lambda = 0.000801$
$\bar{t} = 15.9$ min
Multiple-occupancy dwellings
Average number of deaths per fire ($\lambda\bar{t}$) = 0.012153
$\lambda = 0.000596$
$\bar{t} = 20.4$ min
Source: Fire Statistics, United Kingdom, 1978–1991.

The parameter λ in equation [8.8], as defined in Section 8.7, is the fatality rate or probability of one or more deaths during a unit period of one minute. It is an approximation for the exponential function defined in equation [8.2]. Also, $p(o, t)[= \exp(-\lambda t)]$ is the probability of no death during a period of t minutes such that $[1 - p(o, t)]$ is the probability of one or more deaths during this period as defined in equation [8.3].

For a population of fires, the variable t has an average $\bar{t}$. According to a property of the Poisson distribution, an estimate of $\lambda\bar{t}$ is provided by the overall average number of deaths per fire based on frequency distribution such as in Table 8.7. Assuming an average of, say, 8 deaths for the category 5 or more, calculations would show that $\lambda\bar{t}$ is approximately equal to 0.0127 for single-occupancy dwellings and 0.0122 for multiple-occupancy dwellings. Using the figures for λ given in Table 8.6 approximate values for $\bar{t}$ are 15.9 and 20.4 min, respectively, for these buildings. Figures in Table 8.7 include scenarios in which sudden exposure to risk conditions could have been quickly fatal.

For multiple-occupancy dwellings, if an average of 15 deaths is assumed for the category 5 or more deaths, $\lambda\bar{t}$ would only increase marginally to 0.0123 and $\bar{t}$ to 20.6 min. The values of $\lambda\bar{t}$ and $\bar{t}$ can be thus adjusted, if necessary, to take account of any reasonable average value for the maximum number of deaths. This maximum would depend on the average number, $\overline{N}$, of people at risk in the type of building for which $\lambda\bar{t}$ is estimated from data such as in Table 8.7. If $\bar{t}$ can be estimated from experimental and scientific investigations, the value of λ can be adjusted accordingly for any value of $\overline{N}$.

The average (fractional) number of deaths per fire, $\lambda\bar{t}$, estimated as described above, is simply the ratio between total number of deaths and total number of fires. This parameter is the same as the overall fatality rate per fire given in Table 8.6. The overall rates in Tables 8.6 and 8.7 differ slightly since Table 8.6 does not include some fires for which information on discovery time was not available.

In equation [8.8], both λ and t have been assumed as constants although they may be functions of x. An increase in the number of deaths may be the result of exposure to untenable conditions for a longer duration such that t would increase with x. If equation [8.8] is modified to take account of variations in λ and t, it would lead to a complex Poisson model, an application of which is beyond the scope of this book. Therefore, the simple model in equation [8.8] may be applied to practical problems in which λ and t can be assumed to be constants.

It may be required, for example, to estimate the probability of r or more deaths occurring in a fire if a group of occupants are exposed to untenable conditions for a specified period of t minutes. This probability based on equation [8.8] is given by

$$q_r(x,t) = 1 - p_r(x,t) \quad [8.9]$$

where

$$p_r(x,t) = \sum_{x=0}^{r-1} p(x,t)$$

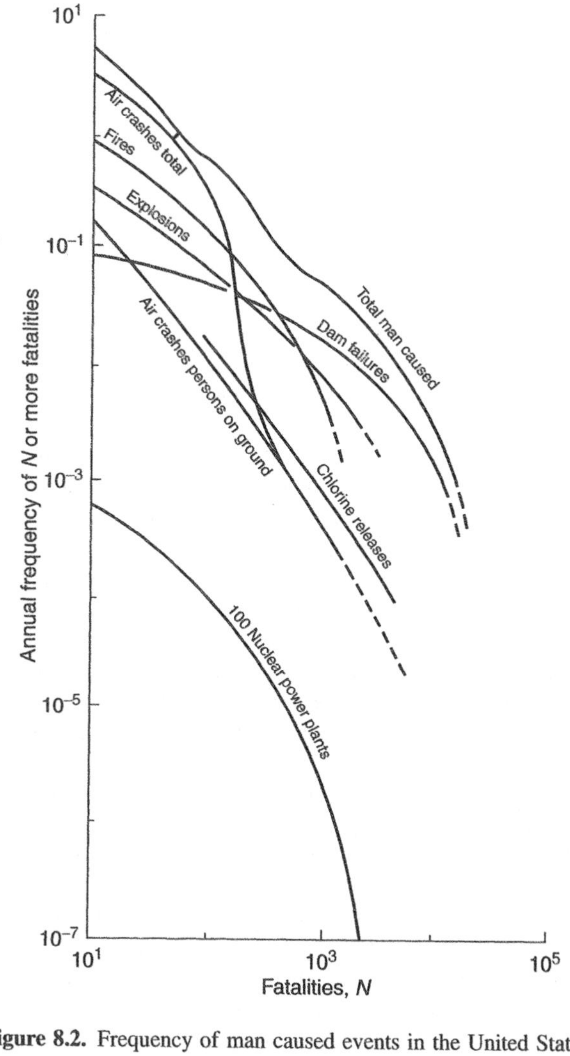

Figure 8.2. Frequency of man caused events in the United States

As one might expect, $q_r(x, t)$ would increase with the duration of exposure denoted by t. In multiple-occupancy dwellings, for example, calculations would show that the probability of 2 or more deaths would increase from 0.00020 for 30 min of exposure to 0.00044 and 0.00060 for exposure of 50 and 60 min respectively.

With $r \geq 2$, an appropriate value of t corresponding to an acceptable level for $q_r(x, t)$ may be selected for designing a large building according to the risk of multiple deaths occurring in a fire. This design value for t will in turn provide the design value for the evacuation time (E) as discussed in the previous section. Subject to the limit imposed on t, the magnitude of E may be adjusted to take account of the presence or absence of fire protection measures such as sprinklers, detectors, and smoke control systems.

In statistical literature $q_r(x, t)$ is known as the *survivor* or *tail* function, which is the complement of the cumulative distribution function $p_r(x, t)$. Rasmussen (1975) used this function, referred to later by Fryer and Griffiths (1979) as $f(N)$ lines, to compare multiple fatalities expected in various types of manmade hazards with hazards of pressurized water reactors in the United States. The figure reproduced in Figure 8.2 has been widely quoted in the subject of quantification of risk. The $f(N)$ relationship has been put forward by several authors for investigating risks due to various types of hazards. Rasbash (1984) has discussed these studies in order to define target probabilities for premises of different sizes (see Table 2.7).

8.10 Other measurements of life risk

The overall fatality rates, single or multiple, given in Tables 8.6 and 8.7 indicate current levels of life risk due to fires on a per fire basis. They denote the probabilities of one or more people dying in the event of a fire occurring in a building of the type considered. These probabilities can be expressed on an annual basis if they are multiplied by the probability of a fire occurring in a building in a year. The occurrence probability per year has been discussed in detail in Chapter 7. This probability would vary depending on the size and type of building as discussed with reference to Table 7.2.

Another factor affecting life risk is the number of occupants (N) in a building. The probability (rate) of death per, say, hundred people would be a useful measurement of life risk particularly for a large building. For any occupancy type, this parameter can be estimated by dividing the fatality rate per fire by the average number ($\overline{N}$) of people in a building. Special surveys may have to be carried out for obtaining information on $\overline{N}$. The fatality rate per person per fire can be expressed on an annual basis by multiplying it by the annual probability of fire occurrence.

Fatal Accident Frequency Rate (FAFR) is another measurement of life risk, which is the number of fatalities that occur during a hundred million man-hours of exposure to an occupation or activity. FAFR has been calculated for various industrial occupations such as nuclear and chemical industries and nonindustrial activities such as traveling by bus, train, car or air, canoeing, and rock climbing. For example, the FAFR for fire safety in the chemical industry as a whole has been estimated to be 4. On this basis, it has been arbitrarily decided in at least part of a major chemical firm that no single activity which any person is carrying out should contribute more than 10% to the FAFR, that is, 0.4 (Kletz, 1976).

As mentioned in Section 7.8, North (1973) has calculated rough values of FAFR for many occupancies in the United Kingdom. These only covered the years 1967 to 1969 and in cases in which deaths were infrequent the estimates had wide confidence limits. For fire deaths in a dwelling, the FAFR was 0.19 and in hotels 3.6. The latter figure was probably distorted by some serious multiple-fatality fires that happened during 1967 to 1969.

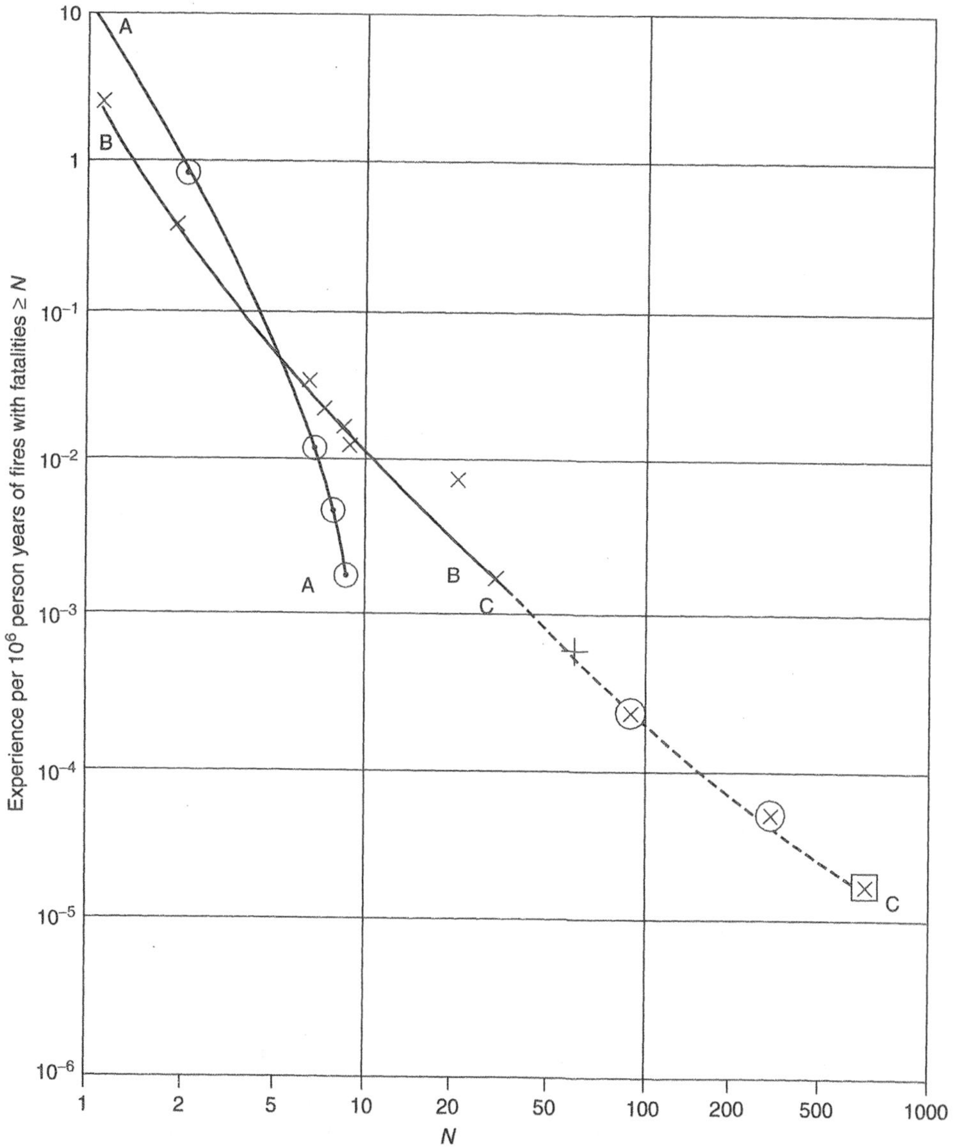

A ⊙ United Kingdom experiences *D* type (dwelling) fires 1963–1973

B × United Kingdom experiences *D* type (non-dwelling) fires 1963–1973

C + Points based on biggest fire in British Isles 1949–1978

C ⊗ Points based on four biggest fires in rich countries 1949–1978 (excl.Japan)

C ⊠ Dubious point based on biggest peace-time type fire in rich countries 1879–1978

Figure 8.3. Experience per 10^6 person years of fires with N or more fatalities.

An American report (Balanoff, 1976) allowed estimates to be obtained for FAFR for firemen on or following activities at the fireground. The report indicated 86 deaths per 100,000 firemen per annum and provided an estimate of 42 for FAFR. About half of the deaths were due to heart failure and 10% each due to building collapse, burns, and smoke inhalation. From these figures, assuming that only 5% of firemen's time was actually spent fire fighting, the FAFR for this activity was about 800 (Rasbash, 1978). Firemen might be expected to endure a substantially high risk during the small fraction of their working time in which they are engaged in the highly risky activity of fighting fires.

Another method of measuring life risk is based on the relationship between the records of actual fire disasters and the population experiencing them. The level of safety that this method reveals has not occurred by chance. It is the result of a continuing process of interaction between the social and technical development of the environment on the one hand and the legislative and regulatory process that has taken place over the years on the other hand. Figure 8.3 produced by Rasbash (1984) is an example of this method, quantifying the measure of safety from multiple-death and catastrophic fires to which people have become accustomed in everyday life under peacetime conditions.

Figure 8.3 omits fires not attended by fire brigades, particularly mine fires and offshore ship fires. It uses information from outside the United Kingdom based on countries with a similar background, particularly Western Europe, North America, and Australia, to extrapolate the UK experience to obtaining estimates of the probability of disasters, of 100 and even 1000 deaths. This extrapolation is rather more speculative but is reasonable because it is a simple extrapolation of the trend already established.

8.11 Impact of product choices on life risk – US method

In the United States, a fire risk assessment method has been recently developed to analyze the impact of changes made to a combustible product on its life-safety risk (see Bukowski *et al.*, (1990)). The method has been designed to calculate the expected severity (in deaths per fire) and the relative likelihood (as fire probability) of each of a large number of fire scenarios that may involve the product as the first item ignited or as a secondary contributor. Briefly, fire risk is measured in terms of both the probability of an event (fire) and the consequence of that event (e.g. deaths resulting from a fire). The problem is to predict how a change in the fire properties of a product (ignitability, heat release rate, toxic potency, etc.) will change the life-safety risk in a given occupancy.

The fire risk assessment method mentioned above combines the likelihood of a fire, based upon fire incident databases, with the expected consequences or severity of a fire, predicted by a computer based hazard calculation method, HAZARD 1, developed at the Centre for Fire Research, National Institute of Standards and Technology, USA. The method thus uses simulation models based on known laws of physics of fire, in combination with information on the behavior of people confronted with fire and the effects of heat and smoke on people, to estimate the severity of a specified fire. The method provides an organized structure for a large series of fire scenarios constructed to represent all the possible ways that a fire might involve the product considered.

A scenario is defined as a detailed description of a specific fire incident related to a building, a fire involving specific items in that building, and the persons occupying the building at the time of the fire. These details are drawn from a review of fire incident data from the national databases, focusing on – buildings and rooms where fires originate, combustible contents of the room, heat sources igniting the first item (flaming or smouldering), final extent of flame damage, factors

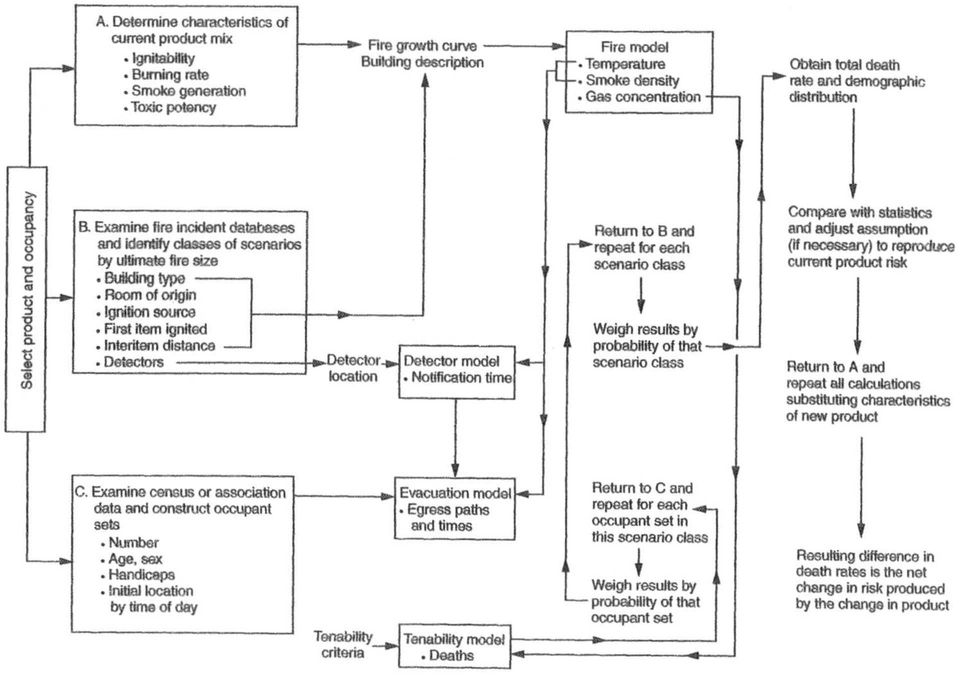

Figure 8.4. Modeling sequence to compute fire risk

contributing to fire spread, and location of any fatalities relative to the fire. Also considered are more specific building characteristics – numbers, types, and layouts of rooms and floors, dimensions of rooms, and the sizes of the openings connecting them.

Figure 8.4 illustrates the sequence of modeling activities used in the risk assessment process. As mentioned earlier, the universe of possible fires is divided into a set of well-defined scenarios. Thereafter, for each scenario, the method follows the sequence described in Figure 8.4 for each occupant set. The deaths-per-fire results for the scenario selected are estimated using the probability for each occupant set. Using the fire scenario probability and the total number of fires, one obtains the number of fires for that scenario. This is combined with the deaths-per-fire estimate for the scenario to obtain the number of deaths. These results are combined with similar results for all other scenarios to produce a sum that gives the estimated risk.

The procedure described above is conducted twice. The first computation is to produce a base line of fire risk associated with the mix of versions of the product in use. This is done either using the product's average characteristics, or, if possible, by conducting runs for the versions of the product in use and weighting the results by the share of product in use. The second computation is done using the characteristics of the new product – its peak rate of heat release, its relative ignitability and so on. A comparison of these two computations then produces a measure of the change in risk achievable by changing to the new product. This process has been demonstrated for each of four case studies – upholstered furniture in residences, carpet in offices, concealed combustibles in hotels, and interior finish in restaurants.

Consider, as an example, the case study (July, 1990), concerned with fires involving upholstered furniture in homes. The output of the method consisted of a series of tables giving fire death rates

Table 8.8. Upholstered furniture in residences – Output from US risk model deaths per 100 fires

	With smoke detectors			Without smoke detectors		
	Day	Evening	Night	Day	Evening	Night
All causes	2.73	1.50	1.00	2.73	1.50	29.74
Causes						
O_2[a]	0.00	0.00	0.00	0.00	0.00	0.00
CT[b]	0.00	0.00	0.00	0.00	0.00	0.00
Convect. heat	2.73	1.50	1.00	2.73	1.50	29.74
Occupants						
Adults	0.13	0.17	0.00	0.13	0.17	0.19
Elderly	0.00	0.00	0.00	0.00	0.00	0.00
Child, 12–18	0.00	0.00	0.00	0.00	0.00	0.00
Child, 3–12	0.00	0.00	0.00	0.00	0.00	0.00
Child, 0–3	0.13	0.17	0.01	0.13	0.17	28.18
Impaired	2.47	1.16	0.99	2.47	1.16	1.08
Drunk	0.00	0.00	0.00	0.00	0.00	0.00

[a]Oxygen deprivation
[b]Concentration–Time product.

in various categories such as age or sex of victim, time of day, type of building, and room of origin. Results were obtained for the class of fire scenarios sharing common characteristics: for example, upholstered furniture first item ignited, ignition sources other than smoking materials, starting in the living room of a ranch house and spreading beyond the living room with bedroom doors open. Table 8.8 is an example for the scenario mentioned above extracted from the report showing a summary of expected deaths per 100 fires with and without smoke detectors for day, evening, and night. These results relate to convected heat as the cause. Oxygen deprivation and toxicity of smoke (measured by the concentration–time product) were the other two causes considered for which the deaths per 100 fires were estimated to be zero. Table 8.8 details how the death per fire numbers are divided among the occupant types. A total of 155 deaths were estimated for this scenario. For the "base" case consisting of living room and bedroom fires, a total of 624 deaths was predicted by considering several scenarios.

Selected studies were conducted to test the sensitivity of the results to changes in key assumptions. These studies covered three input categories – occupant, fire modeling, and building size (volume). The occupant variables examined focused on assumptions related to escape through windows and rescue by persons outside the residence, smoke awareness at night, and location of occupants impaired by alcohol at night. Fire modeling variables included the extent of the smouldering period for upholstered furniture and the impact of breaking of a window in the fire room at flashover, providing an additional source of oxygen.

The "new" upholstered furniture selected had the same fire properties as the "base" case discussed above. However, the materials used in the "new" product would generate smoke with a tenfold increase in toxic potency over the "base" case. Consequently, a comparison of the results indicated a 46% predicted increase from 624 to 909 deaths. Smouldering fire scenarios contributed three-fourths of the increase while flaming scenarios contributed the remaining one-fourth. For the "basic" case, the cause of death was always convected heat. However, for the "new" product, toxicity was the causal factor in 96% of deaths.

Symbols

B	recognition period; gathering phase
D	discovery period
E	elapsed time since the commencement of movement until a safe place is reached inside the building
F	time for a combustion product to attain untenable conditions
FAFR	Fatal Accident Frequency Rate
H	$= (D + B + E)$
K	$= (1 - W)$
N	number of people at risk
p	$= 1 - \exp(-\lambda) \cong \lambda$
$p_d(t)$	probability of one or more deaths during period $t = 1 - \exp(-\lambda t) \cong \lambda t$
$p(x, t)$	probability of exactly x deaths in a fire due to exposure to untenable conditions for t minutes
$p_r(x, t)$	cumulative probability from $x = 0$ to $(r - 1)$ of the extended Poisson distribution $= 1 - \exp(-\lambda t) \cong \lambda t$
$q_r(x, t)$	probability of r or more deaths $= 1 - p_r(x, t)$
r	number of deaths occurring in a fire in the equations for $p_r(x, t)$ and $q_r(x, t)$
t	period spent under untenable conditions $= H - F$
W	$= \exp[-\lambda(B + E - F)] \cong 1 - \lambda(B + E = F)$
x	number of deaths in a fire in the equation for $p(x, t)$
λ	increase in casualty rate per unit time

References

Balanoff, T (1976). *Fire Fighter Mortality Report*, International Association of Firefighters, Washington, DC.

Beard, R E, Pentikainen, T and Pesonen, E (1969). Risk theory. *Methuen's Monographs on Applied Probability and Statistics*, Mehuen & Co Ltd, London.

Berl, W G and Halpin, B M (1976). Fire related fatalities. *An Analysis of their Demography, Physical Origins and Medical Causes*, Fire Standards and Safety ASTM STP 614, A F Robertson (Ed.), American Society for Testing and Materials, Philadelphia, pp. 26–54.

Bowes, P C (1974). *Annals of Occupational Hygiene*, **17**, 143.

Bukowski, R W *et al* (1987). Hazard 1 Vol 1, *Fire Hazard Assessment Method*, National Bureau of Standards, Centre for Fire Research Report NBSIR 87–3602, p. 96.

Bukowski, R W *et al* (1990). *Fire Risk Assessment Method: Description of Methodology*, National Fire Protection Research Foundation, Quincy, MA, USA.

Canter, D (1985). *Studies of Human Behavior in Fire: Empirical Results and their Implications for Education and Design*, Building Research Establishment Report, Fire Research Station, UK.

Chandler, S E (1969). *Multiple Death Fires*, Fire Research Technical Paper No 22, Ministry of Technology and Fire Offices' Committee, Joint Fire Research Organization, UK.

Chandler, S E (1971). *Deaths in Fires Attended by Fire Brigades, 1969*, Fire Research Technical Paper No 26, Ministry of Technology and Fire Offices' Committee, Joint Fire Research Organization, UK.

Chandler, S E (1972). *Deaths in Fires Attended by Fire Brigades, 1970*, Fire Research Technical Paper No 28, Ministry of Technology and Fire Offices' Committee, Joint Fire Research Organization, UK.

Chandler, S E (1979). *The Incidence of Residential Fires in London – The Effect of Housing and Other Social Factors*, Information Paper IP 20/79, Building Research Establishment, Fire Research Station, UK.

Chandler, S E and Baldwin, R (1976). Furniture and furnishings in the home – some fire statistics. *Fire and Materials*, **7**, 76.

Chitty, R and Cox, G (1988). *ASKFRS: An Interactive Computer Program for Conducting Fire Engineering Applications*, Fire Research Station, UK.

Clarke, F B and Ottoson, J (1976). Fire death scenarios and fire safety planning. *Fire Journal*, **70**(3), 20–22, 117–118.

Fahy, R (1986). Fatal Fires and Unsupervised Children *Fire Journal*, 19–24.

Fryer, L S and Griffiths, R F (1979). *United Kingdom Atomic Energy Authority*, Safety and Reliability Directorate, SRD R 149.

Hall, J R (1990). Fire in the U.S.A. and Japan, *International Fire comparison Report 1*, National Fire Protection Association, Quincy, MA, USA.

Harland, W A and Woolley, W D (1979). *Fire Fatality Study*, University of Glasgow, Building Research Establishment Information Paper IP 18/79, Fire Research Station, UK.

Harmathy, T Z and Mehaffey, J R (1985). *Design of buildings for Prescribed Levels of Structural Fire Safety: Fire Safety Science and Engineering*, ASTM STP 882, T Z Harmathy (Ed.), American Society for Testing and Materials, Philadelphia, pp. 160–175.

Karter, M J (1986). Patterns of fire deaths among the elderly and children in the home. *Fire Journal*, **March–April**, 19–25.

Kletz, T A (1976). *The Application of Hazard Analysis to Risks to the Public at Large. World Congress of Chemical Engineering*, Session A5, Amsterdam, July 1976.

Maclean, A D (1979). *Fire Losses – Towards a Loss – Attendance Relationship*, Fire Research Report No 17/79, Home Office Scientific Advisory Branch, UK.

Marchant, E W (1980). *Modelling Fire Safety and Risk. Fires and Human behaviour*, D Canter (Ed.), John Wiley & Sons, pp. 293–314.

Miller, A L and Tremblay, K J (1992). 342 die in catastrophic fires in 1991. *NFPA Journal*, 63–73.

North, M A (1973). *The Estimated Fire Risk of Various Occupancies*, Fire Research Note No 989, Fire Research Station, UK.

Purser, D A (1988). Toxicity assessment of combustion products. *SFPE Handbook of Fire Protection Engineering*, National Fire Protection Association, Quincy, MA, USA, Section 1, Chapter 14, pp. 200–245.

Ramachandran, G (1988). Utility theory. *SFPE Handbook of Fire Protection Engineering*, National Fire Protection Association, Quincy, MA, USA, Section 4, Chapter 8, pp. 64–73.

Ramachandran, G (1990). Probability-based fire safety code. *Journal of Fire Protection Engineering*, **2**(3), 75–91.

Ramachandran, G (1993a). Early detection of fire and life risk. *Fire Engineers Journal*, **53**(171), 33–37.

Ramachandran, G (1993b). Probabilistic evaluation of design evacuation time. *Proceedings of the CIB W14 International Symposium on Fire Safety Engineering*, University of Ulster, Northern Ireland, Part 1, pp. 189–207.

Rasbash, D J (1978). *Statement of the Fire Problem and Formulation of Objectives for Fire Safety*, Lecture at University of Edinburgh, Department of Fire Safety Engineering.

Rasbash, D J (1984). Criteria for acceptability for use with quantitative approaches to fire safety. *Fire Safety Journal*, **8**, 141–158.

Rasmussen, N C (1975). *An Assessment of Accident Risks in US Commercial Power Plants*, WASH 1400.

Sekizawa, A (1988). *Comparison Analysis on the Characteristics of Residential Fires Between the United States and Japan, Summaries of Technical Papers of Annual Meeting*, Architectural Institute of Japan.

Sekizawa, A (1991). Statistical analyses on fatalities. Characteristics of residential fires. *Fire Safety Science – Proceedings of the Third International Symposium*, Elsevier Applied Science, London and New York, pp. 475–484.

Sime, J D (1986). Perceived time available: the margin of safety in fires, *Fire Safety Science: Proceedings of the First International symposium*, Hemisphere Publishing Corporation, New York, pp. 561–570.

Woolley, W D, Ames, S A, Pitt, A I and Buckland, K (1978). *The Ignition and Burning Characteristics of Fabric Covered Foams*, Building Research Establishment, Current Paper CP30/78, Fire Research Station, UK.

9 PROPERTY DAMAGE

9.1 Introduction

As defined in Section 7.2, fire risk in a building is the product of two components – probability of fire starting and probable damage in a fire. Damage to life has been discussed in Chapter 8, while this chapter is concerned with direct material damage to a building and its contents. Note that a fire can also cause indirect/consequential losses (see Ramachandran (1995)). Direct damage can be measured in terms of area damaged, spatial extent of spread, and duration of burning or financial loss. Expected values of these four random variables and their correlations can be estimated approximately by applying the statistical models described in Sections 7.4 to 7.6. A better description of the random variation (uncertainties) governing area damage or financial loss is provided by a probability distribution, with nature and uses in fire protection problems that are discussed in the next section.

Fire losses occurring every year in buildings of any type can be regarded as a sample of observations from a "parent" probability distribution with large losses constituting the "tail" of this distribution. The behavior of the "tail" over a period of years (samples) can be studied with the aid of extreme value distributions (Section 9.3) that have found practical applications in many engineering problems. In the absence of information on small losses, extreme value models can be applied to large losses in order to estimate the average loss in all fires, large and small, in a particular building or group of buildings.

The relative effects of several factors and their interactions considered as "independent" variables can be evaluated simultaneously (not individually) by performing a multiple regression analysis (Section 9.4) with financial loss, or area damaged, as the "dependent" variable. The probability of loss or damage exceeding a specified level or probability of fire spreading beyond, say, the room of origin can also be used as a dependent variable. In such a case, the "logit" transformation is usually applied to the probability to satisfy certain statistical assumptions.

9.2 Probability distribution

Many factors affect the spread of fire in a building and hence the level of damage sustained in a fire is a random variable with a probability distribution. This distribution expresses mathematically the probabilities with which the damage in a fire could reach various amounts. The shape of this

Evaluation of Fire Safety D. Rasbash, G. Ramachandran, B. Kandola, J. Watts and M. Law
 ISBN: 0-471-49382-1

distribution is central for an assessment of fire risk in a building and the determination of fire protection requirements.

Denoting the financial loss in a fire by x, the nature of the probability distribution of x has been investigated in detail by Ramachandran (1974, 1975a) and Shpilberg (1974) and other authors mentioned in these papers. According to these studies, fire loss distribution is skewed (nonnormal or asymmetrical) and in general the transformed variable $z (= \log x)$, that is, logarithm of loss has a probability distribution belonging to the "exponential type." This type, defined by Gumbel (1958) with reference to the limiting (asymptotic) behavior of the random variable at the tail includes exponential, normal, lognormal, chi square, gamma, and logistic distributions.

Among "exponential type" distributions, normal for z, which is the same as log normal for x, has been recommended widely for modeling fire insurance claims. Figure 9.1 is a hypothetical density function $f(z)$ depicting a normal (symmetrical) distribution for z, logarithm of loss x (to base 10). (The density function of x is given by $f(\log x)/x$). The density function, also known as *frequency distribution*, provides an estimate of the relative frequency of observations between two values z_p and z_q. This frequency is given by the shaded area expressed as a proportion of the total area under the frequency curve bounded by the horizontal axis. The relative frequency also corresponds to the probability of logarithm of loss having a magnitude between z_p and z_q.

Figure 9.2 is the curve based on the (cumulative) distribution function of z. This function denoted by $F(z)$ is such that its derivative is the density function $f(z)$ shown in Figure 9.1. The probability of logarithm of loss in a fire being less than or equal to z is given by $F(z)$; this is also the probability of loss being less than or equal to the corresponding value of x. The tail or survivor function $\phi(z)(= 1 - F(z))$ denotes the probability of logarithm of loss (or loss) in a fire exceeding z (or x).

Exponential for z or Pareto for x has also been considered by some actuaries for the distribution of amounts claimed under fire insurance contracts. This distribution can be theoretically justified

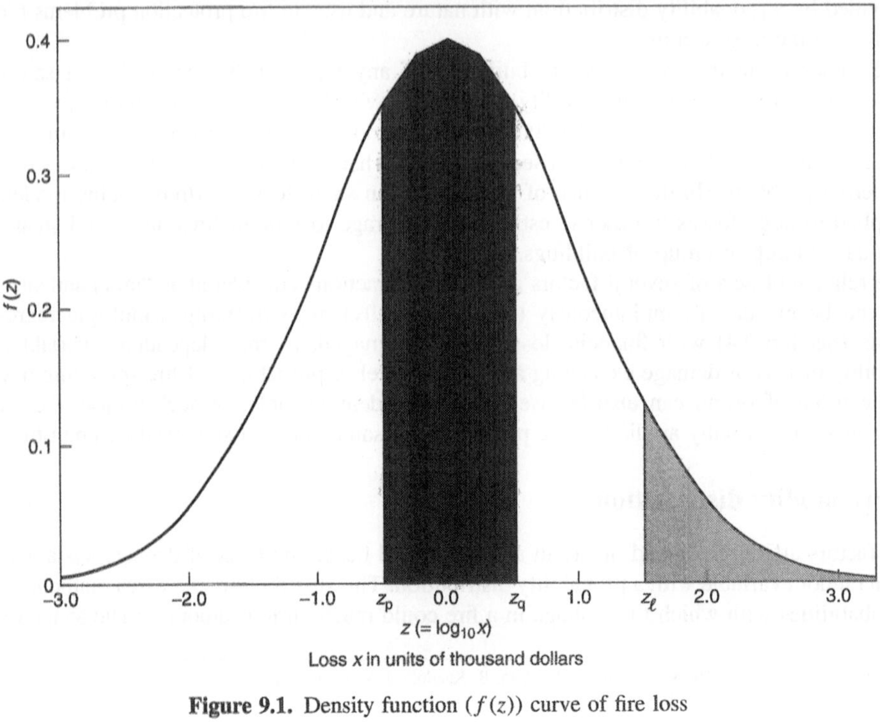

Figure 9.1. Density function ($f(z)$) curve of fire loss

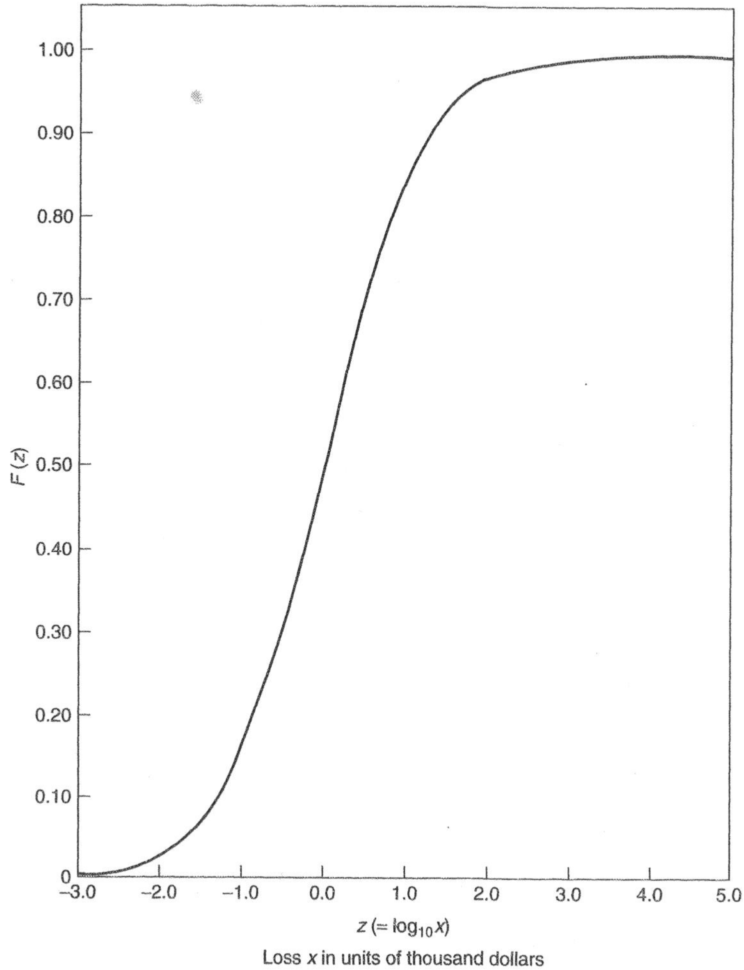

Figure 9.2. Cumulative distribution curve ($F(z)$) of fire loss

by considering fire damage as being the outcome of a Random Walk stochastic process (see Section 15.7). For the exponential distribution

$$F(z) = 1 - \exp(-hz); \quad \phi(z) = \exp(-hz); \quad z \geq 0$$
$$f(z) = h\exp(-hz)$$

such that log $\phi(z)$ has a linear relationship with z. With $z = \log x$, Pareto distribution function for x is given by

$$V(x) = 1 - x^{-h}, \quad x \geq 1$$

with its derivative

$$v(x) = hx^{-h-1}$$

as the density function.

For the exponential distribution, the "failure rate" given by

$$h = f(z)/[1 - F(z)] = f(z)/\phi(z)$$

is a constant independent of z. It is more realistic, however, to treat h as a function of z tracing a curve resembling a "bath tub" (see Section 15.7 and Ramachandran (1975a)).

If the early stages of growth, which are not economically important, are disregarded, then h is an increasing function for large values of z. If the increase is linear,

$$\phi(z) = \exp[-(c_1 z + c_2 z^2)] \quad c_2 > 0$$

where c_1 and c_2 are constants. In this case, the fire loss in the original scale x has the tail probability

$$\phi(x) = x^{-(c_1 + c_2 \log x)} \quad x \geq 1 \qquad [9.1]$$

The distribution according to equation [9.1] and described in Figure 9.3, fits some UK data on fire losses more accurately than Pareto for which $c_2 = 0$. (Ramachandran, 1967). If h is assumed to increase exponentially, x will have one of the forms of Weibull distribution (Ramachandran, 1974, 1975a).

If figures for financial loss are available for all the fires that occurred in a risk category, standard statistical methods or a graphical method can be applied to identify the probability distribution, which would provide the best fit for the data analyzed. But in most countries these data are generally available only for large fires, which in the United Kingdom, for example, are currently defined as fires costing £50,000 or more in property damage. The threshold level that was £10,000 until 1973 has been gradually increased over the years because of inflation and the need to keep the number of large fires to be reported by insurance companies at a manageable level. This led to the development of extreme value statistical models discussed in the next section.

However, a probability distribution can be constructed for area damaged for which, particularly in the United Kingdom, data are available for all sizes of fires. The probability of area damage, d, being less than or equal to d is given by the cumulative distribution function $G(d)$ and the

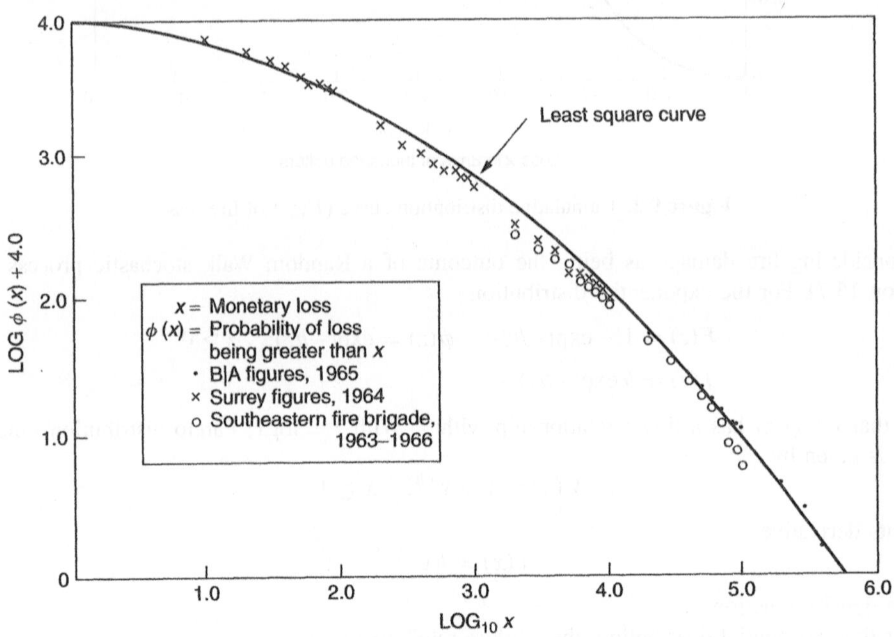

Figure 9.3. Distribution function of loss frequency for fires attended by fire brigades

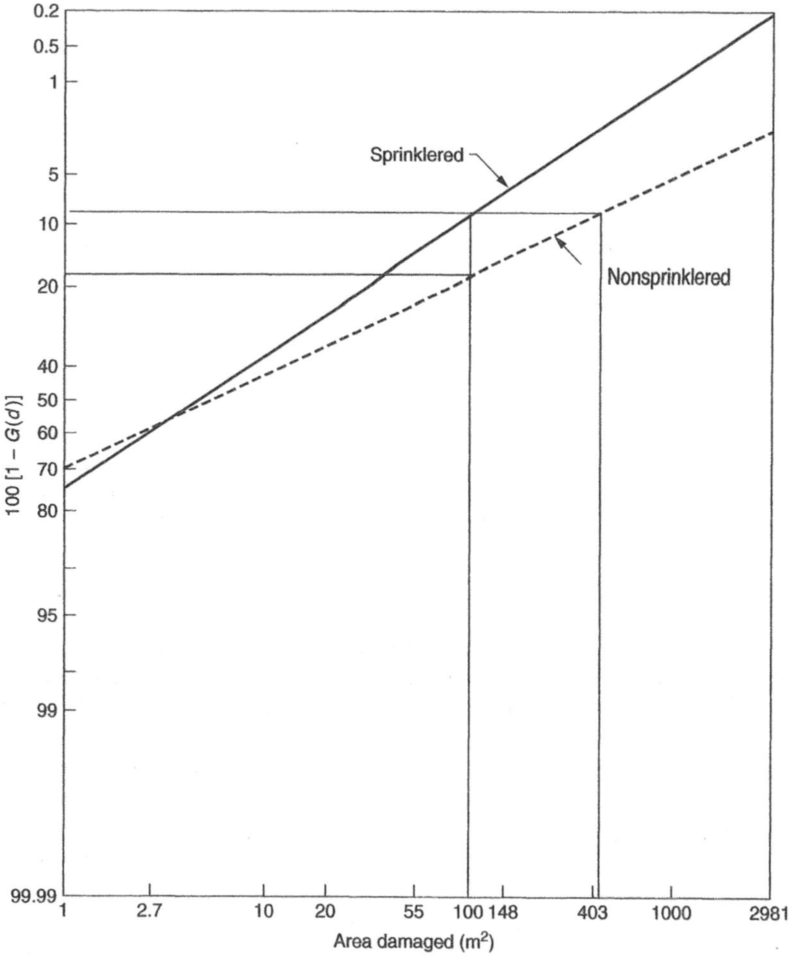

Figure 9.4. Textile industry, United Kingdom – probability distribution of area damaged
$G(d)$ = Probability of damage being less than or equal to d
$1 - G(d)$ = Probability of damage exceeding d

probability of damage exceeding d by $[1 - G(d)]$. Figure 9.4 is an example (textile industry) based on fire brigade data and shows the relationship between d and $[1 - G(d)]$ for a building with sprinklers and a building without sprinklers (Ramachandran, 1988a). The area damage is on a log scale since, as revealed by several statistical studies, this random variable, like financial loss, has a skewed probability distribution such as log normal. The values of the parameters of this distribution vary from one type of building to another and with the effectiveness of fire protection measures.

A log normal was fitted to the data pertaining to Figure 9.4, disregarding fires with damage less than 1 m^2 and following a method appropriate for a "censored" sample. For the range equal to or greater than 1 m^2, the mean (μ_{zd}) of the natural logarithm of area damage was obtained as 0.02 for the sprinklered building and 0.75 for the nonsprinklered case. The standard deviation (σ_{zd}) of the natural logarithm area damage was estimated to be 2.46 and 2.87 for the sprinklered and nonsprinklered cases respectively.

According to statistical theory, the expected (average) value of area damage in the range equal to or greater than 1 m^2 is given by

$$\exp[\mu_{zd} + (\sigma^2{}_{zd/2})]\frac{1 - H(h_o - \sigma_{zd})}{1 - H(h_o)} \qquad [9.2]$$

where

$$h_o = -(\mu_{zd}/\sigma_{zd})$$

and $H(h)$ is the cumulative distribution function of the standard normal variable

$$h = (z_d - \mu_{zd})/\sigma_{zd}$$

with z_d denoting the logarithm of area damage. Using the formula in ([9.2]), it may be calculated that the average damage is 42 m^2 for a sprinklered building and 217 m^2 for a building without sprinklers. Results based on area damage can be converted to provide estimates for financial losses by using an approximate value for loss per square meter (Ramachandran, 1988a).

It appears from Figure 9.4 that an initial damage of 3 m^2 is likely to occur before the heat generated in a fire is sufficient to activate a sprinkler system. It is apparent that, in the range greater than 3 m^2, a successful operation of sprinklers would reduce not only the expected (average) area damage but also the probability of damage exceeding any value. Insurance firms can use such results in determining appropriate reductions in fire insurance premiums for buildings equipped with sprinklers.

Also, according to Figure 9.4, the probability of damage in a fire exceeding, say, 100 m^2 is about 0.18 if the building has no sprinklers and 0.08 if the building has sprinklers that operate in the event of a fire. Hence, the owner of a building with sprinklers can take a risk and opt for a deductible (self-insurance) level equivalent to a damage of 100 m^2. In this case, in addition to the normal reduction in fire insurance premium due to sprinklers, the owner can expect a further reduction in the premium appropriate to the deductible level. A figure such as Figure 9.4 would also enable an insurance firm to calculate reductions in "risk premiums" for various deductible levels for a sprinklered or nonsprinklered building. The "risk premium" would decrease with increasing levels of the deductible. When calculating the premium to be charged for a property, an insurance firm usually adds two types of "loadings" to the risk premium – a "safety loading" and another loading to cover the insurer's operating costs, which include profits, taxes, and other administrative expenses. Ramachandran (1994) has developed statistical techniques for calculating rebates for deductibles and protection measures in industrial fire insurance.

From the point of view of a fire safety code, if a damage of 100 m^2 is acceptable, a sprinklered building should be given some concessions, and increase in the compartment size and/or a reduction in the fire resistance requirement for a compartment. Such concessions are also justified because of the fact that, if a probability level of 0.08 is acceptable, the damage would be 500 m^2 if not sprinklered compared with 100 m^2 if sprinklered. Figure 9.4 can be regarded as applicable to compartments of different sizes since, in probabilistic terms, damage within a compartment constitutes a substantial part of the total damage expected in a building. The probability of a fire spreading beyond a compartment is very small.

Figure 9.5 is an example based on Pareto distribution for area damage. This figure, like Figure 9.4, depicts the relationship between the logarithm of the survivor function $[1 - G(d)]$ and the logarithm of damage d. If d has a Pareto distribution, as mentioned with reference to equation [9.1],

$$1 - G(d) = d^{-W}, \quad d \geq 1 \qquad [9.3]$$

Equation [9.3] has been derived in Section 15.7, following a Random Walk stochastic model (see equation [15.4]).

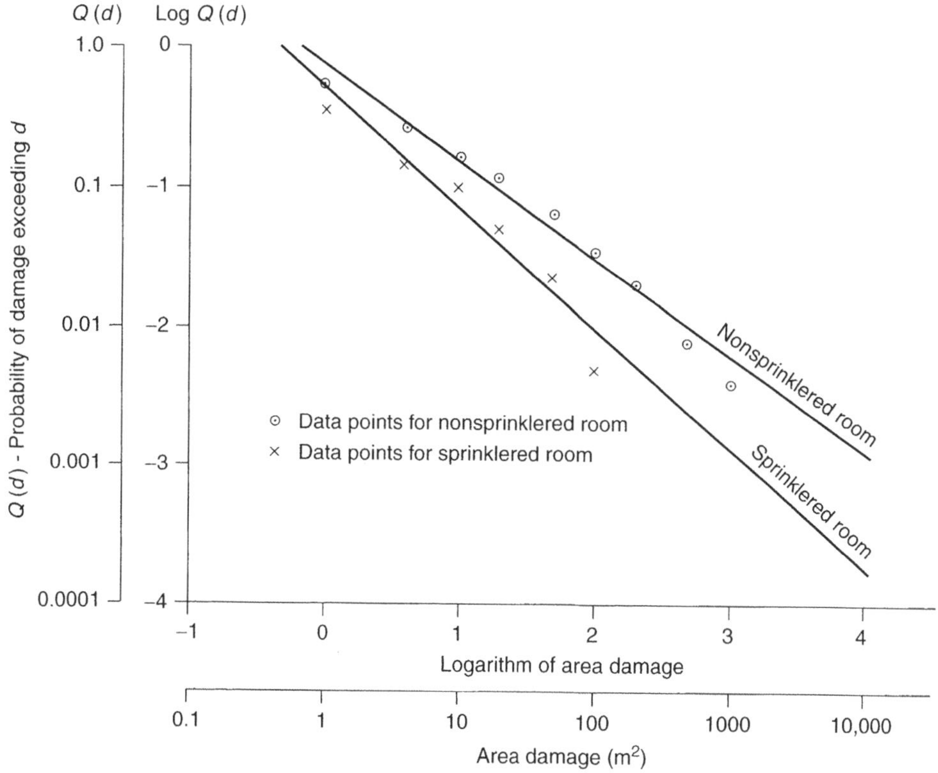

Figure 9.5. Pareto distribution of area damage – retail premises (public areas)

9.3 Extreme value theory

9.3.1 INTRODUCTION

In the context of the extreme value theory, the probability distribution of financial loss or area damage discussed in the previous section is known as the "parent" or "initial" distribution. Large losses fall at the tail of the parent distribution as shown by the shaded portion at the extreme right of the hypothetical distribution in Figure 9.1 with z_l, representing the threshold value. Financial loss figures are generally available only for large fires with losses exceeding a threshold level. These large losses constitute a very small percentage of the total number of fires in a risk category and hence are not amenable to analysis by "standard" statistical methods. Extreme order theory as developed by Ramachandran in a series of papers provides a mathematical framework for making the best use of the information provided by large losses – for a review of these studies see Ramachandran (1982a, 1988b).

When applying the asymptotic theory discussed in this section, the number of fires, n, occurring during a given period, has to be large, say, more than 100. Owing to this requirement, it will be necessary to consider fire losses in a group of buildings with similar fire risks. The asymptotic theory only provides approximate results for three classes of parent distributions discussed by Gumbel (1958). One of the classes is the "exponential type," which includes the exponential, normal, and log normal. Exact results are practically impossible to obtain for large n. For certain parent distributions including normal, exact results are possible for small samples of sizes (n) not exceeding 50.

9.3.2 EXTREME ORDER DISTRIBUTIONS

The logarithms of losses in n fires constitute a sample of observations generated by the parent distribution $F(z)$. If these figures are arranged in *decreasing order of magnitudes*, the mth value in this arrangement may be denoted by $z_{(m)n}$, which is the logarithm of the mth loss. For the largest (maximum) value, the subscript m takes the value of 1 (first rank). The random variables $z_{(m)n}$ $(m = 1, 2, \ldots r)$ are referred to as extreme order statistics. Out of n fires, loss figures are only available for the top r large fires.

The n observations (fires) during a period, say, a year may be considered as a sample in a series of samples each with n observations. The values pertaining to $z_{(m)n}$ in these samples will be distributed according to a probability law. (The loss figures should be corrected for inflation and the corrected figures used in the analysis). For "exponential type" parent distributions, the probability density function of $z_{(m)n}$ has the structure described in equations [9A.1] to [9A.4], if n is large and m has a comparatively small value.

The parameter $b_{(m)n}$ is the modal (most probable) value of $z_{(m)n}$ and $a_{(m)n}$, the value of the "intensity function" of z at $b_{(m)n}$ as given in equation [9A.5]. This function is also known as "failure rate" or "hazard function" in reliability theory and "force of mortality" in actuarial mathematics. The variable $y_{(m)}$, given by equation [9A.2], is known as the "reduced variate" in extreme value theory.

9.3.3 ESTIMATION OF EXTREME ORDER PARAMETERS

Estimation of the values of $a_{(m)n}$ and $b_{(m)n}$ is a simple problem with solutions provided by equations [9A.3] and [9A.4], if the value of n and the exact form of the parent distribution $F(z)$ are known together with the mean and standard deviation of z, the logarithm of loss. Although a normal distribution can be assumed for the variable z, that is, log normal distribution for loss, the mean and standard deviation of z are generally unknown since in most cases data on financial losses are only available for large fires and not for all fires. Estimation of these two parameters of the parent distribution from large loss data is discussed in Section 9.3.6. However, if values of $z_{(m)n}$ are available for N samples (years) each of size n (number of fires) approximately, one of the three methods described in Section 9A.2.2 may be applied for estimating $a_{(m)n}$ and $b_{(m)n}$ (Ramachandran, 1982a). These methods do not require the assumption of any specific form for the parent; it would be sufficient if the parent can be assumed to be of the "exponential type."

It is difficult to assume that the sample size n is a constant since the number of fires would generally vary from year to year. If n varies significantly, the approximate correction discussed in Section 9A.3 of the appendix should be included in the estimation process (Ramachandran, 1974, 1982a).

Consider, as an example, data relating to the top 5 large losses ($m = 1$ to 5) that occurred in a group of buildings during a five-year period. These losses were corrected for inflation by expressing them at the first-year prices with the aid of indices of retail prices. The logarithms of corrected losses (base e) denoted as z are given in Table 9.1. for each extreme (m) and each year together with their ranks (R) in an increasing order of magnitudes. Assuming that the logarithms of (corrected) losses have an "exponential type" parent distribution, the "reduced" extremes (unadjusted) corresponding to ranks R have been obtained using tables of Incomplete Gamma Functions (see Section 9A.2). These extreme values have been denoted as y in Table 9.1.

The corrected reduced extremes for each m denoted as y' in Table 9.1. are based on the following formula:

$$y' = y + \log_e(n_j/n) \qquad [9.4]$$

Table 9.1. Natural logarithms (z) of large losses corrected for inflation and the corresponding reduced extremes (y') adjusted for sample size (n_j)

Extreme (m)	Variable	Year				
		1	2	3	4	5
	z	9.3525(1)	9.9573(3)	10.7529(5)	9.6262(2)	10.0868(4)
1	y	−0.5821	0.3665	1.6998	−0.0950	0.9040
	y'	−0.5964	0.2885	1.7446	−0.0542	0.9848
	z	9.1429(1)	9.3605(2)	10.1955(5)	9.5995(3)	9.7619(4)
2	y	−0.4812	−0.1355	1.0070	0.1751	0.5198
	y'	−0.4955	−0.2135	1.0518	0.2159	0.6006
	z	9.1355(1)	9.3151(2)	9.8309(5)	9.4945(3)	9.6145(4)
3	y	−0.4197	−0.1352	0.7620	0.1147	0.3871
	y'	−0.4340	−0.2132	0.8068	0.1555	0.4679
	z	9.0836(1)	9.2937(2)	9.8050(5)	9.4633(3)	9.5735(4)
4	y	−0.3776	−0.1299	0.6321	0.0853	0.3171
	y'	−0.3919	−0.2079	0.6769	0.1261	0.3979
	z	8.9906(1)	9.2682(2)	9.7231(5)	9.4232(4)	9.3249(3)
5	y	−0.3468	−0.1240	0.5483	0.2728	0.0680
	y'	−0.3611	−0.2020	0.5931	0.3136	0.1488

Notes:

1. The figures within brackets denote the ranks (R) of the variable z in increasing order of magnitude.
2. $y' = y + c$, where c is the adjustment factor for sample size n_j, number of fires, according to the following table with $n = 1200$ fires.

	Year				
	1	2	3	4	5
n_j	1183	1110	1255	1250	1301
$c = \log_e (n_j/n)$	−0.0143	−0.0780	0.0448	0.0408	0.0808

Table 9.2. Extreme value parameters

Extreme (m)	Linear estimation		Moment estimation	
	$a_{(m)n}$	$b_{(m)n}$	$a_{(m)n}$	$b_{(m)n}$
1	1.7556	9.6855	1.7215	9.6800
2	1.5560	9.4630	1.5414	9.4616
3	1.8809	9.3948	1.8703	9.3944
4	1.6074	9.3690	1.5884	9.3681
5	1.5303	9.2816	1.4550	9.2783

where n_j is the number of fires in the jth year and n has the value 1200 to reflect approximately the average number of fires per year during the five-year period. The correction factor $\log_e(n_j/n)$ for each year is shown at the bottom of Table 9.1. Equation [9.4] follows from equation [9A.16].

For each extreme (m), the results in Table 9.2. following Linear Estimation were obtained by fitting the following straight line:

$$z_{(m)n} = b_{(m)n} + (1/a_{(m)n})y'_{(m)n} \qquad [9.5]$$

by the least square method. The values of $b_{(m)n}$ pertain to a sample size of $n = 1200$ fires. The value of $a_{(m)n}$ does not vary significantly with sample size.

Table 9.3. Values of $y_{(m)}$ for selected probability levels

Extreme (m)	Probability levels												
	0.005	0.010	0.025	0.050	0.100	0.250	0.500	0.750	0.900	0.950	0.975	0.990	0.995
1	−1.667	−1.527	−1.305	−1.097	−0.834	−0.327	0.367	1.246	2.251	2.971	3.677	4.601	5.296
2	−1.312	−1.199	−1.024	−0.863	−0.665	−0.297	0.175	0.733	1.325	1.728	2.112	2.600	2.962
3	−1.128	−1.030	−0.878	−0.741	−0.573	−0.268	0.115	0.552	1.002	1.300	1.579	1.929	2.184
4	−1.009	−0.920	−0.784	−0.661	−0.512	−0.245	0.086	0.456	0.830	1.075	1.301	1.581	1.784
5	−0.923	−0.841	−0.717	−0.604	−0.469	−0.227	0.068	0.395	0.721	0.932	1.125	1.364	1.535
6	−0.857	−0.781	−0.665	−0.560	−0.435	−0.213	0.057	0.352	0.644	0.832	1.003	1.213	1.362
7	−0.805	−0.733	−0.623	−0.525	−0.408	−0.201	0.048	0.320	0.587	0.757	0.912	1.100	1.235
8	−0.761	−0.693	−0.589	−0.496	−0.386	−0.191	0.042	0.295	0.542	0.698	0.840	1.013	1.136
9	−0.724	−0.659	−0.560	−0.472	−0.367	0.183	0.037	0.275	0.505	0.651	0.783	0.943	1.056
10	−0.693	−0.630	−0.535	−0.451	−0.351	−0.175	0.034	0.258	0.475	0.612	0.735	0.885	0.990
11	−0.665	−0.605	−0.513	−0.433	−0.336	−0.169	0.031	0.244	0.450	0.579	0.695	0.836	0.936
12	−0.640	−0.582	−0.494	−0.416	−0.324	−0.163	0.028	0.232	0.428	0.550	0.661	0.794	0.887
13	−0.619	−0.562	−0.477	−0.402	−0.313	−0.157	0.026	0.221	0.408	0.526	0.631	0.757	0.846
14	−0.599	−0.544	−0.462	−0.389	−0.303	−0.153	0.024	0.212	0.391	0.504	0.604	0.725	0.810
15	−0.581	−0.528	−0.448	−0.377	−0.294	−0.148	0.022	0.203	0.376	0.484	0.581	0.697	0.778
16	−0.565	−0.513	−0.435	−0.367	−0.285	−0.145	0.021	0.196	0.363	0.467	0.560	0.671	0.749
17	−0.550	−0.500	−0.424	−0.357	−0.278	−0.141	0.020	0.189	0.351	0.451	0.541	0.648	0.720
18	−0.536	−0.487	−0.413	−0.348	−0.271	−0.138	0.019	0.183	0.340	0.437	0.524	0.627	0.700
19	−0.524	−0.475	−0.403	−0.339	−0.264	−0.134	0.018	0.178	0.330	0.424	0.508	0.608	0.679
20	−0.512	−0.465	−0.394	−0.332	−0.258	−0.131	0.017	0.173	0.320	0.412	0.493	0.591	0.658

Following the procedure explained in Section 9A.2.2, the values of $a_{(m)n}$ and $b_{(m)n}$ estimated by the Moment Method are also shown in Table 9.2. for purposes of comparison. In these calculations, since n varied, the mean and variance of the corrected reduced extremes y' were used instead of the unadjusted extremes. If the number of samples (years), N, is large and not small ($= 5$) as in the example considered, and n does not vary, the asymptotic mean and variance of $y_{(m)}$ given by equations [9A.6] and [9A.7] can be used. The estimates provided by the Moment Method are quite close to those given by the Linear Method.

9.3.4 BEHAVIOR OF LARGE LOSSES

The parameter $b_{(m)n}$ is the modal value of $z_{(m)n}$, which is the logarithm of mth loss from the top in a sample of size n. Its value for any other sample size is given by equation [9A.17]. The value of $z_{(m)n}$ for any probability level p is given by

$$z_{(m)n}(p) = b_{(m)n} + (1/a_{(m)n})y_{(m)p} \qquad [9.6]$$

where $y_{(m)p}$ is the value of the reduced variable $y_{(m)}$ corresponding to the probability level p. The values of $y_{(m)p}$ for selected probability levels are given in Table 9.3 for $m = 1$ to 20; these are based on equations [9A.8] and [9A.9].

Consider, as an example, the analysis carried out by Ramachandran (1974) using the top 17 fire losses ($m = 1$ to 17) in the UK textile industry during the 21-year period ($N = 21$) from 1947 to 1967. He applied the linear method (equation [9.5]) including the correction factor in equation [9.4] for estimating $a_{(m)n}$ and $b_{(m)n}$. Base e was used for calculating the logarithms of fire losses expressed at 1947 prices. The results reproduced in Table 9.4. pertained to a sample size of 465 fires, the frequency experienced in 1947. The values of $a_{(m)n}$ indicated an "increasing failure rate" for large values of the variable z, the logarithm of loss.

On the basis of equations [9.6] and [9A.17], Figure 9.6 has been drawn on a log scale. It shows the relationship between the annual frequency of fires in the textile industry and the probable size, at 1947 prices, of the largest ($m = 1$), 7th ($m = 7$) and 16th ($m = 16$) fire in a year in decreasing order of magnitudes. For each of these three ranks the modal (most probable) sizes

Table 9.4. Textile industry, United Kingdom

Extreme (m)	$a_{(m)n}$	$b_{(m)n}$
1	2.247	5.214
2	1.785	4.829
3	1.626	4.534
4	1.460	4.327
5	1.387	4.113
6	1.424	3.988
7	1.239	3.749
8	1.163	3.564
9	1.212	3.448
10	1.034	3.259
11	0.973	3.137
12	0.925	2.972
13	0.886	2.832
14	0.924	2.749
15	0.937	2.680
16	0.950	2.583
17	1.002	2.537

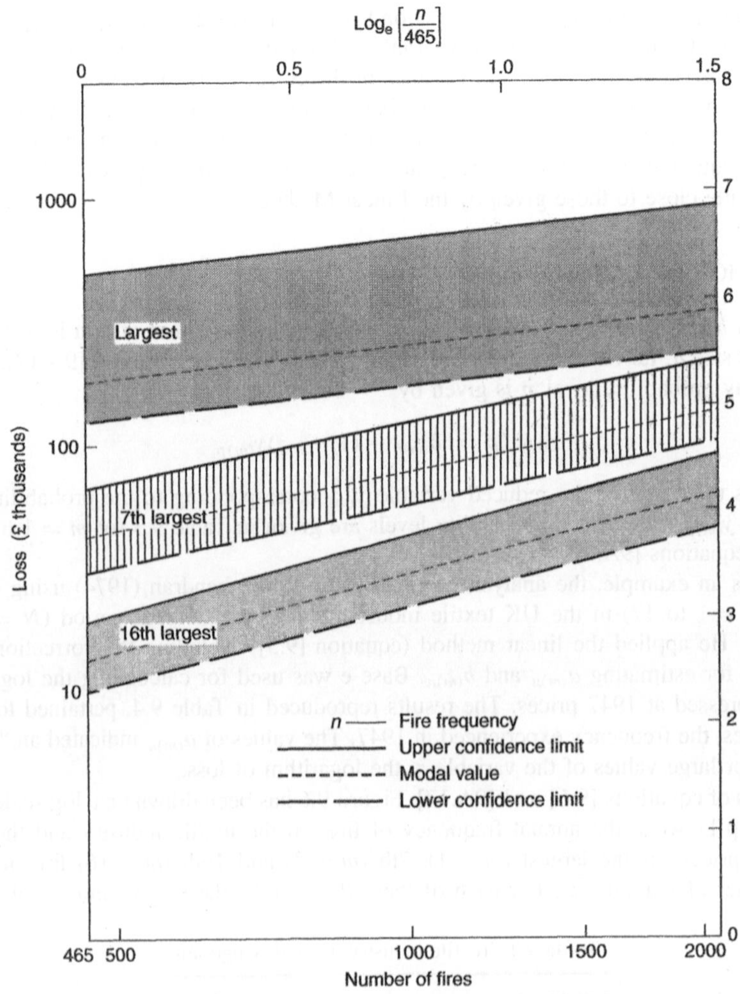

Figure 9.6. Fire frequency and large losses

of the losses are shown with confidence bands. For an estimated number of fires in any year, an ordinate erected at the corresponding point on the horizontal axis would intersect the upper and lower confidence lines at the points giving the corresponding confidence limits. The probability of exceeding the upper limit or falling short of the lower is 1. As an example, if the number of fires expected in a year in the textile industry is 1000, the most probable value of the largest loss would be £260, 000 with upper and lower confidence limits of £700, 000 and £180, 000; all these figures are at 1947 prices.

The confidence lines represent a control chart based on the trend in large losses during 1947 to 1967. The increase in the frequency n of fires in the textile industry may be partly due to an increase in this industrial activity and partly due to the inadequacy of fire prevention measures. In addition, if some or all of the actual large losses corrected for inflation exceeded the corresponding upper limits, it may be concluded that changes in fire fighting and protection methods or in the industrial processes are taking place to alter the picture for the worse. If the corrected losses are less than the lower limits, then the changes are for the better. These arguments and the data on losses and number of fires for 1968 and later years indicated that fire protection and fighting methods were generally coping well with fire outbreaks in the UK textile industry. However, it is necessary to analyze data on fire losses during recent years to assess the current trend. The example described above has been included in this chapter in order to illustrate the application of the extreme value theory.

9.3.5 RETURN PERIOD

For purposes of illustration let us assume that the values given in Table 9.4. are still valid. With an average (n) of about 1000 fires per year in the textile industry over the past few years, equation [9A.17] gives

$$b_{(1)n} = 5.214 + (1/2.247)\log_e(1000/465)$$
$$= 5.555$$

as the logarithm of the modal largest loss in this industry; in the original scale the mode (most probable loss) is £259,000 at 1947 prices.

Under current conditions, annual maximum losses in future years would exceed $b_{(1)n}$ at the rate of 1 in every 1.582 years or 2 in every 3 years approximately. This result is based on the "return period" for the largest value given by

$$\text{R.P.} = [1 - \exp\{-\exp(-y)\}]^{-1} \qquad [9.7]$$

where

$$y = a_{(1)n}(z^{-b}_{(1)n}b_{(1)n}) \qquad [9.8]$$

Equation [9.7] follows from the fact that the (cumulative) distribution function of the largest value is

$$\exp[-\exp(-y)]$$

For the mode $z_{(1)n} = b_{(1)n}$ and hence $y = 0$ such that R.P. = 1.582.

For a maximum loss of, say, £500,000 at 1947 prices, $z_{(1)n} = 6.215$, so that from equations [9.7] and [9.8], the return period for this level of loss is 4.921 years. In other words, this level will be exceeded by maximum losses occurring in future years at the rate of 1 in every 4.9 years. Inversely, one could fix the return period at, say, 10 years such that $y = 2.25$ from equation [9.7]. In this case, with $b_{(1)n} = 5.555$ and $a_{(1)n} = 2.247$, $z_{(1)n} = 6.556$. Hence, maximum losses occurring every year are likely to exceed £704,000 (at 1947 prices) at the rate of 1 in every 10 years.

The calculations mentioned above assume that the number of fires in the UK textile industry would maintain an average rate of 1000 per year. The return periods would be shorter than the estimated values if this average rate increases significantly in future years. Adoption of a fire protection measure such as sprinklers would reduce the logarithm ($b_{(1)n}$) of the modal value of the largest loss. This will in turn reduce the probability (1/R.P.) of maximum loss in any year exceeding any given level of maximum loss and thus lengthen the return period of this level. A discussion about the return period analysis for extreme order statistics smaller than the largest is beyond the scope of this book.

9.3.6 AVERAGE AND TOTAL LOSS

To assess the value of fire protection devices at the industry level, it is necessary to estimate the average and total loss in all fires, large and small, for each industry and for each important class of risk within an industry. The problem is to estimate the average loss making the best use of available data, which is restricted to large values. Using large loss figures it is possible to obtain reasonably good estimates of the parameters of the parent probability distribution if a specific form can be assumed for the parent.

The extreme order parameters $b_{(m)n}$ and $a_{(m)n}$ are related to the location parameter μ and scale parameter σ of the parent distribution $F(z)$ through the following equations:

$$b_{(m)n} = \mu + \sigma B_{(m)n} \quad [9.9]$$

$$a_{(m)n} = A_{(m)n}/\sigma \quad [9.10]$$

where

$$G(B_{(m)n}) = 1 - (m/n) \quad [9.11]$$

$$A_{(m)n} = (n/m)g(B_{(m)n}) \quad [9.12]$$

$G(t)$ and $g(t)$ are the distribution and density functions of the standard variable

$$t = (z - \mu)/\sigma \quad [9.13]$$

$G(t)$ and $F(z)$ follow the same distribution. For example, if $F(z)$ is normal, $G(t)$ is normal. For a normal distribution, μ is the mean (location parameter) and σ the standard deviation (scale parameter) of the variable z. In the case of the exponential distribution

$$F(z) = 1 - \exp(-\lambda(z - \theta)), \quad z \geq \theta \quad [9.14]$$

θ is the location parameter (not mean) and $(1/\lambda)$ the standard deviation (σ) of z. In this case, with $t = \lambda\ (z - \theta)$, the mean of z is given by $\theta + (1/\lambda)$.

In the normal case, the standard variable t has the mean zero and standard deviation unity. For the exponential, both the mean and standard deviation of t have the value unity. The values of $A_{(m)n}$ and $B_{(m)n}$ for the standard normal distribution are given in Table 9.5. for values of (m/n) from 0.001 to 0.100. For the exponential distribution, $A_{(m)n}$ has the constant value unity for all m and

$$B_{(m)n} = \log_e(n/m)$$

As defined earlier, if $z_{(m)n}$ is the mth extreme order statistics from the top, the variable

$$t_{(m)n} = (z_{(m)n} - \mu)/\sigma \quad [9.15]$$

follows the extreme order distribution shown in equation [9A.1] with

$$y_{(m)} = A_{(m)n}(t_{(m)n} - B_{(m)n}) \quad [9.16]$$

and $A_{(m)n}$ and $B_{(m)n}$ as the extreme order parameters instead of $a_{(m)n}$ and $b_{(m)n}$. From equation [9.16]

$$t_{(m)n} = B_{(m)n} + (1/A_{(m)n})y_{(m)} \quad [9.17]$$

It may be seen that the mean of $t_{(m)n}$ is

$$\bar{t}_{(m)n} = B_{(m)n} + (1/A_{(m)n})\bar{y}_{(m)} \quad [9.18]$$

where $\bar{y}_{(m)}$ is the mean of the reduced variable; $y_{(m)}$ given by [9A.6].

From equation [9.15], we can write

$$\bar{z}_{(m)n} = \mu + \sigma\bar{t}_{(m)n} \quad [9.19]$$

Table 9.5. Extreme order parameters of the standard normal distribution

m/n	A	B	m/n	A	B
0.001	3.3700	3.0902	0.051	2.0545	1.6352
0.002	3.1700	2.8782	0.052	2.0463	1.6258
0.003	3.0500	2.7478	0.053	2.0383	1.6164
0.004	2.9625	2.6521	0.054	2.0304	1.6072
0.005	2.8920	2.5758	0.055	2.0225	1.5982
0.006	2.8333	2.5121	0.056	2.0150	1.5893
0.007	2.7843	2.4573	0.057	2.0074	1.5805
0.008	2.7400	2.4089	0.058	2.0000	1.5718
0.009	2.7011	2.3656	0.059	1.9925	1.5632
0.010	2.6650	2.3263	0.060	1.9853	1.5548
0.011	2.6327	2.2904	0.061	1.9782	1.5464
0.012	2.6025	2.2571	0.062	1.9713	1.5382
0.013	2.5754	2.2262	0.063	1.9643	1.5301
0.014	2.5493	2.1973	0.064	1.9575	1.5220
0.015	2.5247	2.1701	0.065	1.9506	1.5141
0.016	2.5019	2.1444	0.066	1.9439	1.5063
0.017	2.4800	2.1201	0.067	1.9375	1.4985
0.018	2.4594	2.0969	0.068	1.9309	1.4909
0.019	2.4395	2.0749	0.069	1.9245	1.4833
0.020	2.4210	2.0537	0.070	1.9181	1.4758
0.021	2.4029	2.0335	0.071	1.9118	1.4684
0.022	2.3859	2.0141	0.072	1.9056	1.4611
0.023	2.3691	1.9954	0.073	1.8995	1.4538
0.024	2.3533	1.9774	0.074	1.8934	1.4466
0.025	2.3380	1.9600	0.075	1.8875	1.4395
0.026	2.3231	1.9431	0.076	1.8814	1.4325
0.027	2.3085	1.9268	0.077	1.8756	1.4255
0.028	2.2946	1.9110	0.078	1.8697	1.4187
0.029	2.2810	1.8957	0.079	1.8641	1.4118
0.030	2.2680	1.8808	0.080	1.8584	1.4051
0.031	2.2555	1.8663	0.081	1.8527	1.3984
0.032	2.2428	1.8522	0.082	1.8471	1.3917
0.033	2.2309	1.8384	0.083	1.8416	1.3852
0.034	2.2191	1.8250	0.084	1.8361	1.3787
0.035	2.2077	1.8119	0.085	1.8307	1.3722
0.036	2.1967	1.7991	0.086	1.8253	1.3658
0.037	2.1857	1.7866	0.087	1.8200	1.3595
0.038	2.1750	1.7744	0.088	1.8148	1.3532
0.039	2.1646	1.7624	0.089	1.8096	1.3469
0.040	2.1543	1.7507	0.090	1.8043	1.3408
0.041	2.1444	1.7392	0.091	1.7992	1.3346
0.042	2.1345	1.7279	0.092	1.7941	1.3285
0.043	2.1249	1.7169	0.093	1.7891	1.3225
0.044	2.1157	1.7060	0.094	1.7840	1.3165
0.045	2.1064	1.6954	0.095	1.7792	1.3106
0.046	2.0974	1.6849	0.096	1.7743	1.3047
0.047	2.0885	1.6747	0.097	1.7694	1.2988
0.048	2.0798	1.6646	0.098	1.7645	1.2930
0.049	2.0712	1.6546	0.099	1.7597	1.2873
0.050	2.0628	1.6449	0.100	1.7550	1.2816

where $\bar{z}_{(m)n}$ is the mean value of the mth observation over N samples (years). If figures for losses are available for, say, r large fires ($m = 1$ to r) and N samples, the straight line in equation [9.19] can be used to obtain rough estimates of μ and σ by plotting the r pairs of values $(\bar{z}_{(m)n}, \bar{t}_{(m)n})$ for $m = 1$ to r and drawing the best line passing close to these points. The intercept on the axis of $\bar{z}_{(m)n}$ will provide an estimate of μ and the slope of the line an estimate of σ. The parameters can also be estimated by the method of least squares. Equation [9.19] can also be used for a single sample ($N = 1$).

An ordinary least square method as mentioned above is somewhat imprecise since the residual errors in equation [9.19] arise from ranked variables and hence vary with the rank (m) and are correlated (not independent). Ramachandran (1974, 1982a) has developed a Generalized Least Square Method involving a variance–covariance matrix based on extreme value theory to deal with this problem. This method will provide the best and unbiased estimates of μ and σ but requires the use of a complex computer program.

Maximum Likelihood (Ramachandran, 1975a, 1982a) is another method that provides comparatively good estimates of μ and σ. For $N \geq 2$, the equations involved in this method have to be solved iteratively, which is tedious and time consuming. But for each sample, with $N = 1$, the solution for σ simplifies to

$$\sigma = \frac{1}{r}\sum_{m=1}^{r} A_{(m)n}(z_{(m)n} - z_{(r)n}) \qquad [9.20]$$

An estimate of μ can be obtained from

$$\mu = \frac{1}{r}\sum_{m=1}^{r} z_{(m)n} - (\sigma/r)\sum_{m=1}^{r} t_{(m)n} \qquad [9.21]$$

The estimate of σ given by equation [9.20] is somewhat biased. If this value is denoted by σ, an unbiased estimate is given by

$$\sigma' = \sigma/\sigma_s \qquad [9.22]$$

where

$$\sigma_s = \frac{1}{r}\sum_{m=1}^{r} A_{(m)n}(t_{(m)n} - t_{(r)n}) \qquad [9.23]$$

Then the σ' given by equation [9.22] may be used in equation [9.21] instead of σ to yield a better estimate of μ. Maximum Likelihood Method is easy to apply for each sample as shown in the example in Table 9.6. based on a normal parent distribution for z. Estimates of μ and σ obtained for different samples (periods) can then be compared to detect changes in the shape of the parent distribution over a period of time.

The parameters μ and σ are the mean and standard deviation of z, the logarithm of loss x. If a normal distribution is assumed for z, that is, log normal for x, the average value of the loss x in all fires in the range 0 to ∞ is given by

$$x = \exp\left(\mu + \frac{\sigma^2}{2}\right)$$

if $z = \log_e x$ or by

$$x = \exp\left\{c\mu + \frac{c^2\sigma^2}{2}\right\}, c = \log_e 10$$

if $z = \log_{10} x$.

Table 9.6. Maximum likelihood estimation of parameters of log normal distribution

Extreme (m)	Loss (x) (£k)	Log_e loss $z_{(m)}$	m/n	$A_{(m)}$	$A_{(m)}$ $(z_{(m)} - z_{(r)})$	$B_{(m)}$	$y_{(m)}$	$t_{(m)}$	$A_{(m)}$ $(t_{(m)} - t_{(r)})$
1	503.7	6.220	0.003	3.0500	9.5081	2.7478	0.5772	2.9370	3.7305
2	411.3	6.0193	0.007	2.7843	8.1154	2.4573	0.2704	2.5544	2.3402
3	212.3	5.3580	0.010	2.6650	6.0053	2.3263	0.1758	2.3922	1.8077
4	131.5	4.8790	0.014	2.5493	4.5235	2.1973	0.1302	2.2484	1.3626
5	72.7	4.2863	0.017	2.4800	2.9306	2.1201	0.1033	2.1618	1.1108
6	47.7	3.8649	0.021	2.4029	1.8269	2.0335	0.0857	2.0692	0.8538
7	32.5	3.4812	0.024	2.3533	0.8863	1.9774	0.0731	2.0085	0.6933
8	31.2	3.4404	0.028	2.2946	0.7705	1.9110	0.0637	1.9388	0.5161
9	25.6	3.2426	0.031	2.2555	0.3113	1.8663	0.0565	1.8913	0.4001
10	25.3	3.2308	0.034	2.2191	0.2801	1.8250	0.0508	1.8479	0.2974
11	24.5	3.1987	0.038	2.1750	0.2047	1.7744	0.0461	1.7956	0.1777
12	23.5	3.1570	0.041	2.1444	0.1124	1.7392	0.0422	1.7589	0.0965
13	22.3	3.1046	0.045	2.1064	0	1.6954	0.0390	1.7139	0
		53.4848			35.4751			27.3179	13.3867

$t_{(m)} = B_{(m)} + (y_{(m)}/A_{(m)})$

$\sigma'(\text{biased}) = 35.4751/13$
$= 2.7289$

$\sigma_s = 13.3867/13$
$= 1.0297$

$\sigma'(\text{unbiased}) = \sigma/\sigma_s$
$= 2.6502$

$\mu'(\text{unbiased})$
$= (53.4848 - \sigma' \times 27.3179)/13$
$= -1.4549$

$$x = \exp\left(\mu' + \frac{\sigma^2}{2}\right)$$
$= 7.822$

Average loss in 291 fires = £7822

Table 9.7. Average loss per fire at 1966 prices (£k)

	Sprinklered single story	Sprinklered multistory	Nonsprinklered single story	Nonsprinklered multistory
Textiles	2.9	3.5	6.6	25.2
Timber and Furniture	1.2	3.2	2.4	6.5
Paper, Printing, and Publishing	5.2	5.0	7.1	16.2
Chemical and allied	3.6	4.3	4.3	8.2
Wholesale distributive trades		4.7	3.8	9.4
Retail distributive trades		1.4	0.4	2.4

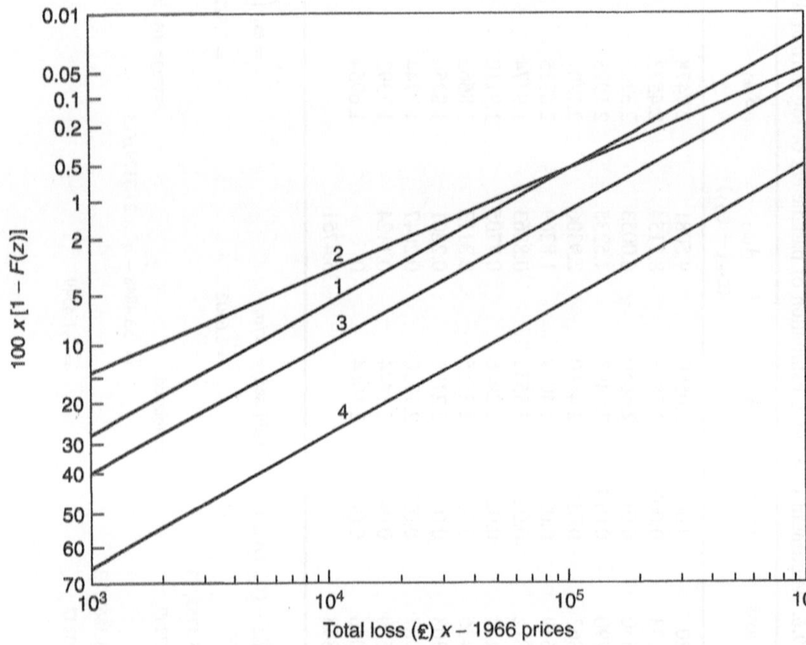

Line	Subpopulation	Parameters
1	Sprink/singlestorey	$\mu = -0.616$ $\sigma = 1.024$
2	Sprink/multistorey	$\mu = -1.419$ $\sigma = 1.340$
3	Nonsprink/singlestorey	$\mu = -0.334$ $\sigma = 1.062$
4	Nonsprink/multistorey	$\mu = 0.401$ $\sigma = 0.992$

$F(z)$ (cumulative distribution for z) = Probability of loss less than or equal to z
$V(x)$ (cumulative distribution for x) = Probability of loss less than or equal to x
Survivor probability = $1 - F(z) = 1 - V(x)$ = Probability of loss exceeding x or z

Figure 9.7. The survivor probability distribution of fire loss for each class in the textile industry

Assuming a lognormal distribution for fire loss and applying the Generalized Least Square Method to large losses, Rogers (1977) has estimated the average losses in all fires in industrial and commercial buildings with and without sprinklers. His results reproduced in Table 9.7. relate to fires that survived "infant mortality;" very small fires were excluded from the total sample size (n). Figure 9.7 is an example based on his investigation.

An investigation (Rogers, 1978) concerned with the application of extreme value theory to normal parent distribution revealed that a high level of accuracy in the estimation of parameters μ and σ can be obtained with about 20 large losses in buildings belonging to any risk category. An economic implication of this conclusion is that there is no need to collect financial loss data for all the fires, which is time consuming and expensive. It will be sufficient for all practical purposes if, for each risk category, data are made available by insurance companies for about 20 large fires, each with a loss greater than a threshold level. This number could arise from fires occurring during two or three years, if necessary, for certain categories.

9.4 Multiple regression

The probability distribution of damage discussed in Section 9.2 provides a method for assessing individually the effectiveness of each fire protection measure such as sprinklers, detectors, and structural fire resistance. In this method, it is assumed that the entire reduction in probable loss or probability of spread is due to the action of the protection device considered. This is not strictly true since part of the reduction could have been due to the operation of other devices, if installed, or other factors and due to interactions between devices and factors. Some factors could also have enhanced the damage, for example, delay in discovering a fire or extinguishing it. A full assessment of fire risk ought to consider all the relevant factors and evaluate their independent contributions to the damage. This is possible by performing a multiple regression analysis.

In a multiple regression model, the financial loss x or area damage d will be the dependent variable. For reasons explained in Section 9.2, the logarithm (z) of x or d should be used in the actual calculations. Factors affecting the financial loss or area damaged are independent variables denoted by the letter v with v_i as the measurement relating to the ith factor. The independent variable pertaining to a qualitative factor such as sprinklers will be assigned the value $+1$ if the factor is present or -1 if the factor is absent. Quantitative factors will be assigned values actually measured in respect to them. In some cases, it may be necessary to use transformed values for quantitative independent variables. Examples of such variables are total floor area of a building and insured value of a building and contents with which damage or loss has a "power" relationship (Section 7.4). For some qualitative variables such as materials, it may be possible to use measurements such as rates of heat release or growth obtained from experimental and statistical studies.

The object of a multiple regression analysis, with say p factors (independent variables) is to estimate the parameters $\beta_i (i = 0, 1, \ldots p)$ of the linear model

$$z = \beta_0 + \beta_1 v_1 + \beta_2 v_2 + \cdots \beta_p v_p \qquad [9.24]$$

for a given category of buildings. Then, for any set of building characteristics specified by the values of v_i ($i = 1, \ldots p$), the logarithm of loss or damage expected in a fire will be predicted by equation [9.24]. The confidence limits for this expected value can be obtained with the aid of "residual error" provided by the analysis. Statistical computer packages are available for estimating this error and the regression parameters β_i. If preliminary studies have shown an interaction between two qualitative factors, this can be taken into account by including in equation [9.24] as an independent variable that will be assigned a value $+1$ if both the factors are present or absent or -1 if one of the factors is absent.

Some data are available for applying a multiple regression analysis as envisaged in equation [9.24] but such an exercise does not appear to have been carried out according to published literature on statistical studies in fire protection problems. However, some attempts have been made to use in a regression model the probability of a fire spreading beyond the room of origin as the dependent variable. Baldwin and Thomas (1968), for example, have used this probability (p_S) as a risk measurement to study the effects of types of building construction. The authors used "logit" transformation to render the effects approximately additive. The "logit" is given by

$$P_s = \tfrac{1}{2}\log[p_s/(1 - p_s)] \qquad [9.25]$$

In a later study, Baldwin and Fardell (1970) used the logit transformation to analyze fire statistics to estimate the influence of various factors on the chance of fire spread beyond the room of origin. According to this study, there were significant differences between buildings used for different purposes and between some multistory buildings and single-story buildings. The biggest single factor affecting fire spread was the time of discovery of the fire, the chance of spread at night being twice that of the day; this was probably because of delays in discovery. The chance of spread was also considerably smaller for modern buildings, particularly in multistory buildings. This was, perhaps, the result of increased building control and safety consciousness. The brigade attendance time had no measurable influence on the spread of fire probably because of the wide range of variation in the size of fire confronting the brigade. There were few differences between buildings in different risk categories adopted by the fire brigade for determining the speed and size of first attendance at a fire.

Shpilberg (1975) used the logit model to quantify the relative effects of types of building construction, number of stories, sprinkler protection, type of fire department, and the subjective Factory Mutual Overall Rating on the probability of loss size. His object was to predict the probability of loss being above or below $10,000, given the particular characteristics of a group of risks. For purposes of illustration, Shpilberg used all fire loss claims in industrial property classified as "Machine Shops" paid by Factory Mutual during 1970 to 1973. In particular, the overall rating adopted by Factory Mutual was found to be of great value for predicting size and degree of loss, that is, fraction of the value of the property that was lost. Sprinklers were also found to be a major factor in determining both expected size and degree of loss.

The method followed by Shpilberg is reasonable if data are available only for number of fires in different loss brackets such as £10,000 to £50,000, £50,000 to £100,000, and so on. It would be wasteful of information to employ this method if figures are available for all individual losses above a threshold, say, of £25,000. In such cases, a multiple regression model based on extreme value theory (Ramachandran, 1975b) will give the expected loss in all fires, large and small, which is more useful than just the probability of loss exceeding the threshold. Using large losses, this model gives estimates of regression parameters equivalent approximately to estimates that would be obtained if loss figures were available for all the fires and were utilized in the calculations.

An application of the extreme value regression model was carried out by Ramachandran (1982b) with reference to possible trade-offs between sprinklers and structural fire resistance. Fires that occurred in each of five industries were classified into eight categories given by the combinations of the following factors:

1. Single story and multistory
2. Sprinklered and nonsprinklered
3. High fire resistance and low fire resistance.

The following simple regression was used for each of the categories:

$$z = \beta_o + \beta \log A \qquad [9.26]$$

where z is the logarithm of financial loss and$A$ the total floor area of a building. For a building of given size, the annual probability of fire starting revealed by another study was multiplied by the expected loss in a fire estimated by the regression model in equation [9.26] to provide an estimate of the annual loss expected in the building. The results obtained showed that, for multistory buildings, sprinklers and/or high fire resistance reduced the annual fire loss to a considerable extent particularly in a large building of total floor area of 1,000,000 sq ft (about 93,000 m^2) compared to a smaller building of 100,000 sq ft (about 9300 m^2).

APPENDIX : PROPERTIES OF EXTREME ORDER STATISTICS

9A.1 BASIC PROPERTIES

If n observations in a sample from an "exponential type" distribution $F(z)$ with density function $f(z)$ are arranged in decreasing order of magnitude, let $z_{m(n)}$ be the mth order statistic. Over repeated samples, for large n, the density function of $z_{(m)n}$ approximates to

$$\frac{m^m a_{(m)n}}{(m-1)!} \exp[-m y_{(m)} - m \exp(-y_{(m)})], \quad -\infty \leq z_{(m)n} \leq \infty \qquad [9A.1]$$

$$y_{(m)} = a_{(m)n}(z_{(m)n} - b_{(m)n}) \qquad [9A.2]$$

where $a_{(m)n}$ and $b_{(m)n}$ are solutions of

$$F(b_{(m)n}) = 1 - (m/n) \qquad [9A.3]$$

$$a_{(m)n} = (n/m) f(b_{(m)n}) \qquad [9A.4]$$

The parameter $b_{(m)n}$ is the modal value of $z_{(m)n}$ and $a_{(m)n} = h(b_{(m)n})$ where

$$h(z) = f(z)/[1 - F(z)] \qquad [9A.5]$$

is the failure rate or intensity function.

The mean $\overline{y}_{(m)}$ and variance $\sigma^2_{(m)y}$ of $y_{(m)}$ are given by

$$\overline{y}_{(m)} = \gamma + \log_e m - \sum_{v=1}^{m-1} (1/v) \qquad [9A.6]$$

where $\gamma = 0.5772$ is Euler's constant.

$$\sigma^2_{(m)y} = \pi^2/6 - \sum_{v=1}^{m-1} (1/v^2) \qquad [9A.7]$$

where $\pi^2/6 = 1.6449 \ldots$

The probability points of $y_{(m)}$ can be calculated by considering

$$u_{(m)} = m \exp(-y_{(m)}) \qquad [9A.8]$$

which has the gamma distribution

$$\{1/(m-1)!\} \exp(-u) u^{m-1} \qquad [9A.9]$$

9A.2 ESTIMATION OF EXTREME ORDER PARAMETERS

Equations [9A.3] and [9A.4] give estimates of $a_{(m)n}$ and $b_{(m)n}$ if the value of n and the exact form of $F(z)$ are known. Otherwise, if $F(z)$ can be assumed to be of exponential type, these two parameters can be estimated if values of $z_{(m)n}$ are available for, say, N samples. In this case, one of the following three methods can be adopted.

9A.2.1 Linear estimation

Let $z_{(m)j}$ be the mth extreme order statistic from the top in the jth sample. (The subscript n has been dropped). If the observations $z_{(m)j} (j = 1, 2 \ldots N)$ are arranged in increasing order and $R_{(m)j}$ is the rank of $z_{(m)j}$, let

$$p_{(m)j} = \frac{R_{(m)j}}{N+1} \qquad [9A.10]$$

Then, from (9A.8), the value $y_{(m)j}$ corresponding to $z_{(m)j}$ is given by

$$u_{(m)j} = m \exp(-y_{(m)j})$$

where $u_{(m)j}$ is estimated by the incomplete gamma integral

$$\frac{1}{\Gamma(m)} \int_{u_{(m)j}}^{\infty} u^{m-1} \mathrm{e}^{-} u = p_{(m)j}$$

The cumulative frequency $p_{(m)j}$ of $z_{(m)j}$ and hence of $y_{(m)j}$ will correspond to the tail value [1 – C.F.] of the gamma distribution. (As $y_{(m)}$ increases from $-\infty$ to $+\infty$, $u_{(m)}$ decreases from ∞ to 0). Using the pairs of values $[z_{(m)j}, y_{(m)j}]$, $j = 1 \ldots N$ in the linear relationship

$$z_{(m)j} = b_{(m)n} + (y_{(m)j}/a_{(m)n}) \qquad [9A.11]$$

$a_{(m)n}$ and $b_{(m)n}$ can be estimated graphically or by the least square method.

9A.2.2 Moment estimation

The estimates in this case are provided by

$$\sigma^2{}_{(m)z} = \text{variance of } z_{(m)j} = \sigma^2{}_{(m)y}/a^2_{(m)n} \qquad [9A.12]$$

$$\bar{z}_{(m)n} = \text{mean of } z_{(m)j}$$

$$= \frac{1}{N} \sum_{j=1}^{N} z_{(m)j} = b_{(m)n} + [\bar{y}_{(m)}/a_{(m)n}] \qquad [9A.13]$$

where the mean $\bar{y}_{(m)}$ and the variance $\sigma^2_{(m)y}$ of $y_{(m)}$ are given by equations [9A.6] and [9A.7].

9A.2.3 Maximum likelihood estimation

Estimates in this case are solutions of

$$\sum_{j=1}^{N} \exp(-y_{(m)j}) = N \qquad [9A.14]$$

$$[1/a_{(m)n}] = m\bar{z}_{(m)n} - \frac{m}{N} \sum_{j=1}^{N} z_{(m)j} \exp(-y_{(m)j}) \qquad [9A.15]$$

and have to be obtained by an iterative process.

9A.3 VARIATION IN SAMPLE SIZE

In the previous sections, it has been assumed that the extremes $z_{(m)j}$ arise from samples with constant size n. This is not likely to be valid or possible in practical situations. The numbers of fires for example, would vary from period to period. In this case

$$z_{(m)n_j} = b_{(m)n} + \frac{y_{(m)j} + \log_e(n_j/n)}{a_{(m)n}} \qquad [9A.16]$$

approximately where $z_{(m)n_j}$ comes from a sample of size n_j. The parameters $a_{(m)n}$ and $b_{(m)n}$ would refer to either a particular sample of size n or a hypothetical sample of average size n (with parameters $a_{(m)n}$ and $b_{(m)n}$).

Equation [9A.16] is based on the relationship

$$b_{(m)n_j} = b_{(m)n} + (1/a_{(m)n}) \log_e(n_j/n) \qquad [9A.17]$$

where $b_{(m)n_j}$ is the modal value of $z_{(m)}$ for samples each of size n_j.

Symbols

A	total floor area of a building
$A_{(m)n}$	failure rate or intensity function of the standard parent distribution of t at $B_{(m)n}$
$a_{(m)n}$	failure rate or intensity function of the parent distribution of z at $b_{(m)n}$
$B_{(m)n}$	modal value of $t_{(m)n}$
$b_{(m)n}$	modal(most probable) value of $z_{(m)n}$
c	adjustment factor for sample size
d	area damaged by fire
$F(z)$	cumulative distribution function for z
$f(z)$	density function of z (derivative of $F(z)$)
$G(d)$	cumulative distribution function for d
$G(t)$	cumulative distribution function of standard variable t
$g(t)$	density function of standard variable t (derivative of $G(t)$)
$H(h)$	cumulative distribution function of h
h	standardized value of z_d
h_0	$= -(\mu_{zd}/\sigma_{zd})$
h_z	failure rate or intensity function of the parent distribution at z
m	ranking of logarithms of losses – decreasing order of magnitude
N	number of samples
n	number of fires in a sample
n_j	number of fires in jth year
P_s	logit transformation of p_s
p	probability level
p_s	probability of spread of fire beyond the room of fire origin
R	ranking of $z_{(m)j}$ – increasing order of magnitude
R.P.	return period
r	number of large fires ($m = 1$ to r)
t	standardized value of z
$t_{(m)n}$	standardized value of $z_{(m)n}$
$V(x)$	Pareto distribution function for x
v	independent variable affecting financial loss or area damaged
$v(x)$	Pareto density function for x (derivative of $V(x)$)

W	parameter in the Pareto distribution of d
x	financial loss in fire
y	reduced extreme value
y'	corrected reduced extreme
$y_{(m)}$	reduced value of $z_{(m)n}$
$y_{(m)p}$	reduced variable $y_{(m)}$ corresponding to p
z	$= \log(x)$
z_d	logarithm of area damage
z_l	threshold value for large loss or large area damage
$z_{(m)n}$	logarithm of the mth loss from the top in a sample of n fires
$\phi(z)$	$= 1 - F(z)$
λ	reciprocal of the standard deviation of z (exponential distribution)
μ	location parameter of parent distribution (mean in the case of normal distribution)
μ'	unbiased (corrected for bias) estimate of μ
μ_{zd}	mean of logarithm of d
θ	location parameter of z (exponential distribution)
σ	scale parameter of parent distribution (standard deviation in the case of normal distribution)
σ'_s	unbiased (corrected for bias) estimate of μ
σ_{zd}	standard deviation of logarithm of d

References

Baldwin, R and Fardell, L G (1970). *Statistical Analysis of Fire Spread in Buildings*, Fire Research Station, UK, Fire Research Note No. 848.

Baldwin, R and Thomas, P H (1968). *The Spread of Fire in Buildings – The Effect of the Type of Construction*, Fire Research Station, UK, Fire Research Note No. 735.

Gumbel, E J (1958). *Statistics of Extremes*, Columbia University Press, New York.

Ramachandran, G (1967). *Frequency Distribution of Fire Loss*, Fire Research Station, UK, Fire Research Note No. 664.

Ramachandran, G (1974). Extreme value theory and large fire losses. *ASTIN Bulletin*, **7**(3), 293–310.

Ramachandran, G (1975a). *Extreme Order Statistics in Large Samples from Exponential Type Distributions and their Application to Fire Loss*, Statistical Distributions in Scientific Work, Vol. 2, D. Reidel Publishing Company, Dordrecht, Holland, pp. 355–367.

Ramachandran, G (1975b). Factors affecting fire loss – multiple regression models with extreme values. *ASTIN Bulletin*, **8**(2), 229–241.

Ramachandran, G (1982a). Properties of extreme order statistics and their application to fire protection and insurance problems. *Fire Safety Journal*, **5**, 59–76.

Ramachandran, G (1982b). *Trade-offs Between Sprinklers and Fire Resistance*, Fire, October 1982, pp. 211–212 (Corrections in Fire, November, 1982, p. 283).

Ramachandran, G (1988a). Probabilistic approach to fire risk evaluation. *Fire Technology*, **24**(3), 204–226.

Ramachandran, G (1988b). *Extreme Value Theory. SFPE Handbook of Fire Protection Engineering*, 1st Edition, National Fire Protection Association, Quincy, MA, USA, Section 4, Chapter 4, pp. 28–33.

Ramachandran, G (1994). *Rebates for deductibles and protection measures in industrial fire insurance. Proceedings of the XXV ASTIN Colloquium*, Cannes, France, September 1994, pp. 299–325.

Ramachandran, G (1995). Consequential/Indirect loss. *SFPE Handbook of Fire Protection Engineering*, 2nd Edition, National Fire Protection Association, Quincy, MA, USA, Section 5, Chapter 7, pp. 63–68.

Rogers, F E (1977). *Fire Losses and the Effect of Sprinkler Protection of Buildings in a Variety of Industries and Trades*, Current Paper CP9/77, Building Research Establishment, UK.

Rogers, F E (1978). *The Estimation of Parameters of Normal Distributions from Data Restricted to Large Values*, Current Paper CP8/78, Building Research Establishment, UK.

Shpilberg, D C (1974). *Risk Insurance and Fire Protection: A Systems Approach, Part 1: Modelling the Probability of Fire Loss Amount*, Technical Report No 22431, Factory Mutual Research Corporation, USA.

Shpilberg, D C (1975). *Statistical Decomposition Analysis and Claim Distributions for Industrial Fire Losses*, Paper for 12th ASTIN Colloquium, Report RC75-TP-36, Factory Mutual Research Corporation, USA.

10 PERFORMANCE OF FIRE SAFETY MEASURES

10.1 Introduction

Under the general title mentioned above, this chapter is concerned with three aspects of fire safety measures – performance, effectiveness, and reliability. The first aspect relates to the operation or behavior of a safety measure in a fire that is affected by several factors such as the location of a fire within or outside the room or compartment of origin, rate of growth of heat or smoke, and environmental factors. Satisfactory operation of a fire protection system, such as automatic detectors or sprinklers, also depends on the reliability factor, the third aspect, which, in the context of this chapter, is the hardware reliability measured in terms of the "failure rate." Some research has been carried out in evaluating the failure rates of components of a fire protection system, for example, detector heads. But, there are practically no published studies on the reliability of an entire fire protection or communication system composed of several components. Statistical models for evaluating the reliability of components and that of "series," "parallel," and mixed systems have been discussed by Bowen (1988).

The second aspect is concerned with the effectiveness of a fire prevention measure in reducing the frequency of occurrence of fires and that of a protection measure in reducing the damage in the event of a fire breaking out. The effectiveness of a safety measure can be assessed in terms of the probable reduction in the damage to life and property. A fire safety measure can produce some level of effectiveness only if it performs or operates satisfactorily in a fire.

Fire prevention activities and protection systems involve expenditure to be incurred by property owners, the government, local and central, and other individuals and organizations. For economic justification, the benefits due to these activities and systems should exceed the costs involved. This aspect relating to cost effectiveness or economic efficiency has been discussed in detail in a recent book by Ramachandran (1998).

10.2 Fire prevention measures

10.2.1 INTRODUCTION

In most countries, fire prevention activities are mainly carried out by fire brigades or departments apart from some efforts by fire protection associations and insurance organizations. These activities aimed at reducing the number of fires are of two basic types – public education and

Evaluation of Fire Safety D. Rasbash, G. Ramachandran, B. Kandola, J. Watts and M. Law
 ISBN: 0-471-49382-1

inspection. Efforts under the first type include visits to schools and community groups, posters, television advertising, fire prevention week, and other publicity campaigns. The second type is concerned with the inspection of hazardous equipments and structures.

Evaluating the effectiveness of public education to prevent the occurrence of fires is difficult since data collected for these activities are usually not amenable to proper statistical analysis. The available evidence suggests that efforts directed at a particular type of fire (e.g. grease or fat pan fire) are most likely to be successful. See, for example, the fire prevention campaigns in Kileen, Texas described in Review of Fire Prevention Education Programmes (National Fire Prevention and Control Administration, Office of Public Education, USA, August 1975). Campaigns on fires caused by fat pans and space heaters carried out in the United Kingdom are described in Sections 10.2.2 to 10.2.4. Several public education programmes are discussed by Swersey *et al.* (1975). There is a growing recognition in the fire service that more effort needs to be devoted to public education. As sufficient evidence of programme effectiveness becomes available, it should be possible to determine the types and amount of public education activities that are most beneficial.

Regulations and related inspections of gasoline distribution systems have contributed to limiting the number of fires in that area. Fire departments are actively involved in inspecting hazardous products and premises. There is widespread agreement that such programmes are necessary and effective although the particular level of effectiveness may be difficult to measure by analyzing available data. Schaenman *et al.* (1976) suggest an approach to measuring the effectiveness of inspection programmes that would relate fire occurrence to inspection effort. It would measure the number of fires that were relatively preventable by inspections per 1000 occupancies. Hall *et al.* (1978) applied the approach using data from 17 cities and one county. They found that "cities that annually inspect all or nearly all inspectable properties appear to have substantially lower fire rates than do other cities."

10.2.2 DWELLING FIRES – HOUSE-TO-HOUSE VISITS

Some measure of the effect of publicity carried out locally may be obtained by comparing the number of fires that occur before and after the publicity with those that occur in a "standard area" or those that occur nationally. Three examples have been analyzed by Chambers in Fire Research Notes 773, 801, and 802 produced by the Fire Research Station, UK. The first deals with the effect of a Fireman's Advice Scheme on dwelling fire frequencies in Worcester City and County. This scheme was in continuous progress for 11 years (1956–1967) before the data collected were analyzed. This fire prevention activity involved a continuing round of visits to all dwellings by firemen giving advice on fire hazards. It was estimated that at the end of the 11-year period, fires were taking place in dwellings at two-thirds the expected national frequency.

The second study by Chambers was concerned with a short campaign for chip pan safety in Exeter. The analysis showed a downward trend in the fire frequency for about 18 months, which was just on the borderline of statistical significance. Chamber's third study was concerned with an intensive publicity campaign in Leicester. This campaign was judged by comparing its performance against that of Nottingham, which was thought to be a city with roughly similar risk tendencies and types of industries. There was a drop in the number of fires that were not extinguished by the fire brigade, and over a period of about six months, there was a drop in the average size of the fires. However, fire frequencies went up as more fires were reported to the brigade and it was found difficult to attribute a significant positive aspect of the results.

10.2.3 CHIP OR FAT PAN FIRES – TELEVISION ADVERTISEMENT

The Home Office, UK, sponsored a television publicity campaign in 1976–1977 on the subject of the prevention and extinction of fat pan fires (see Rutstein and Butler (1977)). One region (Yorkshire) had a high level of television advertising while a second region (Lancashire) had a lower level of advertising. A third region (West Midlands) had no advertising and acted as a control area. One brigade in each region also organized further local publicity, including house-to-house visits.

There were three periods of television advertising in the Yorkshire and Lancashire regions – 12 January to 1 February, 16 February to 7 March, and 19 April to 16 May 1976. The advertisements were shown twice as frequently on Yorkshire television as on Lancashire television. In Yorkshire, there were at least three advertising spots a day in the first period, two to three spots a day in the second period and two spots a day in the third. A reminder campaign was run in Lancashire at the rate of one spot per day during the periods 10 to 20 January and 3 February to 6 March 1977. The television commercials attempted to put across two messages – that chip pans should not be left unattended or filled more than half full, and that the best way to extinguish a fire was to switch off the heat and smother the pan with a damp cloth. There was press advertising in the two areas. There were four insertions in five newspapers in Yorkshire and two insertions in the same publications in Lancashire.

Three brigades, West Yorkshire, Greater Manchester, and West Midlands made house-to-house visits and gave out printed fire warning stickers. The brigades also organized other local publicity. In West Yorkshire, the local radio was used to inform the public which areas were to be visited that day; this generated considerable interest. In Greater Manchester, part of the campaign was conducted in conjunction with the consumer council and local radio was also used. West Midlands brigade made use of the local press and several interviews were presented on local television and radio.

The analysis carried out by Rutstein and Butler (1977) related to fat pan fires starting on the ring or hot plate of domestic cookers. The analysis was necessarily confined to those fires to which the brigade was called. The first problem considered in this study was the effect of publicity on the number of fat pan fires reported to the brigade. The number of fires occurring after the campaign could simply not be compared with the number occurring before the campaign or a year earlier, as the comparison would not be valid if there was a seasonal pattern or trend in the incidence of fat pan fires. It was, therefore, decided to examine the past pattern of fire incidence in each brigade and to forecast the future number of fires, assuming a continuation of this established pattern. The unit of measurement used for this purpose was a four-week period, which provided 13 four-week periods for each year. A seasonal effect was incorporated in the forecast in those brigades in which this seasonal effect was evident. In other brigades, the forecast included only a trend.

In all brigades, the forecast revealed a downward trend possibly due to a shortage of potatoes, which first became apparent in 1975 and affected the 1975 fire incidence and hence the 1976 forecasts. The fire incidence in the control areas in 1976 (in which there was no publicity) closely followed the forecasts, which supported the hypothesis that the forecasts were reliable bases for measuring the effect of the publicity campaign. In all those brigades where there had been television publicity, the actual fire incidence clearly fell below the forecast levels.

In each case, the forecast provided an estimate of the number of fires that would have occurred in 1976 if there had been no publicity campaign. The *forecast number of fires* was then compared with the *actual number of fires* that occurred during and after the campaign to obtain a measure of the effect of the campaign. The change in the number of fat pan fires (actual vs forecast) was expressed as the proportion of the forecast number. Within the high TV and low TV areas, there was no apparent difference between those brigades that had supporting activity and those that did not. In the control areas (West Midlands), there was a random scatter about the zero line with some increase above the forecast levels in 1977. The fire incidence in these areas, which had brigade support activity, was not significantly different from those areas in which there was no publicity at all. The lack of any apparent effect of brigade activity was due to the smallness of any change in fire incidence resulting from this activity.

The effect of brigade activity was, therefore, discounted in the rest of the analysis carried out by Rutstein and Butler. Figure 10.1, reproduced from this report is a summary of the effects of television publicity in the three main areas. The figure shows the change in fat pan fire incidence for the combined brigades in the high TV, low TV, and no TV areas; it reveals the effects of the publicity campaign very clearly. There was a reduction of about 30% in fat pan fires for 6 months following the campaign with both high and low level of TV advertising. Over the year, the reduction was about 20 to 25% in both the cases. The lack of any difference in effect between the two levels of television publicity, high and low, was probably due to the fact that the lower level was sufficient to achieve a saturation effect. The effects of the campaign began to wear off after about six months and the number of fires returned to the original level. Fire incidence in the Lancashire area was again reduced after the reminder campaign and was at least as low as it had been after the initial campaign.

The reduction in fire incidence in most brigades in the TV areas occurred (proportionately) equally with both gas and electric cooker fires. The only exceptions were Lancashire and Merseyside where there was a proportionately greater reduction in the number of electric cooker fires. The proportion of fat pan fires, which were already extinguished by the time the brigades arrived, increased after the publicity campaign from 62 to 67% in the high TV area and 54 to 61% in the low TV area. In the control areas, there was no change. In the 6 months following the campaign, the number of calls to fat pan fires decreased by 30%; the proportion of these fires that had to be extinguished by the brigade decreased by 40%. In both the TV area and the no TV area, there was a slight (not statistically significant) decrease in the average damage caused by fat pan fires following the campaign. For all cooker fires, the number of casualties per 1000 fat pan fires increased, but not significantly, after the main campaign.

In terms of change of attitude, the extinction message came across more strongly than the prevention message. However, in the period immediately following the campaign, more fires were prevented than were successfully extinguished. It appeared that the extinction message was remembered after the prevention message had been forgotten.

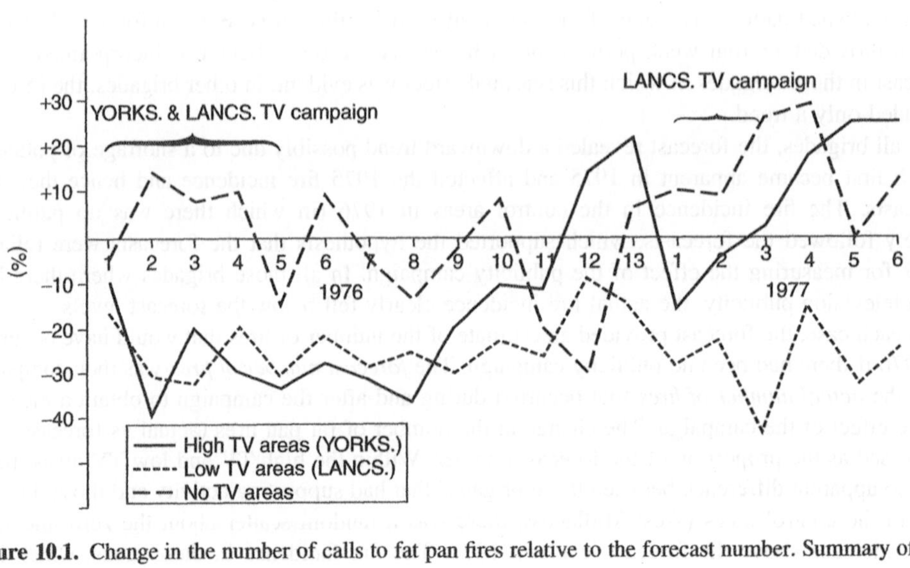

Figure 10.1. Change in the number of calls to fat pan fires relative to the forecast number. Summary of the three television areas

A market research exercise was carried out to provide additional information on the effectiveness of the publicity campaign. Representative samples of housewives were interviewed before, during, and after the campaign. The housewives were questioned on their awareness of the campaign, the contents of the television commercials, their knowledge of the causes of fat pan fires and the correct action to take in the event of such a fire. In general, there was good agreement between the results derived from fire statistics and the market research results.

10.2.4 SPACE HEATER FIRES – TELEVISION ADVERTISEMENT

Following the success of the chip pan fire publicity campaign discussed in the previous section, the Home Office, UK, sponsored a similar campaign in 1978 on the subject of space heater fires. The campaign was run from 1 February 1978 to 25 March 1978 in the Yorkshire Television area. Advertisements were shown once a day during this period. This level of advertising was chosen since the earlier chip pan study had shown that a higher level of advertising did not produce any significant increase in the effect on attitudes or behavior. A control area was introduced in order to provide a check against extraneous factors that might affect the number of space heater fires during the period of study.

The space heater campaign concentrated on the dangers of misusing particularly mobile radiant electric space heaters in the home. The objectives of the campaign were

1. to raise the level of awareness of the potential hazards of portable radiant electric fires,
2. specifically, to increase the awareness of the need to keep such fires at a "safe distance" and to affect behavior in terms of actually placing the heater at a "safe distance."

These objectives were selected on the basis of the results of a social survey carried out before the campaign, which took the form of extended group discussions on the use and perceived dangers of electric space heaters. The television campaign was designed to enliven the preexisting caution expressed in the discussions, with the emphasis on keeping heaters at a "safe distance" (since a large proportion of space heater fires were caused by combustible materials placed too close to the heat source).

To monitor the effects of the campaign, the number and severity of all domestic fires caused by space heaters in general and by electric radiant space heaters in particular were recorded before and after the commencement of the campaign. For this purpose, data for the period 1968 to 1977 extracted from the annual fire statistics were analyzed by Gilbert (1979) who obtained the following relationships by considering the number of fires for each year:

$$\text{All space heaters: } N = 550 - 35.7t - 6.4y$$

$$\text{Radiant electric space heaters: } N = 172 - 13.0t + 0.6y$$

where
t is the mean annual air temperature,
y is the year, counted from 1968 = 1,
and N is the number of fires per million population

The relationships mentioned above were estimated to remove the weather-dependent effects from the recorded annual numbers of space heater fires.

In the second equation, the standard error associated with the coefficient 0.6 for y was sufficiently large that the coefficient was not significantly different from zero. This implied that there was no significant time trend in the incidence of radiant electric space heater fires, although for all space heater fires combined, there was a significant time trend. In order to see the time effects more clearly, the incidence of space heater fires over the period 1968 to 1976 was corrected to a constant annual temperature of 9.9 °C with the aid of the equations mentioned above. This temperature corresponded with the long-term average over the period 1941 to 1970.

Figures for the period 1968 to 1977 corrected for temperature revealed that the incidence of space heater fires in dwellings had been declining in much the same way as for space heater fires generally. One of the reasons for this declining trend could be the increasing use of central heating. A comparison of the incidence of space heater fires in the control and campaign areas did not show any significant change that could be attributed to the campaign. The normal seasonal variation in both the areas would have masked any such difference that might have existed. In order to detect this difference, Gilbert carried out a further analysis. He considered that it was more relevant to estimate the dependence of the number of space heater fires in a four-week period on the temperature in the same period. Figure 10.2, extracted from his study, shows the fire incidence in the campaign and control areas related to the monthly mean air temperature. For each of the four cases considered in Figure 10.2, the relationship between N, t, and y was estimated.

Figure 10.2 does not indicate any discernible reduction in the number of space heater fires that can be attributed to the campaign. Hence, Gilbert used the precampaign relationships to predict the number of fires during the postcampaign period and compared the predicted figures with the actual number of fires that occurred. The actual and predicted numbers of fires were in reasonable agreement and no significant change due to the campaign was detected. The analysis could not also detect any changes attributable to the campaign in

(a) type of space heater fires – type of appliance involved (electric, gas, etc.), cause of fire (appliance fault, drying clothes, etc.), and material ignited (clothing, bedding, etc.)
(b) severity of space heater fires in terms of the method adopted by the brigade to extinguish the fire.

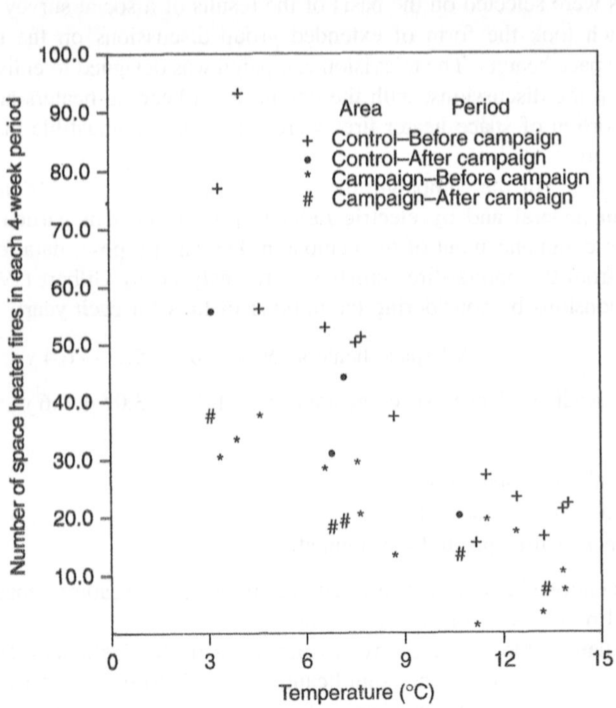

Figure 10.2. Space heater fires versus temperature

10.3 Automatic detectors

10.3.1 PERFORMANCE

Automatic detection systems are designed to detect heat and/or smoke from a fire in its early stages of growth, give an audible signal, and call the fire brigade if directly connected to the brigade. Such a signal would enable first-aid fire fighting to commence early so that the fire could be controlled quickly and prevented from causing extensive damage. Unlike sprinklers, which both detect fires and actively participate in fire fighting, detectors are passive and play no role in fire control. Custer and Bright (1974) have described in detail various types of detectors in a review of the state of the art in detector research. Problems concerned with the design of detection systems and fire alarm audibility have been discussed in detail by Schifiliti (1988).

Although it is possible to calculate from test results the response time of a heat/smoke detector under known conditions of ceiling height, detector spacing and fire/smoke intensity (total heat/smoke release rate), the time of operation of a detector head in an actual fire depends on many factors. The time when a fire product, heat, smoke, or radiation, reaches a detector head depends on the rate of spread of the product, which is controlled by the room/building configuration and environmental conditions. The factors mentioned above cause uncertainties in the performance of a detector, which may or may not operate in an actual fire; if it operates it may do so at a random time. Statistical data are not available for an evaluation of the probability of a detector operating in a fire. Helzer *et al.* (1979), in assessing the economic value of different strategies for reducing upholstered furniture fire losses, assumed that 80% of detectors would respond effectively to a fire. Detectors would fail to operate if the heat or smoke generated is insufficient to activate the system.

The fire statistics compiled by the Home Office, UK, can identify fires discovered first by smoke alarms. The fact that the fire was not discovered first by a smoke alarm does not imply that one was not present, only that the fire was first detected by some other means, usually a person. The number of fires in dwellings, which were discovered by smoke alarm, rose 164% from 1988 to 1991. During this period, according to Home Office research, the number of households owning smoke alarms increased from 15% to over 50%. According to Fire Statistics United Kingdom 1991, published by the Home Office, the proportion of fires discovered under 5 min in dwellings was 69% for fires discovered by smoke alarms and 53% for fires not discovered by smoke alarms or other detectors. In other occupied buildings, 78% of fires that were detected by smoke alarms were detected in 5 min or less, compared to 45% of other fires; this proportion includes fires discovered by other detectors.

According to Bengtson and Laufke (1979/80), operating times for heat detectors range from 2 min in "extra high hazard" occupancies (XHH) such as plastic goods factories, to about 20 min for "extra light hazard" (XLH), which includes flats and other residential premises. The operating times of smoke detectors range from 0.5 min (XHH) to 2.25 min (XLH) for wood materials and 0.75 min (XLH) for polystyrene. Wood materials give out most smoke from glowing fires but polystyrene in flaming conditions produces sufficient smoke for a quicker response time. According to some tests relating to dwelling fires quoted by Custer and Bright (1974), detection times for smoldering upholstery fires are long for both rate-of-rise and fixed temperature detectors. In another test involving a rapidly developing fire in a trash barrel, the rate-of-rise detector operated in 2 min, while the fixed temperature unit responded in 5 min, and the photoelectric detector in 8 min. There is clear evidence of the need for smoke detectors in areas where smoldering fires are likely.

Nash *et al.* (1971) carried out some tests involving high stacked storage using various types of detectors. In a series of similar tests, heat detectors operated between 1 min 16 s and 3 min 58 s of ignition, ionization chamber detectors operated between 1 min 5 s and 4 min 30 s, while optical detectors took over 3 min to operate. Infrared detectors operated in about 3 min and laser beam detectors took about 5 min to operate if well above a fire.

10.3.2 EFFECTIVENESS – PROPERTY PROTECTION

Consider first, property protection, which appears to be the primary reason for installing detectors. According to Baldwin (1971, 1972), the probability of fire spread or of a large fire could both be reduced by early detection or discovery of a fire. Furthermore, the probability of a fire starting during the nighttime and becoming large could be reduced by two-thirds if detected promptly. Figures published in Fire Statistics, United Kingdom, 1991 indicate that among fires in occupied buildings discovered by smoke alarms, 67% of fires are confined to items first ignited and 0.2% spread beyond the building. If smoke alarms were not installed in these buildings, only 36% of fires would be confined to items first ignited and 2.5% would spread beyond the building. For dwellings, the probability of a fire being confined to the item first ignited is 0.68% if the fire is discovered by a smoke alarm and 0.41% if not discovered by a smoke alarm or any other detector.

Cerberus, manufacturers of ionization detectors, maintain a casebook of fires that have occurred in Switzerland in premises protected by their systems. Several items of information are recorded for each fire that has been detected by the fire alarm system and has led, or almost certainly would have led to damage claimable from the insurance company. Fires not resulting in insurance claims are excluded in the calculation of the average insurance loss. The statistics include only fires that occurred in rooms monitored by automatic fire alarm systems. According to an analysis of these statistics for the period 1960 to 1967, the average fire loss (premises and contents) in buildings in Switzerland, monitored by Cerberus fire alarm systems, was only one-third of the average loss in buildings without these systems.

As mentioned in Section 7.6, the exponential fire growth model can be used for assessing the economic value of detectors in reducing property damage. Early detection of a fire through automatic detection systems would reduce the time period T_1 from ignition to discovery of a fire. This will also reduce the control time T_4 since a fire detected soon after ignition will be in its early stage of growth when the fire brigade is called. Hence, the brigade can arrive at the scene when the fire size is small and control the fire quickly. Consequently, the total duration of burning, T, and area damage, $A(T)$, will be reduced considerably. Representing algebraically,

$$T_4 = a + b \cdot T_A \qquad [10.1]$$

where

$$T_A = T_1 + T_2 + T_3$$

$$T = T_A + T_4$$

T_A is the time period from ignition to the arrival of the brigade at the scene of a fire.

The model mentioned above was applied to a pilot study (Ramachandran, 1980) on the economic value of automatic fire detectors for the textile industry. The following values were obtained for the parameters in equation [7.7], and equation [10.1]

$$\theta = 0.0632 \text{ (with a “doubling time” of 11 min)}$$

$$A(\mathrm{O}) = 4.6852\,\mathrm{m}^2$$

$$a = 6.90$$

$$b = 0.83$$

The parameter b expresses the fact that the control time T_4 will increase by 0.83 min for every minute of delay in the arrival of the brigade at the fire scene. Table 10.1 shows the calculations

Table 10.1. Savings due to detectors, textile industry

Case	Average time (minutes)							
	Detection time (T_1)	Call time (T_2)	Attendance time (T_3)	Control time (T_4)	Total duration of burning (T)	Average area damaged (m^2) $A(T)$	Direct loss per fire[a] (£) L	Saving per fire (£)
Detector not connected to the fire brigade	1.0	2.5	5.00	13.96	22.46	19.37	4358	7448
Detector connected to the fire brigade	1.0	0.25	5.00	12.09	18.34	14.93	3359	8447
Fire not discovered at ignition	9.66	2.63	4.84	21.08	38.21	52.47	11,806	–

[a]At the rate of £225 per sq meter at 1978 prices.

involved in estimating the savings due to automatic detectors for the two cases – detector not connected to the fire brigade and detector connected to the fire brigade. The savings are reductions in the damage in a fire not discovered at ignition. An average time of 1 min was assumed for the operation of an automatic fire detector, 2.5 min for calling the fire brigade after discovery of fire by human means, 15 s for the call time of a detector connected to the brigade, and 5 min for the attendance time of the brigade.

The course of the fire is depicted in Figure 10.3 showing the sizes of the fire at the times of fire brigade arrival and control for the three cases considered, and the reduction in damage due to automatic detectors. In the absence of fire brigade attack, a fire can burn for more than 54 min, with damages exceeding 140 m^2. The important time in a fire situation is the first 5 min when the occupants are attempting to escape. During this time, the heat of the fire will be low and hence the smoke temperature will be low and the buoyancy movement sluggish.

Table 10.1 is a general example illustrating the application of the exponential model of fire growth in assessing the economic value of detectors. The input figures in this table can be varied according to factors such as detector type, occupancy type, location of the nearest fire station, and communication systems in a building. For example, instead of 1 min, a different operating time based on experimental results may be assumed. The attendance time for a particular building may be less than or more than five minutes depending on the risk category assigned to it by the brigade. For a particular building, $A(O)$ may be estimated by carrying out a fire load survey but the value of θ may remain the same for any risk category. With particular input values, a figure such as Figure 10.3 can be drawn for any building.

In a further application, Ramachandran and Chandler (1984) applied the following expanded version of the exponential model:

$$A(T) = A(O)\exp[\theta_A T_A + \theta_B T_B] \qquad [10.2]$$

where $T_B = T_4$. The parameter θ denoting the overall rate of fire growth has been split into two parts – θ_A for the growth during the period T_A and θ_B for the growth during T_B. The "doubling time" for these periods can be obtained by substituting θ_A or θ_B for θ in equation [7.8]. The relationship in equation [10.1] was also used with estimated values for a and b. A table similar to Table 10.1 was formed for each group of industrial and commercial premises for estimating the savings due to automatic detectors connected to fire brigade against a loss that might be incurred if a fire was not detected or discovered at ignition. These results, shown in Table 10.2, were only marginally higher than savings due to detectors not connected to the brigade. Fires that were

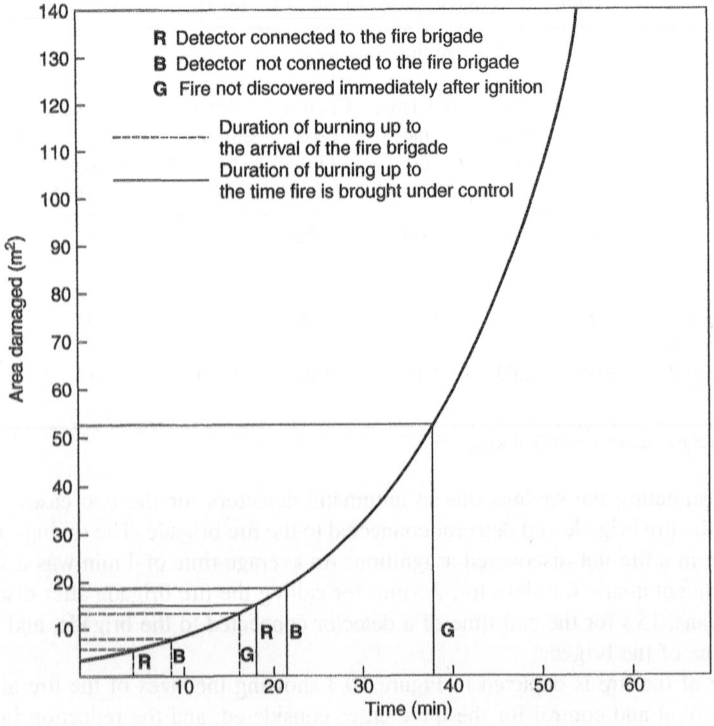

Figure 10.3. Average time and area damaged

Table 10.2. Estimated average savings (£) with detectors connected to fire brigade (Great Britain – 1983 prices)

	Production		Storage	
	First-aid fire fighting?		First-aid fire fighting?	
Occupancy	Yes	No	Yes	No
Food, drink, and tobacco	6730	7980	2170	[a]
Chemicals and allied	120	2750	195	2925
Metal manufacture	60	520	75	295
Mechanical, instrument, and electrical engineering	105	2450	195	1545
Textiles	180	980	580	3390
Clothing, footwear, leather, fur	730	20,710	4280	2185
Timber, furniture, etc.	625	4175	610	2350
Paper, printing, and publishing	60	6625	25	2570
Distributive trades – wholesale	[b]225	[b]2985	120	3880
Distributive trades – retail	[b]15	[b]370	340	730

[a]Estimating equation unstable.
[b]Public/assembly area.

tackled by first-aid fire-fighting methods had shorter discovery times than those that were not tackled. Hence, the additional savings due to automatic detection in such buildings were small. It is apparent that economic benefits due to detectors will be particularly high for buildings with no provision for first-aid fire fighting.

On the basis of equation [7.4], Rutstein and Cooke (1979) used the following formula to estimate the average area damaged in an unprotected building:

$$p_1 \bullet C_1 A^\beta + p_2 \bullet C_2 A^\beta + p_3 \bullet C_3 A^\beta \qquad [10.3]$$

where p_1 and p_2 were proportions of fires that occurred when people were in the room of fire origin and building (not room) of origin respectively; p_3 was the proportion of fires that occurred when there were no people in the building. These proportions were according to the location of the nearest person. For a given value of β, the parameters C_1, C_2, and C_3 were based on the expected damage estimated under certain assumptions. For industrial buildings, for example, values of 1.5, 1.9, and 3.9 were obtained for C_1, C_2, and C_3 respectively. For these buildings, with $\beta = 0.45$, $p_1 = 0.55$, $p_2 = 0.18$, and $p_3 = 0.27$, the value given by equation [10.3] may be seen to be $2.22\mathrm{A}^{0.45}$.

Assuming that the damage was reduced by 60% in "people in building" fires and by 55% in "people not in building" fires, the fire size in a protected building was estimated to be $1.44\mathrm{A}^{0.45}$. This is equivalent to an overall reduction in damage of 35% due to a detection system, whatever the size of a building may be in terms of floor area A (m^2). This reduction was applied to a local alarm system while a reduction of 45% was estimated for a direct line system.

10.3.3 EFFECTIVENESS – LIFE SAFETY

While heat detectors are of economic value for reducing property damage in industrial and commercial buildings, smoke detectors are essential for dwellings where most of the fire deaths occur. Smoke is the leading cause of death in fires occurring in these premises. The economic value of smoke detectors at the national level has been investigated by Helzer *et al.* (1979).

Early detection of a fire would enable the commencement of evacuation of a building soon after ignition. This would increase the chance of occupants reaching a safe place before escape routes become untenable due to heat, smoke, or toxic gases. According to the parameter λ discussed in Sections 8.7 and 8.8, for every minute saved in the evacuation time, the fatality rate per fire would be reduced by 0.0008 for single occupancy dwellings and by 0.0006 for multiple occupancy dwellings.

For fires in single and multiple occupancy dwellings, the average discovery times are 14 and 18 min respectively. If automatic detectors reduce the discovery time to 1 min as assumed in Section 10.3.2, there will be a saving of 13 and 17 min in the discovery time. Hence, the fatality rate will be reduced by 0.01 for both the types of buildings according to the values of λ mentioned above. With about 55,000 fires per year in these buildings, 550 lives can be saved every year if smoke detectors, in particular, are installed in all these buildings. The reduction in life risk is sensitive to the value assigned to the operation time of a smoke detector.

According to the US study mentioned in Section 8.7, detectors would reduce the fatality rate per fire in one- and two-family dwellings by 0.0042. In this connection, it is worth mentioning that, since 1977, every dwelling unit owned by Ontario Housing Corporation, Canada, has been protected by at least one smoke detector. The statistics produced by this organization clearly indicate that smoke detectors save lives. McGuire and Ruscoe (1962) analyzed data on 342 residential fires in Ontario and estimated that smoke detectors could have saved 41% of the victims.

10.3.4 RELIABILITY

As mentioned in the introduction, this section is concerned with the hardware reliability of automatic detectors. Only very limited research has so far been carried out on this aspect of fire safety devices particularly due to lack of data on the "failure rates" of components constituting these systems. However, as discussed by Custer and Bright (1974), some general statements can be made regarding certain critical components of detection systems based on field or laboratory experience and manufacturers' literature. A summary of this study is as follows.

Heat detectors are generally the most reliable in terms of component failure since these devices respond directly to the presence of heat by a physical change in the detector operating elements. Heat detectors may fail due to mechanical damage or abuse after installation or by failure of components or circuitry in peripheral equipment such as power supplies or alarm indicating equipment.

The lamps used in photoelectric type smoke detectors are critical to detector operation. The operational lifetime of incandescent bulbs ranges from about 1 month to about 37 months. Lowering the operating voltage to increase lamp life reduces filament evaporation, one of the causes of failure. Vibration and shock particularly with fragile, aged filaments often lead to lamp failure. Power surges and power failures are also significant factors. The problem of bulb life might be solved through the use of light emitting diodes. These light sources are mechanically stable and should be less prone to damage from vibration. The sensitivity of photocells used in detectors may have a tendency to drift a little with aging. The usual method of counteracting such changes is through the use of compensation photocells in various configurations. These cells act as a reference to maintain balance in the circuit.

The use of batteries as the primary power supply for single station detectors has several problems, which can affect detector reliability. Battery-operated devices are generally required to have a one-year lifetime and an audible signal lasting seven days. Alkaline batteries have a constantly decreasing voltage curve as they wear out. Detectors using these batteries require periodic sensitivity readjustment to maintain the designed alarm threshold. Mercury batteries have a constant voltage throughout most of their life but undergo a rapid drop in voltage at the end of their life. This will reduce the sensitivity and might shorten the operating time of the alarm, or in some cases, prevent its operation.

Detector operation in a ventilation system might result in an alarm when smoke from a failing transformer is detected. The possibility of these failures initiating fires should not be overlooked. Failure of solid-state components might occur due to electrical transients overpowering the built-in transient protection.

Gupta (1984/85) has estimated the hardware failure rates of components of an automatic fire detection (AFD) system at a psychiatric hospital by analyzing the component structure and configuration of the system. The system comprised of a distributed system of ionization type smoke detectors, break-glass units, and heat detectors. All of them were connected to various zone panels, which in turn were connected to a central control unit and three repeater panels. The control unit carried a fire area identity annunciation and was the means for the receipt of the alarm for activating the hospital's audible fire alarm and the transmission of a fire brigade call-out signal. The failure rates of electronic components were obtained from Military Standardization (1974).

For ionization smoke detectors, Gupta estimated the total mode failure rate as 0.057 faults per year of which 0.04 faults per year were of the safe type and 0.017 faults per year were dangerous. Since these figures applied to a first-class environment, factors such as air speed and humidity were taken into account and the failure rate assessed to be 0.46 faults per year, which was eight times greater than that for a first-class environment. For a break-glass unit, the total relevant failure rate was assessed to be 0.032 faults per year of which 0.018 faults per year were of the safe type and 0.014 faults per year were dangerous. The figures did not include the spurious alarm rate due to misuse of the system. For the control unit, Gupta estimated the total failure rate for the unrevealed dangerous type as 0.06 to 0.042 faults per year for indicator control module and 0.018 faults per year for monitor unit.

Studies completed by some reliability engineers (Finucane and Pinkney, 1988) have shown that overall failure rates for control units varied from 0.25 faults per year, up to 1 fault per year, with an unrevealed fail-to-danger rate of typically 0.1 faults per year. According to these authors, the overall failure rates for detectors were 0.1 faults per year with fail-to-alarm failure rates varying from 0.01 faults to 0.1 faults per year.

10.3.5 ADDRESSABLE SYSTEMS

Although conventional detector systems discussed above are still in use in some buildings, new technology addressable systems were introduced in many countries about 15 years ago. In these systems, signals from each detector and each call point are individually identified at the control panel. Each circuit is a form of simple data communication rather than simply an electrical circuit. Within the software of an addressable system, the device identity can be converted into a preprogrammed location, which is then displayed on some form of text such as an LCD or vacuum fluorescent display.

There are essentially four types of addressable systems – two-state, analog, multistate, and multicriteria. In the first type, the detectors themselves make the decision as to whether or not there is a fire. This is not the case with the second type (analog) in which the detectors act as sensors and simply transmit a signal level to the control equipment, representing the amount of heat, smoke, or flame that is being sensed. A "prealarm warning" would be given if the signal exceeds a certain threshold level and a "fire warning" if the signal exceeds a higher threshold level. At a very low threshold level, a fault signal may be given to indicate that the detector has become very insensitive. In a multistate system, each detector is capable of transmitting several states such as fault, normal, prewarning, and fire along with its individual identity. In a multicriteria system, each detector head incorporates more than one sensor and hence is capable of detecting heat, smoke, or flame.

The majority of installed addressable systems are of the analog type. Some of these systems include sophisticated logging facilities, which can provide data for estimating reliability. The unreliability of these advanced detector systems can be expected to be considerably less than that (0.02 per year) of modern conventional systems (see Appleby and Ellwood (1989)).

10.3.6 FALSE ALARMS

Practically, any level of sensitivity to fire, in particular, to signals of combustion, can be achieved with existing technology. However, if sensitivity is set at a high level, a detector may pick up signals given by spurious fires from sources such as cigarette smoking and cooking, which are normal activities. On the other hand, a low level of sensitivity can increase the risk of genuine fires being undetected. Sufficient research has not been carried out so far to close the gap between reliability in detection and rate of unwanted, false alarms. Some studies, for example, by Cholin (1975), have suggested that the gap may be closed to some extent by cross-zoning of detectors where the activation of the alarm is delayed until a second detector is activated. Another approach suggested by Custer and Bright (1974) is the use of multimode detectors requiring signals from several fire signatures before a fire alarm is initiated. Newer generations of detector systems, which employ computers, may be able to check fire signals and spot false or unwanted alarms.

False alarms are a nuisance, and cause wastage of time and money particularly to fire brigades whose response to genuine fires may be delayed due to unnecessary call-outs. Apart from those relating to malicious calls, occupiers of homes and fire brigades may be responding to signals from cooking smoke, bathroom water vapors, tobacco smoke, and other nonthreatening sources of air-suspended particulate. Other causes of false alarms include dust, debris, and insects in the sensing chamber. Apart from high sensitivity levels, location, type of detector, and lack of maintenance can also be reasons for excessive alarms.

Reasons for nuisance alarms arising from smoke detectors in homes have been identified by the National Smoke Detector Project carried out by the US Consumer Product Safety Commission. This information is contained in the Commission's final report (November, 1993) on the first study, "Smoke Detector Operability Survey – Report on Findings." As pointed out in this report, power sources for a high percentage of smoke detectors in homes are intentionally disconnected because of nuisance alarms. The report has also suggested several potential solutions to address this problem. Repeated false alarms for an organization may cause a fire brigade to cancel connection facilities thus exposing the organization to increased fire risks.

The most comprehensive British study of false alarm statistics was that of Fry and Eveleigh (1975) who analyzed data collected in a special survey on detector actuations carried out by fire brigades. On the basis of 5930 reports of fire signals in 1968, only 489 were from genuine fires. There was thus a ratio of 11:1 between false and genuine calls. The ratio was highest with combined heat and smoke detectors at 23:1. The ratio was 11:1 for heat detectors and 14:1 for smoke detectors. Alarms activated by sprinkler systems showed a false to genuine ratio of 10:1, with a ratio of 4.4:1 for manually operated alarms. Mechanical and electrical faults, especially defective wiring or heads, accounted for 46% of the 5441 false calls. Ambient conditions, especially extraneous heat and smoke, accounted for 26% of false calls, with 16.5% being communication faults. The analysis also classified false calls according to occupancy and time of day.

According to Davies (1984), 95 systems of a Swiss manufacturer gave 85 genuine alarms as opposed to 1329 false calls (a ratio of 16:1); of the latter, 1194 were described as "fire stimulating events," for example, blow lamps in fairly normal use. In a letter to "Fire," Bridge (1984) drew attention to 10,000 false alarms a year in New Zealand, of which about 20% were malicious. The reasons for these false alarms are fairly well known due to the inspection systems used by insurers for sprinklers and detectors.

Gupta (1984/85) categorized events that lead to the malfunctioning and proper functioning of automatic detection systems. He divided false alarms into four sub categories:

(i) failure of equipment,
(ii) "nonfire" disturbances,
(iii) "external" effects,
(iv) "unknown" reason for alarm.

Information on failure rates for the first category can be obtained from data banks on hardware reliability such as those maintained by the Atomic Energy Authority (AEA) in the United Kingdom, by a division formally known as SRD (Safety and Reliability Directorate). The second category included causes such as cigarette smoke, steam, dust, and smoke or vapor from cooking. The third category included human error, water from leaks, power supply interruption or surge, electrical interference such as arcing or switching, and birds, animals, and insects.

Gupta analyzed data collected from various sites on time periods between successive events for the categories mentioned above except the first. He fitted Weibull probability distribution to these data in order to understand the statistical behavior of the events. Parameters of the distribution were estimated using maximum likelihood and least squares methods. The mean of the "scale" parameter, interval between successive events, varied between 10 and 42 days for the second category ("nonfire"), 16 and 40 days for the third category ("external effects"), and 14 and 60 days for the fourth category ("unknown"). The value of the "shape" parameter in all the three categories were less than 1.0 (except in "external effects" for one site), which indicated a decreasing failure rate for the occurrence of the events.

On the other hand, the mean duration between two successive false alarms at various sites varied between 5 and 15 days. This duration had little bearing on the type of site, according to results presented in the paper. Little variance in "shape" parameter observed for different sites within a given category indicated a common cause of false alarm. The explanation for this phenomenon was perhaps the attribution to the increasing effectiveness, efficiency, and better maintenance policies of the fire officers of sites. During the visits to various sites, a disturbing number of unsatisfactory features of detection systems were observed such as (1) siting of heads (2) configuration, and (3) maintenance.

10.4 Sprinklers

10.4.1 PERFORMANCE

Although water is not the perfect extinguishing agent for all fires, it is the most commonly used agent in sprinklers, mainly due to the fact that it is widely available, inexpensive, and nontoxic. There are many desirable fire-extinguishing characteristics of water. Even in cases in which water from sprinklers will not suppress the fire, the cooling ability of water spray can protect structural elements of a building, thus containing the fire until it can be extinguished by other means. Such an action can lead to both lower fire temperatures and lower concentrations of smoke and other toxic products in the atmosphere. These less hostile conditions and early attack on a fire would provide more time for escape, rescue, or evacuation.

There are four basic types of sprinkler systems according to Fleming (1988), which differ in terms of the most fundamental aspect of how water is put into the area of fire. A wet pipe system and dry pipe system use automatic sprinklers while a deluge system does not use automatic sprinklers, but rather open sprinklers. The fourth type is similar to a deluge system except that automatic sprinklers are used. There are many other "types" of sprinkler systems, classified according to the hazard they protect (such as residential, in-rack, or exposure protection); additives to the system (such as antifreeze or foam); or special connections to the system (such as multipurpose piping). But all sprinkler systems can still be categorized as one of the four basic types. Fleming has described the basic features of all sprinkler systems. He has also discussed in detail, simple hydraulic calculations for determining water supply requirements, optional calculations that may be performed with regard to hanging and bracing of system piping, and the performance of a system relative to a fire.

Sprinklers are generally required to operate at an average temperature of 68°C, but there are special requirements for certain occupancies and important aspects such as the flow of hot gases in fires, which can determine the siting of sprinkler heads to achieve acceptable operating times. As with detection systems, several factors cause uncertainties in the activation and operating times of sprinklers in actual fires although scientific (deterministic) methods have been developed for estimating the response time (see, for example, Evans (1985)). On the basis of factors such as rate of temperature rise, height of upper fire surface above the floor, and height of the premises, Bengtson and Laufke (1979/80) have estimated sprinkler operation times varying from 2.5 min for XHH occupancies and 16.8 min for XLH occupancies. The operation time of sprinklers in experimental fires have been estimated in several studies carried out, particularly, by the Fire Research Station, UK; Factory Mutual Research Corporation (FMRC), USA; and National Institute of Standards and Technology, USA.

In a sprinklered building, there is a chance that a fire may not produce sufficient heat to activate the system such that it is either self-extinguished or extinguished by first-aid means. This chance of a "small" fire occurring is about 55% according to UK Fire Statistics, which relate to fires attended by or reported to the fire brigades. The remaining 45% are "big" fires requiring intervention by sprinklers with sprinklers operating in 87% of these cases and not operating in 13% of the cases – 39 and 6% respectively of all fires recorded in fire statistics. Some of the fires in which sprinklers operate are extinguished by the system before the arrival of the brigade and some are extinguished by the brigade.

According to an investigation (Rogers, 1977) carried out by the Fire Research Station, UK, several years ago, one-third of fires in sprinklered buildings are extinguished by the system and not reported to the brigade. Hence, fire brigades attend to only two-thirds of fires in sprinklered premises to which 50% should be added to provide an estimate of the total number of fires in these premises. Calculations would show that, out of this total number, 37% are "small" fires as defined earlier. Among 63% of "big" fires requiring intervention by sprinklers, sprinklers operate in 59% of the cases and fail to operate in 4% of the fires. Sprinklers, therefore, operate in 94% (= 59/63) of fires in which their action is required. If required, they might have acted in most of the "small" fires, which were self-extinguished (self-termination) or extinguished by first-aid means.

An analysis as described above needs to be carried out for estimating realistically the performance reliability of sprinklers. A similar analysis based on UK Fire Statistics has been carried out by Rutstein and Cooke (1979) who have estimated that 10% of the fires in which sprinklers operate are not reported to the brigade. Applying this correction, they have obtained a figure of 2.2% for sprinkler failure rate, which denotes a performance reliability of 97.8%. Rutstein and Cooke have also estimated, for various types of occupancies, the percentages of fires in which sprinklers operate satisfactorily, and this ranges from 92 to 97%. This proportion is 95.6% for all industrial buildings. In the remaining 2.2%, the fire is "out of control" and grows into a very large fire in which more than 35 sprinkler heads operate (see Figure 10.4 produced by Baldwin and North (1971)).

Australia and New Zealand compile the most reliable and thoroughly reported statistics on the performance of sprinklers. In these countries, all sprinkler systems are by law directly connected to fire stations and so activation is automatically accompanied by attendance of the fire brigade. In addition, all alarms must be checked on a weekly basis. On the basis of the long-term (from 1886 onwards) statistical data available for these two countries, Marryatt (1988) has estimated a success rate of over 99% for sprinklers. The success rate for sprinklers in the United States was about 96% for the period 1897 to 1964 according to the National Fire Protection Association (NFPA), 85% for the period 1970 to 1972 according to the Factory Mutual Research Corporation (FMRC), and 95% for the period 1966 to 1970 according to the US Navy. The figures mentioned above have been quoted in a study by Miller (1974), who used FMRC experience to estimate success rates of 86% for wet systems, 83% for dry systems, and 63% for deluge systems.

The reasons for sprinkler failure (nonoperation) in the United Kingdom between 1965 and 1969 have been investigated by Nash and Young (1991). The main cause of failure was shut valves, which accounted for 55% of all failures and 87% of failures for which the cause was known. The system was at fault due to problems in design or manufacture in 7% of all failures. Other and unknown causes accounted for 36% of all failures. In the NFPA investigation mentioned above, out of the 4% of failures, 36% were due to system shut down; of these, 85% could probably be attributable to human error.

The success of sprinklers in controlling fire spread can be assessed in terms of the number of heads operating in a fire (see Figure 10.4 produced by Baldwin and North (1971)). The American and British data in Figure 10.4 were not significantly different. Combining these two sets of data, a regression line was fitted by Baldwin and North to $\log q(N)$ and $\log N$ with $q(N)$ as the proportion of fires in which N or more heads operated. According to this analysis, 75% of fires are controlled or extinguished by four heads or less, 80% by five heads or less, and 98% by 35

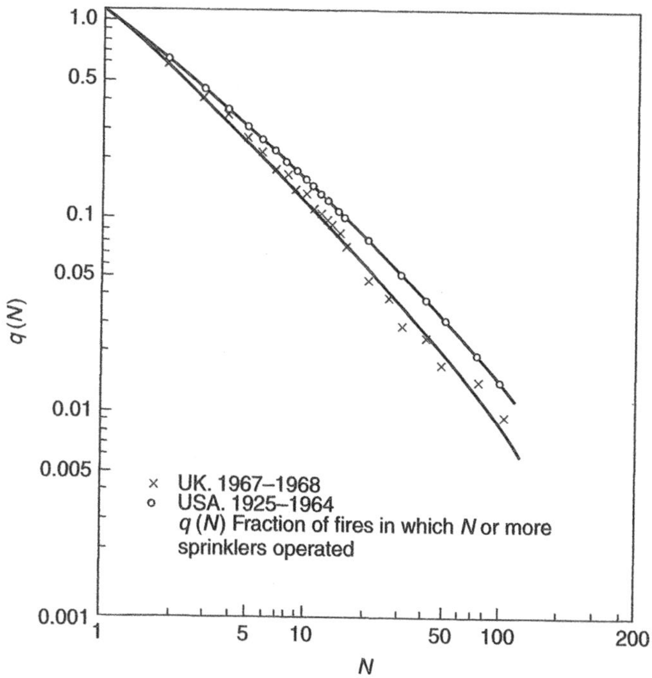

Figure 10.4. Number of sprinkler heads operating in fires in UK and USA

heads or less. The corresponding figures for Australia and New Zealand (Marryatt, 1974) are 90, 92, and 99% respectively. More recent results from America (Rees, 1991) published by FMRC cover the years 1978 to 1987. These reveal that 69% of fires are controlled by 5 heads or less, 83% by 10 heads or less, and 94% by 25 heads or less. The figures mentioned above demonstrate that only sufficient sprinkler heads to control the fire will activate, thus reducing the amount of water damage and fire loss.

Consequences of water damage and accidental leakage have been used as arguments against the installation of sprinklers in certain areas such as computer centers, libraries, and galleries. Reliable data are not available for assessing these losses. According to information provided by some individual fires, losses due to water damage are not appreciable and the chance of a leakage occurring is very small. Additional loss due to water damage is likely to be smaller than that which would result from further fire spread in the absence of sprinklers.

10.4.2 EFFECTIVENESS – PROPERTY PROTECTION

Effectiveness of sprinklers in reducing property damage has been well discussed and established in fire protection literature. Some of the UK studies on this aspect have already been mentioned in Table 7.3, with regard to the extent of fire spread, Figures 9.4 and 9.5, with regard to area damage, and Figure 9.7 and Table 9.7, with regard to financial loss. In a later study, Rutstein and Cooke (1979) have estimated the reduction in average area damage due to sprinklers in various occupancies for a building of size 1500 m^2. The reduction varies between 40% in hospitals and other establishments, and 93% in schools (see Table 10.3). For all industrial buildings, the reduction was estimated to be 73%.

On the basis of Home Office statistics for the years 1981 to 1987, Beever (1991) has found that the probability of fire damage exceeding a given area is very much smaller with sprinkler

Table 10.3. The estimated reduction in fire damage if sprinklers are installed

Occupancy	Average fire size in $1500\,m^2$ building (m^2)		Reduction in damage due to sprinklers (%)
	Without sprinklers	With sprinklers	
All industries	60	16	73
Food, drink, and tobacco	73	6	92
Chemicals and allied	28	12	57
Mechanical engineering	44	5	88
Electrical engineering	64	6	91
Vehicles	56	4	93
Metal goods	34	7	79
Textiles	45	20	56
Timber	112	14	87
Paper	93	17	82
Other manufacturers	140	24	83
Other occupancies			
Storage	157	23	85
Shops	37	6	84
Offices	15	3	80
Hotels, etc.	27	3	89
Hospitals, etc.	5	3	40
Pubs, etc.	33	3	91
Schools	42	3	93

protection than without. According to a research carried out by Morgan and Hansell (1984/85), 1 in 10 fires in offices will exceed $16\,m^2$ in a sprinklered building but $47\,m^2$ in a building without sprinklers. In recent studies, Ramachandran (1993a, 1995) defined a parameter K as the ratio between probabilities of damage being equal to the compartment size in nonsprinklered and sprinklered cases. This probability, as discussed in Section 10.5.1 can be regarded as the probability of flashover or severe structural damage. For office buildings (all rooms) and retail premises (area), the average value of K corresponding to an average room size of $100\,m^2$ was found to be 3.5 and 3.3, respectively; for hotel bedrooms, K was estimated to be 3.3. Sprinklers, therefore, reduce the probability of flashover or severe structural damage by a factor of 3.

One of the most comprehensive sources of data on fire loss in sprinklered versus unsprinklered buildings comes from the FMRC in the United States. Fire loss data compiled by FMRC for the period 1980 to 1989 for a wide range of production, warehouse, and other nonmanufacturing occupancies indicate that the average fire loss for an unsprinklered building is approximately four-and-a-half times greater than that for an adequately sprinklered building (Rees, 1991). An analysis of data by the NFPA covering the years 1980 to 1990 has shown that the reduction in average loss per fire ranges from 43% for stores and offices to 74% for educational establishments (Hall, 1992).

As discussed in Section 7.6, sprinklers would reduce the rate of fire growth in a fire developing beyond the stage of "established burning." The overall rate for a textile industry building is reduced by a factor of 2.7, while the rate for growth within a room is reduced by a factor of 1.7. In a recent investigation, Melinek (1993a) assessed the effectiveness of sprinklers in reducing fire severity expressed in terms of area damage. He estimated that sprinklers would reduce the probability of fire size in industrial and commercial buildings reaching $100\,m^2$ by a factor of 5. He also found that damage to the structure of a building would be reduced by a factor of 2.5. In the study mentioned above, Melinek compared fires that sprinklers extinguish or control with fires they fail to extinguish or control; he did not compare fire size in sprinklered and unsprinklered buildings.

10.4.3 EFFECTIVENESS – LIFE SAFETY

It is a difficult task to estimate the number of lives that could be saved by installing sprinklers in buildings with large numbers of people at risk. Sufficient statistics are not available for analyzing this problem. An estimate can be made only by applying an evacuation model such as the one proposed in Section 8.8. As discussed in that section, sprinklers will reduce both the delay in discovering a fire and the rate of growth of fire and smoke. They also have a high probability of extinguishing a fire during its early stage of growth. Consequently, sprinklers have the potential to reduce the fatality rate per fire in multioccupancy dwellings to 0.0009 from the current level of 0.0122.

Comprehensive data from Australia and New Zealand covering 100 years up to 1986 show that, during that period, there were 11 deaths in 9022 sprinklered fires (Marryatt, 1988). This represents a fatality rate of 0.0012 per fire, which is not much different from the figure of 0.0009 mentioned above. Hall (1992) quotes the results of an analysis of the likely effects of sprinklers on deaths in one- and two-family house fires carried out by the National Institute of Standards and Technology, USA. This was based on laboratory test data, estimates from fire researchers, and available statistics on the likelihood of certain scenarios and the life threat posed by them. According to this study, a 63 and 69% reduction in death rates per thousand fires can be achieved if sprinklers are installed in dwellings that do and do not already have smoke detectors respectively.

Melinek (1993b) has estimated the number of casualties, if all fires were sprinklered, assuming that the average number of casualties per fire depends only on the extent of spread and that the proportion of fires spreading, if all the buildings were sprinklered, would be equal to that in existing sprinklered buildings. He has shown that the number of fatal casualties would be reduced by about half while the number of nonfatal casualties would be reduced by about 20%. He has also shown that sprinklers significantly reduce the number of multicasualty fires.

10.5 Structural (passive) fire protection

10.5.1 FIRE RESISTANCE

Fire-resistant compartmentation has long been a core fire safety measure. A building can be regarded as divided into compartments perfectly isolated from one another and the spread of fire as taking place by successive destruction (or possibly thermal failure) of the compartment boundaries. If the boundaries are of sufficient fire resistance, it is argued, the probability of fire spread beyond the compartment will remain within acceptable limits. Performance of a fire-resistant compartment is assumed to be 100% satisfactory but reliability depends on the level of resistance.

A structural member is said to "fail" if performance criteria relating to stability, integrity, or thermal insulation are violated due to intense heat produced by a fire. For compartments of moderate size, of the order of 100 m^2, it is generally assumed that this heating takes place entirely during the postflashover period. Collapse and destruction relate to stability of the structure, particularly for beams and columns. Floors and walls can also fail where integrity or insulation is lost.

"Failure" would occur if a "limit-state" is reached in the domain of time, temperature, or mechanical strength. Three methods of identifying this extreme condition, on the basis of heat exposure, have been described in the CIB Report (1986). Among these analytical models, the method generally preferred is the one involving the "equivalent time of fire exposure," which is tied up most closely with the stability of the structure. This time (T_e) is estimated by a function of fire load density (q), ventilation factor (w), and a constant (c) associated with the thermal properties of the construction surrounding the element of structure considered.

$$T_e = c \cdot w \cdot q \qquad [10.4]$$

The ventilation factor (w) is based on the CIB equation involving window area, mean window height, bounding surface area, and floor area. The fire resistance period (T) required for the element of structure as measured by ISO 834, is set equal to T_e which is essentially an estimate of "potential" severity likely to be attained in the event of a fire.

$$T = T_e \qquad [10.5]$$

To obtain an estimate of potential maximum severity through equation [10.4], fire load density (q) corresponding to an 80% fractile of the load distribution has been recommended in the CIB Report (1986). By providing fire resistance equivalent to the maximum severity estimated by this fractile value of fire load density, it is deemed that for a complete burnout, the probability of structural success is likely to be 0.8; the probability of structural failure is 0.2.

As discussed in Section 5.9, severe structural damage may be assumed to occur during the postflashover stage when all the objects in the compartment are involved in fire. The effect of fire on the structure will depend on the duration of burning after the occurrence of flashover, which in turn, depends on the ventilation, fire load, and certain other physical parameters. Following the assumption mentioned above, postflashover may be defined statistically (Ramachandran, 1993a) as the stage when most of the floor area of a compartment is affected by fire. Statistical data on area damage can be used to estimate the probability associated with this stage.

Following the definitions and assumptions mentioned above, the probability (C) of structural collapse, termed as *structural failure* for easy reference, is the product of two components – probability (A) of "flashover" and "conditional" probability (B) of structural failure, given "flashover." "Structural failure" would occur if the design fire resistance time is exceeded by severity actually attained in a fire. Because of continuity effects, structural failure of a building framework in an actual fire is not the same as structural failure in a fire resistance test. The probability B, as discussed earlier, depends on the fractile value of the design fire load density.

Expressing numerically,

$$C = A \times B \qquad [10.6]$$

Depending on the consequences in terms of life loss and other factors, an acceptable level can be specified for the product C. If, for example, the value determined for C is 0.02 and if the probability of flashover, A in a compartment is 0.1, then from equation [10.6],

$$B = 0.02/0.1$$

$$= 0.2$$

According to this result, the target value of 0.02 for total fire safety, as defined by the product C, will be met by providing structural fire safety with a fire resistance corresponding to 0.8 ($= 1 - 0.2$) or 80% fractile of the fire load density distribution.

The method described above is a semiprobabilistic approach since uncertainties governing fire severity (S) and resistance (R) in an actual (not experimental) fire have not been taken into account in evaluating the conditional probability B of structural failure, given flashover. The potential maximum severity estimated through equation [10.4] may or may not be attained in a real fire due to several factors affecting its development. Fire resistance of a compartment is not the same as the resistance of any structural element (floor, wall, or ceiling). It is a random variable (not a constant) depending on the type and materials of construction. Fire resistance of a compartment is also affected by weakness caused by penetrations, doors, or other openings in structural barriers.

In a simple probabilistic model on structural reliability, the randomness of fire resistance may be ignored and R regarded as a constant that is set equal to a large value of S. This value is such that the probability of severity exceeding it is a specified small quantity depending on the probability distribution of S. In an expanded probabilistic approach, both R and S are considered as random variables (in units of time) with probability distributions (see CIB W14 (1986), Ramachandran (1990, 1995)). In this approach, it is usual to consider the safety index given by

$$\beta = (\mu_r - \mu_s)/(\sigma_r^2 + \sigma_s^2)^{1/2} \qquad [10.7]$$

where μ_r and σ_r are the mean and standard deviation of R, and μ_s and σ_s, the mean and standard deviation of S. The probabilities of compartment failure for different values of μ_r and μ_s can be evaluated with the aid of the parameter β.

Let us assume, for example, that both R and S have normal probability distributions such that β has a standard normal distribution with zero mean and unit (one) standard deviation. In this case, provision of average fire resistance, μ_r to compartment boundaries with $\beta = 1.96$ in equation [10.7] would ensure that the probability of compartment failure is only 0.025. The probability of failure would be 0.01 if $\beta = 2.33$ and 0.001 if $\beta = 3.09$. Tables of (standard) normal distributions would provide the failure probability associated with any value of β. The probability of failure would be equal to 0.5 if $\mu_r = \mu_s$ with $\beta = 0$, greater than 0.5 if $\mu_r < \mu_s$ with a negative value for β, and less than 0.5 if $\mu_r > \mu_s$ with a positive value for β. A barrier failure analysis on the lines described above has been suggested by Elms and Buchanan (1981).

For convenience, a safety factor θ may be defined by

$$\theta = \mu_r/\mu_s \qquad [10.8]$$

in which case

$$\beta = (\theta - 1)/(\theta^2 C_r^2 + C_s^2)^{1/2} \qquad [10.9]$$

where C_r and C_s are coefficients of variation of R and S:

$$C_r = \sigma_r/\mu_r; \; C_s = \sigma_s/\mu_s$$

Inverting equation [10.9],

$$\theta = \frac{1 + \beta(C_r^2 + C_s^2 - \beta^2 C_r^2 C_s^2)^{1/2}}{1 - \beta^2 C_r^2} \qquad (10.10)$$

Reasonable values for C_r and C_s can be assumed if data are not available for estimating the standard deviations σ_r and σ_s.

If it is further assumed that $C_r = C_s = C$, equation [10.10] reduces to

$$\theta = \frac{1 + \beta C(2 - \beta^2 C^2)^{1/2}}{1 - \beta^2 C^2} \qquad [10.11]$$

In this particular case, from equation [10.7],

$$\beta = (\mu_r - \mu_s)/C(\mu_r^2 + \mu_s^2)^{1/2} \qquad [10.12]$$

For example, if a probability of 0.0014 can be tolerated for compartment failure, $\beta = 2.99$, and if $C = 0.15$ as assumed by Elms and Buchanan (1981), $\theta = 2$ from equation [10.11]. Hence, the desired target is achieved if the mean fire resistance μ_r of the compartment is equal to twice the mean fire severity μ_s likely to be encountered in an actual fire.

An estimate for the mean value for μ_s may be obtained by inserting the mean value of fire load density q in equation [10.4]. From this equation, the coefficient of variation C_s may be estimated

$$C_s^2 = C_w^2 + C_q^2$$

where C_w and C_q are the coefficients of variation of the ventilation factor and fire load density for which data are available for some types of occupancies. Generally, if C_o is the overall coefficient of variation of R or S

$$C_o^2 = C_1^2 + C_2^2 + \cdots$$

where $C_1, C_2, \ldots$ are the coefficients of variation of factors affecting R or S.

Standard deviations and coefficients of variation quantify uncertainties caused by factors affecting fire resistance and severity. An assessment of these uncertainties is of fundamental importance in any structural reliability analysis. Factors and uncertainties would vary depending on the type of structural element considered – steel, concrete, or timber. The fire resistance of these types has been discussed in Chapters 3-6 (J. Milke), 3-7 (C. Fleischmann) and 3-8 (R. H. White) of the SFPE Handbook of Fire Protection Engineering (1988). The fire resistance of a steel member, for example, depends on cross-sectional area of the steel section, perimeter of steel section exposed to fire, specific heat of steel, density of steel, thickness of insulation, thermal conductivity, and specific heat of the insulation. The fire resistance of a structural element is usually based on the standard fire resistance test of ISO 834, which is another source of uncertainty. Fire behavior of a structural frame composed of elements such as beams, columns, and floors differs from the behavior of the elements. Fire severity, as discussed earlier, depends on ventilation factor and fire load, which could vary from compartment to compartment in a building.

Sufficient data are not available for estimating the probability distributions of R and S, although normal distributions have been suggested for the sake of simplicity and purposes of illustrating the application of the probability model. From equation [10.10], it may be seen that a high variability in R implies values of θ that may be unrealistically large. As a better alternative,

$$\beta_{\mathrm{ER}} = \log_e(\mu_r/\mu_s)/(C_r^2 + C_s^2)^{1/2} \qquad [10.13]$$

may be considered as the safety index (see Rosenblueth and Esteva (1971)). In this case, the safety factor is given by

$$\theta_{\mathrm{ER}} = \exp[\beta_{\mathrm{ER}}(C_r^2 + C_s^2)^{1/2}] \qquad [10.14]$$

Equations [10.13] and [10.14] would apply if R and S have lognormal probability distributions (see also CIB W14 (1986)).

The method described in this section would provide "partial safety factors," which constitute a practical design format. These factors based on a probabilistic analysis are more realistic than arbitrary values generally assigned to them. The method based on the safety index in equation [10.7] is considered as "approximately probabilistic" in Reliability Theory, and is usually referred to as First Order (Second Moment) Analysis. A fully probabilistic reliability analysis involves the joint probability density functions of R and S and a "convolution integral." Evaluation of this integral is a complex mathematical problem.

The value of μ_s attained in an actual fire is likely to be different from its potential value estimated through equation [10.4]. The exact relationship between the actual and potential values of μ_s is not known at present. It may be possible to derive an approximate formula for this relationship with the aid of equations [10.4] and [7.12], where severity is defined as the area damaged in a real fire. This would require the estimation of the parameter k in equation [7.12], which relates to severity in a real fire.

For the following reasons, fire statistics do not provide reliable estimates on the effectiveness of a fire-resistant compartment in reducing the probability of fire spread. A "room," as recorded in fire brigade reports, is not necessarily a fire compartment. The figures for the number of fires that spread beyond the room of origin include fires that spread by destruction of structural boundary as well as those that spread by convection through a door or window left open or through some other opening. In the latter case, the boundary elements would still be structurally sound. It cannot be assumed that in all the fires, the spread beyond the room was due to the collapse of the structural boundary. According to fire statistics, structural collapse in a fire rarely occurs.

Between the years 1962 and 1977, fire spread from buildings to adjoining buildings was annually published in United Kingdom Fire Statistics for the whole range of occupied buildings. The figures in the tables relating to this aspect provide for certain occupancies, approximate estimates for the likelihood of failure of separating walls or party walls with prescribed fire resistance and without penetration by doors, ducts, and so on. Two examples are residential houses and retail distributive trades (i.e. shops). Using reasonable assumptions, Rasbash (1994) has carried out an analysis of such data for the three years, 1972 to 1974. He has estimated that among fires involving the structure, 2.8% in houses and 7% of those in shops spread to adjoining buildings. These figures denote approximately the probabilities of structural failure in the two occupancies considered.

10.5.2 COMPARTMENTATION

In Building Regulations, a compartment is defined as a part of a building bounded by fire resisting elements of building construction designed to restrict the size of fire, prevent extensive damage to contents, and reduce the possibility of conflagration. This may require the division of a building into smaller "fire cells" such that a fire in any one cell would be contained in that cell. In contiguous properties, for example, semidetached houses, barriers such as party walls are usually sufficient to prevent fire spread to an adjoining property. But in special cases such as flats, each property may have to be enclosed in a fire cell.

Prevention of conflagrations and the control of fires are interlinked with the fire brigade capabilities, as well as other features such as the separation between buildings and the presence or absence of installed devices like sprinklers. Hence, the fire brigade effectiveness and capability would seem to be a strong factor in determining the fire resistance, size, and number of compartments in a building and the building height. Compartments might ideally contain not more than two or three floors in low-rise buildings, and not more than one floor in high-rise buildings and in basements. Each story in a high-rise building would thus be a fire compartment. The difficulty of fighting fires in high-rise buildings and in basements requires that as much horizontal compartmentation as possible is provided. At present, there is no recognized technique for determining compartment sizes.

As mentioned in the previous section, performance or operation of a compartment in a fire is assumed to be 100% satisfactory. Statistics are not available to assess the effectiveness of compartmentation, that is, the number of compartments in a building. However, a number of fires involving uncompartmented buildings have drawn attention to the absence of subdivision leading to extensive losses. Large storage areas without adequate compartmentation were shown to be a problem by the British Fire Protection Association in their analysis (1979) of large fires in supermarkets, hypermarkets, and warehouses. Lack of compartmentation was also evident at a fire in Donnington Army Depot in the United Kingdom in 1983, which spread rapidly and caused huge financial loss.

10.5.3 MEANS OF ESCAPE FACILITIES

The main object of providing a means of escape in a building is to ensure safe evacuation of all occupants to a place of safety in case of a fire. The escape routes should be available from all parts, remain safe, and effective for the duration for which they are needed, be clearly visible to all users, and be located and sized to meet the needs of all occupants taking account of the use of the building. Means of escape facilities should, therefore, be designed according to the number and characteristics (average age, mobility, etc.) of the occupants.

It is not known how many people are being exposed to fires every year although statistics are available on the number of people escaping or being rescued. If assumptions are made about the average occupancy levels of different types of buildings, these levels can then be compared with the average number of people escaping successfully per fire. This method might give a rough idea of the effectiveness of escape routes. The consequences of inadequate means of escape have been highlighted in a number of incidents in which the absence of properly designed routes, inadequate protection, failure of alarm or warning systems, or some other shortcoming has resulted in serious loss of lives.

According to means of escape provisions in fire safety codes, in the event of a fire, the occupants of a building should be able to reach the entrance of an enclosed staircase or other place of safety within a reasonable period of time. For many large buildings, complete evacuation would be lengthy and difficult; and the codes, recognizing the compartmented nature of buildings, envisage evacuation of only part of the building, usually the fire floor and the floor above. One staircase is assumed to be inoperative due to the effect of the fire. These criteria, together with data on the number of occupants on different floors and their rates of passage through doorways, corridors, and so on, determine the number and widths of staircases required to permit unimpeded flow of people. Formulas used for this purpose contain design evacuation time (E) as an important parameter. This is the time taken by an occupant to reach the entrance of a staircase after leaving his or her place of occupation.

As explained in Section 8.8, the period E is one of three main periods constituting the total evacuation time (H); discovery time (D) and "recognition time" (B) are the other two periods; $H = D + B + E$. For successful evacuation, H should not exceed the time (F) taken by smoke and toxic gases to travel from the place of fire origin and to produce untenable conditions on escape routes. The values of H and F would vary depending on the floor of occupation of a group of escaping population and place of fire origin.

Using the design criterion, $H < F$, a simple (exponential) model has been discussed in Section 8.8, which determines the value of E according to an acceptable level for fatality rate per fire. It has been argued in that section that the value of E can be increased within the acceptable limit for buildings equipped with sprinklers and detectors. As mentioned in that section, the uncertainties associated with H and F should be taken into account by considering the probability distributions of these two variables for any type of occupancy (see Ramachandran (1993b, 1995)).

Depending on the values of H and F, there is a probability of a group of occupants encountering significant amounts of combustion products. If sufficient data are available, this probability can be evaluated by considering a safety index similar to that in equation [10.7]:

$$\beta' = (\mu_f - \mu_h)/(\sigma_f^2 + \sigma_h^2)^{1/2}$$

where μ_f and σ_f are the mean and standard deviation of F, and μ_h and σ_h are the mean and standard deviation of H. The safety factor in this case is $\theta' = \mu_f/\mu_h$. Under normal distribution assumption, the encounter probability denoting evacuation failure would be equal to 0.5 if $\theta' = 1$. It would be less than 0.5 if $\theta' > 1$ and greater than 0.5 if $\theta' < 1$. If an encounter with combustion products occurs, there is the likelihood that one or more escaping occupants could die.

For some occupancies, the variable H may have a lognormal distribution. In a hospital or a building with a high proportion of disabled people, some may be able to escape quickly while some requiring assistance for evacuation may take a longer time to reach a safe place. The variable

F can also have a lognormal distribution depending on the rate of spread of heat, smoke, and other combustion products.

10.5.4 FIRE DOOR

Doors of suitable construction can be said to be one of the most important elements in a building for the safety of life, if a fire should occur in that building. Fire safety codes for means of escape depend upon the fire/smoke check door as an integral part of the escape plan. Without such doors, safety from fire in any building, even a single-story building, may become difficult and often necessitates unorthodox and undesirable means, for example, jumping through a window.

Even if fire doors are of adequate fire resistance, and are strategically sited in a building, they are of no value if they are left open. Even doors for amenity purposes may, if closed, tend to delay the development and spread of fire. Some doors, for example, bedroom doors of parents in a dwelling, may have to be left open, although this would also enable early discovery of the existence of a fire in the kitchen or sitting room in the absence of a smoke detector. It is arguable whether amenity doors should be shut or kept open, but fire doors or smoke check doors used for means of escape purposes must be kept closed if they are to be effective. Doors designed for security purposes are rarely used by large numbers of people. The exception to this is, of course, the sheet entrance doors to flats and maisonettes.

To ascertain how serious the problem of the open door really was in buildings, the Fire Research Station, UK, carried out an analysis of the use of fire-check doors on the basis of information collected by the fire brigades during normal inspection visits. The results of the analysis are contained in a paper by Langdon-Thomas and Ramachandran (1970). The authors found that the frequencies of doors propped open at the time of fire brigade visits ranged from 5% in assembly buildings to 39% in institutional buildings. In storage premises also, the frequency was as high as 37%. In a majority of these instances, it was claimed that the doors were intended to be shut at night or in an emergency. In 1 to 7% of the cases, the doors were found open but obstructed so that they could not be closed.

Langdon-Thomas and Ramachandran also found that the display of a notice asking people to keep a fire door closed, while not completely effective, generally reduced the number of doors propped open. In order to achieve the closing of the door, some type of mechanical closing device may be adopted – single or double action floor spring, overhead door closures, and spring hinges. The authors found that, in regard to closing action, the third type was the most satisfactory automatic closing device. Sixty-three per cent of overhead door closures and 35% of the floor spring type had defective closing action. The authors also suggested the use of a smoke detector coupled with a door retainer.

Fire doors in industrial buildings, if kept in the closed position, can be expected to reduce property damage in a fire. In order to estimate this reduction (saving), Ramachandran (1968) analyzed a small sample of data on large fires, which were available for the years 1965 and 1966. Loss in each of these 17 fires was £10,000 or more. In five of these fires, the doors were in the open position. These included a very large fire in a paper tube factory with loss exceeding a million pounds, in which a number of fire doors were left open. Excluding this fire, the average loss in the remaining 4 fires was £135,000. In 12 fires, the doors were closed and performed their function satisfactorily so that the average loss was only £106,000 even if a very large fire with a loss of £450,000 was included. It was, therefore, estimated that keeping the fire doors closed in industrial buildings could save at least £30,000 per fire (at 1965 prices). In two other cases, closed doors contained the fire – a fire with a loss of £10,000 in a retailed grocers and provision merchants, and another with a loss of £75,000 in a cinema and bingo hall.

10.6 Effectiveness of other fire protection devices

10.6.1 FIRE EXTINGUISHERS

In some fires, occupants of a building use a number of first-aid methods to attack the fires before the arrival of the brigade. Portable fire extinguishers constitute one such method apart from "sundry means" such as buckets of water or sand, garden hose, and smothering. Extinguishers available in the market are mainly of the following types – dry powder, water, carbon tetrachloride, foam, and other vaporizing liquids and carbon dioxide.

Some fires are extinguished or controlled by first-aid means. An initial attack by occupants does reduce the severity of a fire, but there have been some instances in which such an action has led to fatal or nonfatal casualties. In industrial and commercial buildings (Table 10.3), early detection followed by quick action to extinguish the fires by sprinklers, extinguishers, or "sundry" means reduces the damage but this saving would be higher if first-aid fire fighting is not undertaken. This implies that, if a fire is discovered early by human or automatic detection devices, the fire brigade should be called quickly instead of launching an attack by first-aid means.

While early detection followed by a call to the fire brigade appears to be a better option than first-aid fire fighting, some doubts on the effectiveness of fire extinguishers have also been cast by some research studies in the United Kingdom. An analysis of fire brigade data by Ramachandran *et al.* (1972) disclosed that extinguishers were unlikely to be as effective as "sundry" means in attacking dwelling fires. In this study, the effectiveness of first-aid methods was assessed in terms of proportion of fires put out by these methods and the average time taken by brigades to control fires that were not put out. Occupants were able to put out 43% of the fires tackled by "sundry" means but only 27.5% of those in which extinguishers were used. The average control time of a fire brigade was 6.5 min for "sundry" means and 8.9 min for extinguishers. Without further statistical analysis, it was found difficult to judge the significance of this difference in mean control time.

It was possible that extinguishers in dwellings were located at considerable distances from the places of fire origin, for example, cars and garages. An analysis by Sime *et al.* (1981) gave some indication that people have inadequate knowledge of the location of extinguishers. A householder may be more inclined to tackle a fire in an armchair than one in a fat pan. According to Chandler (1978), people were less likely to use extinguishers on small fires, which might partly explain the lower success rate with extinguishers than with other methods in hospitals. This conclusion was confirmed by Canter (1985) who found that the contribution of extinguishers to fire fighting was subject to many constraints in actual fires. Canter also suggested that people (especially staff in hospitals, hotels, etc.) should be made aware of the location of extinguishers, and trained in the use and capabilities of different types and sizes of extinguishers.

As in the case of sprinklers, a number of small fires extinguished by portable fire extinguishers were not reported to the fire brigades. In the United Kingdom, statistics supplied by the Fire Extinguishing Trade Association (Fire Prevention, March 1990) indicate that over 70% of fires were not reported to the fire brigade because they were put out by fire extinguishers. In addition, 17% of the fires reported to the brigade were found to have already been put out by staff or residents using fire extinguishers prior to the brigade's arrival. All the factors and data mentioned above should be taken into account in a detailed statistical analysis for establishing the effectiveness of extinguishers.

10.6.2 VENTILATION SYSTEMS

Fires that occur in a restricted area, such as the kitchen of a dwelling with all the doors and windows closed, may burn out within the confined room area without anyone being aware of the fire's existence. This happens because of a lack of oxygen. In a large uncompartmented building, however, the seepage of air through gaps in the structure is sufficient to prevent the fire being starved of oxygen. As the fire grows, smoke and hot gases will collect at roof level and will in a fairly short time (depending upon the material burning), extend from floor to ceiling if there is no means of expelling them as fast as they are produced. Hot smoky gases can also collect at the upper parts of a compartment that is partially or completely involved in a fire. These gases initially form a stratified layer beneath the ceiling, which, without venting, deepens and after a relatively short time mixes into the clear air beneath. The speed with which an unvented compartment or building can become smoke-logged has been demonstrated in many fires.

Venting is the removal of hot smoky gases from the upper parts of a compartment or building and the introduction of air from outside into lower parts. This process may involve natural convection through openings that occur fortuitously or are provided purposely, or it may involve mechanical (powered) extract or inlet or both. Hinkley (1988) has discussed the basic engineering concepts underlying the design of complete venting systems, including the provision of openings (vents and inlets) or fans, and allied features such as the provision of screens (curtains) to limit the spread of smoke beneath the ceiling. It is important that a venting system be designed as a whole, taking into account other fire safety measures including the provision of structural fire protection, escape routes, and sprinklers.

Venting is provided with one or more of three objectives:

1. To facilitate escape of people by restricting spread of smoke and hot gases in escape routes,
2. to facilitate fire fighting by enabling fire fighters to enter the building and to see the seat of the fire,
3. to reduce damage due to smoke and hot gases.

The extent to which the objectives mentioned above are fulfilled is a measure of the effectiveness of a ventilation system. The effectiveness would vary from one system to another and from one type of building to another.

Experiments carried out in the United Kingdom, the United States, and other countries have revealed that, in unsprinklered fires, vents remove convective heat effectively, greatly retard the buildup of smoke, and improve visibility. Ventilation systems can, therefore, be expected to satisfy the three objectives mentioned above if fires occur in unsprinklered buildings. However, it has been suggested that heat and smoke vents may be less effective after sprinklers have operated. The interaction (Section 10.7.4) between vents and sprinklers has been the subject of intensive debate and investigation for the past few years.

There is a lack of statistical data for assessment of the effectiveness of ventilation systems. However, a small sample of data for industrial buildings analyzed by Ramachandran some years ago did show that vents reduce the control time of fire brigades. In buildings without sprinklers, the average control time was estimated to be 119 min in the absence of vents and 57 min if vents were installed. The sample sizes were 21 and 26 fires, respectively. This result indicated that vents would enable fire fighters to bring fires under control quickly. Sufficient information was not available for an estimation of the reduction in damage due to vents or the interaction between vents and sprinklers. The results of this study were not published.

10.7 Interactions

10.7.1 AUTOMATIC DETECTORS AND FIRE BRIGADE

As discussed in Section 10.3.2, early discovery or detection of a fire would reduce the time required by the fire brigade to control a fire. This is due to the fact that the fire size would be small when the brigade is notified about the fire and when it arrives at the fire scene. Early detection of fire would, therefore, reduce damage to life and property. However, if no change is desired in the level of damage, some delay can be tolerated in the response time for the initial dispatch of men and equipment to an incident.

The trade-off between automatic detection systems and response time has implications for policies concerned with fire department deployment and fire cover provided for various risk categories. This problem, in particular, investment in detector-alarm systems as a substitute for traditional fire department expenditures on manned fire stations and equipments' has been investigated by Halpern (1979) in regard to the protection of single- and double-family dwellings. Halpern has presented a model, which provides a cost-benefit comparison between the productivity of detector-alarm systems and manned fire stations in reducing fire losses. The parameters of the model were based on data collected in Calgary, Canada. Assuming that 10 detectors per home were enough and using a New York study as a general guideline, Halpern concluded that the detector-alarm system was a viable and competitive alternative to additional fire stations.

10.7.2 SPRINKLERS AND FIRE BRIGADE

Sprinklers have a high potential for extinguishing fires. If they fail to put out a fire, sprinklers will reduce the control time by restricting the spread of fire until the arrival of the brigade. Statistical data are available in the United Kingdom for estimating the reduction in control time due to sprinklers. The interaction between sprinklers and fire brigade action should be taken into account in determining the number and size of fire stations.

In the United States, the City of Fresno was, after 1970, the setting for a remarkable experiment on the capability of sprinklers to reduce fire loss. Within the city, all buildings in two separate districts ranging from 1 to 16 storys high were fitted with complete automatic sprinkler protection. This gave 93.5 and 96% sprinkler protection in these two areas. Despite an 8% rise in the number of fires, the fire losses in these areas reduced considerably with the result that a fire station was closed down.

10.7.3 SPRINKLERS AND STRUCTURAL FIRE PROTECTION

It is a well-established fact that sprinklers reduce area damage, extent of spread, probability of flashover, and of course, financial loss and life risk. Hence, subject to any levels specified for damage to property and life loss, one or more of the following concessions can be given to a building fully protected by sprinklers.

(a) reduction in fire resistance
(b) increase in building size
(c) increase in compartment size
(d) increase in design evacuation time.

Consider first, the fire resistance of a compartment. Since sprinklers reduce the probability (A) of flashover, the conditional probability (B) of compartment failure can be increased up to a limit such that the value of the product $C (= A \times B)$ does not exceed any specified level (see equation [10.6]). A simple mechanism for allowing an increase in B is the factor K mentioned in Section 10.4.2. It may be recalled that K is the ratio of the values of A for nonsprinklered and sprinklered compartments. If the increased value of B is denoted by B_s,

$$B_s = K B_0$$

where B_0 is the B value specified for a compartment without sprinklers. The fractile value of fire load density for a nonsprinklered compartment is $F_0(= 1 - B_0)$, and that for a sprinklered compartment is $F_s(= 1 - B_s)$.

If, for example, $B_0 = 0.2$ and $K = 3$, $B_s = 0.6$, and $F_s = 0.4$. Hence, based on the semiprobabilistic approach proposed in Section 10.5.1, fire resistance for a sprinklered compartment can be determined through Equation [10.4] according to fire load density corresponding to 40% (0.4) fractile value. This process reduces the fire resistance for a sprinklered compartment, since the fire resistance for a nonsprinklered compartment is based on an 80% (0.8) value of fire load density. This approach has been suggested by Ramachandran (1993a, 1995) for deriving the "sprinkler factor." The value of B_s can also be determined by following the expanded probabilistic approach discussed in Section 10.5.1.

On the basis of Equation [7.4], the permissible increase in the size of a sprinklered building has been discussed in Section 7.4 with the aid of an example shown in Figure 7.1. Allowable increase in the size of a sprinklered compartment has also been discussed through the example in Figure 7.2. The relaxation in the size of a building or compartment can also be determined according to the probability distributions of area damage in sprinklered and nonsprinklered buildings (Figures 9.4 and 9.7). Relaxation (increase) in the design evacuation time, and hence in the maximum travel distance for buildings equipped with sprinklers and detectors has been discussed in Section 8.8, using the exponential model in equation [8.3]. As suggested in that context, a further increase in the design evacuation time may be permitted for a building equipped with both sprinklers and detectors, on the basis of an interaction between the two fire protection systems.

10.7.4 SPRINKLERS AND VENTILATION SYSTEMS

As mentioned in Section 10.6.2, there is an interaction between sprinklers and ventilation systems, which has so far not been clearly evaluated by fire scientists and engineers. This interaction arises from the fact that water sprays from activated sprinklers remove buoyancy from combustion products and generate air currents that counter the outflow from the vents, and additionally, transport combustion products to the floor. Use of vents may reduce the total water demand by the sprinkler system but this benefit due to vents has not yet been clearly established. There may also be other types of interactions between sprinklers and vents. It may take a few more years of research and experimentation before all the interactions between these two types of fire protection measures are well ascertained and evaluated. Sufficient statistical information also needs to be collected before assessing the effects of the interactions on damage to life and property and the performance of the fire service.

There are arguments for and against operation of a vent before operation of a sprinkler. There are indications from current research that the effect of venting on the opening of the first sprinklers and their capacity to control the fire are likely to be small. There are also indications that the earlier the vents are opened, the more likely that they would be effective in preventing smoke-logging of a sprinklered building. The controversy between sprinklers and vents can, perhaps, be resolved for any type of building by deciding whether property protection or life safety is the main objective. In the initial stage of fire growth, a vent should operate before a sprinkler if life safety is the dominant objective, for example, in hotels, shopping centers, and office buildings. In industrial buildings, the first sprinkler may operate before the opening of any vent.

Symbols

A	floor area *or* probability of flashover
$A(O)$	area originally ignited

$A(T)$	area damage
a, b	constants in equation for control time
B	conditional probability of structural failure *or* recognition time
B_o	B (failure probability) without sprinklers
B_s	B (failure probability) with sprinklers
C	probability of structural collapse *or* coefficient of variation
C_o	overall coefficient of variation of R or S
C_q	coefficient of variation of q
C_r	coefficient of variation of R
C_s	coefficient of variation of S
C_w	coefficient of variation of w
c	constant associated with thermal properties of the construction
D	discovery time
E	design evacuation time
F	time to attain untenable conditions
F_o	$= (1 - B_o)$
F_s	$= (1 - B_s)$
H	total evacuation time $= (D + B + E)$
K	ratio of probabilities of flashover without and with sprinklers
L	direct loss
N	number of fires per million population *or* number of heads operating
p_1	proportion of fires when people are in the room of fire origin
p_2	proportion of fires when people are in the building (not room) of fire origin
p_3	proportion of fires when there are no people in the building of fire origin
q	fire load density
$q(N)$	proportion of fires in which N or more sprinkler heads operated
R	fire resistance in actual fire
S	fire severity
T	total duration of burning *or* fire resistance period
T_1	elapsed time from ignition to detection of a fire
T_2	elapsed time from detection to call of brigade
T_3	elapsed time from call to arrival of brigade
T_4	elapsed time from arrival of brigade to control of fire (control time)
T_A	$= (T_1 + T_2 + T_3)$
T_B	$= T_4$
T_e	equivalent time of fire exposure
t	mean annual air temperature
w	ventilation factor
y	year counted from 1968 = 1
β	safety index; constant in the formula for average damage in an unprotected building
β'	$= (\mu_f - \mu_h)/(\sigma_f^2 + \sigma_h^2)^{1/2}$
β_{ER}	= safety index β for R and S with log normal probability distributions
λ	increase in fatality rate per fire per minute
μ_f	mean of F
μ_h	mean of H
μ_r	mean of R
μ_s	mean of S
θ	overall rate of fire growth *or* safety factor $= \mu_r/\mu_s$
θ'	safety factor $= \mu_f/\mu_h$

θ_A	growth part of θ during period T_A
θ_B	growth part of θ during period T_B
θ_{ER}	safety factor θ for R and S with log normal probability distributions
σ_f	standard deviation of F
σ_h	standard deviation of H
σ_r	standard deviation of R
σ_s	standard deviation of S

References

Appleby, D and Ellwood, S H (1989). *A fire detection system using distributed processing. Proceedings of the 9th International Conference on Automatic Fire Detection (AUBE)*, Duisberg, Germany, 1989, pp. 101–115.

Baldwin, R (1971). *A Statistical Approach to the Spread of Fires in Buildings*, Fire Research Station, UK, Fire Research Note No. 900.

Baldwin, R (1972). *Some Notes on the Mathematical Analysis of Safety*, Fire Research Station, UK, Fire Research Note No. 909.

Baldwin, R and North, M A (1971). *The Number of Sprinkler Heads Opening in Fires*, Fire Research Station, UK, Fire Research Note No. 886.

Beever, P (1991). *How fire safety engineering can improve safety. Presented at Institution of Fire Engineers Meeting*, Torquay, UK, October 1991.

Bengtson, S and Laufke, H (1979/80). Methods of estimation of fire frequencies, personal safety and fire damage. *Fire Safety Journal*, **2**, 167–180.

Bowen, J V (1988). Reliability. *SFPE Handbook of Fire Protection Engineering*, National Fire Protection Association, USA, Chapter 4–5, pp. 34–42.

Bridge, N W (1984). *How New Zealand Deals with its False Alarm Problems*, Letter to Fire, Fire, December 1984.

Canter, D (1985). *Studies of Human Behaviour in Fire: Empirical Results and their Implications for Education and Design*, Report to Building Research Establishment, Fire Research Station, UK.

Chandler, S E (1978). *Some Trends in Hospital Fire Statistics*, Building Research Establishment, Current Paper CP 67/78, Fire Research Station, UK.

Cholin, R R (1975). Reappraising early warning detection. *Fire Journal*, **69**(2), 54–58.

CIB W14 (1986). Design guide: structural fire safety. *Fire Safety Journal*, **10**(2), 75–137.

Custer, R L P and Bright, R G (1974). *Fire Detection: The State-of-the-Art, NBS Technical Note 839, National Bureau of Standards*, Washington, DC.

Davies, D (1984). *Means of Cutting Down False Alarms in Automatic Systems Fire*, June 1984.

Elms, D G and Buchanan, A H (1981). *Fire Spread Analysis of Buildings*, Report R 35, Building Research Association of New Zealand, Judgeford, NZ.

Evans, D D (1985). Calculation of sprinkler actuation time in compartments. *Fire Safety Journal*, **9**, 147–155.

Finucane, M and Pinkney, D (1988). *Reliability of Fire Protection and Detection Systems*, Report SRD R431, Safety and Reliability Directorate, UK, Atomic Energy Authority, Culcheth, Warrington, UK.

Fleming, R P (1988). Automatic sprinkler systems calculations. *SFPE Handbook of Fire Protection Engineering*, National Fire Protection Association, USA, Chapter 3–2, pp. 22–34.

Fry, J F and Eveleigh, C (1975). *The Behaviour of Automatic Fire Detection Systems*, Building Research Establishment, Current Paper CP 32/75, Fire Research Station, UK.

Gilbert, S (1979). *Fire Prevention Publicity – Statistical Analysis of the 1977 Spaceheater Campaign*, Memorandum 5/79, Scientific Advisory Branch, Home Office, UK.

Gupta, Y P (1984/85). Automatic fire detection systems: aspects of reliability, capability and selection criteria. *Fire Safety Journal*, **8**, 105–117.

Hall, J R (1992). *US Experience with Sprinklers: Who Has Them, How Well Do They Work*, National Fire Protection Association, USA, June 1992.

Hall, J, Karter, M, Koss, M, Schainblatt, A and McNerney, T (1978). *Fire-Code Inspections and Fire Prevention: What Methods Lead to Success?* The Urban Institute, Washington, DC.

Halpern, J (1979). *Fire Loss Reduction: Fire Detectors vs Fire Stations*, Working Paper No. WP-08-79, Faculty of Management of the University of Calgary, Calgary, Alberta, Canada.

Helzer, S G, Buchbinder, B and Offensend, F L (1979). *Decision Analysis of Strategies for Reducing Upholstered Furniture Fire Losses*, NBS Technical Note 1101, National Bureau of Standards, Washington, DC.

Hinkley, P L (1988). Smoke and heat venting. *SFPE Handbook of Fire Protection Engineering*, National Fire Protection Association, USA, Chapter 2–3, pp. 33–44.

Langdon-Thomas, G J and Ramachandran, G (1970). Improving the effectiveness of the fire-check door. *Fire International*, **27**, 73–80.

Marryatt, H W (1974). The Australasian experience with automatic sprinklers. *Australian Fire Protection Association Journal*, 2–7.

Marryatt, H W (1988). *Fire: A Century of Automatic Sprinkler Protection in Australia and New Zealand*, Australian Fire Protection Association, pp. 1886–1986.

McGuire, J and Ruscoe, B (1962). *The Value of Fire Detector in the House. Division of Building Research*, National Research Council of Canada, Ottawa, Fire Study No. 9.

Melinek, S J (1993a). Effectiveness of sprinklers in reducing fire severity. *Fire Safety Journal*, **21**, 299–311.

Melinek, S J (1993b). Potential value of sprinklers in reducing fire casualties. *Fire Safety Journal*, **21**, 275–287.

Military Standardization MIL-HDBK-217B (1974). *Reliability Prediction of Electronic Equipment*, Naval Publications and Forms Centre, Philadelphia.

Miller, M J (1974). *Reliability of Fire Protection Systems Chemical Engineering Progress*, **70**(4), 62–67.

Morgan, H P and Hansell, G O (1984/85). Fire sizes and sprinkler effectiveness in offices – implications for smoke control design. *Fire Safety Journal*, **8**, 187–198.

Nash, P, Bridge, N W and Young, R A (1971). *Some Experimental Studies of the Control of Developed Fires in High-racked Storages by a Sprinkler System*, Fire Research Station, UK, Fire Research Note No. 866.

Nash, P and Young, R A (1991). *Automatic Sprinkler Systems for Fire Protection*, Paramount Publishing Ltd, UK.

Ramachandran, G (1968). *Fire Doors and Losses in Large Fires*, Fire Research Station, UK, Fire Research Note No. 690.

Ramachandran, G (1980). *Economic Value of Fire Detectors*, BRE Information Paper IP 27/80, Fire Research Station, UK.

Ramachandran, G (1990). Probability-based fire safety code. *Journal of Fire Protection Engineering*, **2**(3), 75–91.

Ramachandran, G (1993a). Fire resistance periods for structural elements. The sprinkler factor. *CIB W14 International Symposium on Fire Safety Engineering*, University of Ulster, Northern Ireland.

Ramachandran, G (1993b). Probabilistic evaluation of design evacuation time. *CIB W14 International Symposium on Fire Safety Engineering*, University of Ulster, Northern Ireland.

Ramachandran, G (1995). Probability-based building design for fire safety. *Fire Technology*, **31**(3), 265–275 (Part 1) and **4**, 355–368 (Part 2).

Ramachandran, G (1998). *The Economics of Fire Protection*, E. & F. N. Spon, Chapman & Hall, London.

Ramachandran, G and Chandler, S E (1984). *Economic Value of Early Detection of Fires in Industry and Commercial Premises*, BRE Information Paper IP 13/84, Fire Research Station, UK.

Ramachandran, G, Nash P and Benson, S P (1972). *The Use of Fire Extinguishers in Dwellings*, Fire Research Station, UK, Fire Research Note No. 915.

Rasbash, D J (1994). The effectiveness of fire resistant barriers. *Fire Safety Engineering*, **1**(5), 13.

Rees, G (1991). Automatic sprinklers, their value and latest developments. *Fire Surveyor*, **20**, 9–13.

Rogers, F E (1977). *Fire Losses and the Effect of Sprinkler Protection of Buildings in a Variety of Industries and Trades*, Current Paper CP 9/77, Building Research Establishment, UK.

Rosenblueth, E and Esteva, L (1971). *Use of reliability theory in building codes. Conference on Application of Statistics and Probability to Soil and Structural Engineering*, Hong Kong, September 1971.

Rutstein, R and Butler, A J (1977). Cost effectiveness of fire prevention publicity. *Final Report of the 1976/77 Fat Pan Fire Publicity Campaign*, Report No 9/77, Scientific Advisory Branch, Home Office, UK.

Rutstein, R and Cooke, R A (1979). *The Value of Fire Protection in Buildings*, Report 16/78, Scientific Advisory Branch, Home Office, UK.

Schaenman, P, Hall, J, Schainblatt, A, Swartz, J and Karter, M (1976). *Procedures for Improving the Measurement of Local Fire Protection Effectiveness*, National Fire Protection Association, USA.

Schifiliti, R P (1988). Design of Detection Systems. *SFPE Handbook of Fire Protection Engineering*, National Fire Protection Association, USA, Chapter 3-1, pp. 1–21.

SFPE (1988). SFPE Handbook of Fire Protection Engineering (First Edition) Society of Fire Protection Engineers, National Fire Protection Association, Quincy, MA.

Sime, J, Canter, D and Breaux, J (1981). University team studies. Use and success of extinguishers, *Fire*, **73**(909), 509–512.

Swersey, A, Ignall, E, Corman, H, Armstrong, P and Weindling, J (1975). *Fire Protection and Local Government: An Evaluation of Policy-Related Research*, Report R-1813-NSF, The Rand Corporation, Santa Monica.

PART III METHODS OF MEASURING FIRE SAFETY

11 DETERMINISTIC FIRE SAFETY MODELING

11.1 Introduction

Fire is the most complex physical phenomenon occurring in nature encompassing all disciplines of scientific investigation such as thermodynamics, reaction chemistry, combustion, and fluid mechanics, to mention a few. Any modeling approach – physical or mathematical – presents a formidable and challenging task for the investigator.

Various interpretations are put on the term *modeling* with context alluding to meaning. One interpretation may be an object's perceived attributes pertaining to be a "model." Yet, in another, a "model" may be construed only as an abstraction, a limitation, a crude simplification, or semblance of real object. The former belongs to the objective world of aesthetics, while the latter serves as a powerful tool in understanding the phenomenological existence. It is in this context that "models" are used in engineering science to acquire quantitative understanding of nature. Kanury (1987) classifies them into *physical* and *abstract* models. In this chapter, we are concerned with the abstract models (also referred to as *mathematical models*) that are successfully used in the investigation of the fire phenomenon.

A number of theories, backed by experimental evidence, have been formulated to identify and isolate the underlying component processes that give rise to the observed phenomena. Mathematical representations describing the interdependency of determinate variables are the powerful expressions of interrelations between the perceived and the conceptualized phenomenon. In this way, human knowledge may be regarded as only a knowledge of concepts (mental constructs) and not of reality itself. Thus, perceptions, by definition, are always one step removed from reality and do not form a one-to-one relation with it. It is in this respect that all mathematical models, by necessity, require *verification* and *validation* to minimize inherent uncertainties.

Uncertainty is an inevitable consequence of all model construction. The term model itself implies that a semblance of reality is considered. In any modeling process, two types of uncertainties are identified: *statistical* and *state-of-knowledge*. These will be discussed later in Chapter 12. This Chapter deals with models of fire behavior in enclosures (11.2 to 11.6) and models of evacuation from buildings (11.7).

Evaluation of Fire Safety D. Rasbash, G. Ramachandran, B. Kandola, J. Watts and M. Law
 ISBN: 0-471-49382-1

11.2 Models of enclosure fires

Over the last decade or so, there has been considerable activity in development of fire computer codes. These codes have basically been aimed at assessing the hazard (toxic and thermal) associated with a potential fire. The codes vary in complexity from simple slide rule-type deterministic calculations to finite difference field models, employing rigorous mathematical techniques requiring large computer power even to analyze simple fire scenarios. However, because of the advances in computer technology, field models have become more common and are widely used.

Basically, there are two types of models available: *deterministic* and *probabilistic*. Whereas the former allows for a "single possible development," the latter tries to investigate a range of possible developments. Over the years, the former has become more popular among fire safety engineers mainly due to the fact that they readily provide the numbers that a fire safety engineer can work with (more often these are the conservative numbers). These *deterministic models* rely on the basic assumption that for a given set of initial conditions, the phenomenological outcome is entirely determined. That is to say, in the context of fire, that for given conditions, the course of fire development, its characteristic features, and its consequences can be predicted. However, experience shows that no two fires are the same and in this respect, the above assumption is difficult to justify. In contrast, the *probabilistic models* consider a range of possible fire developments, but their practical usefulness is rather limited. In this chapter, we shall confine ourselves to the discussion of deterministic models. For practical applications, these models have proved to be more useful as they can be used to assess when a room or a building is no longer safe to occupy.

In broad terms, there are two types of deterministic models: *zone models* and *field models*. The former rely mostly on empirical correlations between specific variables derived from laboratory scale experiments. Zone models are subdivided into one-layer, two-layer, and HVAC (Heating, Ventilation, and Air Conditioning) models, depending on the type of problem they are attempting to solve (see Figure 11.1). Field models assume fewer empirical relations and attempt to solve the governing conservation equations (mass, momentum, and enthalpy) using numerical techniques.

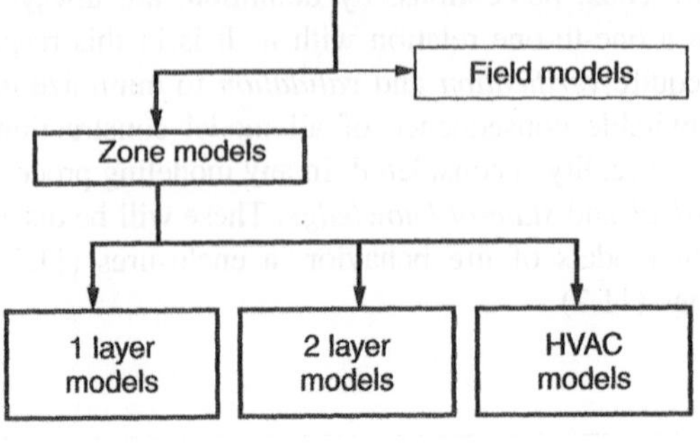

Figure 11.1. Types of computer fire models

One-layer models attempt to calculate smoke movement in regions remote from the fire and can handle large, complex buildings with numerous floors and rooms. Two-layer models, on the other hand, are limited to fires in small enclosures (with no vertical shafts) and consider smoke movement in the immediate vicinity of the fire. The HVAC models calculate smoke spread by HVAC systems and are theoretically similar to one-layer models.

11.2.1 THEORY AND CONCEPT OF ZONE MODELS

The concept of a zone model is very simple and is based on physical phenomena observed in real enclosure fires. In its simplest form, the fire room environment before flashover is assumed to consist of two distinct homogeneous zones or layers: the lower cooler layer at ambient temperature and the upper hot layer at a uniform higher temperature. This two-layer concept is based on the laboratory scale temperature measurements carried out by Thomas *et al.* (Figure 11.2a) and later confirmed by Quintiere *et al.* (Figure 11.2b).

The concept of zones could be further extended to include other identifiable regions such as *flaming zone, thermal plume, thin hotter gas layer, and the walls*. Processes and conditions in each zone, assumed to be uniform throughout, can be approximated by mathematical equations derived empirically and from theoretical considerations. These equations coupled at the boundaries of each domain, then, describe the full behavior of an enclosure fire.

The models are then used to predict various aspects of fire growth and its consequences for the enclosure. These studies predict many variables in the fire growth process such as the hot-layer temperature, hot-layer species concentration, and the hot-layer depth. Over the years, the ideas have been taken further to deal with the fires in multiroom buildings in which similar zones are assumed to exist. These models also help to understand the contribution by materials to fire growth and the impact of fire on its surroundings.

Depending on the aspect of fire under investigation, the models are available for preflashover and postflashover fires. Here we shall describe the theoretical basis of both these approaches. The preflashover models describing the fire growth are generally concerned with personal safety, while the postflashover models deal with the thermal impact on structures.

Before examining these models in more detail, it will be beneficial to review the phenomena of enclosure fires, for it is on this basis that the zone models are derived.

11.3 Dynamics of enclosure fires

11.3.1 HEAT RELEASE

The rate of heat release in a fire is one of the most important factors determining its impact on the surroundings: the structure and the occupants. Because of the radiation feedback on the fuel surface itself, the rate of heat release also determines the rate of fire spread. In order to estimate the potential hazard of a fire, it is necessary to be able to calculate the rate of heat release for a given fuel and the surrounding conditions.

The pyrolysis products (material vapors) released by the heating of a fuel surface react with the oxygen in the air generating heat and flames. The fuel pyrolysis rate depends on the fuel type, its geometry, and on the fire-induced environment, that is, the radiation feedback and the oxygen concentration in the air. The heat release rate is directly proportional to the generation rate of material vapors (volatiles). The heat release rate $\dot{Q}_f$ is generally expressed as

$$\dot{Q}_f = \chi \dot{m}'' A_f \Delta H_c \qquad [11.1]$$

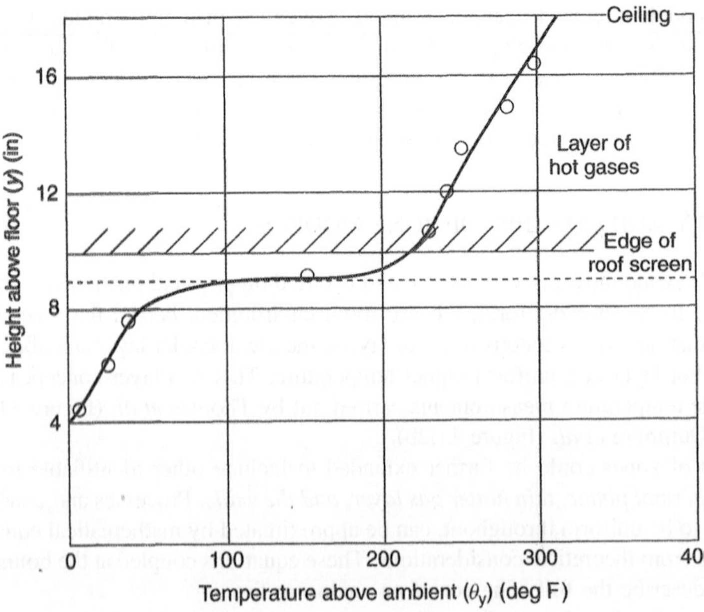

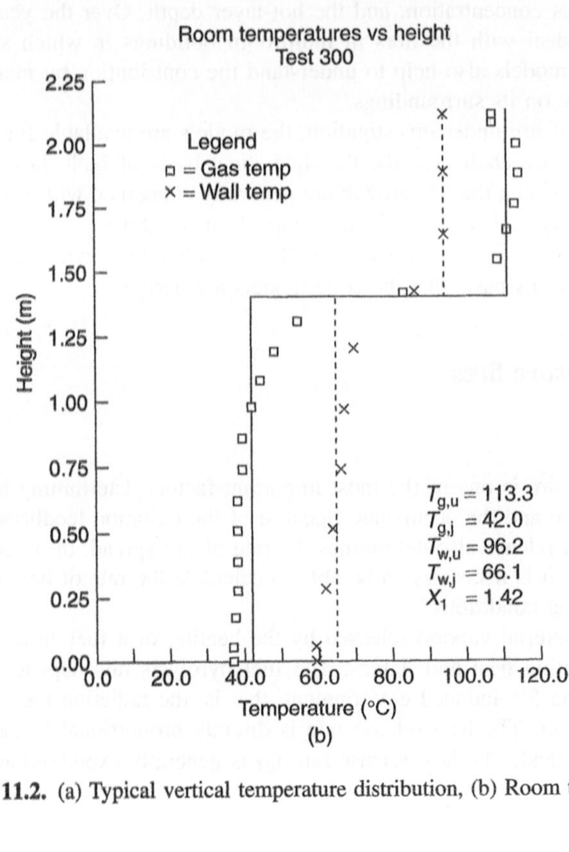

Figure 11.2. (a) Typical vertical temperature distribution, (b) Room temperatures versus height

where A_f is the fuel surface area (m^2), ΔH_c is the heat of combustion of the volatiles (kJ/g), χ is the combustion efficiency (<1.0) and $\dot{m}''$ is the mass loss rate of the material g/(m^{2}s) per unit surface area of fuel.

In real fires, the combustion of material vapors is not always complete and thus the heat of combustion is always lesser than the net heat of complete combustion. Incomplete combustion is usually characterized by the release of unburnt vapors just above the visible flame in the form of soot particles and the production of carbon monoxide. The ratio of heat of combustion to the net heat of complete combustion is known as the *combustion efficiency* (χ) of the material.

The fuel combustion efficiency depends on a number of factors such as the nature of chemical bonds between various atoms in the material, fire ventilation, and mixing of material vapors and air during combustion. It decreases as the supply of fresh air to the fire is reduced or restricted. The combustion efficiency also reflects the fact that not all the amount of fuel leaving the burning surface is taking part in the combustion process. That is to say, the burning rate may be somewhat lesser than the mass loss rate of the material.

In free burning fires (with unrestricted supply of oxygen from the air), the pyrolysis rate and the energy release rate are affected only by the burning of the fuel itself. The primary heating of the fuel is from the flames of the burning item. This type of fire generally is referred to as the *fuel-controlled* fire.

In enclosure fires, the burning rate, and hence the heat release rate, is limited by the amount of available oxygen. The rate of energy release of the fire can, therefore, be related to the inflow of air through openings such as the doors and windows. This type of fire is known as the *ventilation-controlled* fire in which, assuming all the oxygen is consumed in the fire, the heat release rate is given by

$$\dot{Q}_f = \dot{m}_{air} \Delta H_{c,air} \qquad [11.2]$$

where

$\dot{m}_{air}$ is rate of air flow into the enclosure

$\Delta H_{c,air}$ is the heat of combustion in terms of air consumed

Kawagoe (1958) studied a large number of enclosure fires and concluded that for fires nearing flashover and postflashover, the rate of burning is determined by the ventilation and can be expressed as

$$\dot{m}_b = K A_w \sqrt{H} \qquad [11.3]$$

Here, the value of the constant K usually depends on the shape of opening and is generally estimated to be about 0.5 kg/(sm$^{5/2}$). The term $A_w\sqrt{H}$ is sometimes known as the *ventilation factor*.

From the above brief discussion, it is clear that the rate of heat release is determined not only by the type of fuel but also most importantly by the ventilation conditions in which the combustion takes place, that is, the amount of oxygen available for oxidation. In addition, the rate of generation of other products of combustion such as soot, carbon monoxide, and carbon dioxide are also functions of ventilation. It is clear that any fire modeling approach must take these important factors into account.

11.3.2 FIRE-GENERATED FLOWS

As noted earlier, all zone models are based on the observed flow phenomena resulting from enclosure fires. These flows have been under investigation for a long time mostly in Japan (Kawagoe, 1958) and the US (Emmons, Rockett, and Quintiere). The salient features of these flows are summarized by Zukoski (1985).

In the description of fire-generated enclosure flows, the fire is considered to be a point source of heat generating a vertical buoyant plume rising toward the ceiling entraining surrounding air and thereby cooling and increasing in diameter. Whether or not it reaches the ceiling is determined by the physical conditions – the enclosure height – and the size of the fire (i.e. the rate of heat release). In a stable stratified atmosphere in which the temperature and hence the density vary with height, the maximum height to which a buoyant plume can rise is given by (Quintiere, 1983, Heskestad, 1989):

$$z_{max} = 3.79 F_0^{1/4} G^{-3/8} \qquad [11.4]$$

where,

$$G = \frac{g}{\rho_{\infty 1}} \frac{d\rho_\infty}{dz} \qquad [11.5]$$

and

$$F_0 = \frac{g\dot{Q}_c}{\rho_{\infty 1} T_{\infty 1} c_p} \qquad [11.6]$$

Here, F_0 is the flux of buoyancy (or weight deficiency) from source and G is the density stratification parameter; g is the acceleration due to gravity; ρ_1, T_1 the ambient density and temperature at the level of source; c_p is the specific heat of ambient air; z is the height above the fire source of convective heat release rate $\dot{Q}_c 5$; $d\rho_\infty/dz$ is the ambient density gradient.

Although, in the original work reported by Morton *et al.*, the above relationship (equation [11.4]) was derived from stratified liquid experiments, subsequent work shows this to be true also for buoyant plumes from large oil fires (Turner, 1973).

If the enclosure height is greater than z_{max}, no hot layer will be formed below the ceiling. Under these conditions, it is likely that the fire will not spread and that it will have no serious consequences for fire safety in the enclosure. However, if the plume reaches and impinges on the ceiling (i.e. z_{max} greater than the enclosure height), it very quickly forms a growing layer of hot gases below the ceiling that may result in fire and smoke spread to the adjacent areas. Most problems involve fires that are large enough to overcome stratification.

In the initial stages of fire, the plume of gas (combustion products) rises due to buoyancy, entrains the unvitiated compartment air, and impinges on the ceiling in a region directly above the fire where it turns to flow radially along the ceiling. As it spreads – entraining more compartment air – in the form of a ceiling jet, it cools and progressively loses its buoyancy. On reaching the walls, it is turned downward, mixing with the cooler air in the room. This process of mixing and turning eventually results in the formation of a stratified hot layer growing in depth as the fire progresses (Figure 11.3). When the hot and cold layer interface reaches the top of a sidewall opening, such as a door, the hot gases (smoke) flow out into the adjacent areas. A complex flow pattern is set up across this opening with the hot gases flowing out and cooler air trying to enter the fire room.

Thus, based on these observations, the compartment atmosphere could be described as comprising three distinct layers: a relatively *cool bottom layer*, a *deep hot upper layer*, and a *thin hotter layer* at the ceiling. Zukoski (1985) points out a number of implications of this flow pattern for intracompartmental thermal hydraulics. First, convective heat transfer to the ceiling is determined by the temperature of the thin ceiling layer, which is not treated explicitly by a two-layer model. Therefore, at least a rudimentary description of the flow within the upper layer is needed to allow a description of convective and radiative heat transfer to the ceiling. Second, the fire plume (being partially immersed in the hot layer) will entrain gas having composition and temperature that varies with height. This causes the temperature of the thin ceiling layer to increase further when the plume penetrates the upper layer of increasing depth. Third, there may

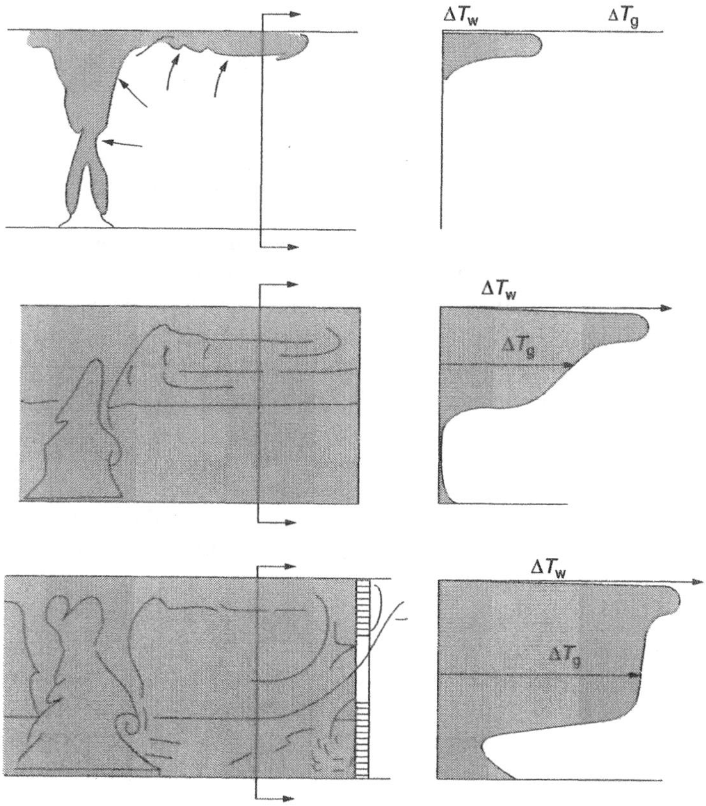

Figure 11.3. The development of the ceiling layer (Zukoski,1985)

be considerable temperature variations in this deep hot layer, which are not accounted for in zone models.

11.3.3 STRATIFICATION

In the zone modeling approach of compartment fires, to be discussed in detail later, only two stratified layers are assumed to exist, that is, the hotter ceiling layer (Zukoski, 1985) is ignored, with the plume acting as the primary source of material for the hot layer. The degree of *stratification* in the ceiling layer will depend in detail on the flow within the layer, changes in the heat release rate of fire, the enthalpy fluxes resulting from the flow of gas into and out of the layer at openings, and heat loss to the walls. An important parameter determining stratification is the Richardson number (Ri) defined as the ratio of the buoyancy forces acting over a height to the momentum flux in the layer, expressed as

$$Ri = \frac{\Delta\rho g h}{\rho U^2} \qquad [11.7]$$

where $\Delta\rho$ is the density difference between the lower and upper layers, g is the acceleration due to gravity, and h, ρ, and U are the vertical depth, density, and velocity of one of the layers.

According to the above definition, two Richardson numbers may be defined, depending on which layer is used for the definitions of h, ρ, and U, and these are appropriate to the dynamics of turbulent layers when the turbulence is generated by motion of one of the layers with respect to a fixed surface. Additional Richardson numbers could be defined when the velocity scale is taken as the velocity difference between the two layers. According to whichever of the two is larger, the Richardson number determines the evolution of the two layers.

When *Ri* is large compared to a number of order unity, the buoyancy forces are dominant and the two layers will remain distinct with little interaction between the two. When *Ri* is small compared to unity, mixing between the two layers will be rapid and they will quickly evolve to form a well-mixed single layer extending between the ceiling and floor. This is what may be expected well away from the smoke source. It is also possible that a single layer may evolve to form two layers, when the single layer is slightly stratified (i.e. not well mixed) and the flow conditions alter so that the Richardson number becomes large compared to unity. Once the *flashover* (discussed elsewhere in the book) has occurred, the two-layer assumption breaks down and the two-layer models no longer apply. It is for this reason that two separate models exist to study the varied flow phenomena – *Preflashover* models and *Postflashover* models. The preflashover models (describing the fire growth) are generally concerned with personal safety. It is at the initial stages of the fire that the detection and fire suppression systems must work and the personnel evacuation completed within the safe period of time. The postflashover fires are, however, studied with a view to predicting the fire resistance required to ensure structural safety. Since the heat output from the fires is at a maximum, these models calculate the heat transfer to structures to predict their performance.

Zukoski (1985) has identified three main processes by which the degree of stratification of the hot layer may be affected. First, the recirculation in the hot layer caused by the entrainment of hot gas into the part of the plume surrounded by the hot layer and the ceiling jet reduces stratification. This recirculation is also responsible for homogeneity and mixing in the layer.

The second process, which also reduces stratification, is the flow of gas from the hot layer to the outside through openings in the walls (such as doors and windows). In such a situation, mixing between the layers may occur.

The third process depends on the Richardson number of the hotter ceiling jet as it impinges on the walls. Thus, it is the flow pattern within the hot layer that will determine the degree of stratification of the hot layer.

Richardson numbers for some typical values of U, h, $\Delta\rho/\rho$ are shown in Table 11.1.

The Richardson number is evaluated for the upper layer. The larger density ratio corresponds to smoke at 1000 K and the smaller density ratio corresponds to smoke at 650 K, both overlying a layer of ambient air at 300 K. A thick layer at 1000 K will cause ignition by radiation of fuel in the lower zone, leading to flashover.

Table 11.1. Typical values of the Richardson number (*Ri*)

U (m/s)	h (m)	$\Delta\rho/\rho$	*Ri*
10	0.35	1.17	0.04
		2.33	0.08
	1.5	1.17	0.20
		2.33	0.40
1.0	0.35	1.17	4.0
		2.33	8.0
	1.5	1.17	18.0
		2.33	35.0

Clearly, the overriding factor in *Ri* seen in Table 11.1 is the velocity of the smoke layer. When the velocity is large, the Richardson number is small and, for the numbers we have chosen, always smaller than unity (though never small compared to 0.01). Smoke velocities can vary substantially throughout a building and depend strongly on the size of the fire and the amount of ventilation. One may conclude, therefore, that it is quite possible that no or little stratification will occur in some cases or parts of a building, and substantial stratification may occur otherwise. This is true for areas remote from the fire.

11.3.4 MIXING

The spread of flame and smoke (hot combustion products) within the fire room and in the adjoining areas is important for fire spread. In addition, it has implications for personnel safety as smoke dilutes and cools because of various entrainment and mixing processes that affect the movement of smoke in buildings. It is for this reason that it is these processes are important to take important that into account for modeling.

To simplify modeling within the enclosure, the hot ceiling layer that grows and descends toward the floor is generally assumed to be homogeneous (in temperature and species concentrations) and strongly stratified. In other words, a two-layer environment is assumed to exist within the fire enclosure with no interchange of fluid across the interface except via the plume itself. However, in practice, this is far from the case. Such a two-layer simplification is complicated by the mixing phenomenon that is found to exist and plays an important role in the spread of smoke.

A number of phenomena have been identified by which the mixing process takes place Quintiere (1984). They are

- *Mixing in doorways caused by the opposing flows* (Zukoski *et al.*, 1985), (Figure 11.4).
- *Interaction between the wall thermal boundary layer arising from the differences in wall temperature* (Jaluria, 1988, Cooper, 1984a) (Fig 11.5).
- *Mixing caused by impingement of plume and ceiling jets on solid boundaries* (Zukoski *et al.*, 1985).
- *Mixing between the upper and lower layers of a two-layer flow* (Zukoski *et al.*, 1985).

Mixing in doorways has been studied experimentally by Zukoski *et al.* (1985). They found that the mixing rate is a function of a Richardson number based on the interface height as the characteristic length and the flow velocity through the vertical opening. They also suggested a modified definition of the Richardson number based on the density and velocity gradients to study the stability of layers.

The hot ceiling layer formed by the combustion products gradually becomes thicker, and then flows out to the adjacent rooms through the open doorways or vents (Figure 11.4a). Fresh air enters the room through the lower part of the door to replace the air entrained into the plume (Figure 11.4b). Near the doorway, the motions of the hot gas and the fresh air are in opposite directions (Figure 11.4c), forming a shear (mixing) layer. The turbulent mixing between the two layers is controlled largely by the inertial and buoyancy forces (see the definition of Richardson number, Equation [11.7]). The mixing process is suppressed by the gravitational forces (high Richardson number) and promoted by the inertial forces (low Richardson number). The experiments of Zukoski *et al.* (1985) show that the mixing layer grows down stream of the incoming airflow, up to a critical distance. Beyond this point, the effects of buoyancy are strong enough to suppress the mixing and entrainment processes, and the mass flux into the mixing layer stops (Figure 11.6).

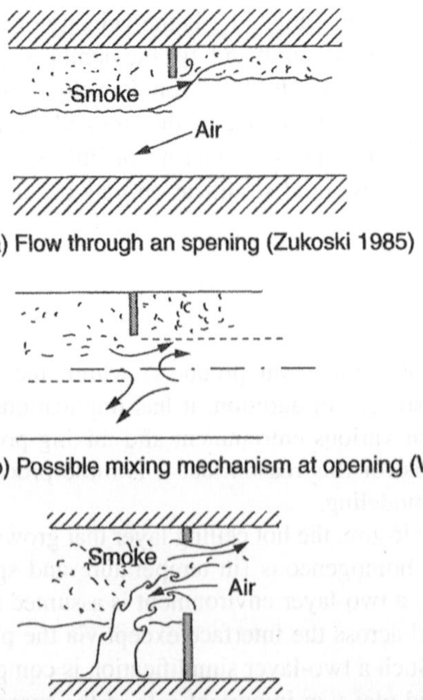

Figure 11.4. (a) Flow through an opening (Zukoski, 1985), (b) possible mixing mechanism at opening (Jones, 1985), (c) mixing in the opening (Zukoski and Kubota, 1980)

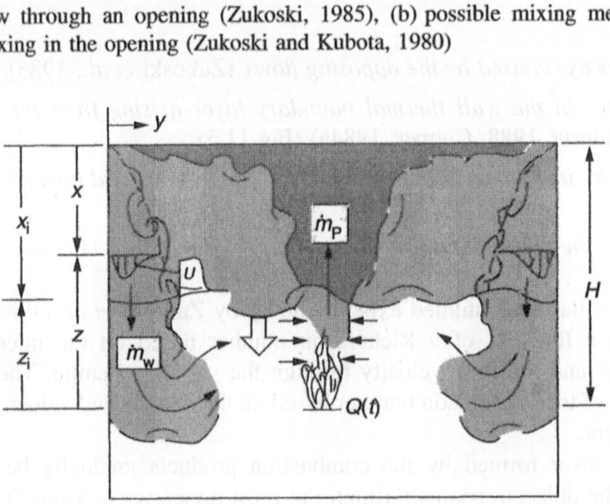

Figure 11.5. Sketch of the significant features of the wall effect at the early stages of a room fire (Cooper, 1984)

In general, the number of configurations by which smoke may mix at an opening is large depending on the orientation of the opening and the positions with respect to the opening of any stratified interfaces that may be present. Mixing is promoted by strong shear layers at openings and by the penetration of buoyant or negatively buoyant plumes or jets into the flow on the opposite side. The simple doorway and window flows shown in Figure 11.4a with the hot smoke and cold air flowing in the opposite directions could be complicated under certain circumstances.

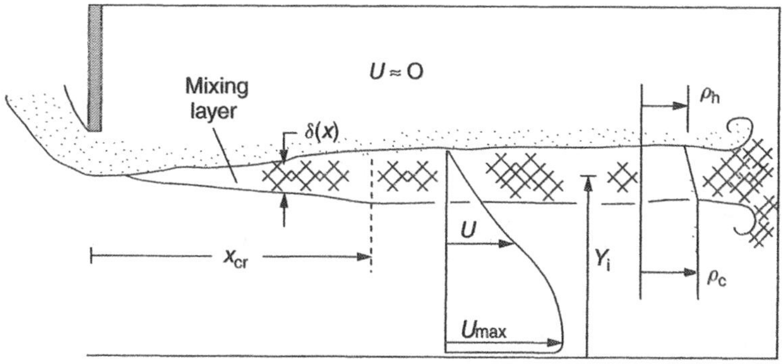

Figure 11.6. Formation of the turbulent mixing layer (Zukoski *et al.*, 1985)

The fresh air from the adjacent room into the lower layer of the fire room can cause entrainment of hot gases into the lower layer in the fire room and thus can lead to substantial contamination of the air in this layer. This mixing process is illustrated in Figure 11.4b and 11.4c.

Zukoski *et al.* (1985) have also reviewed the propagation of buoyant gravity currents along ceilings. They conclude that although the current will not be entrained into the ambient fluid to any significant extent, the converse is not true. That is, ambient fluid will be entrained into the gravity current and they recommend this should be treated. They also conclude that the propagation speed of the current would be at most about 2 m/s and, therefore, is an important factor in determining the time involved in the initial spread of smoke along long corridors.

At solid vertical walls within a compartment, with differences between the local wall and gas temperatures caused by convective and radiative heat transfer, significant buoyancy-induced flows may exist, leading to additional mixing across the layer interface (Jaluria, 1988). In the upper layer, a thermal boundary layer is formed next to the wall resulting in a downward flow that penetrates the interface and becomes upwardly buoyant as it encounters the lower layer. At a certain depth, it begins to rise again toward the interface, causing further mixing (Figure 11.5). Jaluria presents an analytical method for incorporating this mixing phenomenon into the zone models. The zone models discussed in this chapter do not take this phenomenon into account. However, it must be said that for relatively large enclosures, this type of localized mixing may not contribute significantly to the dynamics of upper layer. Though the phenomenon is important, it may be neglected in certain fire situations.

In this brief discussion, we have shown the importance of the mixing processes that occur in fire-generated enclosure environments. We have also shown how the Richardson number may vary widely and may be quite small in certain circumstances. In such situations, vertical mixing will not be suppressed, resulting in the rapid formation of a single layer. Consequently, the usual assumption in smoke movement models that mixing between two layers can be ignored throughout a building may not be valid under certain fire situations.

11.3.5 HEAT TRANSFER AND MASS LOSS TO SURFACES

As discussed, buoyancy is the most important factor causing smoke movement within a building, thus having implications for fire spread and safety. The buoyancy is mainly generated by smoke at an elevated temperature. This is important not only for its effect on the fire compartment but also because the heat transfer through the compartment boundaries may result in the initiation of secondary fires in the adjacent compartments or the failure of the building structure. In addition, the heat loss processes determine the environmental temperature of the compartment. The deleterious effects of smoke increase with temperature and this is an additional reason for taking heat loss into account.

The environment temperature within a compartment is the balance between the rate of heat production by a fire and the rate of heat loss to the surroundings. The prediction of this temperature necessarily involves the calculation of the heat loss phenomena. Heat transfer from a fire source within a building involves heat transfer by radiation and convection to walls, floors, ceilings, and other objects, causing heat transfer by conduction into the body of these objects.

Radiative heat transfer is usually treated because of its importance to heat transfer in compartments within which a fire is present. It is important in two respects: one, the radiation feedback to the fuel surface and two, the radiation heat transfer to the surrounding surfaces (Quintiere, 1984). The former determines the burning rate and fire spread, while the latter has implications for structural fire performance. In general, the radiative heat transfer between the surfaces depends on the geometry, orientation, and temperature (Tien *et al.*, 1988). In practice, most surfaces are assumed to be isothermal. The geometry and orientation of each surface is commonly accounted for in calculation of the configuration factors, which are also variously known as *view factors, shape factors, angle factors* or *geometric factors*. A configuration factor is purely a geometric relation between two surfaces and is defined as the fraction of radiation leaving one surface that is intercepted by the other surface. Prediction of the radiative heat flux from a flame is important in determining ignition and fire spread hazard and the development of the fire detection devices. The actual shapes of flames and other radiating surfaces such as the hot layer are arbitrary and time-dependent, which makes detailed radiation analysis very cumbersome. However, in most cases, some simplifications provide reasonably good answers. In these assumptions, flames are idealized as simple geometric shapes such as planar layers or axisymmetric cylinders and cones (Mudan and Croce, 1988).

A rigorous calculation of radiative heat transfer, though possible, would be costly to undertake in compartments whose ceilings, walls, and floors may each be at different temperatures and within which the atmosphere is separated into two layers of different temperature. This is complicated by the presence of smoke and radiating-active gases such as water vapor and carbon dioxide. The presence of soot in the combustion products markedly alters the radiation properties of the hot layer and the buoyant plume. Soot particles are produced as a result of incomplete combustion and agglomerate to give a distribution of particle size with a mass median diameter on the order of 1 μm (Drysdale, 1985). A method for calculating the heat radiation contribution of the soot particles is provided by Tien *et al.* (1988). Because of these difficulties, simplifications are often made when treating radiation effects in two-layer fire models. Examples of the methods used may be found in Siegal and Howell (1981).

In fire situations, the mechanisms of *convective heat transfer* are usually classified as forced convection, laminar natural convection, and turbulent natural convection. The dominant mechanism depends on the compartment dimensions, the temperature difference between the atmosphere and surfaces, and the flow speeds within the compartment. Under certain conditions, both natural and forced convection may occur simultaneously, resulting in a mixed mode of convective heat transfer. Correlations have been developed for these and are straightforward to implement in integral models (see McAdams, 1954) for details). The correlations apply to idealized flows whereas, in practice, a number of phenomena may occur, which make the assessment less straightforward. Examples of complexities are ceiling, wall, and door jets, and, more generally, any complex flow pattern within a compartment. Usually, these complications are ignored. Forced convection is important in compartments where forced ventilation is provided by the heating and ventilation (HVAC) systems. A study of forced ventilation fires is conducted by Alvares *et al.* (1984), as discussed later. Atkinson and Drysdale (1992) have also made a study of convective heat transfer from fires.

Heat loss from the smoke by conduction depends on the heat capacity, thermal conductivity, and thickness of the structural materials. Because of the complexity of buildings, heat transfer in solid components can be calculated only approximately. The usual assumption made is to treat the heat *conduction* as one dimensional. Thus, components such as walls are treated as slabs, perfectly insulated at their edges, with uniform surface temperatures. In special cases, for example, a component of uniform composition and steady boundary conditions, the heat loss can be determined from a simple analytic expression (see Carslaw and Jaeger, 1980 for details). Otherwise, the heat conduction equation, one for each component, must be solved numerically.

Mass transfer to surfaces is important for the assessment of deleterious smoke effects. The deposition of smoke particles, condensation, and adsorption of chemically reactive vapors, such as hydrogen chloride, are most important. Additionally, smoke deposition in filters is of special interest in nuclear facilities because of the possibility of deleterious effects on filter performance. Deposition is also important for the proper functioning of electrical safety-related equipment.

Smoke particles may settle on surfaces by gravitation, Brownian motion, and thermophoresis (other mechanisms are also possible but they are not important in the present context). Which mechanism dominates depends primarily on the properties of the particles, though thermal-hydraulic factors also play a role (See Friedlander (1977), for details).

Deposition and adsorption of chemically reactive vapors occurs by convective transfer of vapor to surfaces and adsorption of the vapor onto the surface. The former can be treated in an analogous way to convective heat transfer (see Collier, 1981). The transfer rate depends on the vapor pressure at the surface and this is where chemical reaction of the vapor with the surface is important.

In some zone models, (FAST, (Fire growth and smoke transport, for example)) attempts have been made to include the mass transfer phenomenon. How successful this has been is yet to be demonstrated.

11.3.6 FLOW THROUGH OPENINGS

In any building, internal flows between spaces (such as rooms, vertical shafts, and corridors) result as a consequence of building leakage and the pressure distribution. In the normal everyday use of the building, these flows may not be important, but in fire situations, these flows are mainly responsible for the migration of smoke to locations remote from the seat of the fire.

The most obvious leakage paths in a building are the cracks around doors and windows. Ventilation ducts as well as the small cracks in the walls and partitions also provide important routes for smoke flow. Even in the case of a hermetically sealed building, such cracks are unavoidable. Fire in a hermetically sealed room soon develops enough overpressures owing to gas expansion and may cause that the doors and the windows to burst open.

The pressure difference driving these flows is caused by a number of factors. These are: building stack effect, buoyancy of combustion gases, expansion of combustion gases, the atmospheric wind, fan powered ventilation systems, and the elevator piston effect. They are discussed in detail by Klote (1989), who also provides methods for calculating their contribution.

Vertical openings

In general, there are many geometrical relationships of compartment-to-compartment openings and the flow of buoyant – possibly layered – gas between the two. Of particular interest is the motion of a hot layer of smoke through vertical openings that do not extend to the ceiling, such as doorways and windows. The earliest model applicable to this situation was developed by Kawagoe (1958) for the case of an opening separating hot gas on one side and ambient air on the other. Subsequently, several models have been developed to treat a wide range of conditions when the gas flow on either side of the opening is described in terms of two layers (Quintiere and Denbraven, 1978, Steckler *et al.*, 1982, Emmons, 1987). The flow rate depends on not only the pressure difference on either side of the opening but also on the heights of the stratified interfaces either side and the densities of the layers.

Bodàrt and Curtat (1989) have developed the model CIFI (Circulation dans un Immeuble des Fumées de l'Incendie) to predict entrained mass flows through vertical openings in compartments. Until recently, little attention had been paid to the complex flow patterns that occur through these vents. Usually, flow rates are estimated by integrating functions of the static pressure difference over the vent. The modified model discussed by Bodart and Curtat accounts for entrainment between the flows at the vent. The approach is, as the authors note, ad hoc. But it has the virtue that it is easily implemented in a computer code and allows for all possible entrainment configurations without resulting in contradictory flow patterns. The model was compared with experiments reported by Jones and Quintiere (1984). Results for the burn room temperature and the layer height in the burn room are in good agreement with the experimental data, when taking into account the arbitrariness in defining the layer height. The predictions were found to be not sensitive to the details of how the virtual origin of the vent plume was determined.

Recently, Cooper (1988, 1989a) has developed a model for flow through vertical vents that allows for arbitrary (i.e. large) pressure differences across the vent, as may be so when compartments are hermetically sealed. The approach is to develop an analytic prescription for flow rates through vertical openings with parallel sides with two layers on either side, and is an extension of previous models that assume that the pressure difference across the vent is small.

When there is hydrostatic pressure variation with height at the opening and the layers have uniform density, the pressure difference across the opening is a piecewise linear function of height, comprising up to three pieces according to the positions of the layer interfaces with respect to one another and the opening. Since the pressure difference may change sign at the most once for each piece, up to six vertical segments may be defined for which the pressure difference varies linearly with height and does not change sign. The flow rate across the opening is taken to be the sum of the flow rates across each segment (for more details see Section 11.4.9).

To determine the flow rate across a single segment, the flow is assumed to be one-dimensional and isentropic and, at a given height, independent of the flow elsewhere. The gas is assumed to be ideal. This enables the discharge formula in Shapiro (1953) to be used to obtain the mass flow rate as a function of height, which is then integrated analytically for each segment. The discharge coefficient is taken as a linear function of the pressure ratio across the opening (downstream pressure over upstream pressure) lying between 0.68 and 0.85. Penetrating flows are assumed to mix uniformly with the downstream layer at the height of the penetrating flow. Because the flow is assumed to be isentropic, the temperature of the penetrating flow is taken as the stagnation temperature on the upstream side.

Strictly, the formula from Shapiro cannot be applied when the pressure drop across the opening varies with height and so the prescription described above should be regarded as an ad hoc means of obtaining a mass flow rate that is continuous as a function of the independent variables. It would be of great interest to compare it systematically with experiment, to explore satisfactory prediction of flow rates over the range of independent variables of practical interest.

These models treat the general case of two layers on either side of a vertical opening and, consequently, can be applied to special cases such as when only one layer exists on one or both sides.

Horizontal openings

At first sight, it might appear simpler to determine the flow across a horizontal opening rather than a vertical opening, once it is assumed that the pressure field at the vent is horizontally uniform.

Thomas *et al.* (1963) carried out the earliest study of this type for the design of roof venting systems. Kandola (1990) later carried out a study of the wind effects on the performance of roof vents. However, a number of hydrodynamic effects may occur, which complicate the flow.

The flow field at the vent may not be hydrodynamically stable when the density of the fluid below the vent is less than that above and, when it is unstable, counter-current flow may occur through the vent. Cooper (1989a) discusses how this is contradictory with the assumptions usually made for calculating the vent flow rate. In particular, unidirectional flow is assumed, the direction chosen according to the sign of the pressure difference across the vent, and the flow rate is calculated from Bernoulli's formula, which predicts zero flow rate when the pressure difference across the vent is zero (pressure difference does not imply a stable interface).

When the pressure difference is zero, a formula due to Epstein (1988) may be used to calculate the exchange flow through circular vents. Of course, flow is not predicted when the interface at the vent is hydrodynamically stable. Cooper suggests how the usual pressure-driven flow and Epstein's exchange flow formulas may be combined to obtain a uniformly continuous estimate of the flow rate in both directions for any pressure and density difference across the vent.

The flow rate in each direction is assumed to be the sum of the flow rates due to the pressure difference and exchange flow, with the latter modified by a factor that causes the exchange flow to become zero when the pressure difference is larger than a limit referred to as the *flooding pressure*. The flooding pressure corresponds to a pressure-driven flow that overtakes the counter-current exchange flow, causing the flow to be unidirectional. The flooding pressure is estimated from a consideration of exchange flows in inclined vents and, essentially, is obtained as the pressure difference that results in equality of the pressure driven and the exchange flow rates.

Shanley and Beyler (1989) have studied experimentally the motion of gases through a horizontal circular ceiling vent in a rectangular compartment in which two layers of gas are present, the upper layer being less dense than the lower layer. The upper layer comprised a mixture of helium and air, to simulate the lowered density of a hot gas.

It was found that the buoyant layer was stable and the vent discharge coefficient was within 15% of the value for a homogeneous gas mixture when the upper layer depth was similar to, or greater than, the vent diameter. The upper layer became unstable at smaller depths and a weir-type flow developed, resulting in a significant reduction in flow through the vent (by up to a factor of 3) and hence a smaller discharge coefficient.

These results indicate the importance of two-layer dynamics for determining flows through horizontal vents. Note this effect is quite distinct from that considered by Cooper, which is concerned with hydrodynamic instabilities at the vent. It is therefore essential to take these effects into account in zone modeling of fires.

Narrow openings

Flow through narrow openings such as the cracks around doors and windows is determined by the flow Reynolds number (Kandola, 1979). Gross and Haberman (1989) have obtained analytic correlations for determining air flow rates through narrow gaps around door edges. Their results are applicable to steady, laminar flow over a wide range of pressure difference and are shown to improve upon the commonly used correlation

$$Q = AC_d(\Delta p)^n \qquad [11.8]$$

in which the discharge coefficient C_d and the exponent n are not always known, here Q is the volume flow rate and A is the area of the gap. Comparison of measured flow rates for stairwell doors with predictions from their correlations show agreement within 20%. Their correlations are applicable to straight-through, single-bend and double-bend gaps of constant thickness, and in an extended form, can be used for connected gaps of different thickness. For the flow of heated air through simple door gaps, their correlations predict that the flow rate may increase or decrease with temperature according to the gap size and flow regime.

The correlations are based on the dimensionless groups

$$N_Q = \frac{Qd}{L\nu x} \qquad [11.9]$$

and

$$N_P = \frac{\Delta p d^2}{\rho \nu^2}\left(\frac{d}{x}\right)^2 \qquad [11.10]$$

where Q is the volume flow rate through the gap, L is the lateral gap length (e.g. the perimeter of a door), ν is the kinematic viscosity of the fluid, d is the gap hydraulic diameter (which equals twice the gap thickness for rectangular gaps), x is the longitudinal gap length (i.e. in the flow direction), Δp is the pressure difference across the gap, and ρ is the fluid density. Note that in N_Q, the first group of variables on the right is the gap Reynolds number, and in N_P, the first group of variables on the right is a viscous pressure coefficient.

In terms of these groups, the correlation for straight gaps is

$$N_Q = 0.01042 N_P, \quad N_P \leq 250 \qquad [11.11]$$

$$N_Q = -3.305 + 0.2915 N_P^{1/2} + 0.01665 N_P^{3/4} - 0.0002749 N_P, \quad 250 < N_P < 10^6 \qquad [11.12]$$

and

$$N_Q = 0.555 N_P^{1/2}, \quad N_P \geq 10^6 \qquad [11.13]$$

For flow in gaps with sharp 90° bends, they suggest that an additional factor should multiply the correlation for straight gaps. The correction factor is supplied in graphical form in terms of N_P and is unity for N_P below 4000 and is less than unity otherwise, the reduction depending on whether one or two bends are present.

For connected gaps of different thickness, they suggest the pressure drop across each section should be determined, taking into account the intrusion of the boundary layer in upstream sections into adjacent downstream sections when they are wider. They obtain a graphical correlation to determine boundary layer growth at the inlet to wider downstream sections. Using this method, they show that the flow rate can be significantly affected according to the order of narrow and wide gaps.

11.4 Zone modeling of preflashover enclosure fires

11.4.1 CONSERVATION OF MASS AND ENERGY

For multicompartment buildings, the mass and energy conservation equations for the upper and lower layer in each compartment can be written as (Jones, 1990):

Conservation of mass

$$\frac{dm_L}{dt} = \sum_i \dot{m}_{L,i} \qquad [11.14]$$

$$\frac{dm_I}{dt} = \sum_i \dot{m}_{I,i} \qquad [11.15]$$

where the index i refers to the ith source of mass m in the respective layer, the subscripts **L** and **l** refer to the upper and lower layers.

Conservation of energy

The heat balance for the lower layer is written as

$$\frac{\mathrm{d}}{\mathrm{d}t}(c_{\mathrm{v}}m_{\mathrm{l}}T_{\mathrm{l}}) + P\frac{\mathrm{d}V_{\mathrm{l}}}{\mathrm{d}t} = \dot{Q}_{\mathrm{l}} + \dot{h}_{\mathrm{l}} \qquad [11.16]$$

The heat balance for the upper layer is written as

$$\frac{\mathrm{d}}{\mathrm{d}t}(c_{\mathrm{v}}m_{\mathrm{L}}T_{\mathrm{L}}) + P\frac{\mathrm{d}V_{\mathrm{L}}}{\mathrm{d}t} = \dot{Q}_{\mathrm{L}} + \dot{h}_{\mathrm{L}} \qquad [11.17]$$

where $\dot{Q}_{\mathrm{l}}$ is the sensible heat net input rate into the lower zone due to heat of combustion, radiation, and convection, and $\dot{h}_{l,i}$ is referred to as the enthalpy contribution to the lower zone relative to the origin of the incoming mass, c_{v} is the atmospheric specific heat at constant volume and T_{R} is an enthalpy reference temperature, T_{L} is the upper layer temperature and T_{l} is the lower layer temperature. Corresponding definitions apply for the upper zone designated by subscript L.

The terms $P(\mathrm{d}V_{\mathrm{l}}/\mathrm{d}t)$ and $P(\mathrm{d}V_{\mathrm{L}}\mathrm{d}t)$ are included to account for the work done by pressure forces on the moving interface between the two layers. The specific heats of the gases in the room are assumed to remain constant and are evaluated at the initial room temperature.

In addition, a further simplification is made in that the static pressure P in the room is assumed to be constant. This assumption introduces slight errors in the calculation as the static pressure varies due to the hydrostatic effects. Since these effects are comparatively small, of the order of ρgh, the error can be neglected. Zukoski illustrates this by considering a room of height 2.5 m, for which the ratio of the hydrostatic pressure to the static pressure in the room is

$$\frac{\rho gh}{P} < 3 \times 10^{-4} \qquad [11.18]$$

Emmons (1987) also shows that for a 100-m high building, the pressure change at the top of a building is less than 1% of the ambient pressure. It must, however, be noted here that although most of the existing computer codes make the above simplifying assumption, the fire room pressure in reality may vary as a function of time and elevation to affect vent flows significantly. The effect of such variations has been investigated by Brani and Black (1992), who conclude that the time variations of the room pressure are as important. In their model, this is accounted for by considering the conservation of momentum.

With this assumption (i.e. constant pressure), the above conservation equations are closed by the following relations:

The equation of state, assumed to be valid for both the upper and lower layers, gives

$$P = \rho RT = \rho_{\mathrm{L}}RT_{\mathrm{L}} = \rho_{\mathrm{l}}RT_{\mathrm{l}} \text{ and } m = \rho V \qquad [11.19]$$

Since the enclosure volume, V, remains constant,

$$V = V_{\mathrm{L}} + V_{\mathrm{l}} \qquad [11.20]$$

The above governing equations are solved with the prescribed source terms and initial conditions to calculate the layer temperature and its depth as the fire develops with time.

The initial conditions (such as the heating and ventilation and wind) determine the conditions prevailing when a fire starts. The source terms specify the fire itself in terms of its burning characteristics and the associated heat fluxes. They also incorporate ignition, spread of flame, rate of supply of gaseous fuel, charring rate, heat release, melting, smoke release, and so on. (Thomas, 1992). Without proper statement of these conditions, the applications of these models to practical fire problems are rather limited.

Once the source terms are prescribed, the hot-layer temperature and its depth are calculated as follows:

11.4.2 THE HOT LAYER

The temperature, T_L, of the upper layer at any given time is a function of the mass, energy, and specific heat of the layer. The specific heat is assumed to be the same as that of ambient air at 25 °C.

$$T_L = \frac{E_L}{m_L c_p} \quad [11.21]$$

where E_L is the total energy, c_p is the specific heat at constant pressure, and m_L is the total mass of the hot layer (Emmons, 1978).

Ignoring the small changes in molecular weight due to fuel (combusted and not combusted), the layer mass is given by

$$m_L = \frac{\rho_a T_a}{T_L} L W h_L \quad [11.22]$$

where h_L is the layer depth. Combining the above:

$$h_L = \frac{E_L}{L W c_p T_a \rho_a} \quad [11.23]$$

$$\dot{h}_L = \frac{h_L \dot{E}_L}{E_L} \quad [11.24]$$

The total energy of the hot layer at a given time, t, from the start of a fire is given by the integral sum of the rate of change of the layer energy:

$$E_L = \int_0^t \dot{E}_L \mathrm{d}t \quad [11.25]$$

where $\dot{E}_L$ is the total rate of change of internal energy of the hot layer, which is made up of the following component energies:

$\dot{E}_L$	=	$\sum \dot{E}_P$	−	$\sum \dot{E}_V$	+	$\dot{E}_{LW}$	+	$\dot{E}_{LR}$
Hot layer energy change rate		Enthalpy into layer from plumes		Enthalpy out through vents		Convective heat loss rate from layer		Layer energy change from hot surfaces

Terms on the right-hand side of the above equation are calculated from the source terms.

The zones are modeled by simple theories and are linked by setting the outputs from one zone as the inputs to the other. The governing conservation equations (mass and energy) are solved by considering the source terms, described in detail in Sections 11.4.3 to 11.4.10 below. The discussion is based on the pioneering development work carried out by Emmons and Mitler at Harvard University (the Harvard Fire Code series). Some elements of the NIST (National Institute of Standards and Technology) fire and smoke movement code FAST (Walter Jones) are also included.

11.4.3 FLAME AND BURNING OBJECTS SOURCE TERMS

Each room is assumed to contain objects that are specified in terms of their location, size, shape, and material of construction. As the fire involving one of these objects grows, the pyrolysis rate, energy release rates, fire radius, and the flame shape are calculated as functions of time. The pyrolysis rate per unit area is assumed to be proportional to the net impinging surface heat flux, which is taken as the algebraic sum of the radiation and convection from all flames plus reradiation. The flaming region above a burning object is represented as a cone of homogeneous, gray emitting gas with a semiapex angle of ψ (Figure 11.7) calculated from Mitler and Emmons (1981)

$$\tan\psi = \chi \tan\psi_o \frac{|\dot{m}_f|}{\dot{m}_b} \qquad [11.26]$$

where m_b is the mass burning rate (kg/s), m_f is the mass pyrolysis rate (kg/s), χ is the combustion efficiency, ψ_o is the initial value of ψ (default $= 30°$) and the cone temperature is assumed to be 1260 K. The above relationship is assumed to be valid when burning is limited by oxygen supply.

The burning rate *limited by combustion efficiency* is given by

$$\dot{m}_b = \chi |\dot{m}_f| \qquad [11.27]$$

and when *oxygen starvation* occurs, the burning decreases, and is given by

$$\dot{m}_b = \frac{\dot{m}_{air}}{\gamma} \qquad [11.28]$$

where $\dot{m}_{air}$ is the amount of air entrained in the part of the plume below the layer and γ is the air/fuel mass ratio in a free burn of the object.

The actual burning rate, limited by combustion efficiency χ, or by oxygen starvation, is then selected according to

$$\dot{m}_b = \min\left(-\chi\dot{m}_f, \frac{(\dot{m}_p + \dot{m}_f)}{\gamma}\right) \qquad [11.29]$$

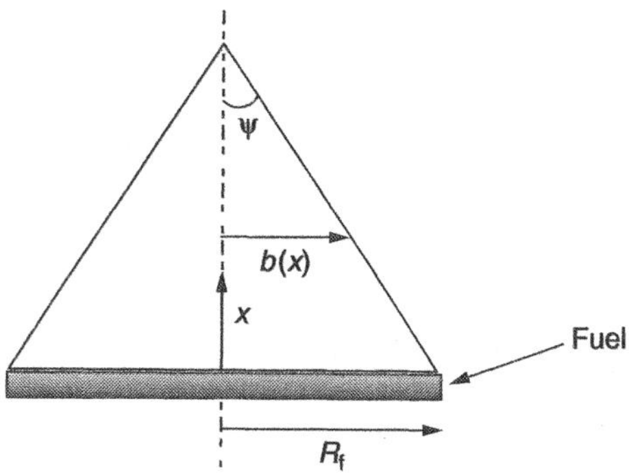

Figure 11.7. Schematic of a flame cone

The pyrolysis rate is taken to be proportional to the net heat flux to the surface and to the instantaneous burning area. It is calculated to allow for the smooth burnout as the fuel is used up.

$$\dot{m}_{\mathrm{f}} = \dot{m}_{\beta}(1 - \mathrm{e}^{(m_{\mathrm{f}}/2\dot{m}_{\beta})}) \qquad [11.30]$$

where m_{f} is the remaining mass of object, given simply by the initial mass (m_{o}) minus the total mass burnt at time t:

$$m_{\mathrm{f}} = m_{\mathrm{o}} + \int_{t}^{0} \dot{m}_{\mathrm{f}}\,\mathrm{d}t \qquad [11.31]$$

and $\dot{m}_{\beta}$ (kg/s) for a fire of radius R_{f} (m), and net radiative heat flux ϕ (W/m^2) and the specific heat of pyrolysis H_{v} (J/kg) is given by

$$\dot{m}_{\beta} = -\frac{\pi R_{\mathrm{f}}^2 \phi}{H_{\mathrm{v}}} \qquad [11.32]$$

Three types of fire are modeled (e.g. in the Harvard Fire series code): *a gas burner, a pool fire,* and *a spreading fire*. The mathematical models are now described for each.

A Gas Burner: The mass pyrolysis rate ($\dot{m}_{\mathrm{f}}$) of the specified fuel is calculated from the gas flow rate input by the user. The gas flow rate is assumed to vary linearly with time between fixed input points entered as a set of discrete points by the user.

A Pool Fire: The pool area, which is the fuel area, is assumed to have a constant radius throughout the fire, R_{max}. The fuel can either be a solid or a liquid. When the entire horizontal surface of the fuel is heated to its ignition temperature, T_{ig}, ignition is assumed to start at a "hot spot" and spread very rapidly over the entire surface. The fire radius is assumed to increase asymptotically from a user-specified initial value to its user-specified maximum value, reaching its maximum size 60 s after ignition. The pyrolysis rate per unit area is assumed to be proportional to the net impinging surface heat flux and is calculated as:

$$\dot{m}_{\mathrm{f}} = \frac{\pi R_{\mathrm{f}}^2 \phi}{H_{\mathrm{v}}} \qquad [11.33]$$

The radius of burning surface after ignition is calculated from

$$R_{\mathrm{f}} = R_{\mathrm{max}}(1 - \mathrm{e}^{-(t - t_{\mathrm{ig}})/2}) \quad \text{for } 0 < (t - t_{\mathrm{ig}}) \leq 60 \qquad [11.34]$$

and

$$R_{\mathrm{f}} = R_{\mathrm{max}} \text{ for } (t - t_{\mathrm{ig}}) > 60.$$

A spreading fire: The pyrolysis rate for a burning object subjected to a heat flux $\dot{q}''$ is given by

$$\dot{m}_{\beta} = \frac{\pi R_{\mathrm{f}}^2 \dot{q}''}{H_{\mathrm{v}}} \qquad [11.35]$$

where the radius of the spreading fire is given by

$$R_{\mathrm{f}} = R_{\mathrm{o}} + \int_{0}^{t_{\mathrm{o}}} \dot{R}_{\mathrm{f}}\,\mathrm{d}t \qquad [11.36]$$

and R_{o} is the initial radius of fire.

The burning rate is related to the pyrolysis rate through the combustion efficiency, χ, defined as

$$\chi = \frac{\dot{m}_\beta}{\dot{m}_f} \qquad [11.37]$$

The net radiant heat flux to the fuel surface is made up as follows:

$\dot{q}''$	=	$(\dot{q}'')_{LF}$	+	$(\dot{q}'')_{WF}$	+	$(\dot{q}'')_{PF}$	−	$(\dot{q}'')_{RR}$
Net heat flux to object		From hot layer to object		Walls + ceiling to object		Flames to object		Reradiation from object

The component radiation heat transfers are modeled as follows (details are given in Mitler and Emmons (1981)):

Hot layer to object $(\dot{q}'')_{LF}$

The hot layer is modeled as a right circular cylinder containing hot, homogeneous gray gas with effective emissivity e_i. The cylindrical layer is assumed to be divided into four quadrants by the point at the center of the object.

Walls to object $(\dot{q}'')_{WF}$

The heat flux from the compartment walls and the ceiling, $(\dot{q}'')_{WF}$, is calculated by dividing the room into four rectangular sections and representing each section as a quadrant of an equivalent cylinder.

Flame to object (horizontal surface) $(\dot{q}'')_{PF}$

The radiation heat flux impinging on a horizontal burning object is calculated for all flame cones in the enclosure and from the flame of the burning object itself. The relative position and the size of each object is taken into account in the calculations. The radiation from a flame cone to an object depends on the temperature and effective mean opacity of the flame cone and on the view factor. More details are given in Mitler and Emmons (1981). The object itself may or may not be burning.

The heat flux at a burning object due to another flame is made up of radiation from its own flame plus radiation from the other flame:

$$(\dot{q}'')_{PF} = (\dot{q}'')_{FO} + (\dot{q}'')_{12} \qquad [11.38]$$

The pyrolysis rate is assumed to be proportional to the net heat flux to the surface, which in turn is determined by the convective heating from the flame minus the reradiation ($\dot{q}_{R''R}$). For small flames, convective heating is important, while reradiation becomes significant as the fire grows. The reradiation to the surface is given by

$$(\dot{q}'')_{RR} = \sigma T_s^4 \qquad [11.39]$$

where T_s is the temperature of the surface of object [K].

In Harvard *V* computer code, this is assumed to be

$$(\dot{q}'')_{RR} = \min(13200, 217000\,\kappa R_f) \qquad [11.40]$$

where κ is the absorption coefficient assumed to be 1.55 m^{-1} and R_f is the fire radius.

11.4.4 FIRE PLUME SOURCE TERMS

The treatment of fire plumes in zone models is based on the pioneering work by Morton *et al.* described earlier in Section 11.3. The earliest application of this work to fire-related problems is found in Thomas *et al.* (1963). For this work, the plume theory was employed to predict the flow through horizontal vents in the roof.

A turbulent buoyant plume is formed above a burning object consisting of gaseous combustion products, soot particles, and unburnt gases. The rising plume spreads radially with entraining surrounding cold air decreasing the centerline temperature. As it impinges on the ceiling, it spreads laterally, forming a hot layer under the ceiling. In this way, the plume transfers mass and energy to the hot layer.

The fire plume model to describe these flows is based on the work done by Morton, Taylor and Turner for point source plumes. In the case of a fire, the point source is taken to be a virtual one below the actual fuel surface such that the plume width at the fuel surface is equal to the fire radius.

For a point source of heat the plume radius, $b(x)$, at any distance x along the plume axis is given by

$$b(x) = 1.2\,\alpha x \qquad [11.41]$$

where α is the entrainment coefficient (assumed to be =0.1) for the plume.

The centerline velocity of the plume is given by

$$u^3(x) = \frac{25\,g\dot{Q}_f}{48\pi\alpha^2 c_p T_a \rho_a} \cdot \frac{1}{x} \qquad [11.42]$$

where $\dot{Q}_f$ is the heat release rate, α is the entrainment coefficient, T_a, ρ_a are the temperature and density of ambient air, c_p is the specific heat of air.

For a plume resulting from a finite fire size, the plume can be thought of originating from a virtual heat source at a distance x_s below the fire given by

$$x_s = \frac{R_f}{1.2\alpha} \qquad [11.43]$$

where R_f is the fire radius

The mass of combustion products flowing into the hot layer is

$$\dot{m}_p = \pi\rho_a(b^2 u - R_f^2 u_f) + |\dot{m}_f| \qquad [11.44]$$

where $\dot{m}_f$ is the fuel pyrolysis rate. The plume radius and the centerline velocity are taken at layer height above the fuel bed, that is, $x = h_p + x_s$.

The enthalpy flow into the hot layer from the plume is calculated from the plume energy balance. The object's flame is included as part of the plume. When an object burns, it produces energy at a rate $\dot{Q}_f$. Some of this energy is lost by radiation from the flame, $\dot{E}_{PR}$. Another portion of the energy produced by the burning of object is used to maintain the object at pyrolysis temperature, T_p, specified by the user as input. Any energy leftover from the combustion of the object plus the energy in the air entrained into the plume is carried into the hot layer.

$$\dot{E}_p = \dot{m}_p c_p T_a + |\dot{Q}_f| - \dot{E}_{PR} \qquad [11.45]$$

The total radiative heat loss from the flame is calculated from the following expression

$$\dot{E}_{PR} = A\sigma T^4 \left[1 - e^{\frac{-4\kappa V_f}{A}}\right] \qquad [11.46]$$

where A is the area of cone (or conical shell) and V_f is the volume of cone (or conical shell).

Apart from the work of Morton *et al.*, improved fire plume models are also available in the work of McCaffrey (1983, 1988) and Zukoski (1985) who developed empirical correlations for the entrainment of air into the plume. According to McCaffrey's model, the flame and plume are divided into three regions; *flaming, intermittent*, and the *buoyant plume* itself. The mass flow rate in each region is expressed in terms of height and fire energy generation rate.

The three correlations are

The Flaming region:

$$\frac{\dot{m}_p}{\dot{Q}_f} = 0.011\left(\frac{Z}{\dot{Q}_f^{2/5}}\right)^{0.566} \qquad 0.0 \leq \frac{Z}{\dot{Q}_f^{2/5}} < 0.08 \qquad [11.47]$$

The Intermittent flaming region:

$$\frac{\dot{m}_p}{\dot{Q}_f} = 0.026\left(\frac{Z}{\dot{Q}_f^{2/5}}\right)^{0.909} \qquad 0.08 \leq \frac{Z}{\dot{Q}_f^{2/5}} < 0.20 \qquad [11.48]$$

The Plume region:

$$\frac{\dot{m}_p}{\dot{Q}_f} = 0.124\left(\frac{Z}{\dot{Q}_f^{2/5}}\right)^{1.895} \qquad 0.20 \leq \frac{Z}{\dot{Q}_f^{2/5}} \qquad [11.49]$$

where $\dot{m}_p$ and $\dot{Q}_f$ are the mass flow rate in the plume at height Z and total heat release of fire, respectively.

Note that if the air entrainment into the plume is $\dot{m}_e$, then

$$\dot{m}_p = \dot{m}_f + \dot{m}_e \qquad [11.50]$$

In other words, the total mass rate of flow in the plume is the sum of the air entrainment rate and the mass rate of production of pyrolysis products. As will be seen later, the McCaffrey correlation is used in the computer code FAST.

11.4.5 THE HOT-LAYER SOURCE TERMS

The buoyant plume gases impinging on the ceiling spread radially to form a hot layer of combustion products under the ceiling. In this treatment, the early stages of layer formation are ignored. As stated earlier, the layer depth and layer temperature (assumed to be uniform) are described by the conservation of mass and energy for the layer.

For energy conservation, the source terms are calculated by taking account of variable emissivity and beam length for the hot layer. The emissivity and absorption coefficient for the hot layer are calculated by taking into account the composition of the layer in terms of combustion product concentrations such as soot and carbon dioxide.

The radiative flux from the hot layer, incident on the extended ceiling (ceiling plus upper portion of walls washed by the layer), is given by

$$\dot{q}''_{LWR} = \sigma \varepsilon T_L^4 \qquad [11.51]$$

where ε is the emittance of the layer given by

$$\varepsilon = 1 - e^{-\left(\frac{\zeta}{1+0.18\zeta}\right)} \qquad [11.52]$$

where ζ is the mean opacity of the layer given by

$$\zeta = \frac{4\kappa V_L}{A_L} \qquad [11.53]$$

and A_L is the bounding area of the hot layer given by

$$A_L = 2WL + 2(W + L).h_L \qquad [11.54]$$

where W and L are the width and length of enclosure
V_L is the volume of hot layer
κ is the absorption coefficient of hot layer gases (m^{-1})
σ is the Stephan–Boltzmann constant (5.67×10^{-8} W/(m^2K^4))

By taking into account the radiative exchange between the hot layer and the walls and the vents, the total heat gain of the layer by radiation is

$$\dot{q}_{LR} = -\varepsilon\sigma[A_L T_L^4 - (WL + A_V)T_a^4 - (A_L - WL - A_V)T_W^4)] \qquad [11.55]$$

where T_L is the mean layer temperature, T_W is the extended ceiling temperature, A_V is the area of part of vent covered by the layer (it is assumed to be zero if outside the layer) and T_a is the ambient air temperature.

To complete the radiation heat transfer for the layer, the radiation absorbed by it from the flames E_a is needed. This is done by calculating the mean energy emitted by the flame cone and multiplying by the view factor (see Mitler (1978)).

The rate of change of energy in the hot layer by radiation from all hot sources in the room is given by

$$\dot{E}_{LR} = \dot{q}_{LR} + \dot{E}_a \qquad [11.56]$$

The convective heat flux from the layer to the extended ceiling is given by

$$\dot{q}_{L''WC} = h_i(T_L - T_W) \qquad [11.57]$$

where h_i is the convective heat transfer coefficient between the layer and the extended ceiling and is assumed to be 5 W/m^2K for a quiescent atmosphere. T_L and T_W are the layer and wall temperatures (K) respectively.

For a growing fire, the heat transfer coefficient is assumed to increase linearly to a maximum value of 50 W/m^2K over a 100-K rise in the layer temperature (Mitler, 1978). The intermediate value of h_i is calculated from

$$h_i = h_s + (h_{max} - h_s)\left(\frac{T_L - T_a}{100}\right) \qquad [11.58]$$

where $h_s = 5$ W/m^2K and $h_{max} = 50$ W/m^2K and T_a is the environment temperature.

Heat loss from external surface of the extended ceiling to the environment is given by

$$\dot{q}_{\mathrm{A''W}} = h_{\mathrm{e}}(T_{\mathrm{a}} - T_{\mathrm{N}}) \qquad [11.59]$$

By taking into account the vent openings in the hot layer, the total power loss by convection from layer to wall is

$$\dot{E}_{\mathrm{LW}} = \dot{q}''(A - A_{\mathrm{V}}) + q''h_{\mathrm{L}}\frac{\dot{E}_{\mathrm{L}}}{E_{\mathrm{L}}}(2L + 2W - b_{\mathrm{V}}^{*}) \qquad [11.60]$$

where

$$q'' = \int_0^t (\dot{q}_{\mathrm{LW}} + \dot{q}''_{\mathrm{AW}})\mathrm{d}t \qquad [11.61]$$

A_{V} is the area of vent and b_{V}^{*} is its width immersed in hot layer.
E_{L}, energy of the layer is given by

$$E_{\mathrm{L}} = (c_{\mathrm{p}})_{\mathrm{L}} T_{\mathrm{L}} \rho_{\mathrm{L}} V_{\mathrm{L}} \qquad [11.62]$$

$\dot{E}_{\mathrm{L}}$ is the rate of change of hot-layer energy.

Flame to extended ceiling

The radiative heat transfer from each flame cone to the extended ceiling (i.e. the ceiling plus the portion of walls wetted by the hot layer) is calculated by approximating the layer as a cylinder. The axially located flame cone is approximated by a point source at its centroid.

11.4.6 HEAT CONDUCTION SOURCE TERMS

The conductive heat transfer to the objects and the extended ceiling surface is calculated using the one- dimensional heat conduction model. The general form of the heat flow equation is

$$\frac{\partial T}{\partial t} = \alpha\frac{\partial^2 T}{\partial x^2} \qquad [11.63]$$

where the thermal diffusivity α is given by

$$\alpha = \frac{k}{\rho c_{\mathrm{p}}} \qquad [11.64]$$

and k, ρ, and c_{p} are the thermal conductivity, density, and specific heat of the material.
The conduction equation is solved numerically using the finite central difference method, where the time interval δt and the subdivision δx are related, to ensure solution stability, by

$$\delta x = \sqrt{(2\alpha\delta t)} \qquad [11.65]$$

The net heat flux on the inside wall, $\dot{q}_{1''}$, is given by

$\dot{q}_{1''}$	$=$	$\dot{q}_{\mathrm{L''WD}}$	$+$	$\dot{q}_{\mathrm{L''WR}}$	$+$	$\dot{q}_{\mathrm{P''WR}}$
		Convective flux from layer to ceiling/wall		Radiative flux from layer to ceiling/wall		Radiative flux from plumes to ceiling/wall

Each of the terms on the right-hand side of the above equation is calculated as described in the following section.

11.4.7 CONVECTIVE HEAT FLUX SOURCE TERMS

Layer to ceiling/wall

This is calculated by assuming the hot layer to be at uniform temperature, T_L. Likewise, the surface of the wall/ceiling (also referred to as *extended wall*) are assumed to be at uniform temperature, T_W, where they are in contact with the hot layer. The convective heat transfer is calculated using Equation (11.57) with $h_i = 5\,W/m^2K$. Similarly, the convective heat loss from the outer surface of the hot walls to the ambient atmosphere is given by equation [11.59] with $h_e = 5\,W/m^2K$. The emissivity ε is given by equation [11.52].

11.4.8 RADIATIVE HEAT FLUX SOURCE TERMS

Layer to ceiling/wall

This term is simply given by equation [11.51]

Flame to ceiling/wall, $\dot{q}_{P''WR}$

This is the radiative heat transfer to the portion of the walls and ceiling contiguous with the hot layer from the flames. Each flame is approximated by a point source located at its centroid.

11.4.9 FLOWS THROUGH VERTICAL VENTS SOURCE TERMS

Natural ventilation

It is assumed that hot gases flow out and cold air flows into the fire room through vertical vents. The pressure drop across a vent is assumed to be piecewise linear function of height. The pieces are strips whose horizontal boundaries are clearly recognizable such as the vent bottom and top, the layer interface, and the neutral planes.

The flow through each strip is separately calculated and summed to give the total outflow and inflow through the vent. A room can have any number of such vents but in Harvard V, a limit of vents is imposed. In order to calculate the net flow through a vent, it is necessary to know

- *the total number of strips in the vent,*
- *the height of each strip, and*
- *the gas density in each strip.*

The pressure at height y on one or other side of the vent is expressed as

$$p_{1/2}(y) = p_0 - g\int \rho_{1/2}\mathrm{d}y \qquad [11.66]$$

where p_0 is the pressure at floor level, g is the acceleration due to gravity, and $\rho_{1/2}$ is the gas density. Subscript 1 refers to the gas in the fire compartment and Subscript 2 to the gas beyond, which may be the outside atmosphere or the air in an adjacent room for a multicompartment fire.

The pressure difference across the vent is then

$$\Delta p(y) = p_1(y) - p_2(y) \tag{11.67}$$

With the assumptions of linear variation of pressure and constant gas density within a strip, the above equation can be written in terms of the pressure differences at the top and bottom of a strip.

$$\Delta p(y) = \frac{\Delta p_{i+1}(y - y_i)}{(y_{i+1} - y_i)} - \frac{\Delta p_i(y - y_{i+1})}{(y_{i+1} - y_i)} \tag{11.68}$$

where subscripts i and $i + 1$ refer to the bottom and top of the ith strip.

The total mass flow rate through a strip of uniform width B is taken as

$$\dot{m}_i = BC_\text{d}(2\rho_i)^{1/2} \int (\Delta p_i(y))^{1/2}\text{d}y \tag{11.69}$$

where C_d is a flow coefficient

Substituting Equation [11.68] into Equation [11.69] gives

$$\dot{m}_i = G_i \frac{|\Delta p_{i-1}|^{3/2} - |\Delta p_i|^{3/2}}{|\Delta p_{i+1}| - |\Delta p_i|} = G_i \frac{|\Delta p_{i+1}| + \sqrt{|\Delta p_i \Delta p_{i+1}|} + |\Delta p_i|}{\sqrt{|\Delta p_i|} + \sqrt{|\Delta p_{i+1}|}} \tag{11.70}$$

where

$$G_i = \pm\tfrac{2}{3}C_\text{d}B(y_{i+1} - y_i)\sqrt{(2g\rho\rho_\text{a})} \tag{11.71}$$

The plus or minus sign is taken according to the sign of Δp_i and ρ is ρ_a (density of atmosphere in the adjacent room) when $\Delta p_i < 0$ and ρ is ρ_i when $\Delta p_i \geq 0$.

The flows are calculated in this way for each vent in the room and then summed to obtain the total inflow and outflow rates.

If only the outflows are considered through a vent, then the mass rate of flow of hot gases through the vents is given by

$$\dot{m}_\text{e} = \sum (\dot{m}_i)_\text{out} \tag{11.72}$$

The total enthalpy flow out of the vents is given by

$$\dot{E}_\text{V} = \dot{m}_\text{e} c_\text{p} T_\text{L} \tag{11.73}$$

As noted above in Section 11.3.4, various mixing processes are important not only in determining the fire spread characteristics but also in contributing, to a considerable degree, to the spread of smoke from one room to another, thus having implications for smoke concentration in these rooms. It has been shown that modeling of mixing is essential to the modeling of enclosure fires.

The mixing phenomenon at the vents is shown to be similar in nature to the entrainment in plumes. Consider a two-room fire situation (Figure 11.8) in which smoke from the fire room (Room 1) hot layer flows out through a vent forming another hot layer in the adjacent room (Room 2). This type of flow through the vent resembles the normal plume flow in that it entrains air from the lower layer of the adjacent room ($\dot{m}_{43}$). Given that $\dot{m}_{13} > 0$, the entrainment can be calculated assuming that the plume is formed by a fictitious source at a distance Z_o below door jet center level such that the flow rate given by McCaffrey plume model becomes equal to the door jet flow rate at the level (Tanaka, 1983). The fictitious source position Z_o is calculated by assuming the source strength (equivalent heat flux) to be given by

$$\dot{Q}_\text{E} = c_\text{p}(T_1 - T_4)\dot{m}_{13} \tag{11.74}$$

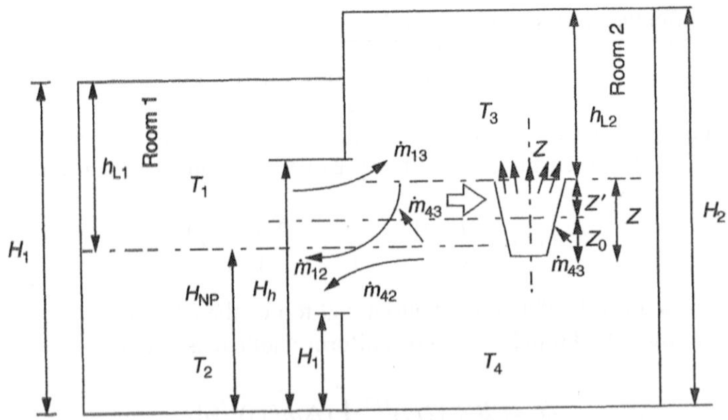

Figure 11.8. Flow through a vent opening between room 1 and room 2 (fire room)

Then from the position of the fictitious source (Z_0), the room dimensions and layer interface height in each room the total flow into the upper layer of room 2 is calculated by substituting m_{13}, Q_E and Z_0 for m_p, Q_f and Z in the plume equations [11.47–11.49].

$$Z' = (H_2 - h_{12}) - \left[\frac{\min(H_h, H_2 - h_{12}) - \max(H_{NP}, H_p H_l - h_{L1})}{2}\right] \quad [11.75]$$

$$Z = Z' + Z_0 \quad [11.76]$$

These dimensions are illustrated in Figure 11.8. Note that the distance Z represents the distance from the virtual source to the hot-layer interface in room 2.

Then using equations [11.49–11.51], the mass flow into the layer can be calculated. From Figure 11.8, it is clear that:

$$\dot{m}_p(z) = \dot{m}_{13} + \dot{m}_{43} \quad [11.77]$$

From this, the entrainment $\dot{m}_{43}$ is then calculated.

The other type of mixing is much like the inverse plume and causes contamination of the lower layer, $\dot{m}_{12}$. This term is calculated by assuming that the incoming cold plume behaves like the inverse of the door jet discussed above. The entrainment for this descending plume ($\dot{m}_{13}$) is then calculated following the above procedure. On this, Jones (1990) remarks that it is possible that the entrainment is overestimated in this case, since buoyancy, which is the driving force, is not nearly as strong as for the usually upright plume.

Forced ventilation

In some respects, forced ventilation modeling is mathematically simpler to implement in the zone models than natural ventilation. This is because the flow can be controlled (directionally as well as in flow rate) with a fan of known characteristics (i.e. pressure drop and the volume flow rate). In the case of natural ventilation, as shown above, however, the pressure difference (hence the flow rate) varies continuously with the local conditions.

Mitler (1984) has identified two main cases of forced ventilation flow:

(i) *Fan operating in the extract mode* ($\dot{V} > 0$)

(ii) *Fan operating in the blower mode* ($\dot{V} < 0$)

where $\dot{V}$ is the volume flow rate

These cases are implemented in the Harvard V and Harvard VI codes. Each case is now considered in detail.

Extract mode ($\dot{V} > 0$)

With respect to the hot layer and the vent position, there are three possible configurations of flow as shown in Figure 11.9. Here, three distinct stages of layer development are identified. Figure 11.9a represents the case when the fan extracts from the lower colder layer. As the hot layer develops, it partially covers the vent, in which case, hot gases together with the cold air are extracted by the fan (Figure 11.9b). And, finally, a stage is reached when the hot layer completely covers the vent and only the hot-layer gases are extracted (Figure 11.9c). The associated mass flow rates, derived by Mitler, for each of the cases are given by

Case a:

$$\dot{m}_{\mathrm{l}} = \rho_{\mathrm{l}}\dot{V},\ \dot{m}_{\mathrm{u}} = 0 \quad \text{for } h_{\mathrm{L}} \le H_{\mathrm{t}} \qquad [11.78]$$

Case b:

$$\dot{m}_{\mathrm{l}} = (1-\xi)\dot{V}\sqrt{(\rho_{\mathrm{l}}\overline{\rho})},\ \dot{m}_{\mathrm{u}} = \xi\dot{V}\sqrt{(\rho_{\mathrm{u}}\overline{\rho})} \quad \text{for } H_{\mathrm{t}} < h_{\mathrm{L}} < (H_{\mathrm{v}} + H_{\mathrm{t}}) \qquad [11.79]$$

Case c:

$$\dot{m}_{\mathrm{l}} = 0, \dot{m}_{\mathrm{u}} = \rho_{\mathrm{u}}\dot{V} \quad \text{for } h_{\mathrm{L}} \ge (H_{\mathrm{t}} + H_{\mathrm{v}}) \qquad [11.80]$$

where

$$\xi = (h_{\mathrm{L}} - H_{\mathrm{t}}), \overline{\rho} = [\xi\sqrt{\rho_{\mathrm{u}}} + (1-\xi)\sqrt{\rho_{\mathrm{l}}}]^2 \qquad [11.81]$$

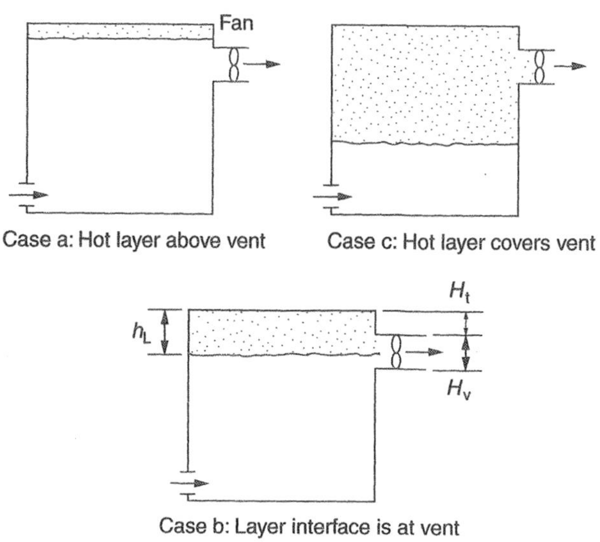

Figure 11.9. Three subcases of Case I, forced outflow. Case a: Hot layer above vent; Case b: Hot layer covers vent; Case c: Layer interface is at vent

Blower mode ($\dot{V} < 0$)

This is the case where the fan blows air into the fire room. Again, three cases are considered, depending on the relative density of the incoming gases. For a single enclosure, the incoming gases may be the outside air but for a multiroom building, these gases may contain smoke from adjacent fire rooms.

Cases a and b (Figure 11.10): If the assumption of two distinct layers is to be satisfied, a further assumption is needed, such that the gas entering the upper or lower layers mixes completely with the existing gas. In other words, the assumption implies that the inflow in no way interferes with the two layers, that is, the two layers remain stratified.

With these assumptions, we can write

$$\dot{m}_{\text{in}} = \rho_{\text{in}}|\dot{V}| \quad \text{for} \quad \rho_{\text{in}} > \rho_{\text{l}} > \rho_{\text{L}} \text{ or } \rho_{\text{in}} < \rho_{\text{L}} \qquad [11.82]$$

where ρ_{in} is the density of the incoming gas.

When the incoming gas is lighter than the upper layer gas, it mixes completely with the upper layer. And for the case when the incoming gas is denser than both the upper layer and lower layer gases, it must sink into the lower layer and mix with it to lower its temperature.

Case c (Figure 11.10): In this case, the incoming gas density lies between the upper and lower layer densities. The incoming fluid is assumed to mix partly with the upper layer and partly with the lower layer. Accordingly, provided the two-layer stratification is maintained, the masses

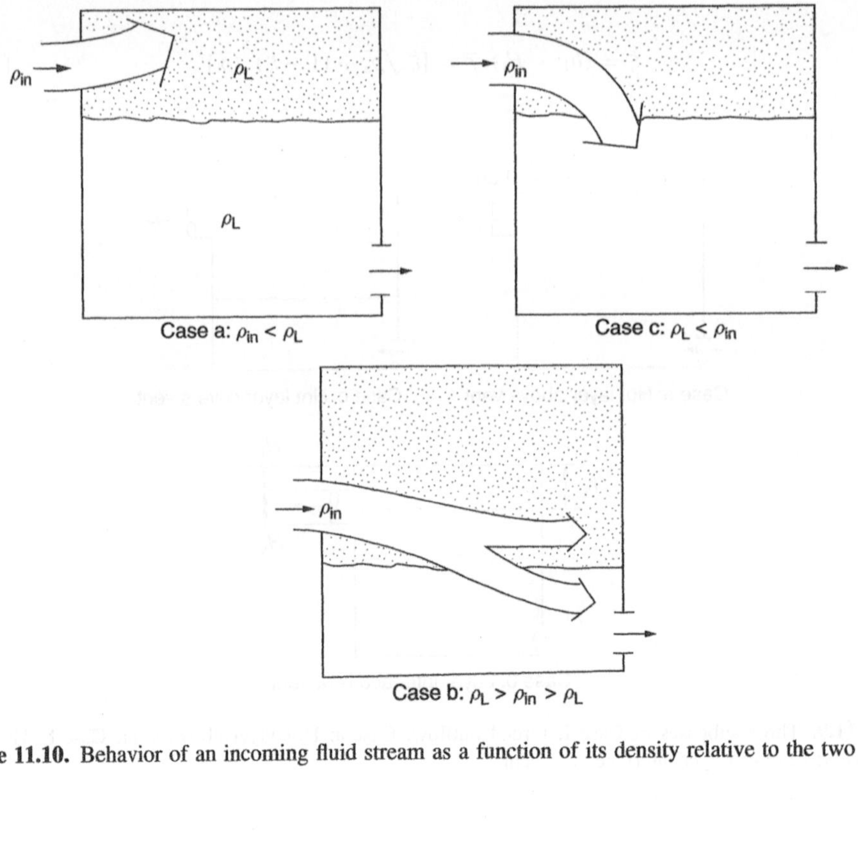

Figure 11.10. Behavior of an incoming fluid stream as a function of its density relative to the two layers.

entering the upper ($\dot{m}_L$) and lower ($\dot{m}_l$) layers are determined by the relative temperatures of each layer as well as the temperature of the incoming gas, thus

$$\dot{m}_L = \dot{m}_{in}\left(\frac{\rho_l - \rho_{in}}{\rho_l - \rho_L}\right) = \dot{m}_{in}\frac{T_L}{T_{in}}\left(\frac{T_{in} - T_l}{T_L - T_l}\right) \qquad [11.83]$$

$$\dot{m}_l = \dot{m}_{in}\left(\frac{\rho_{in} - \rho_L}{\rho_l - \rho_L}\right) = \dot{m}_{in}\frac{T_l}{T_{in}}\left(\frac{T_L - T_{in}}{T_L - T_l}\right) \qquad [11.84]$$

The above temperature relations are implemented in the Harvard V computer code. Here, it must be noted that in deriving these relations, the mixing processes are ignored, as the fluid plunges from one layer to another.

Mechanical ventilation modeling in FAST

In FAST, the modeling of mechanical ventilation is based on the nodal network approach. In this approach, each room is regarded as a node of uniform pressure interconnected to other nodes through a network of flow paths. The rules governing these nodes are the same as Kirchoff's rules for electrical circuits. In the treatment of forced ventilation, following assumptions are made:

1. *Flow is unidirectional through a flow path or duct*
2. *Effects of gas expansion are negligible*
3. *Effects of pressure change within a duct are negligible.*

The mass conservation for each node is written as

$$\sum_j \dot{m}_{ij} = 0 \qquad [11.85]$$

where the index j is the flow path and i is the node number.

At each connection to a compartment, the pressure is specified and the above mass conservation equation (Equation [11.85]) is solved for mass flow rate for a given flow path.

The mass flow rate is given by

$$\dot{m} = G_1\sqrt{\Delta p} \qquad [11.86]$$

where G_1 is the flow path conductance and is calculated for the known duct geometry and fluid properties and

$$G_1 = A_o\sqrt{\left(\frac{2\rho}{C_d}\right)} \qquad [11.87]$$

where A_o is the area of duct inlet, C_d is the flow coefficient (usually about 0.65).

11.4.10 PRODUCTS OF COMBUSTION SOURCE TERMS

In order to assess smoke hazards (toxicity, visibility) throughout a building, the concentration build up of the combustion products needs to be calculated for the given enclosure or room. The burning conditions of fuel are identified and appropriate semiempirical relations are used to calculate mass fractions of oxygen and other products in the hot layer.

For complete burning with unlimited supply of oxygen, the rate of change of oxygen mass in the layer can be shown to be given by Jones (FAST documentation) as

$$\dot{m}_{ox} = 0.23[\dot{m}_p - (\gamma_s\chi + 1)\dot{m}_f - \dot{m}_e Y_{O_2}] \qquad [11.88]$$

where $\dot{m}_p$ is the mass flow from plume to layer; γ_s is the stoichiometric air/fuel ratio; χ is the combustion efficiency; $\dot{m}_f$ is the fuel pyrolysis rate; Y_{O_2} is the mass fraction of oxygen in the hot layer; and $\dot{m}_e$ is the mass flow out of vent.

For the case when insufficient oxygen is available for combustion, despite the fact that some of the air is also entrained by part of the plume covered by the hot layer, the rate of change of oxygen mass in the layer is given by

$$\dot{m}_{ox} = 0.23(\dot{m}_p + \dot{m}_f)\left(1 - \frac{\gamma_s}{\gamma}\right) - Y_{O_2}\left(\dot{m}_{a'}\left(\frac{\gamma_s}{\gamma}\right) + \dot{m}_e\right) \qquad [11.89]$$

where $\dot{m}_{a'}$ is the mass of air entrained by the part of plume covered by the layer.

The concentrations of other products of combustion (such as smoke, carbon dioxide, carbon monoxide, and unburnt hydrocarbons) are calculated from the empirical yield values for the components. Smoke in this context is considered to consist of soot and hydrocarbons.

The rate of change of carbon monoxide mass in the layer is given by

$$\dot{m}_{CO} = -\dot{m}_f Y_{CO} - \dot{m}_e Y_{CO}^{(L)} \qquad [11.90]$$

The rate of change of carbon dioxide mass in the layer is given by

$$\dot{m}_{CO_2} = 5.0 \times 10^{-4}(\dot{m}_p + \dot{m}_f) - \dot{m}_f Y_{CO_2} - \dot{m}_e Y_{CO_2}^{(L)} \qquad [11.91]$$

The rate of change of smoke mass in the layer is given by

$$\dot{m}_s = -\dot{m}_f Y_s - \dot{m}_e Y_s^{(L)} \qquad [11.92]$$

where Y_{CO}, Y_{CO_2}, Y_s are the measured yield values of carbon monoxide, carbon dioxide, and smoke respectively.

$Y_{CO}^{(L)}$, $Y_{CO_2}^{(L)}$, $Y_s^{(L)}$ are the mass fractions of carbon monoxide, carbon dioxide, and smoke respectively in the layer.

The main purpose of FAST is to estimate the transportation of combustion species between the spaces (or room). The treatment of the combustion products is therefore more detailed than, for example, can be found in Harvard V. The program predicts concentrations of species away from the fire source. The species treated are

- *carbon dioxide* (CO_2),
- *hydrogen cyanide (HCN),*
- *hydrogen chloride (HCl),*
- *nitrogen* (N_2),
- *oxygen* (O_2),
- *soot,*
- *total unburned hydrocarbons (TUHC), these are assumed to result from the lack of oxygen availability,*
- *water.*

The model assumes fuel to be entirely composed of carbon, hydrogen, and oxygen. The species production rates are supplied, by the user, to the program in the form of input data at specified intervals. These can change from one time interval to the next, but are constant within a time interval. With HARVARD V, the evolution fractions of combustion products are constant throughout a run.

On the basis of the user-supplied information, in terms of the species yield, the program calculates the species concentrations as functions of time by considering mass balance in three regions, namely, portion of the plume in the hot layer, portion of the plume outside the hot layer, and in the region of the vent.

The species conservation equation is written as

$\dot{m}_{i,u}$	$=$	$Y_i \dot{m}_f$	$+$	$\sum_j f_{i,j} \dot{m}_j$
Rate of change of mass of species i in the upper layer		Rate of production of species i in a fire		Net rate of flow of species i out of upper layer through surface j

Smoke obscuration

In a fire, the smoke obscuration or the reduction in visibility only presents an indirect hazard in that it impedes escape, thereby prolonging exposure to toxic products of combustion. From the victim's point of view, obscuration is a trap, and toxicity and heat are the actual killers.

Generally, smoke resulting from a fire is associated with reduced visibility owing to the light scattering and absorption properties of the solid particle components of smoke (soot particles). The optical density of smoke is a measure of the light obscuration properties of these solid particles. The increase in the number distribution of particles (i.e. particle mass density) in a given volume results in increased optical density and reduced visibility. The optical density is defined as

$$\frac{I}{I_o} = 10^{-DL} \qquad [11.93]$$

where I_o is the intensity of light in the absence of smoke, I is the intensity of light in the presence of smoke, L is the optical path length, and D is the optical density. When the above equation is expressed in terms of natural logarithm, e, it becomes

$$\frac{I}{I_o} = e^{-KL} \qquad [11.94]$$

The quantity K is then called the *extinction coefficient*. From the above definitions, the optical density and the extinction coefficient are related as follows:

In general, the visibility is expressed in terms of a distance, S, at which an object is clearly visible to an observer under the room lighting conditions. Visibility depends on many factors such as the optical density of smoke, the illumination in the room, and whether the object is light-emitting or light-reflecting. The relationship between visibility and the extinction coefficient is expressed as

$$D = 2.3K \qquad [11.95]$$

$$K.S = B \qquad [11.96]$$

where B is a constant.

Experiments (Jin, 1975) show that for light-emitting signs, $B = 8$ and for light-reflecting signs, $B = 3$.

The extinction coefficient per unit path length, K (or D), is an extensive property and can be expressed as a function of the mass concentration of the smoke aerosol, C_s.

$$K = K_m C_s \qquad [11.97]$$

where C_s is in gm^{-3} and K_m (log to the base e, Be) is the specific extinction coefficient of smoke Be (m^2/g) or sometimes referred to as the *particle optical density*.

In FAST, the following relation is used to calculate the smoke optical density:

$$D = 3300C_s \qquad [11.98]$$

where the constant 3300 (log to the base 10, B) has the units of (B) m^2/kg and C_s kg/m^3.

In FAST, smoke obscuration is used only in the evacuation model. The walking speed of an occupant is adjusted according to the smoke density; at low density, the person walks faster while high density slows his progress. Also, high-density smoke will cause the person to seek an alternative escape route.

"Within the hazard calculation in FAST, smoke obscuration is accounted for only within the evacuation model. That is, smoke density is used to adjust the walking speed of an occupant. (A little smoke makes the person walk faster, and a greater amount slows his progress.) Smoke also represents a psychological barrier to an occupant entering a room. In the latter case, excessive smoke will cause the person to seek an alternate route and can result in the occupant being trapped in a room without a safe exit (door or window)." (Jones' technical guide to Hazard I).

Toxic hazards

In the treatment of toxic effects, the so-called *N-Gas* (N being number of gas components in the smoke) model is used within FAST. This model simply states that the total observed effect of a mixture of gases is equal to the sum of the effects of each of the component parts, for example, 50% lethal dose of carbon dioxide together with the 50% lethal dose of, say, HCN will result in death.

On the basis of experimental evidence, the most commonly found fire gases, carbon monoxide, carbon dioxide, hydrogen cyanide, and oxygen, are considered in the FAST toxicity model ($N = 4$). To assess the cumulative toxic effects of these gases, the concept of *Fractional Effective Dose* **(FED)** is used. Here, the term dose is defined as the time integral of gas concentration (i.e. area under the concentration-time curve). Accordingly, an exposure to a time-varying concentration $C(t)$ for time t would give

$$\text{Dose} = \int_0^t C(t).\mathrm{d}t \qquad [11.99]$$

The FED is then defined as

$$\text{Dose} = \frac{\text{dose received at time } t}{\text{effective dose to cause effect}} \qquad [11.100]$$

In FAST, the FEDs for each component gas are calculated from the species concentrations in the hot layer. For N-Gas components, the total FED is obtained from

$$\mathbf{(FED)}_{N\text{-G}} = \mathbf{(FED)}_{O2} + \mathbf{(FED)}_{CO} + \mathbf{(FED)}_{CO2} + \mathbf{(FED)}_{HCN} \qquad [11.101]$$

If the value of $(\text{FED})_{N\text{-G}} >= 1$, then the program predicts incapacitation or death.

Here, $(FED)_{O2}$ refers to the incapacitation due to oxygen depletion.

Two FED models are available: one based on the work at NIST and the other on the work by Purser.

FAST also provides an alternative method for the assessment of toxicity. This method considers the *smoke* as consisting of all the combustion products and uses the results of Hartzell from studies on exposure to smoke from real fires. In this approach, the dose $(C.t)$ is calculated in the upper hot layer by taking the cumulative mass of fuel lost and distributing it into upper layers in each of the rooms according to the calculated mass flows through the openings (vents). The calculated concentration is then integrated over time to produce a concentration-time product, the dose. This is then compared to the threshold lethal dose of $900\,\text{g.min.m}^{-3}$ (Hartzell). Death or incapacitation is assumed to occur if this value is exceeded.

The toxic effects are not dealt with in Harvard V computer code.

11.5 One-Zone modeling of postflashover fires

The growth of fire in an enclosure is marked by a specific event called the *flashover*. Although there are some arguments about the precise definition of flashover, it is agreed that once this event has occurred, the hot gas layer covers the entire enclosure volume and that the highest internal temperatures are reached after this event. Flashover marks the transition from a localized burning of an individual item of fuel to the simultaneous ignition of all combustibles in the enclosure. It can therefore be said that the study of the postflashover fire is important for structural performance (or fire endurance) of fire barriers and walls, while personal safety and fire detection are important in the preflashover stage.

Compared with preflashover fires, the analysis of postflashover fires is easy to deal with, as in this case, the enclosure is assumed to consist of a single "hot" and well-mixed layer. As the fire burns in the vitiated air, its burning characteristics and hence the rate of heat release changes. For the purposes of theoretical analysis, the following assumptions are made:

- The enclosure environment is well mixed such that the gas temperature is only a function of time (i.e. uniform gas temperature). The temperature variations near the enclosure boundaries and in and around the fire plume are ignored.
- The exchange of gas between the inside and outside occurs through a single vertical opening with no interaction between the opposing flows. This flow is caused by the stack effect alone.

11.5.1 THEORETICAL MODEL

The compartment temperature can be estimated from the heat balance equation, which can be described as (neglecting the pressure and dissipation terms):

$$\dot{Q}_f = \dot{Q}_c + \dot{Q}_W + \dot{Q}_R + \dot{Q}_B \quad [11.102]$$

where $\dot{Q}_f$ is the rate of heat release from the fire, $\dot{Q}_c$ is the rate of heat loss by convection through vents, $\dot{Q}_W$ is the rate of heat loss by conduction through compartment boundaries, $\dot{Q}_R$ is the rate of heat loss by radiation through vents, and $\dot{Q}_B$ is the rate of change of heat stored in the gas volume.

The heat balance is illustrated in Figure 11.11. The rate of change of heat stored in the gas volume is associated with the change of temperature of the gas, T_g, and since the process is *quasi-steady*, this term is very small and can be dropped for simplification.

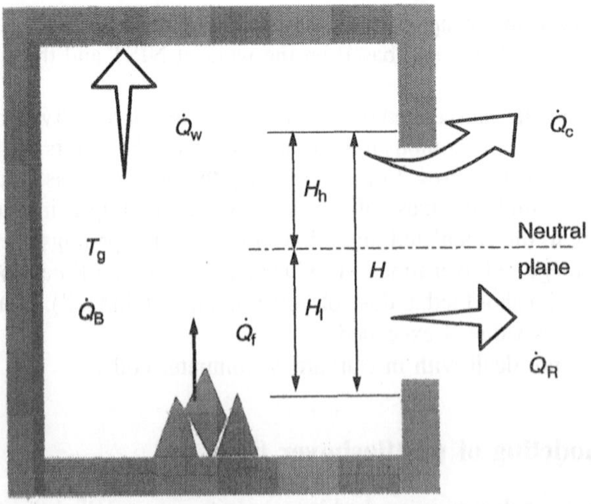

Figure 11.11. Fully developed enclosure fire

Terms on the right hand side of equation [11.102] are functions of the gas temperature T_g, and by calculating the heat release rate of a given fire, the equation can be solved for temperature, T_g.

Quintiere (1983) has presented a simple empirical correlation for the calculation of hot gas temperature rise ($\Delta T = T_g - T$). In his calculation procedure, the heat release rate is either estimated or derived from experimental results. On this basis and a number of simplifying assumptions, he concludes that the following expressions for temperature rise predicts temperatures within 100 °C (at worst). This is illustrated by the application of this formula to a number of experimental cases.

$$\frac{\Delta T}{T_\infty} = 1.6\left(\frac{\dot{Q}_f}{\sqrt{g}c_p\rho_\infty T_\infty A_w\sqrt{H}}\right)^{2/3}\left(\frac{h_k A_{WS}}{\sqrt{g}c_p\rho_\infty T_\infty A_w\sqrt{H}}\right)^{-1/3} \quad [11.103]$$

where
T is the ambient temperature (°K)
A_{WS} is the wall surface area (m^2)
A_w is the area of opening (m^2)
h_k is the wall conductance
H is the height of opening (m)

The wall conductance h_k is given by the following approximate expression:

$$h_k = \sqrt{\frac{k_W\rho_W c_W}{t}}, \quad t \le t_p \quad [11.104]$$

$$h_k = \frac{k_W}{\delta} \quad t \ge t_p \quad [11.105]$$

k_W is the thermal conductivity of wall (kW/mK)
ρ_W is the density of wall material (kg/m^3)
c_W is the specific heat of wall material (kJ/kgK)
δ is the wall thickness (m)
t is time (s)

The solid thermal penetration time is given by

$$t_{\text{p}} = \left(\frac{\rho_{\text{W}} c_{\text{W}}}{k}\right)\left(\frac{\delta}{2}\right)^2 \quad [11.106]$$

Note that the term $h_{\text{k}}A_{\text{WS}}$ is computed by taking into account all of the materials used in the construction of floor, walls, and ceiling. Thus,

$$h_{\text{k}}A_{\text{WS}} = \Sigma(h_{\text{k}}A_{\text{WS}})_i \quad [11.107]$$

where i is the wall region

Quintiere's formula is an oversimplication of the problem. It does not, for example, take into account radiative losses, which in some cases can be significant, and only treats conduction and convection in a rudimentary fashion. However, after the application of this formula to experimental measurements, he concludes that this empirical relation could provide a simply derived estimate of thermal effects. Figure 11.12 shows a comparison of measured temperatures with those calculated using the above formula. In this example, the fire consisted of a typical furnished bedroom with a closet and a single doorway opening to the outside. The floor material was plywood, and the walls and ceiling were sheetrock. The computed temperature was compared to two measured temperatures –0.24 m from the ceiling between the bed and the doorway, and 0.13 m from the top of the doorway.

Alternative approach has been offered by Kawagoe and Sekine in which the heat loss terms are expressed explicitly and calculated with greater accuracy. Some parameters are, no doubt, derived from experimental measurements, but this approach does represent a more realistic analysis of the problem. The heat generations and heat loss terms involved in Equation [11.102] are now discussed in detail.

Heat release rate from fire

This has already been dealt with in Section 11.3.1. It has been shown that the rate of heat release depends on the amount of oxygen available in the room, that is, the rate of flow of air into the room. By combining equations [11.2], [11.3], and [11.29], it can be shown that

$$\dot{Q}_{\text{f}} = \gamma K A_{\text{w}}\sqrt{H}.\Delta H_{\text{c,air}} \quad [11.108]$$

Values for γ, and $\Delta H_{\text{c,air}}$ are known from experiments for fires involving given fuel. Kawagoe and Sekine recommend 5.2 kg air/kg fuel and 10,780 kJ/kg for wood respectively.

Convection heat loss

The convective heat losses from the room are calculated from the mass rate of flow of hot gases to the outside. If the outside temperature is $T(K)$ and mass rate of flow of hot gases $\dot{m}_{\text{g}}$ (kg/s), then the heat loss can be written as

$$\dot{Q}_{\text{c}} = \dot{m}_{\text{g}} c_{\text{p}} \Delta T \quad [11.109]$$

where

c_{p} is the specific heat of gases (1.15 kJ/kgK)

$\Delta T = (T_{\text{g}} - T)$

The above equation shows that if the mass rate of flow of hot gases from the room is known, the convective heat loss can be calculated.

From the considerations of stack effect and ignoring the hydrostatic pressure differences, Kawagoe and Sekine have suggested the following expression for the mass flow rate:

$$\dot{m}_g = 1042\,WC_d(H_h)^{3/2}\left[\frac{1}{T_g}\left(\frac{1}{T_\infty}-\frac{1}{T_g}\right)\right]^{1/2} \qquad [11.110]$$

where
W is the width of opening and
C_d is the orifice discharge coefficient (approximately = 0.68)
H_h is position of neutral plane (see Figure 11.11).

By assuming the internal temperature to be in the range 600 °C and 1000 °C and the mass rate of flow of air into the room between one and two times stoichiometric, Kawagoe and Sekine have estimated the position of the neutral pressure plane to be

$$H_h = \frac{2H}{3} \qquad [11.111]$$

where H is height of opening (m).

Conduction heat loss

To calculate conduction losses, the one-dimensional heat conduction model is used on the assumption that temperature variations along the wall surfaces are small. This way, the walls can be represented as plates of infinite surface area. The conduction equation is then solved with the internal gas temperature and ambient temperature as the boundary conditions.

The heat loss through the wall consists of radiative and convective terms:

$$\dot{Q}_W = A_{WS}\lfloor \sigma\varepsilon_{eff}(T_g{}^4 - T_W{}^4) + h_w(T_g - T_W)\rfloor \qquad [11.112]$$

where
A_{WS} inside surface area of the walls
ε_{eff} is the effective emissivity of fire gases and the wall surface
h_w is the convective heat transfer coefficient

From equation [11.112], it is clear that in order to calculate $\dot{Q}_W$, we need to know the internal wall temperature T_W. This is done by solving the heat conduction equation of the form:

$$\rho_W c_{pW}\frac{\partial T_W}{\partial t} = \frac{\partial}{\partial x}\left(k_W\frac{\partial T_W}{\partial x}\right) \qquad [11.113]$$

where
x coordinate is such that $x = 0$ is the inside wall surface and $x = L$ is the outside wall surface
L is the wall thickness (m)
k is the thermal conductivity of wall material (kW/m/K)
ρ_W density of the wall material (kg/m^3)
c_{pW} is the heat capacity of the wall material (kJ/kg/K)
t is time (s)
T_W is the wall temperature as function of x, t.

This equation is solved subject to the following initial and boundary conditions.

Initial conditions

$$t = 0, \quad 0 \leq x \leq L, \quad T_W = T_\infty \qquad [11.114]$$

Boundary conditions

$$x = 0, \quad t \geq 0 \quad -k\frac{\partial T_W}{\partial x} = h_W(T_g - T_W) + \varepsilon_{eff} .\sigma(T_g{}^4 - T_W{}^4) \qquad [11.115]$$

$$x = L, \quad t \geq 0 \quad -k\frac{\partial T_W}{\partial x} = h_\infty(T_W - T_\infty) + \varepsilon_{eff} .\sigma(T_W{}^4 - T_\infty{}^4) \qquad [11.116]$$

Radiation heat loss

The radiation losses from the window opening are calculated, treating the opening as a blackbody (emissivity is equal to unity). Then,

$$\dot{Q}_R = A_w\sigma(T_g^4 - T_W^4) \qquad [11.117]$$

where
A_w is the area of the opening
σ Stefan–Boltzmann constant (5.67×10^{-11} kW/m^2/K^4)

11.5.2 NUMERICAL SOLUTION PROCEDURE ASSUMPTIONS

Using the heat balance equation [11.102] together with the above-defined quantities and the wall conduction equation, a solution is found for the wall temperature T_W and the room gas temperature T_g. To do this, a number of further simplifications are made.

Effective emissivity, ε_{eff}

In order to calculate heat transfer to the inside wall from the hot gases and the heat losses to the ambient from the outer wall surface, emissivities for the wall and the hot gas mixture are required. Kawagoe and Sekine make the simplest of assumptions and use only the emissivity value for concrete ($\varepsilon_{eff} = \varepsilon_W = 0.7$). This is an oversimplification.

It is important to remember when evaluating the heat transfer between hot gas and a solid surface that within the enclosure, radiative heat transfer is more dominant than the convective heat transfer – hence the importance of emissivity. While outside the enclosure, both convective and radiative heat transfer are important.

Babrauskas (1981) made a number of improvements by explicitly treating the emissivity of the gases and the wall material. The expression used is

$$\varepsilon_{eff} = \frac{1}{\dfrac{1}{\varepsilon_g} + \dfrac{1}{\varepsilon_W} - 1} \qquad [11.118]$$

where
ε_g is the emissivity of the hot gases
ε_W is the emissivity of the wall material

The gas emissivity, ε_g, is further refined with contributions from the following two components:

- emissivity from band radiation of CO_2 and H_2O, ε_{gb}
- emissivity from soot, ε_{gs}.

The resulting gas emissivity is given by

$$\varepsilon_g = \varepsilon_{gs} + (1 - \varepsilon_{gs})\varepsilon_{gb} \quad [11.119]$$

where

$$\varepsilon_{gs} = 1 - e^{(-k\delta_f)} \quad [11.120]$$

k is the absorption coefficient and δ_f is the flame thickness.

Convective heat coefficient, h_W

Kawagoe and Sekine use a constant value for the convective coefficient ($h_W = 23\,\mathrm{W/m^2K}$).

The natural convective flows in a room depend on the mixing processes and the turbulence generated by the interaction of plumes and ceiling jets with horizontal and vertical walls (Jaluria). Disruption of the wall thermal boundary layers under actual fire conditions adds further uncertainties in the calculation of convective coefficient. Babrauskas (1981) discusses some of these issues and provides a useful way of calculating h_W for horizontal and vertical surfaces.

11.6 Field modeling of enclosure fires

It is clear from the above discussion of zone fire models that a knowledge of fire physics in terms of fire geometry (fire plume) and smoke spread (two layers) is a necessary and sufficient condition for the validity of these models. Indeed, if clearly identifiable physical processes do exist in a given fire scenario, it would be futile and unnecessary to attempt to calculate fire conditions everywhere in the domain of interest. If, from physical observations of real or model scale fire tests, the underlying assumptions could be justified with acceptable degree of accuracy, the validity for the use of such models could be assured. If we know from such observations of enclosure fires that temperatures higher than the ambient only prevail in the "hot" smoky layer and in the "fire plume," it would be unwise to attempt to calculate temperatures throughout the fire room. In many fire situations, therefore, if individual fire processes such as fire plumes, hot layers, wall plumes, spill plumes, ceiling jets, and so on could be distinctly identified, the task of fire modeling to calculate temperatures, heat, and species concentrations for hazard assessment purposes, becomes simple.

Clearly, such a methodology relies heavily on the use of verifiable, empirically derived correlation. Herein lies both the strength and weakness of zone modeling techniques. Strength, because the governing simplified equations are easy to solve, requiring only minimal computing power. Weakness, because such simplifications rely on a priori knowledge of the physical processes. In other words, the physical processes and their interactions are not predicted but are assumed to exist.

However, real fire situations are not so simple. Fire plumes may interact with, for example, other fire plumes, the building structure, and the environmental conditions to produce complex situations in which hot layers and simple plumes become impossible to distinguish. The processes become inter dependent and interlinked. "Hot" and "cold" layers, and simple plumes may not exist. Under these conditions, zone modeling approach becomes inadequate and invalid, for there are no zones.

What is needed is a new approach in which the relevant physical processes are described mathematically on the "microscopic" scale and the governing differential equations solved over time and space for the "field" variables. Such an approach would, hopefully, make fewer empirical assumptions and be applicable to all situations irrespective of the physical complexity. This approach is called *field modeling* or *Computational Fluid Dynamics* (CFD) modeling.

The Navier-Stokes equations, conservation equations for each of the flow constituents, the energy equation, equation of state, and initial and boundary conditions are the equations that govern the fire phenomena. Field models attempt to solve these equations. In principle, this approach is exact, but some approximations are made to incorporate complicated phenomena like turbulence, which still defies accurate description. A number of field models have been developed worldwide.

Yamamuchi (1986) describes results of numerical simulations of smoke movement and coagulation in order to determine the response of a smoke detector within an enclosure. He assumes that coagulation of the smoke is dominantly due to Brownian motion and that the particle size distribution has a time-independent shape. From these assumptions, he derives closed field equations for the volume concentration and number concentration of particles as a function of position that are solved numerically.

Habchi *et al.* (1988) have developed a computer code for fire protection and risk analysis of nuclear plants. It is primarily intended for the study of enclosures such as cable rooms and control rooms but has the flexibility to enable it to be used in more complex situations. The differential equations that express heat, mass, and momentum conservation are solved on a three-dimensional grid. The k-ε model (discussed later) is used to treat the effects of turbulence. Chemical reactions may be modeled with instantaneous, one-step or multistep kinetics and radiation is treated with a six-flux model. The code has been applied to a nuclear power plant fire test simulation done by Factory Mutual Research. The test was conducted in a large rectangular, ventilated enclosure with a gas burner inside.

Morita *et al.* (1989) have used field methods for calculating heat flow in fire compartments. They take the governing equations to be those for a viscous, heat conducting, compressible fluid and model the effects of turbulence with the k-ε model. The equations are solved with an implicit, upwind method. They conclude that a time step of 10 ms and a grid spacing of 5 cm are the upper limits needed to obtain numerically accurate results in high Reynolds number flows.

Huhtanen (1989) has modeled an oil pool fire in a two-unit turbine hall with the PHOENICS computer model, utilizing a turbulence-dependent burning rate model. PHOENICS is a three-dimensional field model that utilizes the k-ε turbulence model and includes the effects of heat radiation and heat transfer to structures. It was found to be difficult to choose the proper physical models to obtain the essential features of the situation being studied. However, final results were found to be in qualitative agreement, with observations of a similar fire that took place in the Hanasaari power plant near Helsinki in 1986. A grid of 8000 cells was used (which in three dimensions and a plane of symmetry results in about 25 cells in each dimension) required 225 h of computing time on a MicroVAX II for a six-minute transient.

Boccio and Usher (1985) have used PHOENICS to assess fire behavior in complex nuclear power plant enclosures. Code predictions were compared with cable fire tests conducted by Factory Mutual Research on behalf of the Electric Power Research Institute, and enclosure fire tests conducted by Sandia National Laboratories on behalf of the US Nuclear Regulatory Commission. Additionally, fire scenarios for the control room of the La Salle plant and a lubrication oil fire in the center of the bottom of a PWR (Pressurized Water Reactor) containment dome were simulated. Qualitative agreement was obtained in the comparisons, but difficulties in determining the heat release rate in the former experiment and inaccuracies in temperature measurements in the latter experiment could have prevented better agreement.

Cox *et al.* (1986) discuss validation studies with the field model JASMINE, a development of the PHOENICS model, which treats fire and smoke behavior. The geometry they consider is relatively simple, being a closed room and a tunnel with forced ventilation. They conclude that the model predicts smoke behavior reasonably well except in the vicinity of the fire. Further work was recommended to improve the modeling of turbulence and chemistry in the vicinity of the fire to improve the reliability of predictions for fire spread.

In a later study, JASMINE was used by Pericleous *et al.* (1989) to model smoke movement and the distribution of temperature caused by a 2-MW methanol pool fire in a large air-supported sports arena. The results were compared with experimental measurements done in a one- sixth scale model. Body-fitted coordinates were used in order to match the smooth contours of the dome of the sports arena. 1200 computational cells were used, the number selected being a compromise between accuracy and economy. The fire was treated parametrically as a source of heat. Although not detailed, the comparison showed that the temperature histories of the thermocouple rakes could be reproduced reasonably well. Qualitatively, the predicted flow patterns were similar to those observed. Calculations with a polar grid with 2860 cells failed to reproduce the observed ceiling jet, and instead predicted the appearance of a well-mixed layer. Both grids were coarse relative to the dimensions of the fire and its plume and consequently failed to predict details of the temperature field near the fire. With time steps varying between 2 and 30 s during the 7-min transient starting when the fire was first lit, the calculations required between 5 and $7\frac{1}{2}$ h computing time on a concurrent 3280 super mini. Results were not shown to demonstrate grid and time step-size independence.

Galea (1988) has discussed the merits of field models and notes that they are very good at resolving the details of complex flows, particularly flows in single compartments with unusual geometry. This was admirably demonstrated by the calculations done for the Kings Cross fire (Hamer, 1988) in which a previously unknown mode of fire propagation up the well of an escalator (the "trench effect"), promoted by the flow of gases in the well, was predicted and subsequently confirmed experimentally.

11.6.1 THEORY AND CONCEPT OF FIELD MODELS

The instantaneous velocity, pressure, enthalpy, temperature, and density components can be written as

$$u = \overline{u} + u', \quad v = \overline{v} + v', \quad w = \overline{w} + w', \quad p = \overline{p} + p' \qquad [11.121a]$$

$$H = \overline{H} + H', \quad T = \overline{T} + T', \quad \rho = \overline{\rho} + \rho' \qquad [11.121b]$$

With these definitions, it can easily be shown that for two flow variable Φ and Ψ the following relations hold:

$$\overline{\Phi\Psi'} = 0, \quad \overline{\overline{\Phi}\Psi} = \overline{\Phi}\,\overline{\Psi}, \overline{\Phi + \Psi} = \overline{\Phi} + \overline{\Psi} \qquad [11.122]$$

In reality, two types of turbulence are encountered: stationary and nonstationary. A stationary turbulence is one in which an average is independent of the time at the midpoint of the averaging period (t_1). If, however, it does depend on t_1 as in the case of a nonstationary turbulence, the *ensemble average* over N identical measurements is used (see Figure 11.12). The ensemble is defined as

$$\overline{\Phi}(x_o, t_o) = \frac{\sum_{1}^{N} n\Phi_n(x_o, t_o)}{N} \qquad [11.123]$$

For the purposes of this discussion, we confine ourselves to the time averages only.

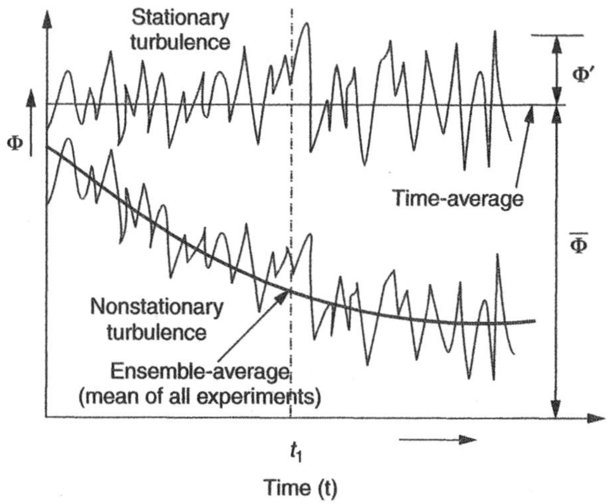

Figure 11.12. Stationary and nonstationary turbulence

Using the above definitions for instantaneous values of flow components, it can be shown that the conservation equations for mass, momentum, and energy (enthalpy) can be written as shown in the following section.

Conservation equations

Mass conservation:

$$\frac{\partial \overline{\rho}}{\partial t} + \frac{\partial(\overline{\rho u_j})}{\partial x_j} = 0 \qquad [11.124]$$

Momentum conservation:

$$\frac{\partial}{\partial t}(\overline{\rho u_i}) + \frac{\partial}{\partial x_j}(\overline{\rho u_i}\overline{u_j}) = -\frac{\partial \overline{p}}{\partial x_i} + \frac{\partial}{\partial x_j}\left[\mu_{\text{eff}}\left(\frac{\partial \overline{u}_j}{\partial x_i} + \frac{\partial \overline{u}_i}{\partial x_j}\right)\right] + (\overline{\rho} - \rho_o)g \qquad [11.125]$$

Energy conservation:

$$\frac{\partial}{\partial t}(\overline{\rho H}) + \frac{\partial}{\partial x_j}(\overline{u}_j\overline{\rho H}) - \frac{\partial}{\partial x_j}\left(\Lambda_{\text{H}}\frac{\partial \overline{H}}{\partial x_j}\right) = \frac{\partial \overline{p}}{\partial t} \qquad [11.126]$$

Species conservation:

$$\frac{\partial}{\partial t}(\overline{\rho} m_\alpha) + \frac{\partial}{\partial x_j}(\overline{u}_j\overline{\rho} m_\alpha) - \frac{\partial}{\partial x_j}\left(\Lambda_\alpha\frac{\partial m_\alpha}{\partial x_j}\right) = S_\alpha \qquad [11.127]$$

All these are transport equations that express the transport of a quantity (momentum, or enthalpy) through the faces of an infinitesimal control volume. The transport processes are governed by the mean flow, turbulent diffusion, and molecular diffusion. The transported quantity may be produced within the control volume by external body forces or pressure gradients or by exchange with another quantity (for the production of turbulent energy from the mean flow kinetic energy). It may be destroyed either by further exchange (e.g. dissipation of turbulent kinetic energy into thermal kinetic energy) or by direct loss (e.g. heat loss by conduction).

In addition, all components of the fluid are assumed to obey the perfect gas law:

$$p = \frac{\rho_\alpha}{W_\alpha} RT \qquad [11.128]$$

with the mixture density, ρ, being given by

$$\frac{1}{\rho} = \sum_{\alpha=0}^{N} \frac{m_\alpha}{\rho_\alpha} \qquad [11.129]$$

$$\Lambda_{\mathrm{H}} = \left(\frac{\mu_{\mathrm{T}}}{\sigma_{\mathrm{H}}} + \frac{\lambda}{c_{\mathrm{p}}}\right), \quad \Lambda_\alpha = \left(\frac{\mu_{\mathrm{T}}}{\sigma_\alpha} + \Gamma_\alpha\right) \qquad [11.130a]$$

where Λ_α is the molecular diffusion coefficient of α and S_α is the volumetric source/sink term representing generation rate of α per unit volume (may be specified by the user).

μ_{eff} is the effective viscosity defined by

$$\mu_{\mathrm{eff}} = \mu + \mu_{\mathrm{T}} \qquad [11.130b]$$

with μ_{T} being the turbulent viscosity,

λ is thermal conductivity, c_{p} is the specific heat at constant pressure, σ_{H} is the turbulent Prandtl number for enthalpy, m_α is the mass fraction of species α such that $1 \le \alpha \le N$, N is the number of species, Γ_α is the molecular diffusivity, σ_α is the turbulent Schmidt number for species α, H is the enthalpy, W_α is the molecular weight, R is the universal gas constant.

These conservation equations describe the turbulent motion in terms of the mean and turbulent components (turbulent components appearing implicitly in the definition of turbulent viscosity).

Turbulent viscosity

The effective viscosity term, μ_{eff}, appearing in the momentum equation (Equation [11.125]), is made up of molecular viscosity and turbulent viscosity (Equation [11.130b]). Whereas the former is a property of the fluid, the latter results from the fluid flow.

Turbulent viscosity arises from the fact that the fluctuating component produces additional stresses (in addition to those resulting from pressure and molecular viscosity), which are generally referred to as *Reynolds stresses*. These stresses appear in the full momentum transport equation in the form $-\rho\overline{u'u'}2$. Like molecular viscosity, it has been suggested by Boussinesq (1877) that the turbulent shear stress is related through eddy viscosity (or turbulent viscosity) to the velocity gradient:

$$-\rho\overline{u_{i'}u_{j'}} = \mu_{\mathrm{T}}\left(\frac{\partial \overline{u}_i}{\partial x_j} + \frac{\partial \overline{u}_j}{\partial x_i}\right) \qquad [11.131]$$

It can, therefore, be seen that in turbulence, a correlation between the fluctuating quantities creates new unknowns (such as Reynold's stresses) for which the conservation laws in terms of known quantities are not well established. An iterative process for constructing the conservation laws for unknown correlation lead to higher order unknown correlation. Thus, the complete set, which can provide a detailed description of the turbulent motion, ends up in an infinite set of partial differential equations, making the solution impossible. To get over this difficulty, a semiempirical approach is used to model turbulence. Various such models are available, but in this chapter, we shall confine our discussions to the so-called k-ε model, which has been found to be applicable to a large number of practical problems. In most field modeling, computer codes options are available to select any one of these models. A detailed discussion of turbulence models is given by Kumar (1983).

Turbulence modeling

As noted above, the appearance of the effective viscosity term, μ_{eff}, in the momentum conservation equation – made up of molecular viscosity and turbulent viscosity – introduces difficulties in solving these equations. There are more unknowns (turbulent stresses and heat flux quantities) than the number of available equations. To proceed, we need to find additional equations involving the new unknowns or make assumptions regarding the relationship between the new apparent turbulent quantities and the time-mean flow variables. This is known as the closure problem, which is most commonly handled through turbulence modeling techniques.

In a Newtonian viscous fluid, viscosity is a property of the fluid and depends only on temperature and on the pressure. Virtually all gases and most liquids are closely Newtonian (Bradshaw, 1985). The effective viscosity as stated above is a function of turbulent and molecular viscosity.

It is generally considered that turbulence is generated mainly from the shear of the mean flow and from the buoyancy (i.e. external body force). On the microscopic scale, it consists of a highly disordered array of eddies of widely different sizes. These eddies are considered as a tangle of vortex elements (or lines) that are stretched in a preferred direction by the mean flow and in a random direction by one another. This "vortex stretching" mechanism ultimately leads to the breakdown of large eddies into smaller ones (Hinze, 1975). This process takes the form of "energy cascade." Since eddies of comparable size can only exchange energy with one another, the kinetic energy from the mean motion is extracted from the large eddies. This energy is then transferred to neighboring eddies of smaller scales continuing to smaller and smaller scales (larger and larger velocity gradients), the smallest scale being reached when the eddies lose energy by the direct action of viscous stresses that finally convert it into internal thermal energy on the smallest size eddies. It is the larger eddies that determine the rate at which the mean flow kinetic energy is fed into turbulent motion, and can be passed on to smaller scales and finally dissipated. The larger eddies are thus responsible for the transport of momentum and heat, and hence need to be properly simulated in a turbulence model. Because of the direct interaction with the mean flow, the large-scale motion depends strongly on the boundary conditions of the problem under consideration (Kumar, 1983).

In the k–ε model the turbulent viscosity is assumed to be related to the turbulence kinetic energy (k) and to its dissipation rate (ε) from large eddies to smaller ones:

$$\mu_{\mathrm{T}} = \frac{C_\mu \rho k^2}{\varepsilon} \qquad [11.132]$$

The turbulence kinetic energy is obtained by solving a transport equation of the form:

$$\frac{\partial \rho k}{\partial t} + \frac{\partial}{\partial x_j}(\rho \bar{u}_j k) - \frac{\partial}{\partial x_j}\left[\left(\mu + \frac{\mu_{\mathrm{T}}}{\sigma_{\mathrm{k}}}\right)\frac{\partial k}{\partial x_j}\right] = P + G - \rho\varepsilon \qquad [11.133]$$

where $\underline{P}$ is the shear production term and G is the buoyant production term.

The transport equation for ε is added to close the system:

$$\frac{\partial \rho \varepsilon}{\partial t} + \frac{\partial}{\partial x_j}(\rho \bar{u}_j \varepsilon) - \frac{\partial}{\partial x_j}\left[\left(\mu + \frac{\mu_{\mathrm{T}}}{\sigma_\varepsilon}\right)\frac{\partial \varepsilon}{\partial x_j}\right] = C_1\frac{\varepsilon}{k}(P + C_3 G) - C_2 \rho \frac{\varepsilon^2}{k} \qquad [11.134]$$

where C_1, C_2, C_3 are empirical constants, σ_{k} is the Prandtl number for kinetic energy ($\sim$1.0), σ_ε is the Prandtl for the dissipation rate.

These modeling equations for the transport of k and ε are only valid in the fully turbulent regime, that is, away from any wall damping effects. For wall flows, the boundary conditions are dealt with by the use of wall functions (Anderson *et al.*)

Fire modeling

Fire modeling (the source term) in field models remains rather rudimentary. The user is required to put in the heat release rate as a function of time to specify the fire. Similarly, the increase in fire area with time may also be input by the user in order to model fire spread.

Alternatively, a choice of combustion models is also provided. In these models, an assumption is made that if fuel and oxidant are simultaneously present at the same point, then an instantaneous reaction occurs producing combustion products. The following combustion models are available: Mixed-is-burnt Model, Eddy breakup model. These models are not generally used for fire applications, as they have not been fully validated. In these models, complete combustion is assumed to take place such that

$$1\ \text{kg fuel} + i\ \text{kg oxidant} = (1 + i)\ \text{kg products}$$

Here, combustion models are not discussed in detail. However, further information can be found in various text books and in the CFDS-FLOW3D user guide (CFDS, 1994).

11.6.2 METHOD OF SOLUTION

The nonlinear differential equations give a description of the physical phenomena in their entirety, however complex it might be. This complexity and the lack of understanding of the constituent physical phenomena such as turbulence makes analytical solution almost impossible except in cases of little practical importance. A numerical solution is the only answer.

A numerical solution of a differential equation consists of a set number from which the distribution of the dependent variable ϕ can be constructed (Patankar, 1980). The method starts by dividing the domain of interest into a finite number of grid points, thus replacing the continuous information contained in the exact solution of the differential equation with discrete values. This process of discretization of differential equations (Anderson *et al.*, 1984) enables standard methods to be used for the solution of these equations; by replacing differential equations with simple algebraic equations. Because a discretized equation is derived from the differential equation, it still expresses the same physical information as the differential equation. This means that, among other things, properties are assumed to vary linearly over the space of any given grid cell; the calculated value for that cell being an average. The method assumes the equations to be linear during each iteration cycle by taking certain quantities (coefficients and sources) to be constant. These quantities are "upgraded" after each iteration.

The number of grid cells used may affect the results. However, as the number of grid points becomes very large, the solution of the discretized equations is expected to approach the exact solution of the corresponding differential equation. This does not, of course, mean that going from a coarser grid to a finer grid necessarily produces results that are closer to experimental observations. In principle, it is more desirable to use a finer grid rather than a coarser one. The main reason for using coarser grids seems to be much greater time and expense entailed in using finer grids.

The CFDS-FLOW3D solution method uses finite-volume solution method in which all variables are defined at the center of each control volume (cell), which fills the physical domain being considered. Each equation is integrated over each cell to obtain a discrete equation that connects the variable at the center of the control volume with its neighbors. The discretized equations are solved subject to prescribed boundary conditions.

Boundary conditions

The above equations are solved subject to the appropriate boundary conditions.

Wall boundary condition: The conditions at a solid wall surface may be specified by the user in terms of the velocity or the surface shear stress. The most commonly used condition is the no-slip condition in which the surface velocity, k and ε are assumed to be zero.

Wall functions for turbulent flow: Because of the existence of a surface boundary layer, many of the flow variables change rapidly in the near-wall regions. Extremely fine grids may be required to resolve flow details in this region. Such an approach may take up lot of computer time and memory space. Accordingly, wall functions are usually specified to overcome this problem. One simple example of a wall function is the description of the wall shear stress in terms of the turbulence kinetic energy for a fully developed boundary layer over a stationary wall. This is given by

$$\tau_k = \rho C_\mu^{1/2} k$$

The equation for the turbulence kinetic energy k is solved in the control volume immediately adjacent to the wall. From this, the value of the wall shear stress may be obtained.

The wall boundary conditions for heat transfer and scalars are treated in a similar way. In this case, the enthalpy or scalar (e.g. concentration) in the wall layer is specified. In the case of the heat transfer equation, the wall temperature is specified.

Flow boundaries: A flow boundary is defined as a boundary where fluid can enter or leave the domain. Three types of such boundaries are identified

Inlets: For subsonic flows, as is the case for most of the fire-generated flows, all variables (u, k, ε, T, and scalars) except pressure are specified upstream. The inlet velocities are specified in such a way that the flow enters the domain. The pressure boundary condition is extrapolated from downstream.

Mass flow boundaries: The flow leaving or entering the domain is specified in terms of the total or some fraction of the mass flow rate. Mass flow rates may be specified in two ways:

- As fraction of the total flow rate into the domain through inlets by ensuring that mass conservation is satisfied.
- May be directly specified by the user again, ensuring that the global mass conservation is satisfied.

Mass flow boundaries are not used for compressible transient flows, as they do not satisfy the global mass conservation on each time step.

Pressure boundaries: Pressure is specified together with the condition that the velocity gradient normal to the flow direction is zero.

At inflow, the entrance velocity is restricted to subsonic. The user specifies temperature and additional scalars. k and ε are extrapolated from downstream.

At outflow, temperature, additional scalars, k and ε are all extrapolated from upstream.

Planes of symmetry: The boundary conditions at the symmetry planes are such that all variables are mathematically symmetric with thus no diffusion across the boundary. However, this requires that the component of velocity normal to the boundary together with the Reynolds shear stress and Reynolds flux be antisymmetric.

In conclusion, the purpose of the above discussion is to introduce the concepts of field modeling. The intent, clearly, has not been to present a detailed exposition but rather to outline the rationale and the basic theoretical model. For practical use, detailed information can be found in the appropriate references and user manuals accompanying the software packages.

11.7 Evacuation modeling

Having calculated the building environment in terms of temperatures and combustion product concentrations (toxicity and visibility etc.) the next step in the evaluation of fire safety is to estimate the evacuation time for a given building and occupancy. Building layout, in particular the escape routes, and the occupant behavior together with their ability to move are some of the important factors that may influence evacuation time. Also, at the building design stage, this information is essential for the design and location of escape routes as well as the requirements protection and detection measures. According to Marchant (1976), the escape time in an emergency situation has three main components:

$$t_{\text{escape}} = t_{\text{p}} + t_{\text{a}} + t_{\text{rs}} \qquad [11.135]$$

Where t_{p} is time from ignition to perception of fire, (perception time), t_{a} is time from perception to the start of escape action, (action time) and t_{rs} is time taken to move to a place of safety (i.e. travel time)

The *perception time* and *action time* largely depend on a person's state of awareness and the alarm system characteristics and performance. For example, early detection can reduce perception time considerably. Trained personnel, accustomed to regular drills, may be expected to react promptly to an alarm, thereby reducing the action time.

Travel time, t_{rs}, on the other hand, is influenced by a number of important factors such as the following:

- Number of people – travel time increases with the increase in the number of people;
- Speed of movement – the higher the travel speed, the lower the travel time;
- Crowd density – higher the density, the higher the travel time;
- Widths of doors and escape routes – wider doors decrease the travel time;
- Distance to a place of safety – travel time increases with distance;
- Overall geometry – large complex buildings often increase the travel time;
- Familiarity with the building layout – unfamiliarity increases travel time;
- Hazard progression – faster spread of fire or smoke may cause people to seek other routes resulting in increased travel time.

Primary aspects of crowd movement

The speed and flow characteristics of crowd movement have been analyzed by a number of different researchers over the past 40 years. The primary speed and flow characteristics of crowd movement are illustrated in Figure 11.13 and 11.14. These graphs are derived from data presented by Fruin, Hankin and Wright, Ando and Predtechenskii and Milinskii.

The flow rate figures used in the exit "arcs" of most network-node evacuation models are derived from a "maximum sustained flow rate." The maximum flow rate is normally observed in crowd densities of greater than 2 persons/m^2, but values differ greatly between different researchers and between different cultural and psychological differences in the test groups. Most building regulations adopt a figure of between 1.25 and 1.4 persons/m (exit width)/s as a "safely sustainable flow rate." The assumption that exit flow rate is directly proportional to exit width seems to hold true for door widths greater than 1 m, but flow rates can become erratic and disproportionately congested for narrow doorways (for which there is little data available).

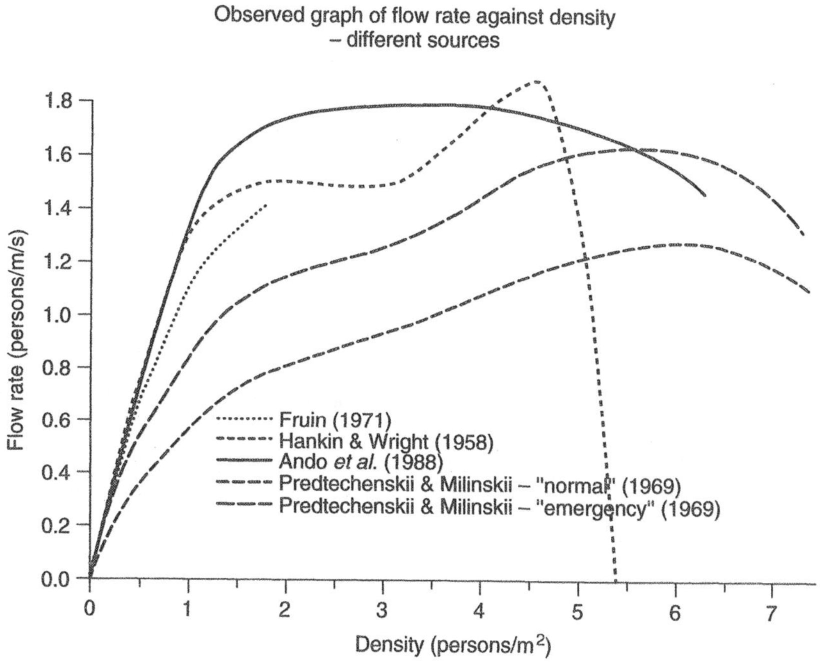

Figure 11.13. People flow rate against density

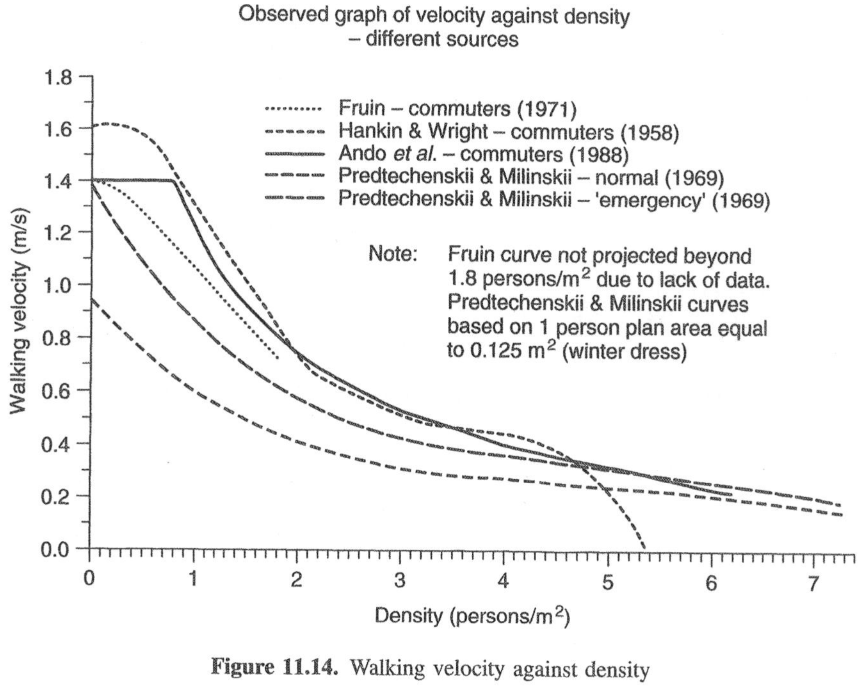

Figure 11.14. Walking velocity against density

The graph illustrated in Figure 11.15 is derived from data presented by Thompson (1994), which was obtained by analyzing digitized video footage of the movement of fairly calm people through different building areas in Edinburgh. When both people walk in similar directions, the interperson distance to the person in front affects the velocity of the "obstructed" person. The term "Interperson Distance" is defined as the linear distance between the centerpoints of two pedestrians (plan-view) where one pedestrian is walking directly within the forward path of another pedestrian, and hence becomes a potential obstruction. This graph was obtained specifically for use in Simulex (Thompson and Marchant, 1996).

The building population

The building population may be defined one person at a time by the user, or as large groups, spread over specified areas of the building. Single people or groups of people with associated population density or finite number of occupants can be inserted into the building space.

The physical characteristics that are assigned to the defined people, are governed by options, for example, the SIMULEX model given in Tables 11.2 and 11.3. The physical characteristics are defined by using "occupant type" categories. Each occupant type is made up of percentages of

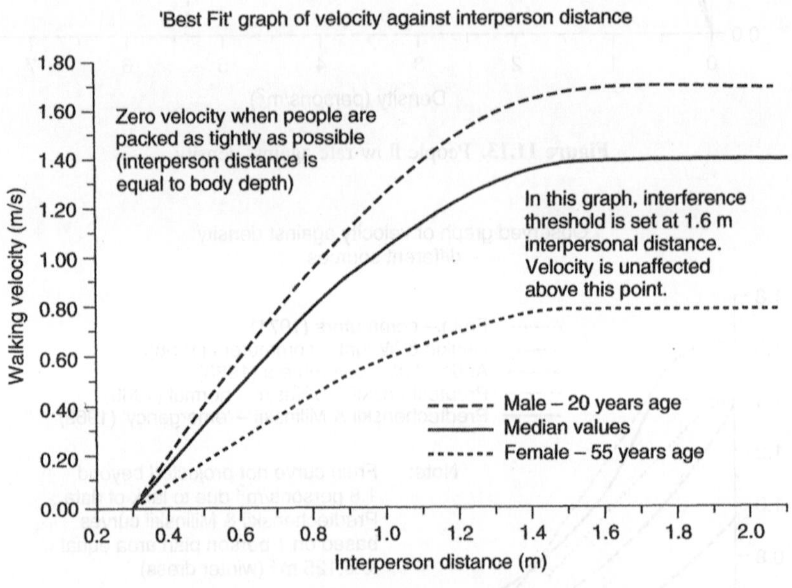

Figure 11.15. Walking velocity against interperson distance (Thompson, 1994)

Table 11.2. Defining different types of individual person

Defined individual	Body breadth (m)	Body depth (m)	Normal walking speed 'V' (m/s)	+/– limits for 'V' distribution (m/s)
"Average"	0.50	0.30	1.3	0.0
Adult male	0.54	0.32	1.35	0.2
Adult female	0.48	0.28	1.15	0.2
Child	0.42	0.24	0.9	0.3
Elderly	0.50	0.30	0.8	0.3

Table 11.3. Specification for occupant types

Occupant type	Percentage of defined individuals
Office staff	60% Adult male, 40% adult female
Commuters	50% Adult male, 40% adult female, 10% children
Shoppers	35% Adult male, 40% adult female, 15% children, 10% elderly
School population	3% Adult male, 7% adult female, 90% children
All male	100% Adult male
All female	100% Adult female
All children	100% children
All elderly	100% elderly

different types of predefined individual. The different "physical" types of individual are described in Table 11.2 (controlling body size and unimpeded walking speed), and the distribution of the different individual types within occupant categories are described in Table 11.3. These tables are based on data from Predtechenskii and Milinski and Fruin.

A variety of computer models attempt to calculate the evacuation time from buildings. Some are confined to the calculation of travel time (EVACNET+, EGRESS) while others also try to take into account behavioral characteristics (EXITT), albeit in a rudimentary fashion. However, no single model has the capability to deal with evacuation in a comprehensive and rigorous manner. All models have strengths and weaknesses and are valid only for very limited situations.

Most evacuation models use the network approach in which a building is represented by nodes (or rooms) linked by escape paths. People move from node to node at predetermined speeds along these paths. The evacuation time is determined by the travel speed and the distance between nodes. Some models (e.g. EXITT) attempt to account for behavioral responses by allowing for delays and response times. Such models do not attempt to describe the building in detail.

Another approach generally referred to as *cellular automata* is a considerable improvement on simple network technique. In this approach, the floor plan is divided into a grid on which people are allowed to move according to randomly selected weighted functions that define their speed and direction of travel. In this way, people interactions are modeled by appropriately modifying the weighted functions.

In this chapter, we summarize the underlying features of both these approaches: EXITT (Bukowski *et al.*, 1987) representing the network approach and EGRESS (Ketchell) the cellular automata approach.

11.7.1 EVACUATION MODEL – EXITT

EXITT (Levin, 1987), is a deterministic evacuation model designed to simulate occupant decisions and actions in fire situations. This code forms part of the HAZARD I package (Bukowski *et al.*) and is designed to interact with the smoke movement model FAST (Jones) to take account of escape route blockage or change of walking speed due to smoke density or toxicity. The movement model uses a nodal representation of the building and a tree-searching algorithm to determine the shortest path to target node (i.e. exit) for each person.

The building is represented as a network of nodes (rooms) and escape routes (links) between the nodes. The occupants are allowed to move from one node to another at a travel speed that is a function of their assigned travel speed and smoke density. This is further modified by considering whether or not they are assisting other occupants. The decision rules, which are based on experimental case studies, are used as the basis for path selection. The occupant movements and decisions can be displayed on the screen or saved in file to be printed later.

The model can be used for up to 12 rooms and a total of 35 nodes. It cannot be used for large buildings with many people. The model is deterministic and relies heavily on some fairly arbitrary figures for delays and functions for psychological impact. It has only been validated for a limited number of cases and fire scenarios. The time taken to reach the target node is independent of other people and therefore cannot account for bottlenecking and crowding.

11.7.2 EVACUATION MODEL – EGRESS

The EGRESS evacuation model (Ketchell *et al.*, 1993 and Webber *et al.*, 1993) is designed to be an integrated hazard assessment tool for evaluating hazards to people resulting from growing fires. The input to the model requires a description of the building or a structure as well as the way the hazard (fire or smoke) escalates within the building. The building design and layout of escape routes influence people movement as well as determining how the hazard spreads.

In general, as seen above, the simplest calculation of the time required for travel to a place of safety is based on the travel distance and the expected speed of movement (network approach). The procedure is refined by taking into account movement restrictions (such as doorways) and other delays.

In EGRESS, however, this approach has been improved considerably. The building is described in the form of two-dimensional floor plans divided into a hexagonal grid that, in principle, provides six degrees of freedom for movement. People are modeled moving from one cell to another, once their attributes and specified locations on the grid have been assigned. This way, the method allows the progress to be traced as an individual moves toward a set goal, that is, a place of safety outside the building. Different groups of people can be assigned different goals.

The movement of each individual (or "Automaton") located on a hexagonal grid is based on the weighted probabilities between each hexagonal cell (each cell having the projected area of a person). These probabilities represent the decision-making attributes and responses of people to the situation. The decision of which cell to move to is based on these probabilities of moving to six adjacent cells and a probability of remaining in the same cell (i.e. no movement). The relative weights of these probabilities determine the mean speed of motion in terms of cells traversed. The probabilities are calibrated against experimental and real-life data. In certain cases, the model can vary the probabilities for cells to reflect changes in the event as it progresses. For example, regions of the structure can become blocked by smoke or radiation heat as a fire progresses. These blockages result in the "people" having to find alternative escape routes, which is represented by the variation in the cell probabilities. A person's movement to adjacent cells is restricted to cells that are not part of the structure or occupied by another person or in the case of fire, for example, blocked by hazardous conditions (fire or smoke).

This type of movement model is classified as *cellular automaton*. The methodology has the ability to model a number of key issues relating to evacuation:

1. Persons with impaired mobility can be modeled by defining different groups of people having different movement speeds;
2. Assigning different goals for different groups allows certain people to move against the evacuation flow, for example, fire brigade personnel may wish to head toward the fire;
3. In principle, there is no upper limit to how many people (automata) can be modeled.

The model provides a flexible and robust method for calculating movement time. The method allows the times to fire hazard, determined prior to evacuation modeling, to be taken into account as the evacuation progresses. It enables the effects of "bottle necks" to be highlighted. As the code is PC-based and faster to run, it provides a useful way of exploring different escape route strategies at the building design stage. The code has been considerably validated for large populations and a range of complex buildings (see Chapter 12).

11.7.3 EVACUATION MODEL – SIMULEX

Simulex is a PC-based evacuation model, designed to model large, geometrically complex buildings, with multiple floors and staircases (Thompson and Marchant, 1996). Large building populations can be accommodated, and the user can view the movement of each individual in the building at any time during the evacuation. A text file that contains detailed information about the evacuation process is produced at the end of a simulation.

The movement algorithms in Simulex are based on a combination of data obtained from video analyses of individual movement (Thompson, 1994) and the crowd movement data collected by Fruin (1971), Hankin and Wright (1958), Ando *et al.* (1988), and Predtechenskii and Milinskii (1978). The progression of each person through the building space is modeled in time steps

Figure 11.16. Distance map for a simple two-story building

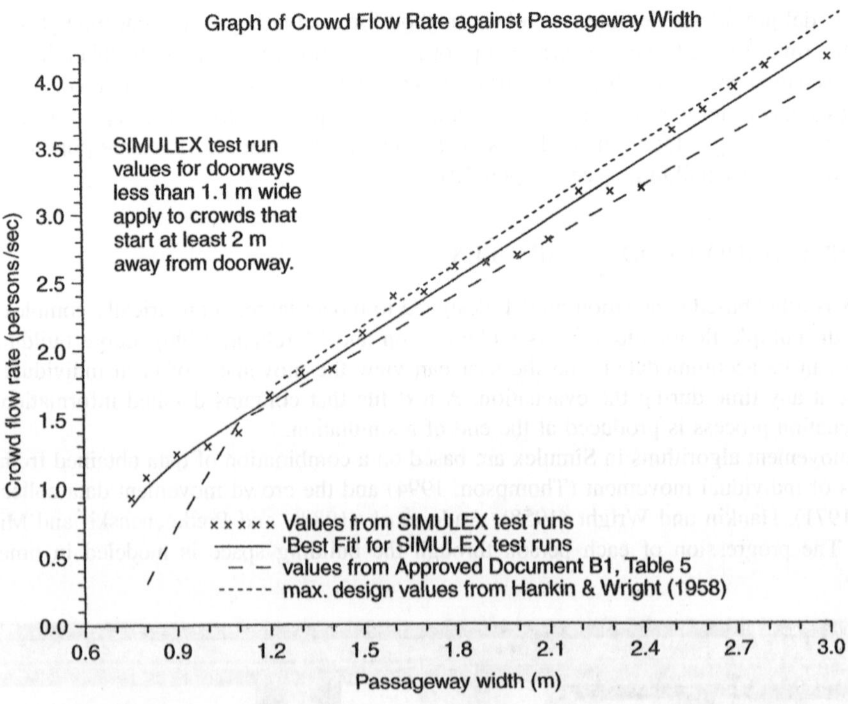

Figure 11.17. Comparison of modelled and 'real-life' exit flow rates

of 0.1s. Positions and distances are calculated in meters. Types of movement that are modeled include normal unimpeded walking, the reduction of walking speeds due to the proximity of other individuals, overtaking, body "twist", and sidestepping.

Each person assesses his own direction-to-exit by the use of "distance maps," which map out the distance to an exit from any point in the building. Different distance maps are stored in memory, describing routes to different final exits. Simulex models a number of psychological aspects including choice-to-exit and response time to an alarm. Further improvement of these psychological factors is part of the ongoing development of the model.

The output is known as a "distance map," which is illustrated in Figure 11.16. Shaded bands represent the values of distance-to-nearest-exit where the shading is graduated to represent 1 m travel contours. Distance maps are very useful when assessing travel distances in a building, and are also used by the route-finding functions during a simulated evacuation.

The linear relationship between flow rate and exit width was observed in the SIMULEX test runs for widths greater than or equal to 1.1 m, at the average sustainable rate of 1.40 persons/meter width/sec (Figure 11.17). This value compares well with data presented by Hankin & Wright (1958) and The Building Regulations (1991), and may be regarded as a reasonably realistic value.

At the time of writing this book, large-scale tests of Simulex (modeling and comparison to full-scale evacuations) are being carried out at the University of Belfast. The evacuation of the building populations from large department stores is being simulated and analyzed, with particular emphasis on different exit choices and walking speeds. The results of these tests should be available soon.

Nomenclature

A	area of gap (m^2); area of flame cone (m^2); area of extended ceiling (m^2); [11.8; 11.46; 11.60]
A_f	fuel surface area (m^2) [11.1]
A_L	bounding area of the hot layer (m^2) [11.54]
A_o	area of duct inlet (m^2) [11.87]
A_V	area of vent covered by the hot layer (m^2) [11.55]
A_w	area of vertical ventilation opening (m^2) [11.3]
A_{WS}	area of wall surface (m^2) [11.107]
b	plume radius (m); body depth (m); [11.44]
$b(x)$	plume width at height x (m) [11.41]
b_V^*	vent width immersed in layer (m) [11.60]
B	width of strip at a vent (m); = K.S; [11.69; 11.96]
c_p	specific heat at constant pressure for air (kJ/(kg K)) [11.6]
c_v	specific heat at constant volume for air (kJ/(kg K)) [11.16]
c_W	specific heat of wall material (kJ/(kg K))
C_d	discharge coefficient; minimum contact distance (m); [11.8]
C_s	smoke aerosol concentration (g/m^3)
d	hydraulic diameter of gap (m) [11.9]
D	height between top of fuel surface and bottom of hot layer (m); optical density of smoke (m^{-1}) [11.93]
e_i	effective emissivity of the hot layer
$\dot{E}$	total emission rate from flame (W)
$\dot{E}_a$	energy absorbed by layer (W)
$\dot{E}_L$	rate of change of hot-layer energy (kW) [11.24]
$E_{L:}$	energy of the hot layer (kJ) [11.25]
$\dot{E}_{LR}$	layer energy change from hot surfaces (kW) [11.25]
$\dot{E}_{LW}$	convective heat loss rate from layer to wall (kW) [11.25]
$\dot{E}_{PR}$	total radiative heat loss from flame (kW) [11.46]
$\dot{E}_V$	enthalpy flow out through vents (kW) [11.25]
$\dot{E}_p$	enthalpy flow into hot layer from plume (kW) [11.25]
$\dot{E}_{12}$	power absorbed by flame cone 1 from radiation from flame cone 2 (W)
E_2	nth exponential integral for flame
f	projected area of person (m^2)
F_o	flux of buoyancy from fire source (m^4/s^3) [11.6]
FED	fractional effective dose [11.101]
g	acceleration due to gravity (m/s^2) [11.5]
G	density stratification parameter (s^{-2}) [11.5]
G_i	parameter defined in Equation [11.71]
G_1	flow path conductance $(kgm)^{1/2}$ [11.87]
h	layer depth (m); distance between the opposite sides of a hexagon [11.7]
h_e	heat transfer coefficient by convection to external environment ($W/(m^2K)$) [11.59]
h_i	heat transfer coefficient by convection to extended ceiling ($W/(m^2K)$) [11.57]
h_k	wall conductance (kW/m^2K)
h_L	depth of the hot layer (m) [11.23]
$\dot{h}_l$	enthalpy contribution to the lower layer (kW) [11.16]

$\dot{h}_L$	rate of change of hot-layer depth (m/s); enthalpy contribution to the upper layer (kW) [11.24; 11.17]
h_{max}	maximum value of h_i; [11.58]
h_p	layer height above fuel bed (m) [11.44]
$h_{s'}$	minimum value of h_i; [11.58]
h_w	convective heat transfer coefficient (kW/(m^{2}K))
H	($H_p - R_f$); height of vertical ventilation opening (m); height of layer interface above the flame centroid; height of conical flame [11.3]
H_B	room height (m)
H_p	height of plume (m) [11.75]
H_t	distance between top of vent and ceiling (m) [11.79]
H_v	specific heat of pyrolysis (J/kg); vent height (m) [11.33; 11.80]
ΔH_c	heat of combustion of the volatiles (kJ/g) [11.1]
$\Delta H_{c,air}$	heat of combustion in terms of air consumed (kJ/g) [11.2]
I	intensity of light in the presence of smoke
I_0	intensity of light in the absence of smoke
k	absorption coefficient (m^{-1}); thermal conductivity (kW/(m K)) [11.64]
k_W	thermal conductivity of wall (kW/(m K))
K	extinction coefficient (m^{-1}); a constant [11.94; 11.3]
K_m	specific extinction coefficient [11.97]
L	length of enclosure (m); lateral gap length (m); optical path length (m), wall thickness (m) [11.9,11.93]
m	mass (g) [11.14]
$\dot{m}''$	mass loss rate per unit surface area (g/(m^{2}s)) [11.1]
$\dot{m}'_a$	mass of air entrained by part of plume covered by layer (kg/s)
$\dot{m}_{air}$	air entrainment rate into part of plume below the hot layer (kg/s); rate of air flow into the enclosure (kg/s) [11.28; 11.2]
$\dot{m}_b$	fuel mass burning rate (kg/s) [11.3]
$\dot{m}_{CO}$	rate of change of CO mass (kg/s) [11.90]
$\dot{m}_{CO_2}$	rate of change of CO_2 mass (kg/s) [11.91]
$\dot{m}_e$	mass rate of flow of hot gases through a vent (kg/s); mass rate of air entrained in plume (kg/s) [11.72; 11.50]
m_f	remaining mass of fuel (kg) [11.30]
$\dot{m}_f$	fuel pyrolysis rate (kg/s) [11.26]
$\dot{m}_g$	mass flow rate of hot gases to the outside (kg/s)
$\dot{m}_i$	mass flow rate through a vent at strip i (kg/s) [11.70]
m_l, m_L	mass of the cool layer (lower layer), hot layer (upper layer) respectively (kg) [11.15, 11.14]
$\dot{m}_l$	mass flow rate from lower layer through vent (kg/s) [11.78]
$\dot{m}_{l,i}$	mass flow rate due to the ith source in the lower layer (kg/s) [11.15]
$\dot{m}_{L,i}$	mass flow rate due to the ith source in the upper layer (kg/s) [11.14]
m_o	initial mass of fuel (kg)
$\dot{m}_{ox}$	rate of change of oxygen mass in the hot layer [11.88]
$\dot{m}_p$	mass flow rate of combustion products into hot layer (kg/s); mass flow rate in plume (kg/s) [11.44; 11.47]
$\dot{m}_s$	rate of change of smoke mass in layer (kg/s) [11.92]
$\dot{m}_u$	mass flow rate from upper layer through vent (kg/s) [11.80]
$\dot{m}_\beta$	pyrolysis rate $= \chi \dot{m}_f$ (kg/s); $= -\pi R_f^2 \phi / H_v$ (kg/s) [11.32]
$\dot{m}_{13}$	mass flux from room 1 to room 3

M	number of automata (population)
N	number of attempted moves
N_P	dimensionless number defined in eq [11.10]
N_Q	dimensionless number defined in eq [11.9]
p_f	pressure at floor level in a room (Pa)
p_o	pressure at floor level in a room (Pa) [11.66]
p_1	pressure at vent inside fire compartment (Pa) [11.67]
p_2	pressure at vent outside fire compartment (Pa) [11.67]
$p_{1/2}(y)$	pressure difference at height y in vent connecting rooms 1 and 2 (Pa) [11.66]
Δp	pressure difference (Pa) [11.8]
$\Delta p(y)$	pressure difference at position y across a vent (Pa) [11.67]
P	pressure at a point (Pa) [11.16]
P_i	probability of movement to an adjacent cell i;
$P(1)$	probability of moving one step toward the exit
$P(-1)$	probability of moving one step away from the exit
$P(0)$	probability of not moving
$\dot{q}_{LR}$	total heat gain of layer by radiation (W) [11.55]
$\dot{q}''$	heat flux per unit area (W/m^2) [11.35]
$\dot{q}''_N$	heat flux to outside wall (W/m^2)
$\dot{q}''_1$	heat flux to inside wall (W/m^2)
$\dot{q}''_{12}$	radiant flux at base of burning object from another flame (W/m^2)
$\dot{q}''_{AW}$	heat loss from surface of ceiling to environment (W/m^2) [11.59]
$\dot{q}''_{FO}$	radiant heat flux from flame to fuel surface (W/m^2)
$\dot{q}''_{LW}$	convective flux from layer to extended ceiling (W/m^2) [11.57]
$\dot{q}''_{LWD}$	convective flux from layer to ceiling/wall (W/m^2) after
$\dot{q}''_{LWR}$	radiative flux from layer to ceiling/wall (W/m^2) [11.51]
$\dot{q}''_{PWR}$	radiative flux from plumes to ceiling/wall (W/m^2) after
$(\dot{q}'')_{FI}$	radiant heat flux from flames to object (W/m^2)
$(\dot{q}'')_{FO}$	radiant flux from flame cone to object (W/m^2)
$(\dot{q}'')_{LF}$	radiant heat flux from hot layer to object (W/m^2)
$(\dot{q}'')_{PF}$	radiant heat flux from flames to object (W/m^2)
$(\dot{q}'')_{RR}$	reradiation from object (W/m^2)
$(\dot{q}'')_{WF}$	radiant heat flux from walls and ceiling to object (W/m^2)
Q	volume flow rate (m^3/s) [11.8]
$\dot{Q}_B$	rate of change of heat stored in the gas volume (kW) [11.102]
$\dot{Q}_c$	convective heat release rate (kW); rate of heat loss by convection (kW) [11.6]
$\dot{Q}_f$	rate of heat release from a fire (kW) [11.1]
$\dot{Q}_l$	sensible heat net input rate into the lower layer (kW) [11.16]
$\dot{Q}_L$	sensible heat net input rate into the upper layer (kW) [11.17]
$\dot{Q}_R$	rate of heat loss by radiation through vents (kW) [11.102]
$\dot{Q}_W$	rate of heat loss by conduction (kW)
r	equivalent ceiling radius = $\sqrt{(WL/\pi)}$ (m)
R	universal gas constant (J/kg.mol); equivalent radius of room (m) [11.19]
R_f	fire radius (m) [11.43]
$\dot{R}_f$	rate of change of fire radius (m/s)
R_i	Richardson number [11.7]
R_i	effective radius of ceiling quadrant (m)
R_{max}	maximum fire radius (m) [11.34]

R_0	initial fire radius (m) [11.36]
R_2	radius of fire cone 2 (m)
S	visibility distance (m); psychological impact of smoke [11.96]
t	time (s)
t_a	time from perception to the start of escape action (action time) (s) [11.135]
t_d	threshold distance (m)
t_{escape}	escape time [11.135]
t_i	time from ignition (s)
t_{ig}	ignition time (s) [11.34]
t_p	time from ignition to perception of fire (perception time) (s); thermal penetration time (s) [11.110; 11.135]
t_{rs}	time taken to move to a place of safety (travel time) (s) [11.135]
Δt	time step (s)
T	absolute temperature (K)
T_e	environment temperature (K)
T_{ig}	ignition temperature (K) [11.34]
T_g	temperature of hot gases (K) Figure 11.11
T_l	lower layer temperature (K) [11.16]
T_L	mean hot-layer temperature (K); upper layer temperature (K) [11.55; 11.17]
T_N	outside wall temperature (K)
$T_N(t)$	back surface temperature at time t (K)
T_p	pyrolysis temperature (K)
T_R	enthalpy reference temperature (K) text below
T_s	temperature of object surface (K) [11.39]
T_W	wall temperature (K); extended ceiling temperature (K) [11.55]
T_∞, T_a	ambient temperature (K); [11.103; 11.23]
$T_{\infty 1}$	ambient temperature at level of fire source (K) [11.6]
ΔT	$= (T - T_\infty)$ (K) [11.109]
u	plume centerline velocity at layer height above the fuel (m/s) [11.44]
u_f	fuel velocity (m/s) [11.44]
$u(x)$	velocity at distance x from source (m/s) [11.42]
U	velocity (m/s) [11.7]
v	impeded walking speed
V	volume of enclosure (m^3) [11.20]
$\dot{V}$	volume flow rate (m^3/s) [11.78]
V_f	volume of flame cone (m^3) [11.46]
V_L, V_l	volume of upper, lower layer respectively (m^3) [11.20]
V_u	unimpeded walking speed
W	width of enclosure (m); width of opening (m) [11.23; 11.116]
x	longitudinal gap length in flow direction (m); distance along plume axis (m); length of differential cone originating from the flame centroid (m) [11.9; 11.41]
x_o	position at which turbulence is measured
x_s	virtual source distance below the fire (m); [11.43]
Y_{CO}	yield of carbon monoxide (kg) [11.90]
$Y_{CO}^{(L)}$	mass fraction of carbon monoxide in the hot layer [11.90]
Y_{CO_2}	yield of carbon dioxide (kg) [11.91]
$Y_{CO_2}^{(L)}$	mass fraction of carbon dioxide in the hot layer [11.91]

Y_{O_2}	mass fraction of oxygen in the hot layer [11.88]
Y_s	yield of smoke (kg) [11.92]
$Y_s^{(L)}$	mass fraction of smoke in the hot layer [11.92]
z	vertical height above a fire source (m) [11.5]
z_{max}	maximum plume rise above a fire source (m) [11.4]
Z	plume height (m); distance from vertical source to hot-layer interface; [11.47; 11.76]
Z_o	fictitious source position [11.76] Greek symbols

Greek symbols

α	entrainment coefficient; thermal diffusivity (m^2/s) [11.43; 11.63]
χ	combustion efficiency
δ	wall thickness (m) [11.105]
δ_f	flame thickness (m) [11.120]
ε	emittance of the hot layer [11.51]
ε_{eff}	effective emissivity of fire gases and wall surface [11.118]
ε_g	emissivity of hot gases [11.118]
ε_{gb}	emissivity from band radiator [11.119]
ε_{gs}	emissivity from soot [11.119]
ϕ	radiative heat flux (W/m^2) [11.32]
ϕ_{LO}	flux from four ceiling quadrants (W/m^2)
ϕ_{WO}	radiation from hot and cold walls (W/m^2)
γ	air/fuel mass ratio in free burn [11.28]
γ_s	stoichiometric ratio air/fuel ratio [11.88]
κ	absorption coefficient (m^{-1}) [11.40]
Λ	molecular diffusion coefficient
ν	kinematic viscosity (m^2/s); pressure (Pa) [11.9; 11.121a]
$\overline{\nu}$	unimpeded mean travel speed (m/s) [11.121a]
θ	$\min(h_L/2, 1/\kappa)$
θ_o	mean acceptance angle for flame (by the hot layer) $= \tan^{-1} r/(H + \theta_1)$; $R/(H + h_L/2)$
θ_1	$\min(H, \theta)$
ρ	gas density (kg/m^3) [11.7]
$\Delta\rho$	density difference (kg/m^3) [11.7]
ρ_∞, ρ_a	ambient density (kg/m^3) [11.5; 11.23]
$\rho_{\infty l}$	ambient density at level of fire source (kg/m^3) [11.5]
ρ_l, ρ_L	density of lower and upper layer respectively (kg/m^3)
ρ_{in}	density of incoming gas (kg/m^3) [11.82]
σ	Stefan–Boltzman constant (5.67×10^{-8} $W/(m^2K^4)$)
σ_ν	standard deviation in the direction of travel
τ	mean optical depth of flame; effective mean opacity of flame cone
ω	view factor of extended ceiling
ξ	$= (h_L - H_t)$ (m) [11.81]
ψ	semiapex angle (Figure 11.7) [11.26]
ψ_0	initial value of ψ [11.26; 11.130a]
ψ_2	semiapex angle of fire cone 2
ζ	mean opacity of the hot layer [11.52]
Λ	view factor

Φ	Instantaneous value of a quantity
$\overline{\Phi}$	Mean value of Φ [11.122]
Φ'	Fluctuating component of Φ
n	The number of sampling point variable [11.123]
N	Total number of points sampled, number of species [11.123; 11.129]
u_i	The value u in the i direction ($i = 1$, 2 or 3) [11.125]
x_i	Spatial distance in the i direction ($i = 1$, 2, or 3) [11.125]
u_j	The value u in the j direction ($j = 1$, 2 or 3) [11.125]
x_j	Spatial distance in the j direction ($j = 1$, 2, or 3) [11.125]
μ_{eff}	effective viscosity [11.130b]
μ	fluid viscosity [11.130b]
μ_{T}	Turbulent viscosity [11.130b]
$\overline{H}$	Mean enthalpy [11.126]
Λ_{H}	Molecular diffusion coefficient
Λ_α	Molecular diffusion coefficient of α [11.127]
$\overline{p}$	Mean value of pressure [11.126]
m_α	mass fraction of species α [11.127]
S_α	Volumetric source/sink term representing generation rate of α per unit volume [11.127]
W_α	Molecular weight of species α [11.128]
R	Universal gas constant [11.128]
T	Absolute temperature [11.128]
σ_{H}	Turbulent Prandtl number for enthalpy [11.130a]
σ_α	Turbulent Schmidt number for species α
σ_{k}	Prandtl number for turbulent kinetic energy [11.133]
σ_ε	Prandtl number for the dissipation rate [11.134]
σ_{v}	Standard deviation in the direction of travel
σ_{t}	Standard deviation traverse to the direction of travel
k	Turbulent kinetic energy [11.132]
ε	Dissipation rate for turbulent kinetic energy (k) [11.132]
C_μ	A constant used in the k-ε turbulence model [11.132]
C_1, C_2, C_3	Empirical constant used in the k-ε turbulence model [11.134]
P	Shear production term [11.133]
G	Buoyancy production term [11.133]

References

Alvares, N J, Foote, K L and Pagni, P J (1984). Forced ventilated enclosure fires. *Combustion Science and Technology*, **39**, 55–81.

Anderson, D A, Tannehill, J C and Pletcher, R H (1984). *Computational Fluid Mechanics and Heat Transfer*, Hemisphere Publishing Corporation.

Ando, K, Ota, H and Oki, T (1988). Forecasting the flow of people, (in Japanese). *Railway Research Review*, **45**(8), 8–14.

Atkinson, G T and Drysdale, D D (1992). Convective heat transfer from fires gases. *Fire Safety Journal*, **19**, 217–245.

Boccio, J L and Usher, J L (1985), *The Use of a Field Model to Assess Fire Behavior in Complex Nuclear Power Plant Enclosures: Present Capabilities and Future Prospects*, NUREG/CR-4479.

Bodart, X E and Curtat, M R (1989). Prediction of entrained mass flows through vertical openings in a multi-room fire model. *Fire Safety Science – Proceedings of the 2nd International Symposium*, Hemisphere, pp. 917–926.

Bradshaw, P (1985). *An Introduction to Turbulence and its Measurement*, Pergamon Press.

Brani, D M and Black, W Z (1992). Two-zone model for a single-room fire. *Fire Safety Journal*, **19**, 189–216.

Bukowski, R W, Jones, W W, Levin, B M, Forney, C L, Steifel, S W, Babrauskas, V, Braun, E and Fowell, A J (1987). *HAZARD I – Volume I: Fire Hazard Assessment Method*, National Institute of Standards and Technology, Center for Fire Research, NBSIR 87–3602.

Carslaw, H S and Jaeger, J C (1980). *Conduction of Heat in Solids*, 2nd Edition, Oxford University Press. CFDS, AEA Technology, Harwell, CFDS-FLOW3D User Guide, Release 3.3, June 1994.

Collier, J G (1981). *Convective Boiling and Condensation*, 2nd Edition, McGraw-Hill.

Cooper, L Y (1984a). On the significance of wall effect in enclosures with growing fires. *Combustion Science and Technology*, **40**, 19–39.

Cooper, L Y (1988). *Calculating Flows Through Vertical Vents in Zone Fire Models Under Conditions of Arbitrary Cross-Vent Pressure Difference*, NBSIR 88–3732, May 1988.

Cooper, L Y (1989a). *Calculation of the Flow Through a Horizontal Ceiling/Floor Vent*, NISTIR 89–4052, March 1989.

Cox, G, Kumar, S and Markatos, N C (1986). Some field model validation studies. *Fire Safety Science, Proceedings of the 1st International Symposium*, Hemisphere, pp. 159–172.

Drysdale, D D (1985). *An Introduction to Fire Dynamics*, Wiley.

Emmons, W. H (1978). The prediction of fires in buildings. *17th Symp. on Combustion*, University of Leeds, August 20–25, 1978, pp. 1101–1111.

Emmons, H W (1987). *The Flow of Gases Through Vents*, Home Fire Project Technical Report 75, Division of Applied Sciences, Harvard University, Cambridge, MA.

Epstein, M (1988). Buoyancy driven exchange flow through small openings in horizontal partitions. *Journal of Heat Transfer*, **110**, 885–893.

Friedlander, S K (1977). *Smoke, Dust and Haze*, John Wiley & Sons.

Fruin, J and Riedlander, S K (1971). Designing for pedestrians: a level-of-service concept. *Highway Research Record*, **355**, 1–15.

Galea, E (1988). Smoking out the secrets of fire. *New Scientist*, 44–48.

Gross, D and Haberman, W L (1989). Analysis and prediction of air leakage through door assemblies. *Fire Safety Science, Proceedings of the 2nd Int. Symp.*, Hemisphere, pp. 169–178.

Habchi, S D, Przekwas, A J and Singal, A K (1988). *A Computer Code for Fire Protection and Risk Analysis of Nuclear Plants*, Report CFDRC 4025/1, CFD Research Corporation, prepared under contract NRC-04-87-374 for the US Nuclear Regulatory Commission.

Hamer, M (1988). The night that luck ran out. *New Scientist*, 29–31.

Hankin, B D and Wright, R A (1958). Passenger flows in subways. *Operational Research Quarterly*, **9**, 81–88.

Heskestad, G (1989). Note on maximum rise of fire plumes in temperature-stratified ambients. *Fire Safety Journal*, **15**, 271–276.

Hinze, J O (1975). *Turbulence*, McGraw-Hill.

Huhtanen, H (1989). Numerical fire modelling of a turbine hall. *Fire Safety Science, Proceedings of the 2nd International Symposium*, Hemisphere, pp. 771–779.

Jaluria, Y (1988). Natural convection wall flows. *SFPE Handbook of Fire Protection Engineering*, National Fire Protection Association, Boston, pp. 1-116–1-129.

Jin, T (1971). Visibility through fire smoke. *Report of Fire Research Institute of Japan*, **2**(33), 31–48; **5**(42), 12–18, 1975.

Jones, W W (1985). A multi-compartment model for the spread of fire, smoke and toxic gases. *Fire Safety Journal*, **9**, 55.

Jones, W W (1990). *Refinement of a Model for Fire Growth and Smoke Transport*, NIST Technical Note 1282.

Jones, W W and Peacock, R D Refinement and experimental verification of a model for fire growth and smoke transport. *Fire Safety Science, Proc. of the 2nd Int. Symp.*, pp. 897–906.

Jones, W W and Quintiere, J (1984). Prediction of corridor smoke filling by zone models. *Combustion Science and Technology*, **35**, 239–253.

Kandola, B S (1979). *Comparitive Studies of Some Aerodynamic Aspects of Smoke Flow in Buildings*, Ph.D. Thesis, Edinburgh University, Edinburgh.

Kandola, B S (1990). Effects of atmospheric wind on flows through natural convection roof vents. *Fire Technology*, **26**(2).

Kanury, A M (1987). On the craft of modelling in engineering and science. *Fire Safety Journal*, **12**(1).

Kawagoe, K (1958). *Fire Behaviour in Rooms*, BRI Report No. 27.

Kawagoe, K and Sekine, T (1958). *Estimation of Fire Temperature-time Curve in Rooms*, Report No. 27, Building Research Institute, Ministry of Construction, Tokyo, Japan.

Ketchell, N, Hiorns, N, Webber, D M and Marriott, C A (1993). When and How will people muster? A simulation approach. *Response to Offshore Incidents' Conference Proceedings*, Marriott Hotel, Aberdeen, June 1993.

Klote, J H (1989). *Considerations of Stack Effect in Building Fires*, National Institute of Standards and Technology, NISTIR 89–4035.

Kumar, S (1983). Mathematical modelling of natural convection in fire- a state of the review of the field modelling of variable density turbulent flow. *Fire and Material*, **7**(1).

Levin, B M (1987). *EXITT – A Simulation Model of Occupant Decisions and Actions in Residential Fires: Users Guide and Program Description*, National Bureau of Standards, Gaithersberg, USA.

Marchant, E W (1976). Some aspects of human behaviour and escape route design. *5th International Fire Protection Seminar*, Karlsruhe, September 1976.

McAdams, W H (1954), *Heat Transmission*. 3rd Edition. McGraw-Hill Book Company, New York.

McCaffrey, B J (1983). Momentum implications for buoyant diffusion flames. *Combustion and Flame*, **52**(149).

McCaffrey, B (1988). Flame height. *SFPE Handbook of Fire Protection Engineering*, National Fire Protection Association, Boston, pp. 1-299–1-305.

Mitler, E (1978), *The Physical Basis for the Harvard Computer Code*, Home Fire Project Technical Report Number 34, Harvard University, Harvard.

Mitler, E (1984). Zone modeling of forced ventilation fires. *Combustion Science and Technology*, **39**, 83–106.

Mitler, H E and Emmons, H W (1981). Documentation for CFCV, The Fifth Harvard Computer Fire Code, NBS Report, NBS Grant, 67–9011.

Morita, M, Oka, Y and Hirota, M (1989). Numerical analysis and computational methods of heat flow in fire compartments. *Fire Safety Science, Proceedings of the 2nd International Symposium*, Hemisphere, pp. 189–198.

Morton, B R Taylor, G and Turner, J S (1956). Turbulent gravitational convection from maintained and instantaneous sources. *Proceedings of the Royal Society*, **234A**, 1–23.

Mudan, K S and Croce, P A (1988). Fire hazard calculations for large open hydrocarbon fires. *SFPE Handbook of Fire Protection Engineering*, National Fire Protection Association, Boston, pp. 2-45–2-87.

Patankar, S V (1980). *Numerical Heat Transfer and Fluid Flow*, Hemisphere Publishing Co.

Pericleous, K A, Worthington, D R E and Cox, G (1989). The field modelling of fire in an air supported structure. *Fire Safety Science, Proceedings of the 2nd International Symposium*, Hemisphere, pp. 871–880.

Predtechenskii, V M and Milinskii, A I (1978). *Planning for Foot Traffic Flow in Buildings*, Amerind Publishing Co, New Delhi, translated from original in Russian, 1969.

Quintiere, J G (1983). *A Simple Correlation for Predicting Temperature in a Room Fire*, NBSIR 83–2712.

Quintiere, J G (1984). A perspective on compartment fire growth. *Combustion Science and Technology*, **39**, 11–54.

Quintiere, J G and Denbraven, K (1978). *Some Theoretical Aspects of Fire Induced Flows Through Doorways in a Room Corridor Scale Model*, National Bureau of Standards, NBSIR 78–1512.

Quintiere, J G Steckler, K and Corley, D (1984). An assessment of fire induced flows in compartments. *Fire Science and Technology*, **4**(1).

Shanley, J H and Beyler, C L (1989). Horizontal vent flow modeling with helium and air. *Fire Safety Science, Proceedings of the 2nd International Symposium*, Hemisphere, pp. 305–313.

Shapiro, A H (1953). *The Dynamics and Thermodynamics of Compressible Fluid Flow*, Vol. 1, Roland Press.

Siegal, R and Howell, J R (1981). *Thermal Radiation Heat Transfer*, McGraw-Hill, New York.

Steckler, K D, Quintiere, J G and Rinkinen, W J (1982). Flow induced by fire in a compartment. *19th International Symposium on Combustion*, The Combustion Institute, Pittsburgh, pp. 913–920.

Tanaka, T *A Model of Multiroom Fire Spread*, NBSIR 83–2718.

The Building Regulations (1991). *Approved Document B*, (Section B1, 1992 ed.), HMSO Publications, London, pp. 9–40.

Thomas, P H (1992). Fire modelling: a mature technology? *Fire Safety Journal*, **19**, 125–140.

Thomas, P H, Hinkley, P L, Theobald, C R and Simms, D L (1963). *Investigations Into the Flow of Hot Gases in Roof Venting*, Fire Research Technical Paper No. 7, London.

Thompson, P A (1994). *Developing New Techniques for Modelling Crowd Movement*, Ph.D. Thesis, University of Edinburgh, Edinburgh.

Thompson, P A and Marchant, E W (1996). Modelling evacuation in multi-storey buildings with simulex. *Fire Engineers Journal*, **56**(185), 6–11.

Tien, C L, Lee, K Y, and Stretton, A J (1988). Radiation Heat transfer. *SFPE Handbook of Fire Protection Engineering*, National Fire Protection Association, Boston, pp. 1-92 – 1-106.

Turner, J S (1973). *Buoyancy Effects in Fluids*, Cambridge University Press.

Webber, D M, Ketchell, N, Hiorns, N and Marriott, C A (1993). *Modelling human movement in emergency evacuations. ESReDA/CEC Seminar on Operational Safety*, Lyon, 14–15 October, 1993.

Yamauchi, Y (1986). Numerical simulations of smoke movement and coagulation. *Fire Safety Science, Proceedings of the 1st International Symposium*, Hemisphere, pp. 719–728.
Zukoski, E E (1985). Fluid Dynamic Aspects of Room Fires. *Fire Safety Science – Proceedings of the First International Symposium*, Hemisphere Publishing Corporation, 1986.
Zukoski, E E and Kubota, T (1980). Two-layer modelling of smoke movement in building fires. *Fire and Materials*, **4**(1), 17–27.
Zukoski, E E, Kubota Toshi and Lim, C S (1985). *Experimental Study of Environment and Heat Transfer in a Room Fire, Mixing in Doorway Flows and Entrainment in Fire Plumes*, NBS-GCR-85-493, May 1985.

12 MODEL VALIDATION

12.1 Introduction

Model validation is a necessary condition for all model development, more so in fire modeling because of the complex nature of fire phenomena and the inevitable use of empirically derived submodels. Validation is what gives the model credibility for its application to practical engineering problems. This condition is all the more important in the realm of safety. Validation is necessary because, as we have seen above, even the most sophisticated of models (e.g. field models) make simplifying assumptions to first, formulate the physical processes (empirical correlation) and second, facilitate the solution (numerical methods) of governing conservation equations. Since these models, by definition, are a semblance of reality – not reality itself – it is the validation that provides the link between reality and the conceptual model.

All models and submodels are mental constructs by definition, having a basis in the observed physical phenomena; they are abstractions and idealizations with inherent uncertainties that are both qualitative (pertaining to conceptual assumptions) and quantitative (numerical). The task of a systematic validation process is to assess these uncertainties and ascertain their acceptability or unacceptability in so far as the practical application of these models is concerned. The models do not "predict" the phenomenon as such; rather they serve to highlight the trend of certain dependent parameters that in themselves are abstractions and mere manifestations of reality. The instruments with which they are measured have associated with them experimental uncertainties and limitations (Beard, 1992).

This does not mean that such models serve no useful purpose or that they are devoid of any sense of reality, but their limitations as regards their use for practical applications must be established by the process of validation.

In its simplest form, validation involves a comparison of model calculations with measurements of quantitative variables in a real case. It seeks to establish the accuracy or otherwise of a given model in relation to the real world. In practice, such a comparison is fraught with difficulties associated with experimental and numerical uncertainties. Beard (1992) and Davies (1985) discuss these in detail.

A validation process should determine the degree of accuracy of calculation of fire parameters for hazard analysis purposes. It should also highlight the strengths and weaknesses of the

Evaluation of Fire Safety D. Rasbash, G. Ramachandran, B. Kandola, J. Watts and M. Law
 ISBN: 0-471-49382-1

underlying physical models and identify the justification and range of validity of the underlying assumptions.

In fire modeling, as discussed earlier in this book, two types of modeling techniques are common: "zone" models and "field" models. Our discussion on validation is therefore confined to these two approaches.

12.2 Validation of zone models

As noted above, the experimental validation of computer models is carried out for a number of reasons. The underlying objective is to determine the degree of trust that can be placed in the calculated dependent variables such that the results could be used for practical fire risk assessment and the design of protection and detection systems to safeguard life, property, and the environment. Because of the issues of safety, validation is of paramount importance in fire computer code development. Validation not in terms of predicting the physical phenomenon but to facilitate effective and efficient design of safety systems and to improve their performance and reliability.

Empirical models use data from laboratory scale experiments combined with general physical principles to model the development of real fires in enclosures. In all models, the descriptions of natural phenomena are incomplete and approximate. They aim to highlight only those features that are important in the context of what is being investigated. For example, fire phenomena will only be described in terms of heat generation if the concern is thermal impact on people or on critical safety equipment. Similarly, smoke movement is considered in order to calculate toxic effects and visibility levels for evacuation strategies and escape route design.

In zone models, a comparison of calculated and measured parameters present particularly difficult problems. These stem from the uncertainties associated with the measured values as well as the definitions of some of the parameters. For example, in practice the position of layer interface height is difficult to quantify. This is particularly true of rooms away from the fire room, because of the temperature variation with height that is more gradual than in a fire room in which a distinct stepwise variation is observed to exist.

One difficulty that arises from the comparison of temperatures is that, for the purposes of modeling the room fire, the generated environment is assumed to consist of two well defined layers ("hot" and "cold") having uniform temperatures and a well-defined interface. However, in a real fire this takes the form of a vertical profile measured using a vertical array of thermocouples. These measurements are then idealized into a two-layer situation for comparison purposes. This is done either in terms of percentage temperature rise of a maximum value or in terms of the total enthalpy of the layer. Clearly, two methods will generate slightly different results and the subsequent comparison with the calculated results will only be approximate.

Generally, the objectives of zone model validation are to examine the theoretical basis and interactions of the model parameters (e.g. plume model and heat transfer models), assess the validity and applicability of the underlying assumptions, and examine the sensitivity of empirically derived input data (e.g. heat transfer coefficients).

With these objectives in mind, developers of various computer codes have carried out a number of validation studies. In this chapter we discuss, in detail, three such studies involving experimental validation of Harvard V (FIRST), Harvard VI and FAST computer codes.

12.3 The Harvard V zone model (first)

In 1981, the Lawrence Livermore National Laboratory (LLNL) carried out tests (Alvares and Foote, 1981) to assess the behavior of ventilated enclosure fires. We discuss here Harvard V

(Mitler and Emmons, 1981), one of several models evaluated against the measurements obtained in these tests.

12.3.1 EXPERIMENTAL SETUP

The test configuration is shown in Figure 12.1. The burn room was 4.5 m high, 4.0 m wide, and 6.0 m long. Clean air was introduced along the floor through four horizontal rectangles (0.5 m long and 0.12 m high) with horizontal center lines 0.1 m above the floor. Combustion products flowed through an exit opening (0.65 × 0.65 m) on the vertical centerline of the west wall with its center 3.60 m above the floor, being extracted by a fan. The volumetric inflow and outflow rates were measured.

12.3.2 THE FIRE

Three types of fire were installed: burner, spray, and pool fires.

For the burner fire, bottled methane was burnt in a 0.28-m diameter rock filled pan on the floor of the test cell.

The spray fires were supplied with liquid fuel from a pressurized reservoir to a jet nozzle located in the center of a 0.91-m diameter steel pan with a 15-cm lip on the floor of the test cell. The atomized spray, ignited by an electric arc, burned before it came in contact with the pan surfaces.

For the pool fire, approximately 40 L of liquid fuel were burned in a 0.91-m diameter steel pan. The effects of the pan walls on the fuel-burning rate became asymptotic for lip heights greater than 7 cm from this size pool. The mass pyrolysis rate was determined using a calibrated load cell.

12.3.3 MEASUREMENTS

Instrumentation of the enclosure and ventilation circuits was extensive and consisted of gas and surface thermocouples, calorimeters, radiometers, combustion gas and oxygen detectors, fuel and ventilation flow sensors, and a video camera for recording the fire shape.

12.3.4 COMPARISON OF RESULTS

Alvares *et al.* (1984) carried out comparisons of measurements and predictions for the three tests shown in Table 12.1.

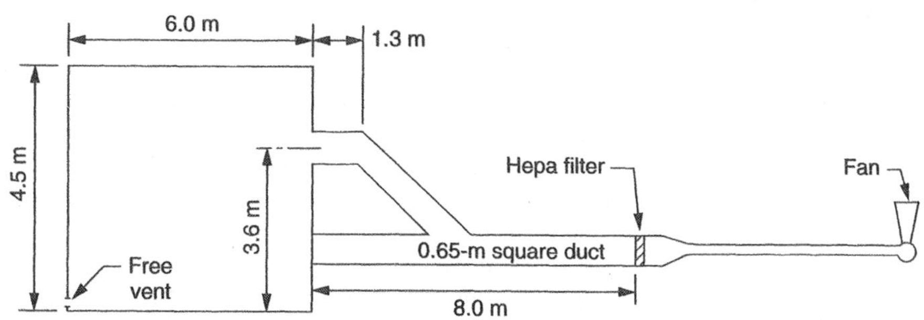

Figure 12.1. Sketch of the Livermore test arrangement

Table 12.1. Experimental fire test details (Alvares *et al.*, 1984)

Test	Fire strength (kW)	Fire type	Ventilation (l/s)
MOD 8	400	Spray	500
MOD 9	800	Spray	500
MOD 27	400	Spray	250

Table 12.2. Comparison of model predictions with test data

Parameter	Upper-layer temp. (°C)	Fire strength (kW)	Heat loss to walls (kW)	O_2 Conc. (% Vol)	Lower-layer height (m)	Wall temp. (°C)	Lower-layer temp. (°C)
(a) Test MOD 8							
Expt.	232	400	300	14.0	1.3	135–180	112
Harvard V	152	395	182	12.1	0.36	113	21
(b) Test MOD 9							
Expt.	299	700	600	6.5	1.29	170–270	146
Harvard V	175	482	231	7.6	0.33	126	34
(c) Test MOD 27B							
Expt.	252	330	270	11.0	1.22	180–210	136
Harvard V	158	341	165	6.8	0.28	123	18

Table 12.2 contains quasi-steady state test data and long-time model calculations for the three comparison tests.

The measured temperature profiles for the MOD 27B fire scenario are shown in Figure 12.2. They show that between 50 and 100 s a discernible hot "upper layer" begins to emerge in the enclosure. Between 200 and 250 s this layer may be thought of as being fully established and it remains virtually unchanged throughout the fire duration. However, looking at these results it cannot be said that two distinct, stratified layers having no mixing between them are formed. The temperature variation is not a stepwise function of height but varies continuously. This means that the interface between "hot" and "cold" layers is not well defined. The lower-layer temperature also increases, be it slowly, from room temperature to about 150 °C in about 250 s. These measurements help to show that the two-layer assumption made in zone models is an idealization. For these reasons, Steckler's two-layer equivalency technique (Mitler, 1984) is used here to make the comparison valid.

Measurements show that for this fire scenario the upper-layer temperature varies between 200 to 250 °C. A comparison of calculated (using Harvard V) and measured upper-layer temperatures (Figure 12.3) shows that the calculations are consistently lower than the measurements despite the fact that the trend is captured quite well. This means that there is more energy going into the hot layer than what is accounted for in the Harvard V code. Where is this energy coming from? To answer this question Mitler (1984) has examined some of the underlying assumptions made in the formulation of the theoretical basis for the code.

In Harvard V, an assumption is made that the lower-layer temperature remains unchanged at ambient. In view of the experimental measurements (Figure 12.2), it is reasonable to say that the heating of the floor by radiation results in the heating of the lower layer by convection. This way some of the energy from the hot floor is radiated back to the lower layer raising its

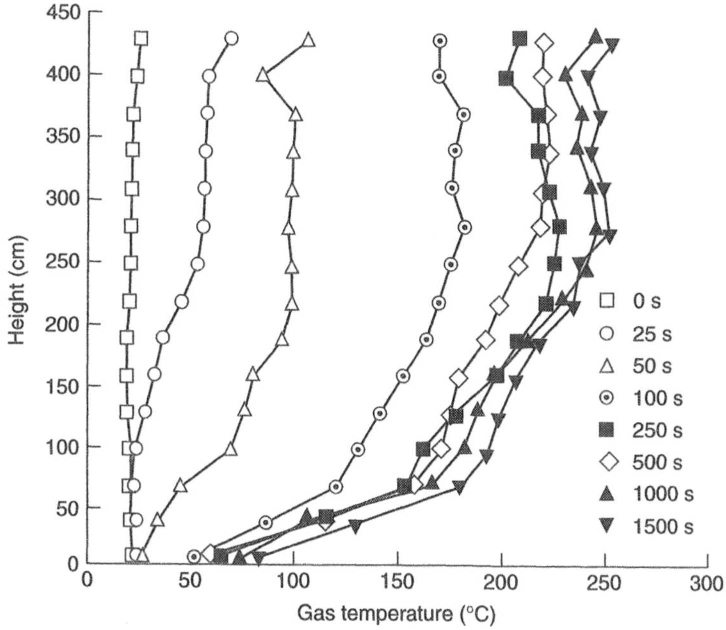

Figure 12.2. Actual gas temperature profiles for MOD 27B at early times

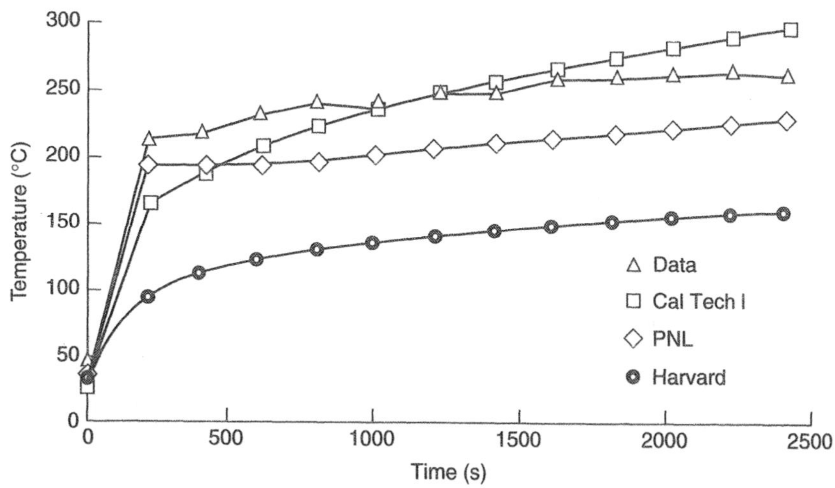

Figure 12.3. Comparison of several models – gas temperature predictions for MOD 27B

temperature. Heating of the lower layer reduces entrainment into the plume (reducing the cooling effect), and moreover because the entrained air is hot, the plume gas temperature is increased, thereby increasing the hot layer temperature. Hence, if this assumption is improved the code can be expected to generate more accurate results as it already gives good qualitative agreement with the measurements. Mitler further suggests that radiation feed back to the layer from the heated floor can raise the upper-layer temperature by 100 to 200 °C. This estimate is confirmed

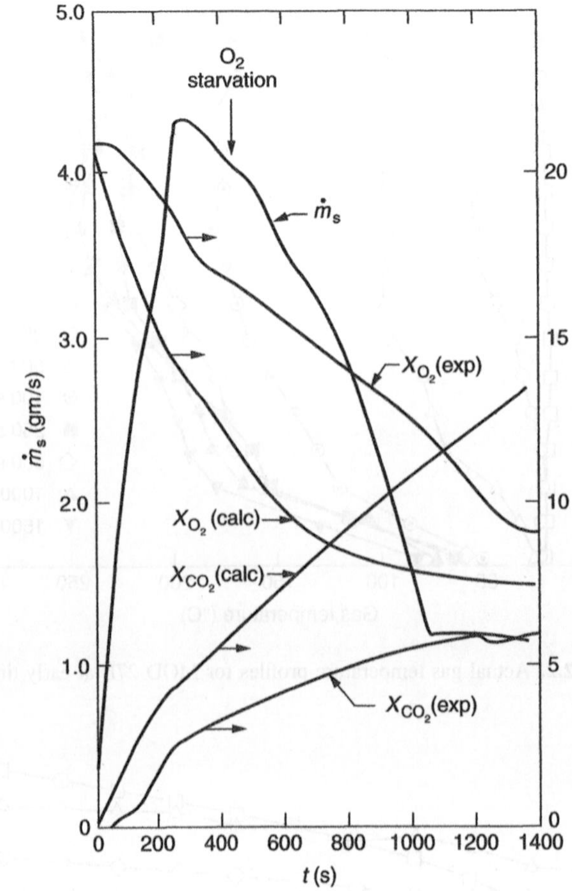

Figure 12.4. Measured versus calculated species concentrations for the MOD 20 case

by the measurements shown in Figure 12.2. This of, course, very much depends on the interface height above the floor, but for a small enclosure this effect could be significant. This explanation is further confirmed by the data summarized in Table 12.2. It can be seen that the measured lower-layer temperature for each of the tests is much higher (over 100° higher) than the ambient temperature assumed for the Harvard V calculations.

Mitler (1984) has analyzed a set of forced ventilation pool fire results (MOD 20, Alvares *et al.*, 1984.). A comparison of measured and calculated results for species concentrations is shown in Figure 12.4. These results also confirm the conclusions arrived at from the above discussion of hot layer temperature comparisons. From Figure 12.4 it can be seen that the calculated values of carbon dioxide (CO_2) concentrations are consistently higher than the measured results. In contrast, the calculated oxygen (O_2) concentration is lower than the measured values. This implies that the theoretical plume model does not entrain enough air.

12.4 Harvard VI zone model

Cooper *et al.* reported in (1982) some full-scale multiroom fire tests. Rockett *et al.* (1987, 1989) have compared Harvard VI calculations with the measurements in these tests. Harvard VI with up to five rooms at the same level, is a multiroom version of the single-room compartment fire model Harvard V. It simulates the dynamic environment generated by fires. It allows fire to exist in more than one room and it includes a database of material properties.

Otherwise, the basic physics remains the same as for Harvard V.

12.4.1 EXPERIMENTAL SETUP

Measurements were carried out with varied combinations of compartment configurations, fire sizes (heat output rate), and area of variable door openings in a 2.0 × 1.07-m doorway that was between the fire room and a variable length corridor. Each test used the same fire room, with a floor area of 14.0 m^2 (Figure 12.5).

12.4.2 THE FIRE

The fire was simulated with a 0.3 by 0.3-m methane diffusion burner positioned centrally at 0.24 m above the floor. The fuel supply to the burner was varied to give four fire sizes: constant heat output of 25, 100, and 225 kW; and a time-varying heat output given by

$$Q(t) = 30\ t$$

t is time in minutes ($t > 0$) and $Q(t)$ in kW.

Artificial smoke was introduced into the ceiling jet above the fire room to get a visual record of smoke spread.

12.4.3 MEASUREMENTS

During each test run, detailed measurements of temperature, pressure, and hot layer depths were carried out.

Five model parameters chosen for comparison with Harvard VI calculations are discussed in the following section.

12.4.4 DISCUSSION OF RESULTS

In general, the Harvard VI calculations were found to be in agreement with the measurements. However, in some cases as discussed below, the two-layer idealized approximation was found to be not so accurate; the transition region between the two layers varied from being relatively narrow to relatively broad. This accounted for some of the discrepancies in the calculated and measured comparisons. Further, in some tests the lower "cooler" layer was contaminated by

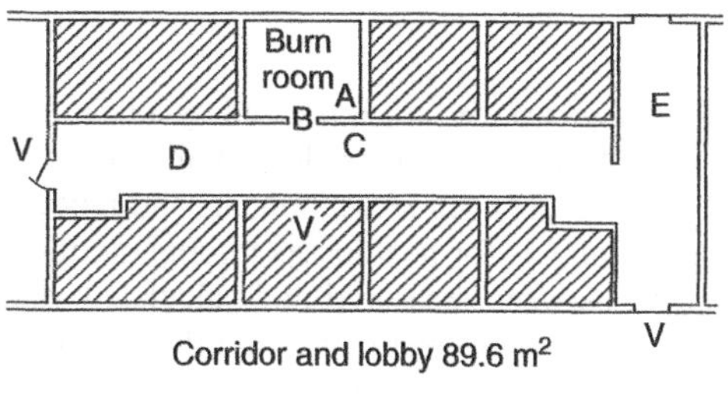

Figure 12.5. Fire test configuration (Rockett *et al.*, 1989)

smoke and warmer than the ambient temperature – indicating mixing between the layers as well as in the doorways.

Rockett *et al.* (1987) have presented detailed results of comparisons for all 19 tests. For the purposes of the present discussion, we confine ourselves to the results of a 100-kW fire, and full corridor and lobby test configuration with full room-to-corridor doorway (Figure 12.5); where appropriate reference will be made to the results of other fire scenarios to illustrate a point, without recourse to actual numerical data. The reader is referred to the original reference for details.

Pressure difference

Figure 12.6a shows the comparison of calculated and measured pressure difference for the 100-kW fire. The results indicate that the calculated values are consistently higher than the measured data. However, the results for smaller fire sizes indicated that this discrepancy is reduced. Also for the 100-kW fire, subdivision of the corridor into smaller volumes (rooms) markedly reduced this discrepancy. This is to be expected. The measured pressure difference reflects the temperature difference between the fire room and the corridor on either side of the door. The measuring points being close together means that the temperature difference is small, while the calculated pressure difference is based on the average hot layer temperature in the corridor, giving high temperature difference and therefore higher pressure difference. The reduced temperature difference for small fires reduces this discrepancy. Given this, the calculated and measured pressure differences are satisfactory.

Average temperature rise

The measurements and predictions for average temperatures in a number of locations (i.e. burn room, corridor, and lobby) are in good agreement for 25- and 100-kW fires, and for the first 200 s or so of the larger fires (225 kW) in small spaces. Results for the 100-kW case are shown in Figure 12.6b. For larger fires beyond 200 s, the model calculates that the average temperature will start to rapidly decrease, whereas the data show it that continues to increase. The authors attribute this discrepancy to the inadequate treatment of "oxygen-limited burning" in Harvard VI. The model assumes that combustion cannot take place in the upper layer because of oxygen depletion. As a result, the calculated temperature begins to decrease with the decrease in oxygen concentration in the upper layer. This has the implication that in cases in which the hot layer descends to the burning object and the object continues to pyrolyse, the model will not accurately calculate the layer temperature. In other words, the model does not adequately simulate environments in which burning takes place in the hot layer. Average temperatures in areas away from the fire room (e.g. corridor and lobby) are calculated satisfactorily (Figure 12.6b).

Overall heat losses

Overall heat losses to the bounding surfaces are expressed by a heat transfer parameter, λ, defined as the fraction of fire heat output that is lost by radiation and convection to these surfaces. The parameter shows general agreement between measurements and model after an initial period. The authors state that the calculation of λ is for heat losses throughout the entire space. Thus, for any given test run, good agreement between measured and calculated values could be obtained, for example, even if predicted losses were overestimated significantly in the fire room and underestimated in the other spaces. The authors state that uncertainties involved in the derivation of λ from experimental measurements and its calculation using the model may limit significantly its

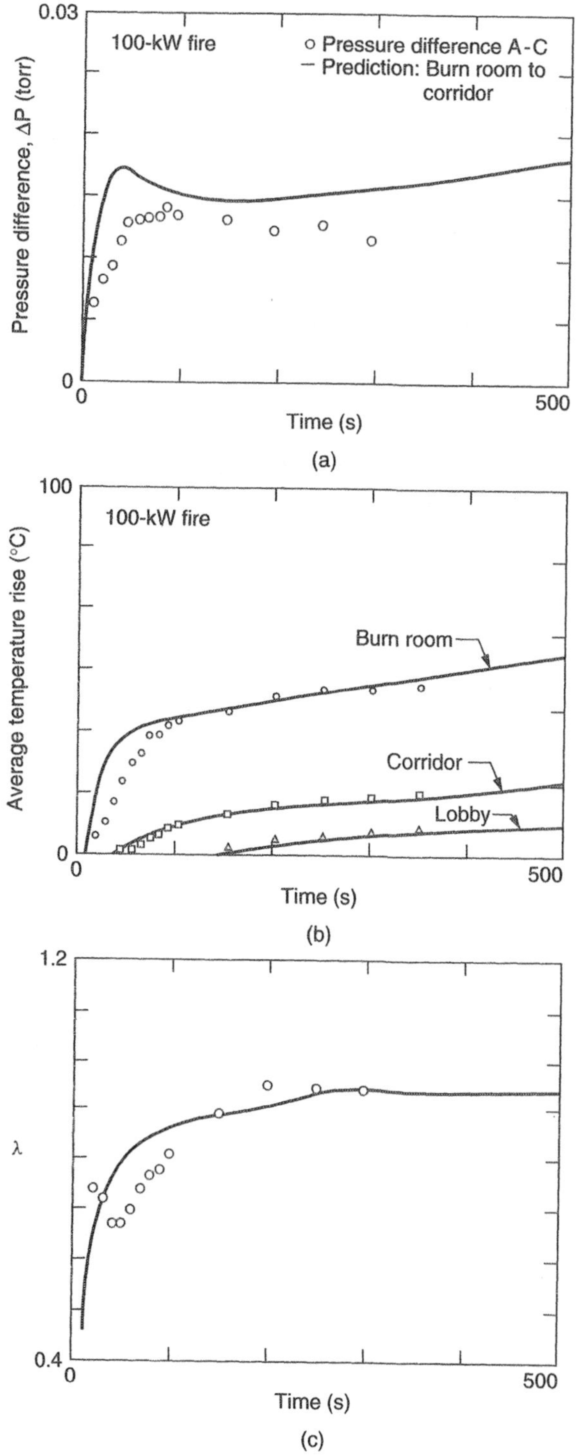

Figure 12.6. Corridor and lobby configuration. (a) Pressure differences, (b) Average temperature rises, (c) λ

accuracy. These comparisons therefore need to be treated with caution. Results for the 100-kW fire are given in Figure 12.6c.

Vertical temperature profiles and layer depths

Figure 12.7 presents the calculated and measured results for temperature profiles for 200 s in the three spaces: the fire room, corridor, and the lobby.

For the fire room (Figure 12.7a), the measured results clearly show two nearly isothermal gas layers with a relatively small (~20 cm thick) transition region between them. The calculated layer depth is in satisfactory agreement with the measured data. However, this agreement deviated considerably when the door width was decreased. Two plausible explanations are provided for this discrepancy.

First, the simulation model does not account for mixing at the doorway between the hot gas exiting from the room and cool gas entering. In reality, this mixing, which more pronounced for narrow openings, causes the lower layer to heat up, which, in turn, raises the hot layer temperature further. Second, the overall modeling of convective heat transfer uses a single heat transfer coefficient based on the average upper-layer and ceiling surface temperatures. The actual convective heat transfer is governed by the differences between plume-driven ceiling/wall boundary flow temperatures and ceiling/wall surface temperatures – which are expected to vary significantly.

The results for the corridor and lobby show that the two-layer assumption that works well for the fire room seem to be unsatisfactory for other rooms. As the results of Figures 12.7b and 12.7c indicate, a clearly defined and identifiable transition region does not seem to exist (even if it does, it is very broad). Experimental data indicate that temperature increases progressively than in a stepwise fashion from floor to ceiling. For this reason, a more detailed transient model for the layer growth may be more appropriate for rooms other than the fire room. Even so, the two-layer model does seem to give adequate overall results of the temperatures and layer depth in these spaces.

12.4.5 CONCLUSIONS

Given the underlying assumptions and simplifications in the modeling process it can be concluded that the multiroom Harvard VI model can be used with confidence to simulate fire generated environments similar to those studied here. In most fire scenarios, the trends found experimentally were reproduced by simulations, although some of the numerical values varied considerably. In particular, the model could be improved by taking into account burning in the hot layer, mixing at the doorways, and better modeling of convective heat transfer.

12.5 FAST zone model

Jones and Peacock (1989) present a limited set of experimental results for validation of the FAST model. In the experiments, fire was simulated as a constant source of heat of about 100 kW in a three-room configuration. Comparisons of hot-layer temperature, interface height, and vent flows were carried out.

FAST is a zone model, in which the gases inside a compartment are treated as residing in two well-mixed zones, an upper hot layer and a lower cool layer. The governing equations for the transport and behavior of gases within a multicompartmented building are the equations for conservation of mass and energy within each zone, and Bernoulli's equation applied to the flow boundaries between compartments. Using Bernoulli's equation avoids the need to solve the differential form of the momentum equation. In common with other zone models, when

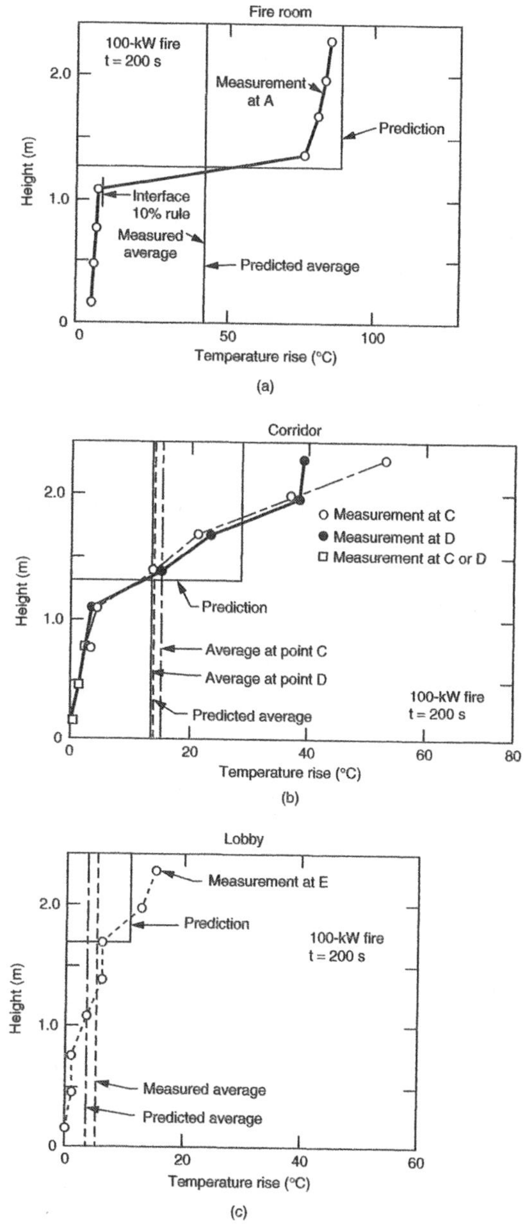

Figure 12.7. Corridor and lobby configuration, vertical temperature profiles (a) fire room, (b) corridor, (c) lobby

considering conservation of mass and energy, the pressure within a compartment is assumed to be spatially constant, thus simplifying the treatment within the compartment. Terms in the equations involving temperature are consistently evaluated relative to ambient temperature. The treatment of vertical vent flows has been improved to account for the variation in hydrostatic pressure on either side of each vent. Depending on the relative densities of the layers either side of the vent, and the heights of the interfaces with respect to the top and bottom of the vent and one another, the flow may reverse up to three times between the top and bottom of the vent. Generally, the flow will reverse at least once at vents adjacent to a room within which a fire is present, the inflow being cold, oxygenated air and the outflow being the hot combustion products.

12.5.1 EXPERIMENTAL SETUP

The arrangement consisted of a burn room opening into a 12-m corridor and a target room on one side. The fire source was a diffusion flame burner supplied with natural gas. Various combination of fire sizes and corridor configuration were tested.

For the purposes of validation, three quantities are discussed in the section below.

12.5.2 COMPARISON OF RESULTS

Upper-layer temperatures

Figure 12.8 shows the upper-layer temperature as a function of time for the three rooms (Room 1 – Fire room, Room 2 – Corridor, Room 3 – End room). It is clear from this figure that FAST overestimates the upper-layer temperatures in the three rooms. The authors attribute this disagreement to errors in the modeling of two effects: heat transfer and plume entrainment. It has been shown in experimental measurements (see Section 12.3.4) that floor heating by radiative heat transfer is significant in enclosure fires. This in turn heats up the lower layer. Since this effect is not incorporated in FAST, the calculation of layer temperature may not be accurate. In addition, further errors are introduced in the modeling of heat losses to walls and convective losses through vents.

As these effects become less significant in rooms away from the fire room, the discrepancy between the calculated and measured layer temperatures is reduced. This is clear from the results for Rooms 2 and 3 in Figure 12.8. However, it must also be stated that in these rooms the layer

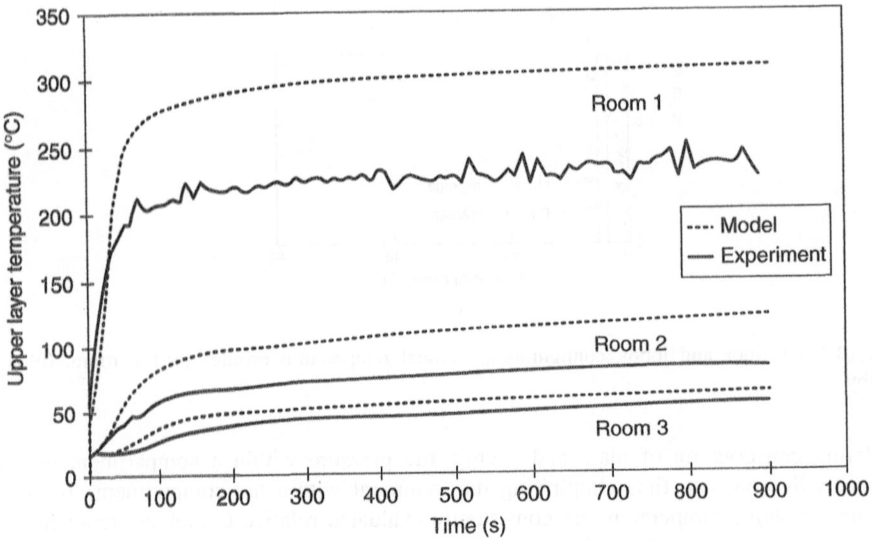

Figure 12.8. Comparison of predicted and measured upper-layer temperature in a three-room test facility

interface is not as sharply defined as in the case of a fire room, because in these rooms the temperature rise with height is more gradual than for the fire room. Greater experimental errors may therefore be expected in such cases, making the comparison difficult.

Layer interface heights

In contrast, the layer interface height is calculated fairly well as Figure 12.9 shows. The interface height is primarily affected by entrainment by the fire plume in the fire room and entrainment at the vents in other rooms. However, it should be noted that in the initial stages of the layer development the interface height is not accurately predicted, at least in rooms away from the fire room. The authors attribute this to the use of a circular plume model for the vent flows. In reality, the plume emerging from a vent is an extended flat plume for which there is no reliable correlation for this configuration.

Mass flow rates

The measured and predicted inflow and outflow at the exit vent in the corridor are in good agreement except for an increasing under prediction of the inflow later in the experiment (Figure 12.10a). The experimentally measured pressures in the burn room and corridor show no evidence of the initial expansion peak predicted by the model (Figure 12.10b).

From these comparisons it can be concluded that FAST calculates fairly well the fire conditions in multiroom configurations. When using this model for hazard-assessment purposes, the underlying modeling assumptions must be kept in mind when interpreting the results.

12.5.3 ANOTHER VALIDATION OF FAST

Gandhi (1993) has presented results of a validation study of FAST for a corner fire in a single room. The experimental measurements were carried out in a room 3.66 × 2.44 × 2.44 m high

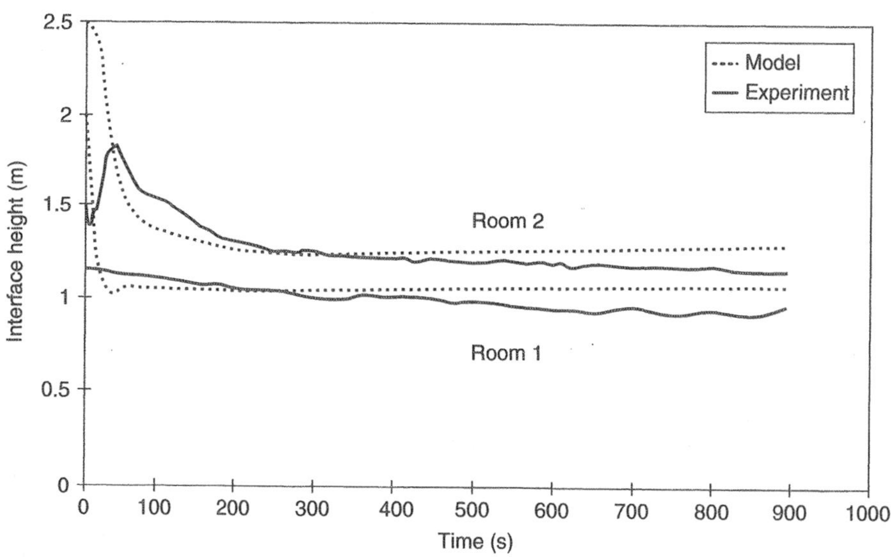

Figure 12.9. Comparison of predicted and measured layer interface in a three-room test facility

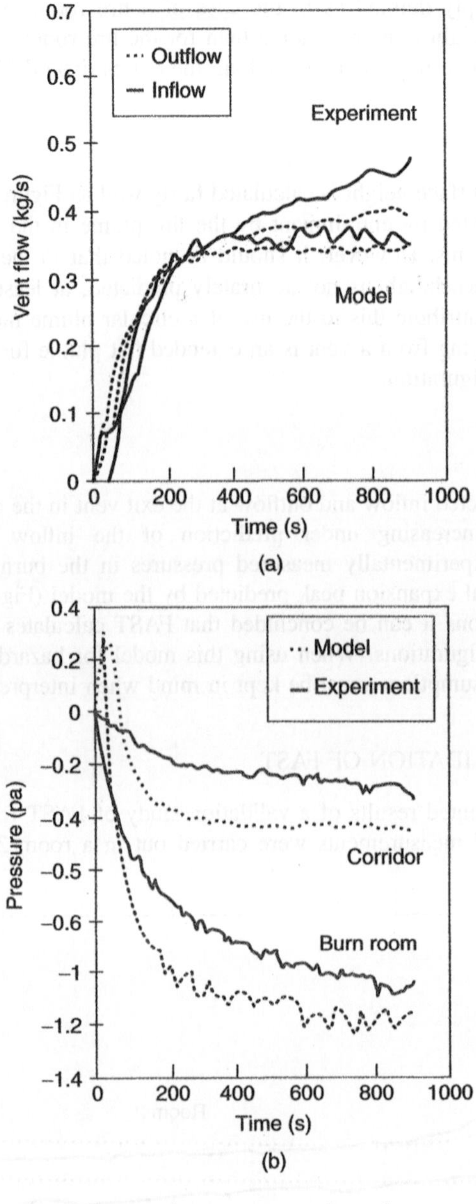

Figure 12.10. Comparison of predictions and measurements in a three-room test facility (a) vent flow, (b) pressure

with a single doorway of 0.76 × 2.03-m height. The tests were performed using three different types of wall lining materials.

The FAST computer calculations were performed using two fire configurations: fire in the center of room and in the corner. The results for both these runs were compared with the experimental measurements. The comparisons of interface height and hot layer temperature are shown in Figures 12.11 and 12.12 respectively. From these results it can be seen that FAST predicts the

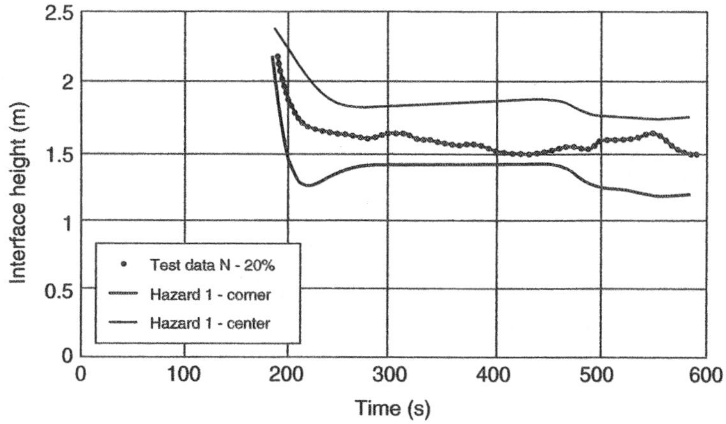

Figure 12.11. Interface height for material C, Test 2

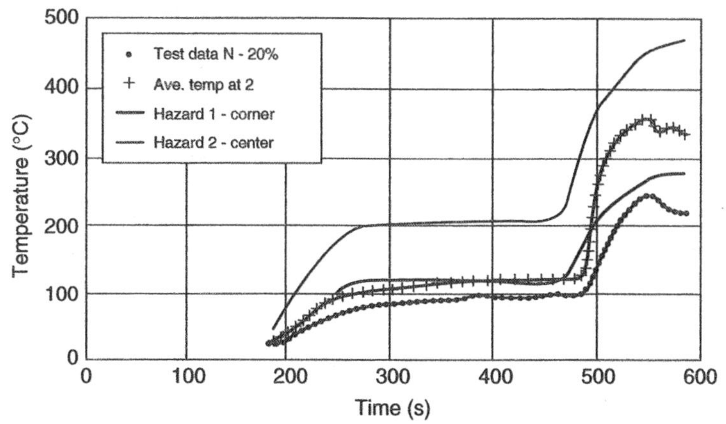

Figure 12.12. Temperature for material C, Test 2

temperature increase trends fairly well, but the actual calculated values are consistently higher than the measured values. The reasons for this are likely to be the way heat transfer is modeled in FAST. The results for the interface height show that the experimental measurements lie between those calculated for the corner fire and fire located in the center of the room.

12.6 CFAST

CFAST is a multiroom zone model developed by merging two previous models FAST and CCFM.VENTS. The underlying physics and the governing equations are similar to those described earlier. The main differences and improvements include the following:

- Treatment of multiple fires in one or more rooms. The interaction of such plumes is not modeled.
- Calculation of lower-layer temperature.
- Burning in the plume, in the upper layer and in a door jet is taken into account.

- Vertical flows through horizontal vents (e.g. hole in a ceiling or floor) are modeled.
- Volumetric expansion of gases in the fire room and the resulting vent flows are calculated.
- Modeling of heat conduction is improved by allowing different material properties to be specified for ceiling, floor, and walls of each room.
- Species conservation equations are solved by taking into account the burning in the plume, in the layer, and at the vents.

12.6.1 CFAST VALIDATION

Peacock *et al.* (1993) present validation studies for CFAST computer code. They selected five different real scale fire tests for comparison.

The fire tests consisted of the following:

1. A single-room fire with upholstered furniture as the fuel. The peak fire size was about 2.9 MW with a total room volume of 21 m^3.
2. A single-room fire similar to that of 1 in which the phenomenon of wall burning was also included. Peak fire size was 7 MW with a total room volume of about 21 m^3.
3. A three-room configuration fire test in which fire was simulated using a gas burner. The fire size was about 100 kW with a total volume of 100 m^3.
4. A four-room test with a gas burner, where the rate of heat release varies with time. The fire size was up to 1 MW with a total volume of 200 m^3.
5. This fire scenario consisted of a series of full-scale experiments conducted in a seven-story hotel building with multiple rooms on each floor and a stairwell connecting all floors. The fire, simulated in a room on the second floor, had a peak fire size of 3 MW with a total building volume of 140,000 m^3.

12.6.2 DISCUSSION OF RESULTS

Upper-layer temperature and interface height

As expected, the comparison of calculated and measured results for the single-room tests showed remarkably good agreement. Upper-layer temperature rise and the layer interface height were predicted with acceptable degree of confidence. However, the burning wall scenario produced some expected deviations. The authors identified modeling of heat conduction or lack of modeling of leakage as reasons for these deviations.

Comparisons of results for multiroom configurations showed that, in general, the calculated values for the upper-layer temperature and interface height were consistently higher than the corresponding measured values. Conversely, the lower-layer temperatures were found to be lower than the measurements. The reasons for these deviations are explained in terms of the underlying assumptions. For example, if modeling of the radiative exchange to the lower layer were included, it would reduce the upper-layer temperature and increase the lower-layer temperature. An improved representation of entrainment at the vents could give improved estimates of interface height. Typical results for upper- and lower-layer temperatures, and the interface height as a function of time are shown in Figures 12.13a to c.

Gas species concentration

The calculation of gas concentrations is basically a reflection of the accuracy of the flow models. The species concentrations are calculated on the basis of user-specified species yield values (as

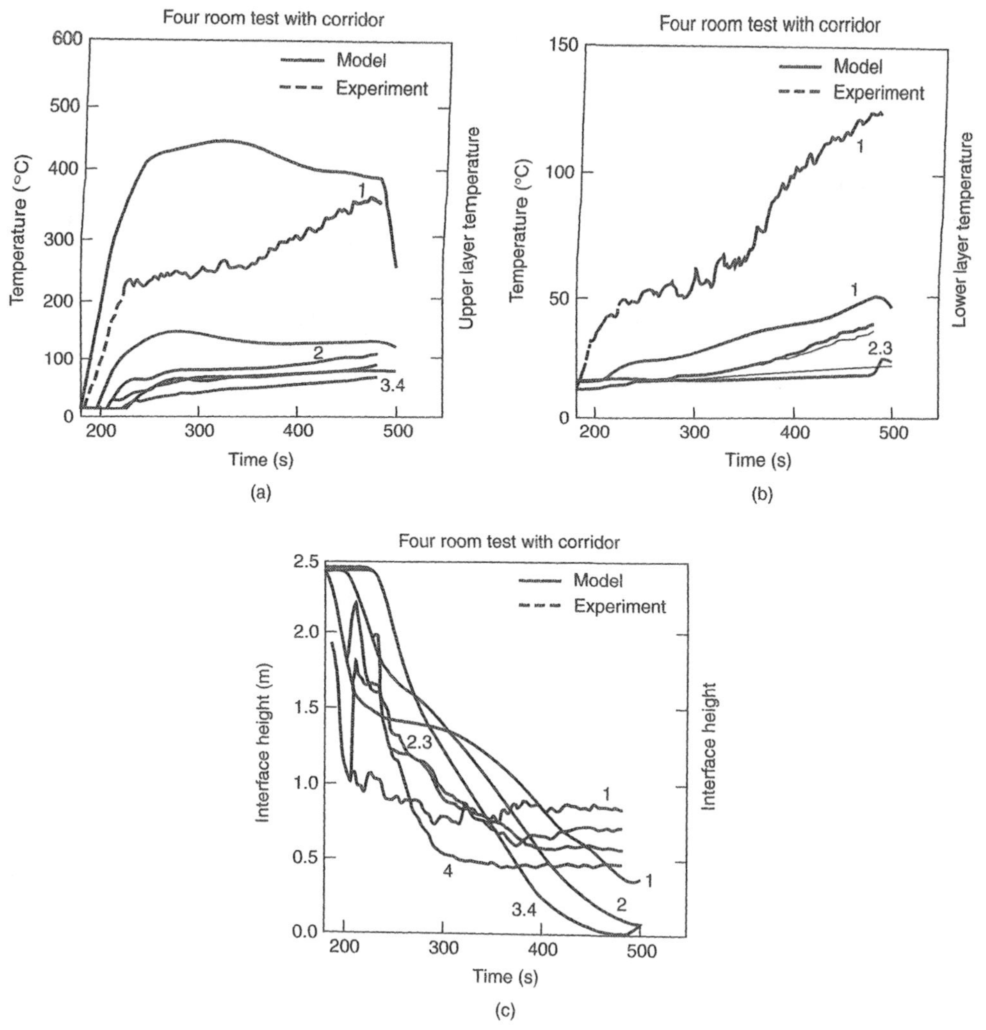

Figure 12.13. Four-room test with corridor (a) Upper-layer temperature, (b) Lower-layer temperature, (c) Interface height

in FAST). Results for the single-room tests show that the calculated concentrations are lower than the measured values. The treatment of oxygen-limited burning in the model is probably the reason for this discrepancy. In contrast, the agreement for the four-room scenario is quite good. However, the results for the multistory building are not so encouraging. The calculated values are far lower than the measurements. Poor estimate of building leakage and the use of estimated species yield values might be the main contributing factors for this discrepancy.

Vent flows

The mass rate of flow through vents is somewhat underpredicted in all the fire tests. In CFAST, the conventional circular cross-section plume model estimates the flow of hot gases through a vent. In reality, however, this flow in the form of a door jet is more akin to an extended flat plume rather like a waterfall. This way the entrainment is wrongly calculated resulting in errors in the calculation of mass flow rate accounting for discrepancy in the calculated and measured mass flow rates.

12.7 FAST and HARVARD VI

Levine and Nelson (1990) present comparisons of measured results with those calculated using two multiroom models: FAST and HARVARD VI.

This experimental study of a domestic incident was carried out to investigate why three people in the first-floor bedrooms were so quickly overcome by smoke, despite being awakened by one of the smoke detectors. The fire started in the ground-floor kitchen. Postmortem examination showed that two of the victims died from carbon monoxide (CO) poisoning, having 91% carboxyhemoglobin in their blood. (The 70% level is generally considered to be fatal). The third victim, who died later in hospital, was also badly burned.

12.7.1 EXPERIMENTAL SETUP

The NIST Centre for Fire Research (CFR) "townhouse" two-story facility was used for the fire tests (Sharon 2 tests). It consists of two upstairs bedrooms (Rooms 6 and 7) connected by a hallway (Room 5) leading to the stairway (Room 4). One of the upstairs bedrooms was fitted with a window, which formed the only vent to the outside at this floor level. It is interesting to note here that from the point of view of internal airflows at this level, the two bedrooms and the hallway may act as one single volume with inflow through the stairway and outflow through the window opening in one of the rooms. As we shall see later, this was crucial in determining the lethal smoke conditions in the bedrooms.

On the ground floor are the kitchen (Room 1) and two other rooms (Rooms 2 and 3) with a doorway from the kitchen into the drawing room. The kitchen had one window opening to the outside.

The fire load, wood cribs, and plywood panels, were designed to cause flashover in the kitchen.

Extensive temperature profile and smoke concentration measurements were made in each of the rooms.

12.7.2 DISCUSSION OF RESULTS

Before discussing the results of this test, it will be instructive to recall the sequence of events that led to the fatalities.

The fire, which started in the early hours of the morning, was mostly confined to the ground-floor kitchen, although there was a evidence to suggest that the ceiling had burned through to the bedroom above causing burn injuries to the occupant of that room. The three victims were asleep in the upstairs bedrooms. At flashover, the kitchen window broke, providing extra fresh air to sustain combustion. The main door opening from the kitchen caused smoke to spread to the rest of the building very quickly. Incomplete combustion produced large quantities of CO. Windows and doors were partially open, apparently by the occupants' desperate efforts to escape.

The fire scenario could be described as follows:

Once flashover had occurred, large quantities of CO-rich smoke had the tendency to flow to the upper floor. This was mainly due to smoke buoyancy and to the fact that doors and a bedroom window on the upper floor were partially open. A flow path was setup, with inflow through the kitchen window on the ground floor and outflow through the upper floor bedroom window. Because there was only one window (i.e. to the outside) opening on the upstairs floor, the hallway as well as the bedrooms (doors open) acted as one single volume. This way the layer depth in each of the rooms descended very quickly to the floor level with high concentration of carbon monoxide. The turbulent nature of the flow ensured that the bedroom environments were well mixed with the resulting smoke concentration in each room very nearly uniform. In fact, this was one of the interesting findings of the tests.

From the comparison of test measurements, observations at the fire scene, and the calculations performed using two of the zone models (FAST and HARVARD VI), the following conclusions could be drawn.

Gas concentration

As Figure 12.14 shows, the two models perform well in calculating the upper-layer average concentration of CO in rooms remote from the fire.

The data agree fairly well for the first 200 s. After this, the FAST calculations indicate that concentrations continue to increase while the measured data show a decrease. The apparent discrepancies can be explained in terms of the growth of layer heights. The models work on the basic assumption that in a room two distinct layers exist across which there is no mixing. In fact, from the measured results it is clear that after 200 s the hot layer seems to occupy the entire room. Therefore, the two-layer assumption breaks down and the calculated results no longer represent a realistic situation. In addition, the calculated values are dependent on the input data: fire growth rate and the yield of CO from the fire. In view of these assumptions, it can be concluded that the models calculate well the onset of the toxic hazard.

Upper-layer temperature

The measured upper-layer temperature in each room is shown in Figure 12.15a. Flashover occurred at 134 s. The temperatures can be compared with the values calculated using FAST (Figure 12.15b) and HARVARD VI (Figure 12.15c).

The comparisons with HARVARD VI results show that the calculated burn room temperature is higher while the other temperatures are lower than the measurements. This is also true for the

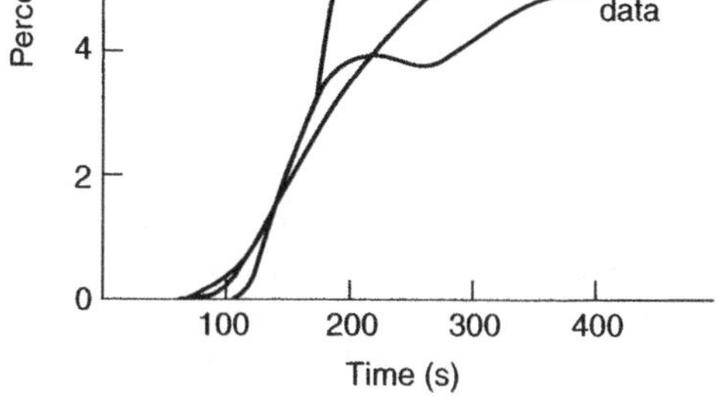

Figure 12.14. CO concentrations versus time from data and two calculation models

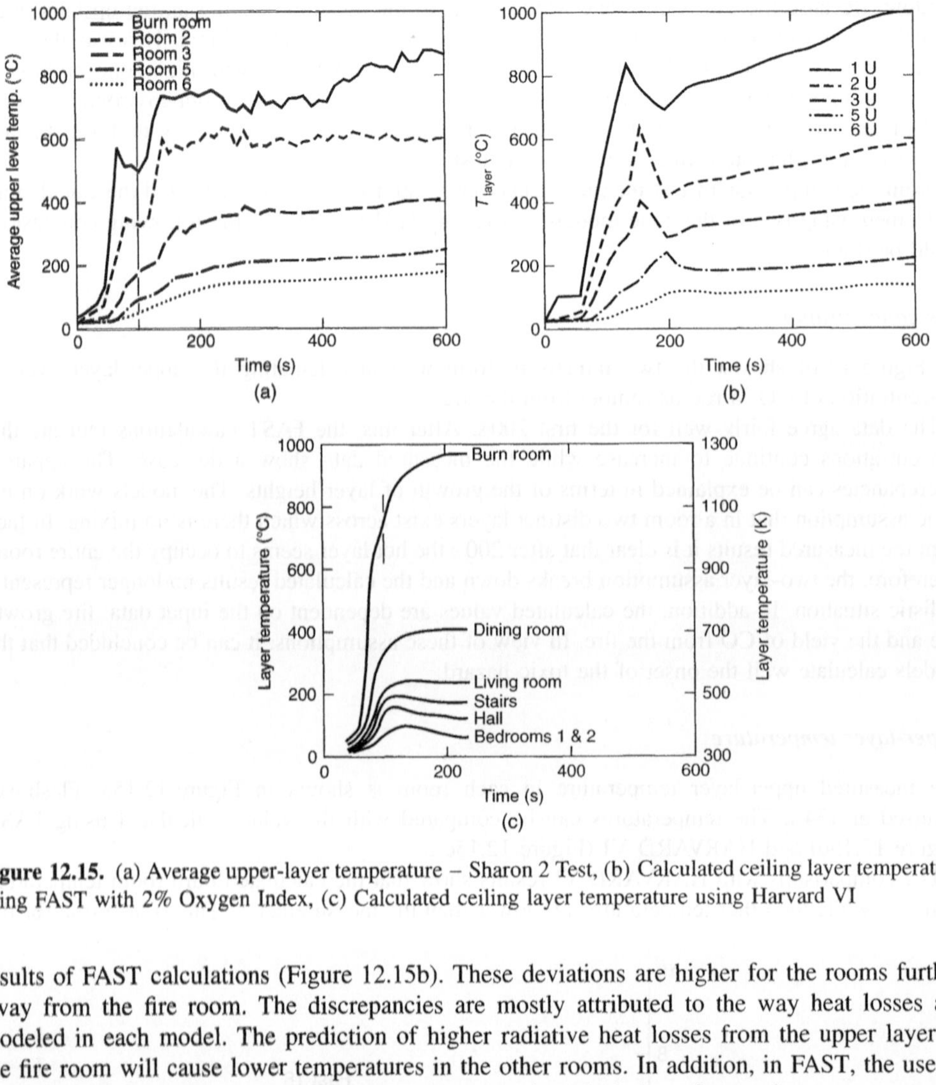

Figure 12.15. (a) Average upper-layer temperature – Sharon 2 Test, (b) Calculated ceiling layer temperature using FAST with 2% Oxygen Index, (c) Calculated ceiling layer temperature using Harvard VI

results of FAST calculations (Figure 12.15b). These deviations are higher for the rooms further away from the fire room. The discrepancies are mostly attributed to the way heat losses are modeled in each model. The prediction of higher radiative heat losses from the upper layer in the fire room will cause lower temperatures in the other rooms. In addition, in FAST, the use of "Limiting Oxygen Index" causes further problems to the calculation of upper-layer temperatures. The use of default value of 6% resulted in very poor agreement. The authors conclude that for the present case a value of 2% is more appropriate and the results of Figure 12.15a are based on this value.

Layer height

Originally, zone models were developed to calculate the hot layer depth and temperature in a fire room. It is therefore not surprising that both models calculate layer heights with an acceptable degree of accuracy. As the rooms further from the fire room do not have discrete thermal layers, the calculation of layer height is less accurate. In fact, even in the case of measurements, the task of defining an interface (between the hot and cold layer) becomes a difficult one. In such situations, layer heights can be defined on the bases of temperature rise or upper-layer enthalpy, the two models perform differently. On the basis of temperature, FAST does best, while HARVARD VI performs well on the basis of enthalpy.

12.8 Validation of field models

The idealization process in field models assumes large number of zones (or "cells") throughout the space of concern with each cell having "field variables" (temperature, gas concentration etc.) associated with it. Although uniformity of conditions is assumed to exist both for cells (in field models) and zones (in zone models), there are fundamental conceptual differences between the two. Cells are purely arbitrary geometric shapes while zones, on the other hand, are determined by the observed physical phenomena. In this respect zone, models assume the course of fire development that the field models attempt to predict or calculate. Zone models are necessarily empirical. In principle, there are no limitations on the size, shape, and the number of cells. The number of zones, on the other hand, is small, usually not more than two. Also, in principle, field models can calculate values of the field variables at a very large number of places (typically thousands) throughout a building. Accordingly, from the point of view of validation, this presents an intriguing problem not encountered in zone model validation. That is, there are too many calculated values and too few measurement points. This is quite the opposite in zone model validation in which there are too few calculated values and too many measurement points. The matter is complicated further when we remember that in field models the precision of a calculated value increases with the number of cells. In other words, the cell size is determined by the physical scale lengths, which are being modeled. So, the dilemma is that accurate comparison requires a large number of cells, and the corresponding experimental measurements present impossible practical difficulties.

Despite these problems, some field model validation studies have been successfully undertaken in the past. In this section, we present a discussion of such studies involving two of the widely used field models: JASMINE and CFDS-FLOW3D.

In principle, as discussed earlier, field models try to solve the governing equations of motion from first principles. But some empiricism is inevitably incorporated because of inadequate mathematical description of physical phenomena such as turbulence and the use of approximate numerical methods to solve these equations. Other areas in which empiricism introduces problems include the modeling of combustion, fire spread, and heat transfer.

The task of field model validation is made further difficult because of the lack of a systematic approach. The majority of the fire experiments have been carried out with the view to zone model validation, in which detailed experimental measurements are not necessary. Such data are insufficient and inadequate for comprehensive field model validation. For example, measurements of temperature, velocities, and mass flow rates through openings have provided detailed data that are adequate for comparison. However, measurements of variables such as heat fluxes, gas, and particulate concentration have not been adequate throughout the domain of interest.

Notwithstanding these difficulties, a variety of validation studies are available. Some of the studies are discussed here.

12.9 JASMINE field model

JASMINE is a field model developed by the Fire Research Station (FRS), United Kingdom, which uses PHOENICS as the equation solver. JASMINE has been used extensively by FRS on in-house research and code validation projects, and is not commercially available.

12.9.1 ENCLOSURE FIRES WITH NATURAL VENTILATION

Kumar *et al.* (1992) have used the JASMINE field model to examine the effects of relative fire location and heat radiation on the enclosure fire thermal hydraulics. Later, Kerrison *et al.* (1994) used the same data to conduct CFDS-FLOW3D validation studies (see Section 12.10 for details).

The data are from tests by Steckler *et al.* (1982a,b) who investigated fires of various sizes placed at several locations within a naturally ventilated compartment (Figure 12.16) Kumar *et al.* (1992) have used the corner fire test results (fire 'B' in Figure 12.16) to demonstrate the capability of JASMINE to handle complex fire scenarios.

Experimental setup

The fire tests were conducted in a compartment measuring 2.8 × 2.8 × 2.18 m high. A 0.3-m diameter porous plate diffusion gas burner was used to simulate the fire. Vertical columns of thermocouples and bidirectional velocity probes provided measurements of temperature and velocity profiles in the enclosure as well as within the doorway opening. The results for a doorway 0.74 m wide by 1.83 m high and a fire of 62.9-kW heat release rate are presented here.

Field model simulations

To examine the effects of heat radiation two sets of numerical simulations were carried out:

1. For the first set of runs, the radiation exchange in the gas phase was ignored. Heat losses to the wall boundaries were calculated using the empirical heat transfer coefficients (radiative and convective components being lumped together), similar to the approach taken in zone models (for example, Harvard V).
2. The radiation exchange within the gas phase was included using a simple six-flux radiation model (for details see Cox, 1995).

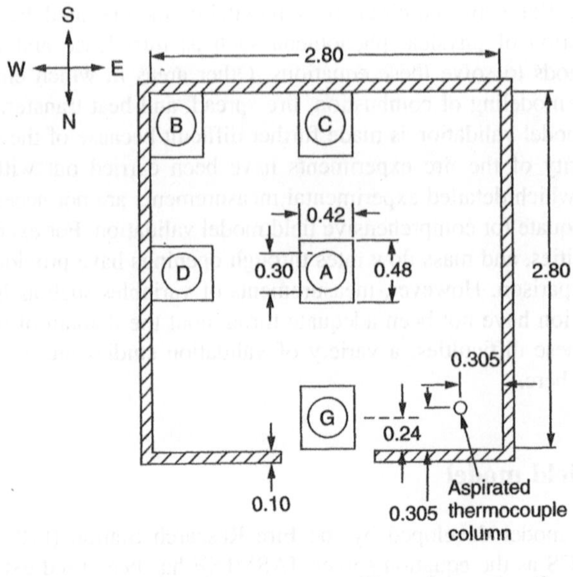

Figure 12.16. Plan of compartment with gas burner locations

The resulting doorway centerline velocity and temperature profiles are shown in Figures 12.17a and 12.17b respectively. They clearly show that the six-flux radiation model makes a marked improvement to the calculated results, especially in the region of hot–cold layer interface. The results confirm the experimental observations that the lower layer does not remain at a constant ambient temperature but is slowly heated by radiation from the upper hot layer. The temperature rise across the layer interface is gradual and not a stepwise function as is generally assumed in zone model formulations. This is further confirmed by the comparisons of measured and calculated temperature profiles inside the compartment as shown in Figure 12.18. The results indicate that radiation redistributes the thermal energy between the lower and upper layer by transferring some of this energy from the hot to the cold layer by raising its temperature. As a consequence, the upper-layer temperature is reduced and seen to fall well below the measured values. The authors do not provide a credible explanation for this discrepancy. However, it is likely that inadequate modeling of heat transfer to the walls is the main reason here. Remember also that the results are compared for a location very close to the enclosure walls where such heat transfer effects are important. Analysis of Jaluria (1988) shows that the layer temperature near the wall surfaces is influenced by the downward flow of hot gases. In this way, the thermocouple measurements in the corner are a reflection of complex flow conditions near the walls and *not* in the rest of the hot layer. With this in mind, the temperatures calculated using the radiation model compare reasonably well with the measurements.

The calculated and measured mass flow rates through the open doorway are given in Table 12.3. These results further confirm the importance of including some kind of a radiation model.

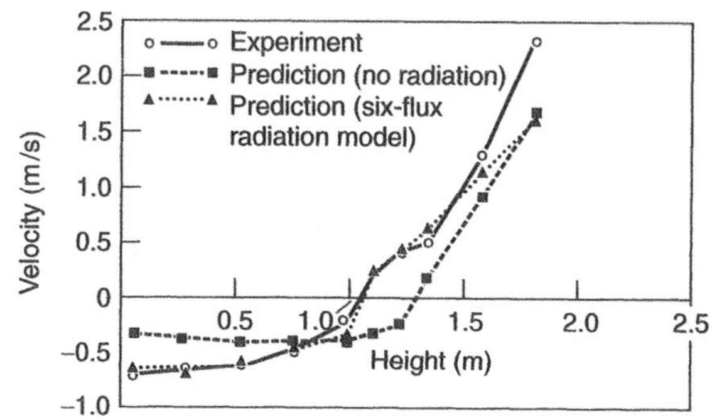

(a)

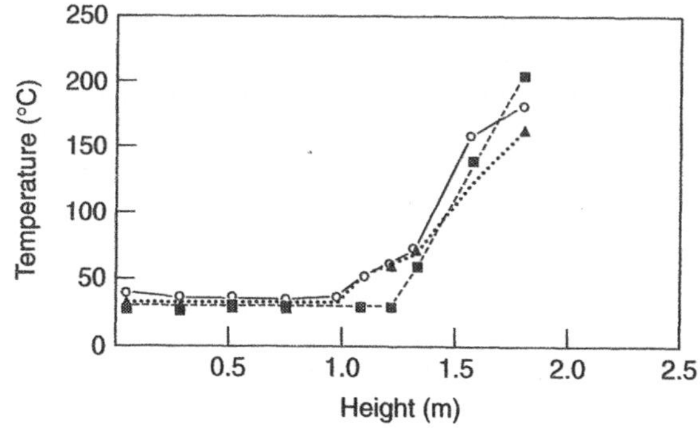

(b)

Figure 12.17. Doorway centerline velocities and temperatures as a function of height for fire B

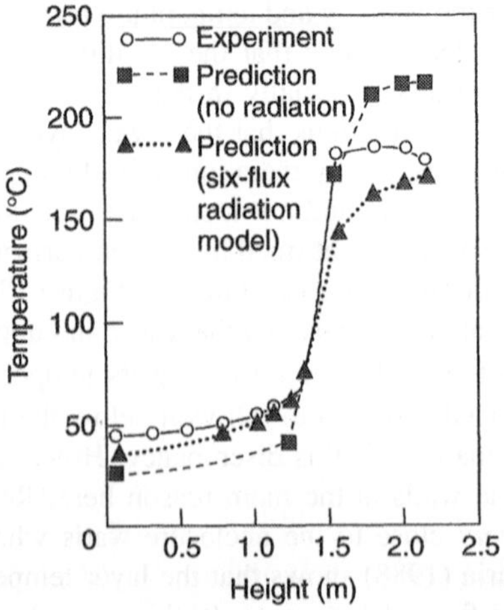

Figure 12.18. Vertical temperature profiles in corner of compartment for fire B

Table 12.3. Comparison of calculated and measured mass flow rates (kg/s) through the doorway for a 62.9 kW corner fire at B

Flow direction	JASMINE run		Experimental
	Without radiation	With radiation	
Inflow	0.303	0.423	0.440
Outflow	0.304	0.424	0.439

Although the measurements are subject to experimental uncertainties, especially in the case of velocity measurements (where a probe might not be aligned with flow streamlines), the inclusion of a radiation model makes a considerable improvement to the calculated results.

Discussion of results

From this study, the authors conclude that the field modeling approach is capable of accurately estimating the enclosure fire phenomena. The simple radiation model used in this study illustrates the importance of radiation and that the use of an improved radiation model could further improve the predictive capabilities of field models. Other calculations (not presented here) confirm that it is the field modeling approach that enables the calculation of plume entrainment and plume tilt caused by ventilation flow through openings such as doors and windows. These considerations are important in hazard analysis and safety implications of protection measures (e.g. the siting of fire detection and suppression systems).

12.9.2 ENCLOSURE FIRES (FORCED VENTILATION)

Enclosure fires under forced ventilation conditions are important because forced ventilation modifies plume entrainment, giving enhanced burning rate, rapid fire spread, and plume interaction with enclosure boundaries. Simple zone modeling techniques do not lend themselves to the analysis of such complex fire situations. It is for these reasons that validation of a field model is of crucial practical importance for such scenarios.

Cox and Kumar (1987a) present results of a JASMINE validation study of enclosure fires under forced ventilation conditions, using the experimental data of Alvares *et al.* (1984). For the purposes of this comparison, the fire scenario designated by Alvares *et al.* (1984) as MOD8 was considered.

Experimental setup

The fire enclosure measures 4.0 × 6.0 × 4.5 m high. A rectangular duct 0.65 m^2, centered 3.6 m above the floor of the compartment provided forced ventilation by an axial extract fan. The air inlet was at lower level through slots in a cylindrical duct close to one face of the compartment. The fire was centrally located at floor level. It was a natural pool fire formed from a spray of isopropyl alcohol from opposing jet nozzles located in the center of the a steel pan of diameter 0.91 m. On the assumption of efficient combustion, the total heat release rate of the fire was estimated to be 400 kW. The fan extracted 500 L/s of ambient air.

The measurements consisted of temperature profiles in the enclosure as well as gas concentrations at the exit.

The fire scenario was modeled using a simple combustion model in which a unit mass of fuel was assumed to combine with the stoichiometric mass requirement, s, of oxygen to give $(1 + s)$ mass units of product. Forced ventilation was modeled by specifying a fixed volume flux with no frictional losses in the duct or across the fan. Turbulence was modeled using the k-ε model. Heat losses to the walls (both radiative and convective combined) were modeled by using a single heat transfer coefficient (25 W/m^2K). The one dimensional heat conduction equation was solved for the wall boundaries.

The comparisons of calculated and measured enclosure gas temperatures and wall and ceiling surface temperatures are shown in Figures 12.19 and 12.20 respectively. The results indicate that the overall agreement is fairly good except for some discrepancies at distances close to the ceiling and floor surfaces. This, as the authors explain, may be due to inadequate modeling of heat transfer to these surfaces. Remember, only an averaged heat transfer coefficient is used rather than the more sophisticated heat radiation model. The higher calculated values near the surfaces indicate that heat lost by convection and radiation is more than can be accounted for by the use of an average heat transfer coefficient. A more detailed modeling would improve the results significantly.

Some of the other calculated and measured overall flow properties are compared in Table 12.4. The results indicate that the general agreement is fairly good except for the gas concentrations and the exit pressure.

Table 12.4. Comparison of calculated and measured flow properties for a forced ventilation fire

Property	JASMINE calculated	Alvares *et al.* measured
Mass outflow rate (kg/s)	0.269	0.240
Mass inflow rate (kg/s)	0.257	0.300
Exit gas temperature (°C)	249	275
Exit heat flow (kW)	66	68
Exit pressure increase due to fire alone (Pa)	14.8	9
Exit oxygen concentration (dried gas) (%)	10.4	14
Exit carbon dioxide concentration (dried gas) (%)	7.5	5.5

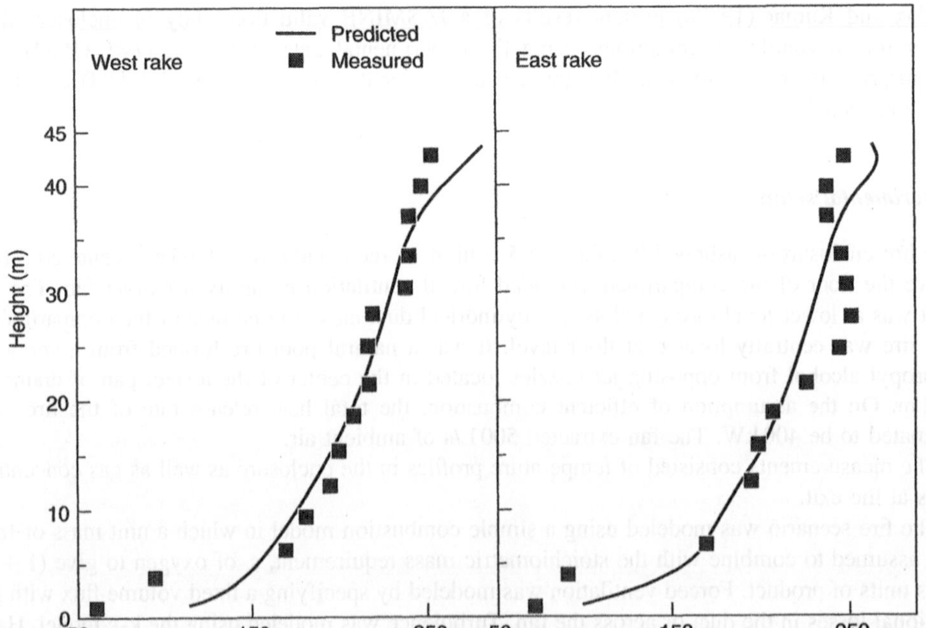

Figure 12.19. Gas temperature with height at two thermocouple rakes

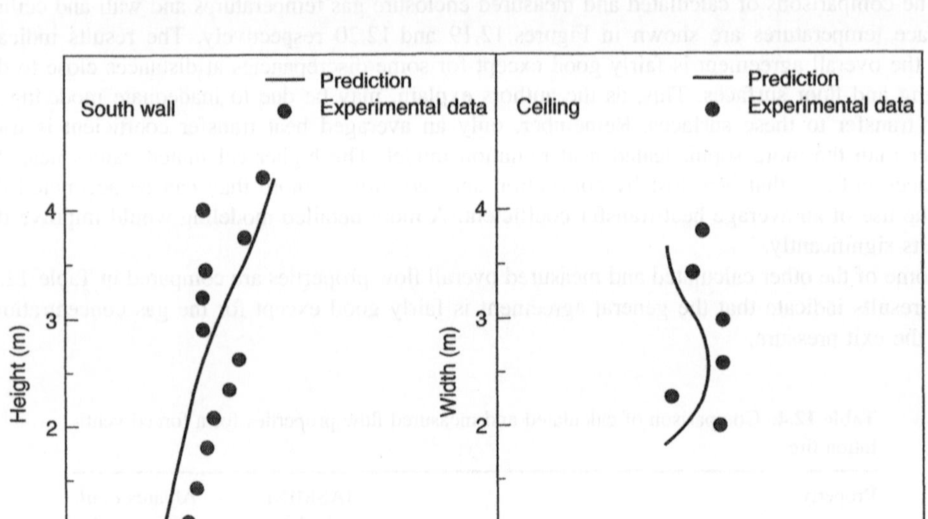

Figure 12.20. Surface temperature with height on south wall and ceiling

Discussion of results

The calculation of gas concentrations is highly dependent on the source terms, that is, on the species concentration of fire products. The combustion model used for this analysis is a very simple one that makes the assumption of efficient combustion. In reality, however, this may not be true due to the complex nature of enclosure ventilation and the consequent variations in the rate of production of gaseous products. A more realistic combustion model would give greatly improved results. As regards the discrepancies in the calculation of exit pressures, it must be noted that the numerical simulation did not take into account the frictional losses in the fan or in the ductwork. This may have contributed significantly to this discrepancy. Better results could be obtained by improving on this assumption.

The authors conclude from this study that further improvements to the calculated results can be made by improving on some of the submodels, such as a better treatment of the heat transfer to boundaries, the use of a more realistic combustion model, and the use of a finer grid in the vicinity of the fire source and near the wall surfaces.

12.9.3 CEILING JET FOR CONFINED AND UNCONFINED CEILINGS

Kumar and Yehia (1994) have conducted a validation study of JASMINE involving the investigation of ceiling jet characteristics for both the confined and unconfined ceilings. The calculated results are compared with the experimental measurements conducted by Montevalli and Marks (1991) for the unconfined ceiling, and Montevalli and Ricciuti (1992) for the confined ceiling jets. Comparisons of vertical temperature and velocity profiles are made for both steady state and transient conditions.

Experimental setup

The experimental measurements were carried out using a premixed methane-air burner of 0.27 m in diameter capable of producing fire strengths in the range of 0.5 to 2.0 kW. The ceiling jet was produced by placing a large ceiling of diameter 2.13 m directly above the fire source at distances of 0.5 m and 1.0 m above the floor. The ceiling was insulated on the top. The velocity and temperature measurements were made using arrays of measuring probes placed at radial distances of 0.26-m and 0.75 m from the point of plume impingement. The JASMINE comparisons are presented for a 2-kW fire and the ceiling height of 1 m for both the confined and unconfined configurations.

Discussion of results

The results of this study show that in the region of plume impingement significant differences are found between the calculated and measured results. This difference reduces considerably as the jet becomes fully developed and approaches steady state conditions at increasing radial distances (compare Figures 12.21a and 12.21b). From these results it is also clear that the model tends to underestimate the jet velocity and overestimate the jet temperatures. In the main, these differences are attributed to both the modeling simplifications and experimental uncertainties, in particular:

- inadequate description of the fire source;
- inadequate modeling of the radiative heat transfer;
- in the impingement region, the discrepancies are most likely due to turbulence modeling and uncertainties associated with the measurement of velocity (probe not in alignment with flow streamlines).

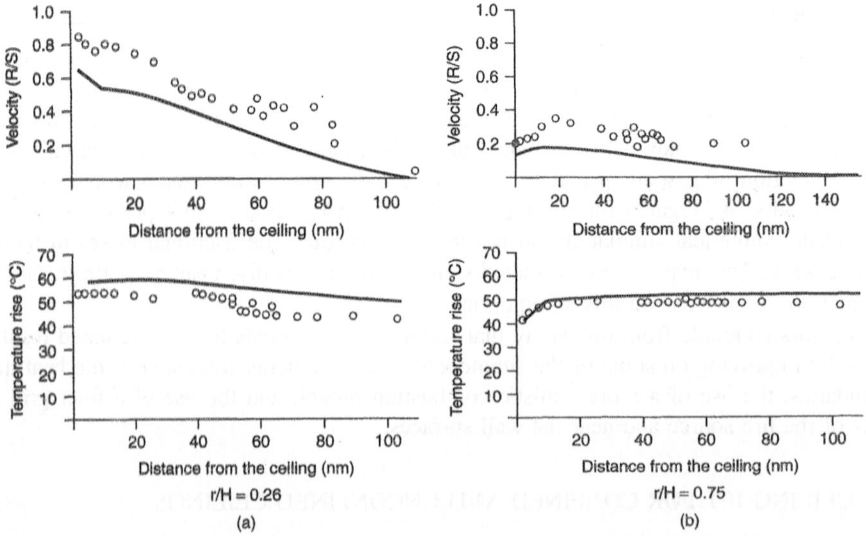

Figure 12.21. Comparison of the predicted and measured vertical profiles of the unconfined ceiling jet at steady state

For the unconfined case, the comparison of maximum temperature and velocities as function of the radial distance is shown in Figures 12.22a and 12.22b. From these results it can be seen that the maximum flow velocity and the maximum temperature are estimated reasonably satisfactorily. This indicates that the field modeling approach can confidently be used for hazard analysis calculations in which the maximum temperatures are critical. However, at regions close to the plume impingement, the results may not be reliable.

12.9.4 TUNNEL FIRES

Kumar and Cox (1985) and Cox and Kumar (1987b) present results of a JASMINE validation study for fires in road tunnels. The experimental data collected from fire tests in the Zwenberg tunnel were used for comparison. In these tests, liquid pool fires were burnt under a variety of conditions of natural and forced ventilation. Gas temperature, gas composition, and visibility were measured.

Experimental setup

The tunnel geometry and measurement stations along the tunnel are shown in Figure 12.23a. The tunnel itself is 390 m long with a sealed south portal and open north portal. The forced ventilation fan at the sealed end provided airflow rates of 2 m/s and 4 m/s from the south toward the north end. In the tests used for this comparison, fire was simulated in a 2.6-m square tray placed at 108 m from the south end containing 200 L of petrol fuel placed at 108 m from the south end.

Field model simulations

The combustion model used in the numerical simulations assumed that the products were only carbon dioxide and water giving the following simple reaction (assuming fuel to be hexane):

$$C_6H_{14} + 9.5O_2 \rightarrow 6CO_2 + 7H_2O$$

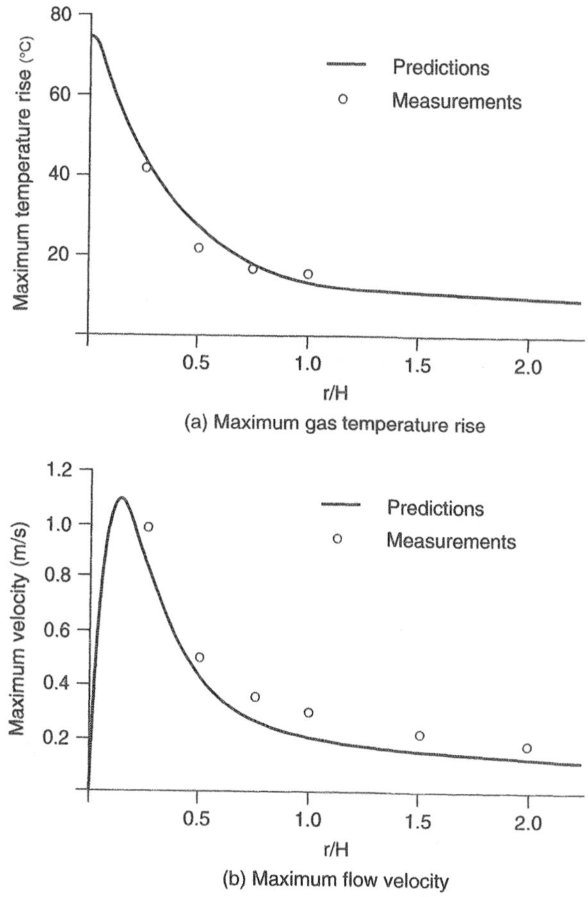

Figure 12.22. Comparison of the predicted and measured radial profiles of the unconfined ceiling jet at steady state

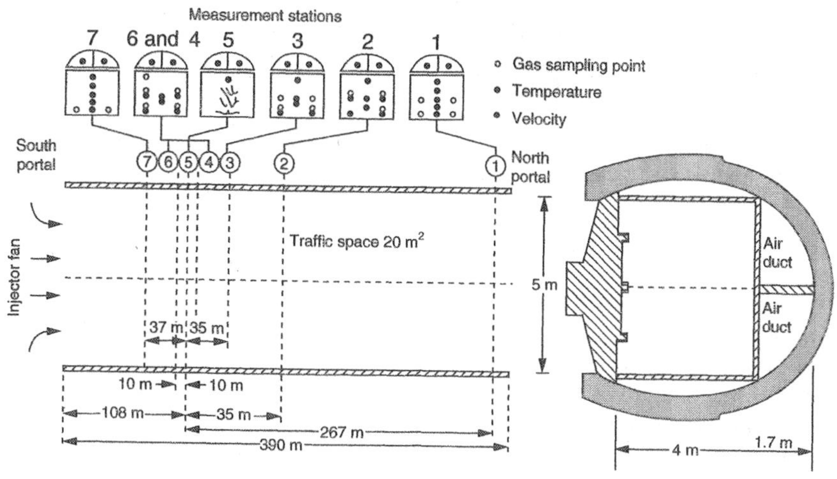

Figure 12.23. The Austrian Zwenberg railway tunnel experiment

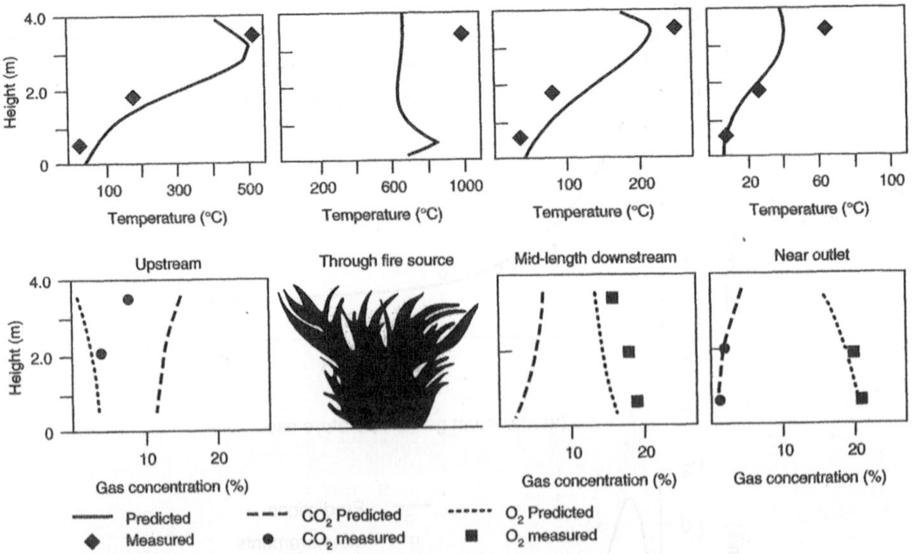

Figure 12.24. Comparison of predictions with measurements – natural ventilation

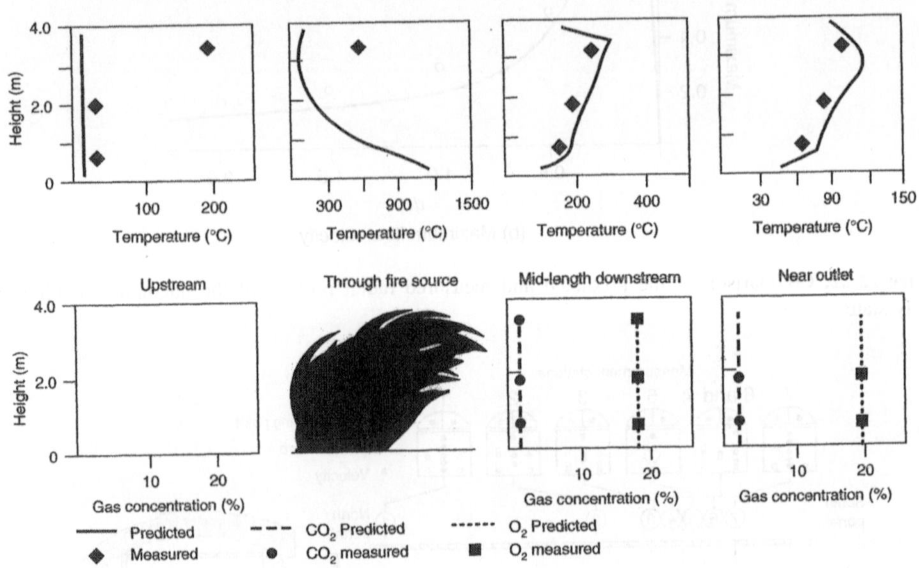

Figure 12.25. Comparison of predictions with measurements – 2 m/s

The rate of heat release was calculated on the basis of the experimental data and the appropriate ventilation rates. This gave heat release rates of 14.45 MW for the natural ventilation case, and 20.25 MW and 24.95 MW for 2 m/s and 4 m/s ventilation rates respectively. Heat losses through the tunnel walls were calculated from the known wall conductivity and thickness, and the computed temperature gradients. The convective and radiative losses were obtained by using a combined local empirical transfer coefficient.

The authors compared the calculated and measured values for three different heights on the centerline of the tunnel at each measurement station indicated in Figure 12.23. These are at 0.5 m below the ceiling at head height (1.8 m above the floor) and 0.5 m above the floor. These comparisons were made for the three ventilation conditions: natural, forced (2 m/s), and forced (4 m/s) ventilation. The results are shown in Figures 12.24, 12.25, and 12.26. Table 12.5 gives a summary of some of the calculated and measured results.

From the results it is clear that the overall agreement is reasonably good except for the position directly above the fire. At this position radiant heat transfer that was not explicitly included in the calculations dominates the measurements. Calculation of velocity profiles (not shown here) further highlighted some important features of the flow. In the case of natural ventilation, recirculation and mixing between the hot and cold layers at the closed south portal was predicted. For the 2-m/s ventilation rate, considerable inflow of air was predicted at the open north portal end that was not present for the 4-m/s ventilation rate.

Table 12.5. Calculated and measured results for the tunnel fire

	Position 7 37 m upstream	Position 5 above the fire	Position 2 85 m downstream
Natural ventilation			
Measured °C	210	1000	255
JASMINE °C	248	664	215
2 m/s Forced			
Measured °C	14	510	250
JASMINE °C	10	35	258
4 m/s Forced			
Measured °C	12	176	220
JASMINE °C	10	12	198

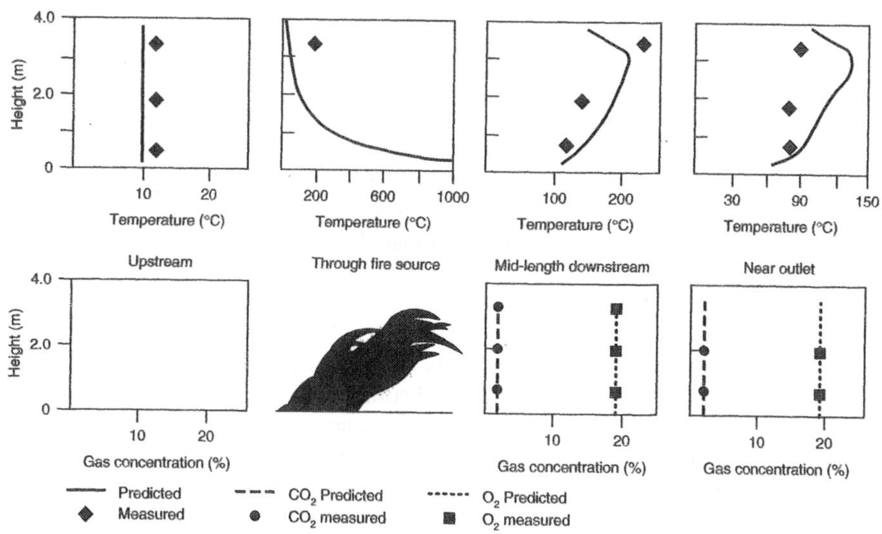

Figure 12.26. Comparison of predictions with measurements – 4 m/s

From these studies, the authors conclude that in common with other validation studies the field modeling approach seem to estimate the "far-field" conditions reasonably accurately. However, close to the fire source more accurate modeling of heat transfer and turbulence-chemistry interaction is required to achieve an acceptable degree of accuracy for the calculated results.

12.10 CFDS-FLOW3D field model

12.10.1 SINGLE-ENCLOSURE FIRES WITH NATURAL VENTILATION

Kerrison *et al.* (1994) have reported results of a validation study based on the room fire tests carried out by Steckler *et al.* (1982a). The test setup is described in Section 13.9.1.

The CFDS-FLOW3D simulations were carried out within the compartment with three door widths (0.24, 0.74 and 0.99 m) and height 1.83 m, and two fire sizes (31.6 and 62.9 kW). Combustion and radiation were ignored and the fire was simulated as a simple source of heat. To account for heat losses through walls and via radiation, the fire source heat release rate was modified. The fire was modeled as a rectangular burner having the same surface area as the round burner used in the tests.

In the absence of detailed experimental measurements – that are a necessary prerequisite for meaningful field model validation – the authors primarily confined themselves to the examination of overall level of agreement between the experimental and calculated values. With this aim in mind, the measured as well as calculated results were averaged or reduced to conform to the physical two-layer idealization model. Before going on to discuss detailed comparisons, it will be prudent to examine the way these average values were determined.

The upper-layer temperature was estimated by averaging the thermocouple measurements in the corner. Here, it must be noted that this is not a particularly good thermocouple position to measure layer temperature. As the enclosure walls become hot (from radiation and convection), the natural convection wall flows give rise to a distinctly different environment near the walls than in the rest of the hot layer. Analysis of Jaluria (1988) shows that layer temperature near the wall surfaces is influenced by the downward flow of hot gases. In this way, the thermocouple measurements in the corner are a reflection of flow conditions near the walls and *not* those prevailing in the rest of the hot layer. It is therefore misleading and incorrect to deduce hot layer temperature from thermocouple measurements in the corner.

The neutral plane height was deduced from velocity measurements in the doorway with zero velocity indicating the neutral plane. The turbulent nature of flow present difficulties in the measurement of flow velocity. The accurate measurement is only possible if the velocity probes are aligned to the flow streamlines. Instead, the measurements were only carried out with the probe axes parallel to the floor. This way significant errors were introduced to the measurement of velocity, as well as to the measurement of mass flux through the doorway.

With these provisos, the results can now be compared.

Discussion of results

Comparisons of measured and calculated velocity profiles in the doorway are given in Figures 12.27a and 12.27b for fires located at the center and corner of the room respectively. As expected, the results indicate that the comparisons for the corner fire are not as good as those for the centrally located fire. Also with the narrow door width (0.24 m), there is poor agreement for both the fire scenarios. The reasons for these discrepancies, as discussed by the authors, are due to significant errors introduced by experimental measurements as well as those due to the modeling assumptions.

The interaction of the fire plumes with the enclosure boundaries determines the thermal hydraulics and heat transfer characteristics in an enclosure fire. In the case of a corner fire,

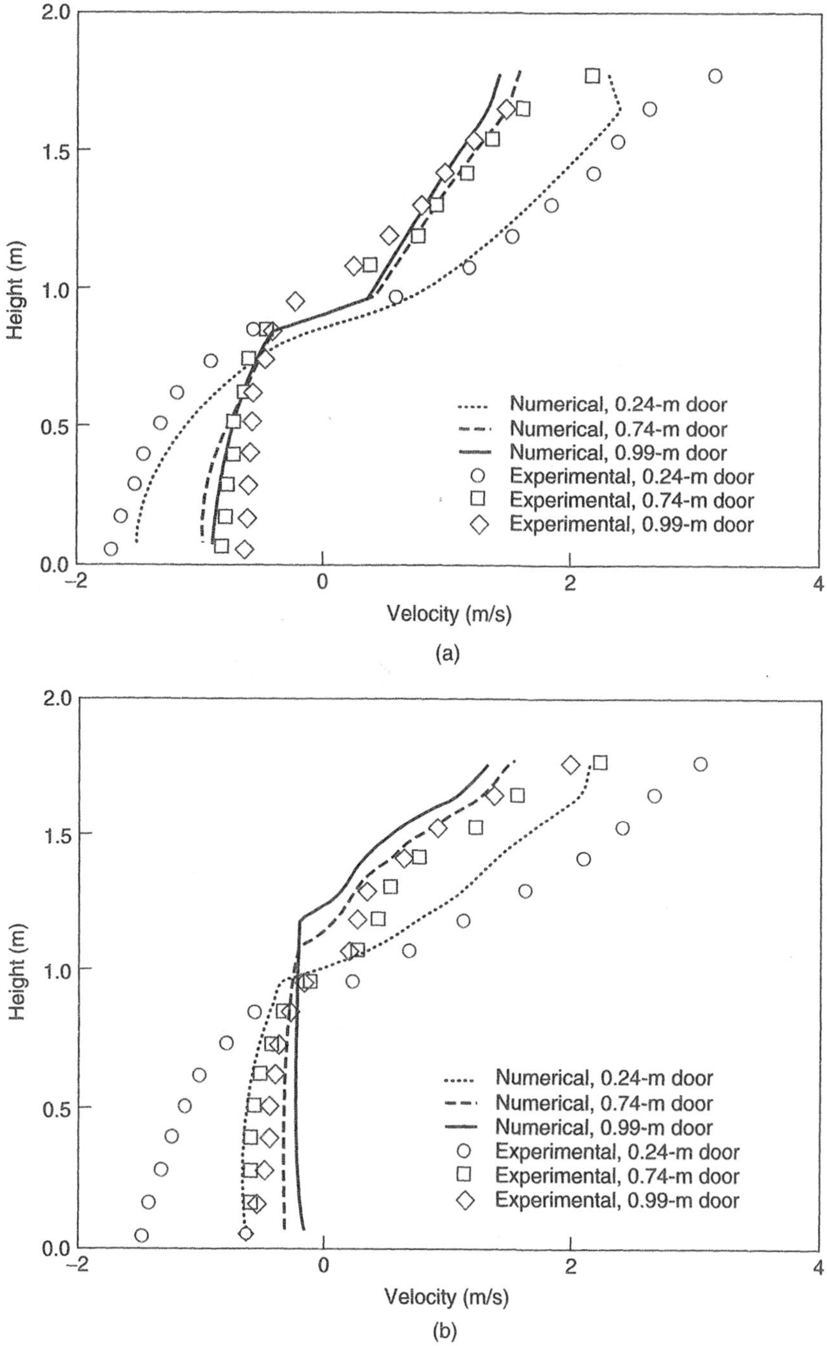

Figure 12.27. Predicted and measured door center vertical velocity profiles (a) Fire centrally located, (b) Fire in the corner of the room

the plume entrainment and the ceiling jet flowing toward the doorway further complicate the inflow and outflow through the doorway. As the door width is decreased, the flow becomes strongly three dimensional introducing greater measurement errors due to probe misalignment. In addition, the heat transfer characteristics, which influence the calculation of flow velocities, are more significant for a corner fire than for a centrally located fire. In order to minimize errors resulting from wall-fire plume interaction and to capture flow details, the calculated results can be improved by refining the grid, at least near the walls and near the fire source. The effect of using a refined grid is shown in Figure 12.28. The results indicate that the use of a finer grid improves the calculation of flow velocities quite considerably.

Given these experimental uncertainties and modeling assumptions, the results of this validation study for room fires show that the overall trends are captured fairly well. As expected, the corner fire case produces the worst correlation while the fire in the center of the room gives the best comparison. In addition, the model gives poor correlation of temperatures for adiabatic boundary condition than for the isothermal boundary condition (Compare (a) and (b) of Table 12.6). This is to be expected, as heat transfer characteristics are important in determining the enclosure flow conditions.

Because a field model is capable of better resolving the intracompartment flow details, it is also possible to calculate the plume tilt caused by the inflow of outside air through various openings such as doors and windows. Such information could be useful, for example, in positioning safety-related equipment or fire detectors. Figure 12.29 shows how in this centrally located fire, the plume is tilted slightly toward the wall facing the door. Such analysis is not possible with zone modeling techniques.

For fires located adjacent to the walls, the wall boundary conditions were found to be critical. The authors concluded that a full treatment of wall heat losses is required rather than using either the isothermal or adiabatic approximations.

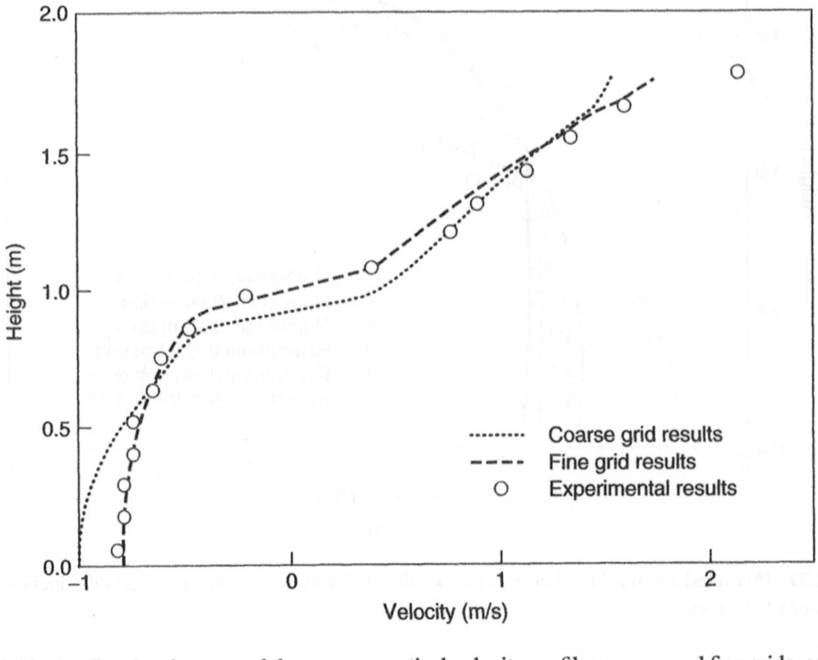

Figure 12.28. Predicted and measured door center vertical velocity profiles, coarse and fine grids, central fire

Table 12.6. Comparison of experimental results and CFDS-FLOW3D calculations for a 62.9-kW fire and 0.99-m door width. $H_{n/p}$ – neutral plane height, H_{door} – door height

Scenario	$H_{n/p}/H_{door}$	Mass flow rate (kg/s)		Hot layer temperature (°C)
		In	Out	
(a) Central fire				
Isothermal	0.515	0.732	0.741	108
Adiabatic	0.508	0.761	0.769	108
Experimental	0.582	0.653	0.701	109
(b) Corner fire				
Isothermal	0.694	0.298	0.299	208
Adiabatic	0.685	0.290	0.290	293
Experimental	0.586	0.513	0.491	172

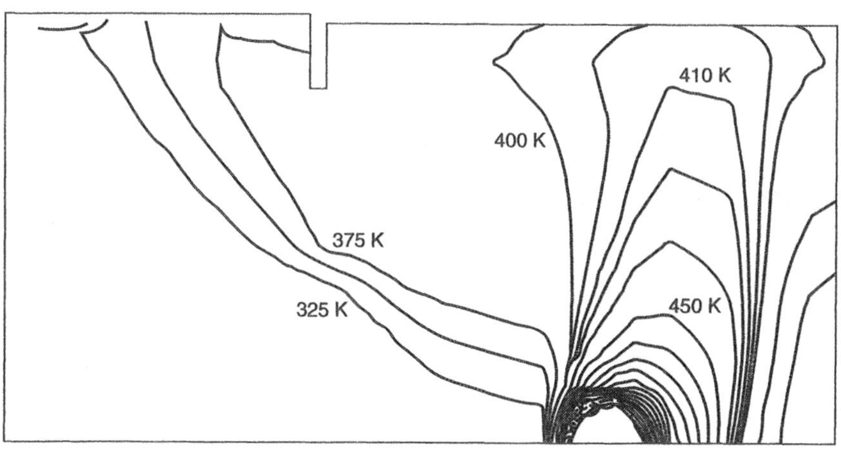

Figure 12.29. FLOW3D predicted temperature distribution through the center of the compartment passing through the center of the door, fire at center

From this validation study, it can be concluded that in view of some of the modeling assumptions and scarcity of the detailed experimental data, field models are very good at capturing some of the complex underlying physical phenomena. However, quantitative validation presents some difficulties both experimental as well as numerical modeling. For a reliable validation study, it is imperative that a detailed programme of field model validation is conducted with experiments specifically designed for such purposes. As this study has shown, the use of existing test data is inadequate.

12.10.2 MULTIROOM ENCLOSURE FIRES WITH NATURAL VENTILATION

Davis *et al.* (1991) present results of CFDS-FLOW3D validation studies for two sets of fire tests involving single-room and multiroom fire configurations. For the single-room fire scenario a two-dimensional grid was used to calculate the flow field, while a three-dimensional grid was utilized for the multiroom case. In each case the following simplifying assumptions were used.

- The fluid was assumed to be air and fully compressible;
- The k-ε turbulence model was used;
- The walls were assumed to be adiabatic, but for the multiroom case, the effect of a conducting ceiling was also examined;
- Radiation losses were accounted for by reducing the measured heat release data by 35%. This way 65% of the heat, that is, the convective component, was assumed to heat the gases.

For the single-room fire, the results were also compared with a zone model calculation. This was done to show that for simple fire situations zone models could readily be used to perform hazard analysis calculations instead of the more complex and relatively more time-consuming field models. However, for complex fire scenarios in which flow physics is uncertain, field models are the only choice for analysis.

Experimental setup

The experimental studies reported by Cooper and Stroup (1988) were used as the basis for the single-room fire scenario validation. The single-room fire tests were conducted in a room measuring 2.44 m wide, 3.66 m long, and 2.44 m high. The door to the fire room was 0.76 m wide and 2.03 m high. The total heat release rate of the fire, which was located next to one of the walls, was estimated from the oxygen consumption measurements just outside the door. Temperature profile in the center of the room was measured from a thermocouple tree.

The multiroom fire configuration is shown in Figure 12.30. It consisted of a three-room layout in which a long corridor (Room 2) connected two smaller rooms (Room 1 and Room 3). A door

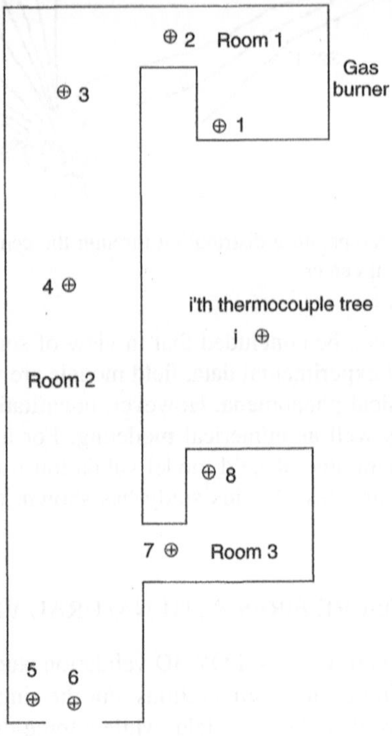

Figure 12.30. Three-room experimental layout and thermocouple location

at one end of Room 2 connects it to the outside. A set of eight thermocouple trees was used to measure temperature profiles at different locations as shown in Figure 12.30. The fire was simulated by a gas burner located in Room 1 giving a heat release rate of 100 kW.

Discussion of results

For the single-room fire, the calculated ceiling jet temperatures at 150 s from the start of fire were compared with the experimental measurements as shown in Figure 12.31. The results indicate that both the field and zone models give a good representation of the ceiling jet temperatures. The qualitative drop in temperature, from ceiling down toward the floor, is captured quite well by both the models. However, quantitatively both models consistently underestimate. The reason for this discrepancy lies in the modeling of heat loss characteristics. As noted earlier, the wall surfaces and the ceiling are modeled as adiabatic and consequently the radiative heating of gases from these surfaces is not taken into account. This assumption results in calculated temperatures that are lower than the measured values. The authors conclude that a more detailed modeling of heat transfer would improve the calculated results significantly.

Similar trends were found in the multiroom fire case. A comparison of calculated and measured temperature profiles in each room showed that the field model underestimated the gas temperatures in the fire room while overestimating in rooms away from the fire room. This was again due to the lack of radiative heating of the ceiling and other wall surfaces in the model. The overestimation of temperatures in rooms away from the fire room result from the fact that heat losses to the boundary walls (including the ceiling) are not accounted for in the calculations. This was clearly demonstrated by the apparent decrease in gas temperature that occurred when the ceiling was made conducting rather than adiabatic as the results of Figure 12.32 show. The authors again suggested that significant improvements in the calculated results would be realized if all the boundary surfaces were allowed to be conducting.

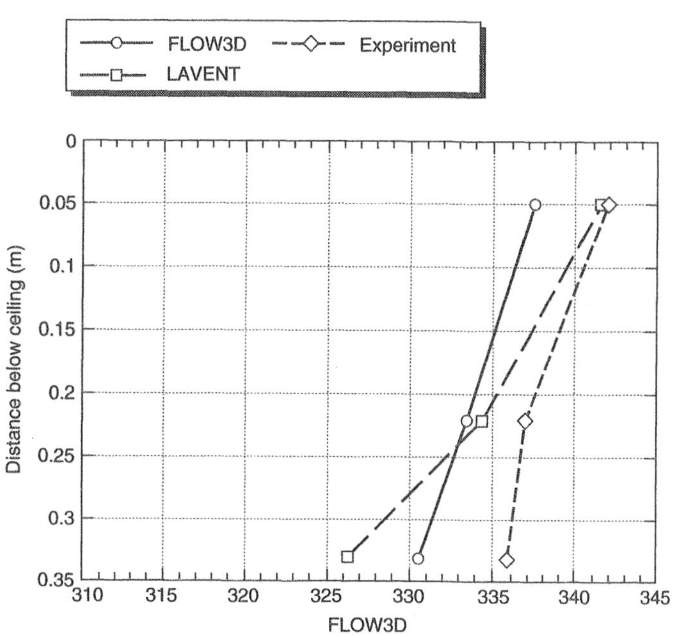

Figure 12.31. Ceiling jet temperatures at 150 s

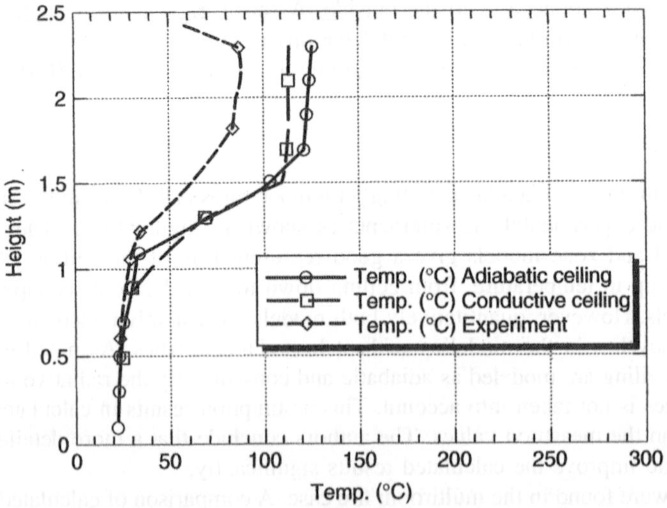

Figure 12.32. Temperature profiles for thermocouple tree 5 at 500 s

From these two sets of validation studies, the authors conclude that the calculation of enclosure gas temperature should include the effects of radiation from the fire source as well as from the hot wall surfaces. In addition, a detailed modeling of heat conduction through solid boundaries is required for accurate estimation of enclosure temperatures.

12.10.3 LARGE SINGLE-CELL ENCLOSURE FIRE

For safety considerations, fires in large spaces such as high bay warehouses, hospital wards, aircraft hangars, and exhibition halls present some difficult and challenging fire protection problems. Early fire detection and activation of suppression systems is important to limit its growth and spread. For the design of such protection systems, an accurate method of hazard analysis in terms of smoke movement and its temperature for determining detector response and design fire size is required. Zone models are inadequate because of their inherent assumptions and their inability to resolve flow details that are required for this type of analysis. Field models should therefore be used. Notarianni and Davis (1993) present results of a validation study using CFDS-FLOW3D for a large single-cell building.

For this study, Notarianni and Davis have used the experimental data gathered from fire tests carried out in an aircraft hangar measuring 389 × 115 × 30.4 m high. The hangar contained draught curtains spaced approximately 12.5 m apart, extending vertically down from the ceiling a distance of 3.7 m. These curtains were designed to prevent the spread of smoke across the ceiling.

The fire was simulated using technical grade isopropyl alcohol as fuel contained in an array of nine different pans forming a total fire area of 7.5 m^2. The fire was located on the floor in the center of the building. The average burning rate of the fire was 0.036 $kg/m^2/s$, giving an average heat release rate of 8250 kW.

Extensive temperature measurements were carried out using arrays of thermocouples positioned at radial distances from the fire centerline. The fire plume centerline temperature profile was also measured using an array of thermocouples placed directly over the fire source.

As a significant departure from other validation studies, the computer runs were made using several sets of parameters for the k-ε model as summarized in Table 12.7. The parameter C_3 represents the production/destruction of turbulence by buoyancy. (All k-ε parameters are defined in the FLOW3D manuals.) The radiation losses were accounted for by assuming that 35% of the total heat release rate was radiated from the plume at the fire source. The remaining 65% was used as the convective heat driving the fire plume flow.

Discussion of results

The plume centerline temperatures were calculated and compared with the experimental measurements for the four sets of k-ε parameters as summarized in Figure 12.33. The results show that for the case of zero buoyancy (K34) contribution to turbulence, the centerline plume temperatures are consistently underestimated. However, including the buoyancy term produces even larger discrepancy, at least for the lower portion of the plume. Near the ceiling calculations K36 and K38

Table 12.7. k-ε parameters used in the analysis of the hangar fire

Run #	C_1	C_2	C_3	C_μ	Prandtl number for enthalpy	CAPPA
K34	1.44	1.92	0	0.09	0.9	0.419
K35	1.44	1.92	1	0.09	0.9	0.419
K36	1.44	1.92	1	0.18	0.85	0.419
K38	1.44	1.92	1	0.15	0.85	0.419

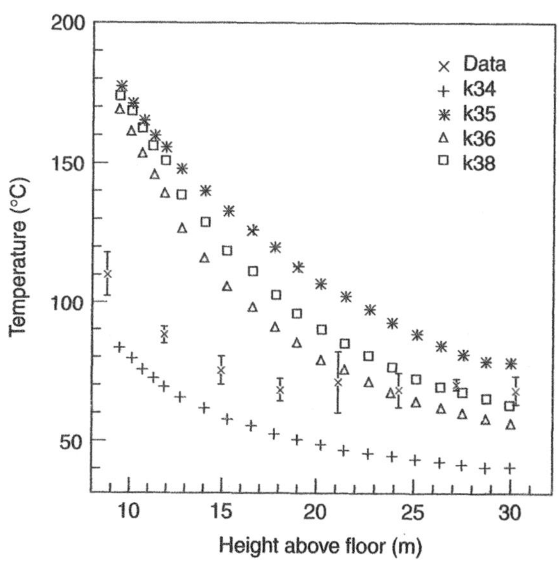

Figure 12.33. Comparison of CFD calculations using four different sets of k-ε constants with the measured centerline temperatures

provide better agreement. The authors identify two reasons for this discrepancy: experimental and modeling.

The experimental contribution comes from the misalignment of the thermocouple array with the plume centerline. In practice, due to the swaying of the thermocouple tree, it is extremely difficult to align the thermocouples with the plume centerline. The temperature measurements are therefore not truly the measurements at the centerline.

The modeling difficulties arise from the fact that heat losses near the ceiling and the effects of beams are not modeled. The presence of additional beams and the consequent heat losses are reflected in the measurements, which indicate that the temperature remains constant at height above 20 m.

Figure 12.34 shows an improvement in the calculated results when the grid was refined near the ceiling and the ceiling was assumed to be conducting and not adiabatic.

Temperature measurements along the ceiling (at 0.15 m below the ceiling) and the corresponding FLOW3D calculations are shown in Figure 12.35. The results indicate that at distances beyond the draught curtain, temperature drops substantially for both the experimental measurements and the calculations indicating good agreement.

From this study the authors conclude that, for a large building the modified k-ε parameter (K36) provide good agreement with experimental measurements. With detailed modeling of ceiling structures, the plume centerline and radial temperatures could be calculated with acceptable degree of accuracy.

12.11 Conclusions concerning validation of zone and field models

In the above discussion of validation studies concerning zone and field models, it is clear that there are many difficulties (conceptual as well as experimental) in carrying out exhaustive and unambiguous validation of fire models. No matter how many different sets of such comparisons between the calculated and experimental values are carried out, no model can be *proved to be*

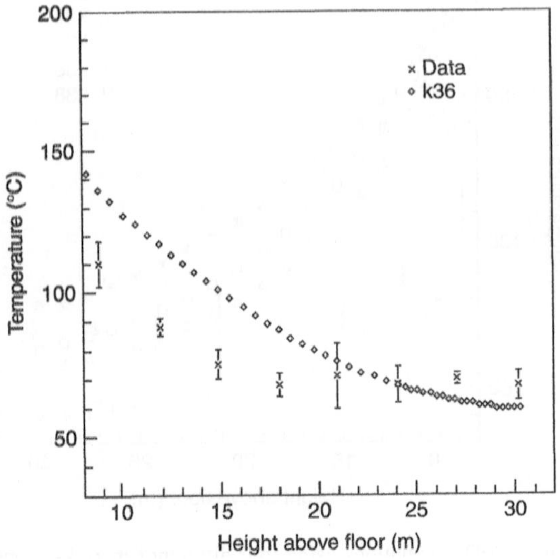

Figure 12.34. Comparison of a CFD calculation using the turbulence constants of k36 and a more detailed grid with the measured centerline temperatures

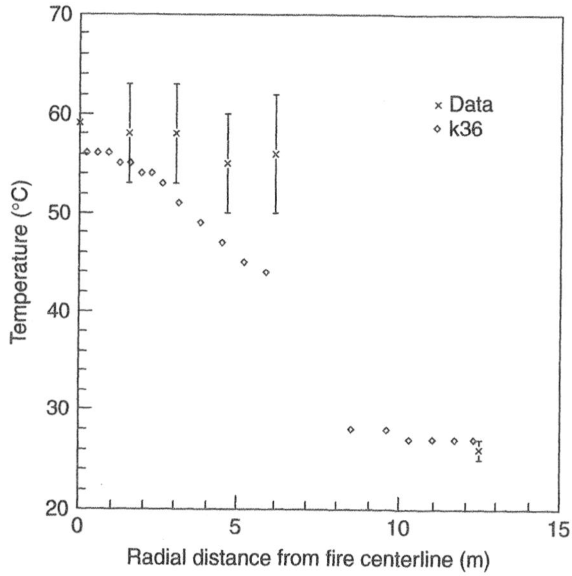

Figure 12.35. Comparison of the CFD calculated ceiling temperatures using a detailed grid and the k36 turbulence constants with the measured temperatures at various radial distances from the fire centerline and 0.15 m below the ceiling

correct for *all* fire situations. All modeling approach inherently consists of systematic variances between experiment test and model estimates. These are termed *the residual differences* (Davies, 1985). The purpose of validation should not be to prove the universality of application of a given model, rather to aim at quantifying residual differences between measured and calculated values.

For this to be done satisfactorily, comprehensive and exhaustive sensitivity studies are necessary for a given model (Beard, 1992). In addition, a large number of such comparisons are needed to gain a degree of confidence in the calculated results. Here, the issues such as "openness" of computer codes and the independence of people involved in the validation process need to be addressed.

There is an urgent need for establishing reliable data sets for different cases to ensure consistency of input data across various models and to make the comparison of results between different models possible.

Above all, it is important that the use of these models for practical applications must not be left to nonexperts. The models cannot be regarded as "black boxes." The user must be aware of a model's limitations and underlying assumptions (both qualitative and quantitative). The interpretation of numbers resulting from a model must be done from a full knowledge and through grasp of the physical phenomena involved, as well as from a complete understanding of the implications of the use of a particular set of input data. In this way, valuable information can be obtained for the purposes of engineering design. Without this understanding, it is possible to draw misleading conclusions.

Symbols

c	specific heat (J/kg K)
H	ceiling height

H_{door}	door height
$H_{n/p}$	neutral plane height k thermal conductivity (W/m K)
$Q(t)$	heat output at time t (kW)
r	radial distance
s	stoichiometric mass requirement of oxygen
t	time (min)
t_0	time at start of fire
α	proportionality constant
λ	fraction of fire heat output lost by radiation and convection to surrounding surfaces
ρ	density (kg/m^3)

References

Alvares, N J, Foote, K L and Pagni, P J (1984). Forced ventilated enclosure fires. *Combustion Science and Technology*, **39**, 55–81.

Alvares, N J and Foote, K L (1981). *Contrast Between Natural and Experimentally Controlled Fires in Forced Ventilated Enclosures*, LLNL Report.

Beard, A (1992). Limitations of computer models. *Fire Safety Journal*, **18**, 375–391.

Cooper, L Y and Stroup, D W (1988). Test results and predictions for the response of near-ceiling sprinkler links in a full-scale compartment fire. *Proc. of the 2nd Int Symp. Fire Safety Science*, pp. 623–632.

Cooper, L Y, Harkleroad, M, Quintiere, J G and Rinkinen, W (1982). An experimental study of upper hot layer stratification in full scale multi-room fire scenarios. *Journal of Heat Transfer*, **104**, 741–50–NOTES.

Cox, G (1995). *Combustion Fundamentals of Fire*, Academic Press.

Cox, G and Kumar, S (1987a). Field modelling of fire in forced ventilated enclosures. *Combustion Science and Technology*, **52**, 7–23.

Cox, G and Kumar, S (1987b). *A Numerical Model of Fire in Road Tunnels, Tunnels and Tunnelling*, March 1987, pp. 55–60.

Davies, A D (1985). *Applied Model Validation*, NBSIR 85-3154-1.

Davis, W D, Forney, G P and Klote, J H (1991). *Field Model Modelling of Room Fires*, NISTIR 4673, November 1991.

Gandhi, P D (1993). *Validation of FAST for room corner fire tests. Proceedings of the 6th International Fire Conference, INTERFLAM '93*, pp. 331–341.

Jaluria, Jogesh (1988). Natural convection wall flows *The SFPE handbook of Fire Protection Engineering*, Society of Fire Protection Engineers, Boston, USA.

Jones, W W and Peacock, R D (1989). Refinement and Experimental Verification of a Model for Fire Growth and Smoke Transport. *Fire Safety Science, Proc of the 2nd International Symposium*, pp. 897–906.

Kerrison, L, Galea, E R, Hoffmann, N and Patel, M K (1994). A comparison of a FLOW3D based fire field model with experimental room fire data. *Fire Safety Journal*, **23**, 387–411.

Kumar, S and Cox, G (1985). Mathematical modelling of fires in road tunnels. *5th Int. Symp. on The Aerodynamics & Ventilation of Vehicle Tunnels*, Lille, France, May, 1985, pp. 61–76.

Kumar, S, Gupta, A K and Cox, G (1992). Effects of thermal radiation on the fluid dynamics of compartment fires. *Proc. of the 3rd Int. Symp. of Fire Safety Science*, pp. 345–354.

Kumar S and Yehia, M (1984). (1994). Mathematical field modelling of transient ceiling jet characteristics in fires, *Proc. of the 1st ISHMT-ASME Heat and Mass transfer conference*, Bombay, India, 5–7 January, 1994.

Levine, R S and Nelson, H E (1990). *Full Scale Simulation of a Fatal Fire and Comparison of Results with Two Multiroom Models*, NISTIR 90-4268, August 1990.

Mitler, H E (1984). Zone modelling of forced ventilation fires. *Combustion Science and Technology*, **39**, 83–106.

Mitler, H E and Emmons, H W (1981). *Documentation for CFCV, The Fifth Harvard Computer Code*, NBS Report, NBS Grant, 67-9011.

Montevalli, V and Marks C H (1991). *3rd Int. Symp. on Fire Safety Science*, pp. 301–312.

Montevalli, V and Ricciuti, C (1992). Technical Report, NIST-GCR-92-613.

Notarianni, K A and Davis W D (1993). *The Use of Computer Models to Predict Temperature and Smoke Movement in High Bay Spaces*, NISTIR 5304, December 1993.

Peacock, R D, Forney, G P, Reneke, P, Portier, R, Walter, W J (1993). *CFAST, The Consolidated Model of Fire Growth and Smoke Transport*, NIST Technical Note 1299.

Rockett, J A, Morita, M. and Cooper, L Y (1987). *Comparisons of NBS/Harvard VI Simulations and Full-Scale, Multi-Room Fire Test Data*, NBSIR 87-3567.

Rockett, J A, Morita, M and Cooper, L Y (1989). Comparisons of NBS/Harvard VI simulations and full-scale, multi-room fire test data. *Fire Safety Science, Proceedings of the 2nd International Symposium*, Hemisphere, pp. 481–490.

Steckler, K D, Quintiere, J G and Rinkinen, W J (1982a). Flow induced by fire in a compartment. *19th International Symposium on Combustion*, The Combustion Institute, Pittsburg, pp. 913–920.

Steckler, K D, Quintiere, J G and Rinkinen, W J (1982b). *Flow Induced by Fire in a Compartment*, NBSIR 82-2520, Washington, September 1982.

13 POINT SYSTEMS – A SINGLE INDEX

13.1 Introduction

A point system produces a number or an index that is the sum total or product of points allocated to various attributes of a system. It is designed to characterize an overall effect within a scale of such effects. For example, the number may indicate the standard of comfort in a weather system. Meteorologists have realized that temperature alone does not represent the coldness of a winter day. Therefore, they created the wind-chill index from a combination of temperature and wind speed to measure the cooling effect of the wind. Such point systems have been widely used in fire safety evaluation. The fire processes and mitigating effects of protection measures are modeled and expressed in points in order to give a simple and rapid evaluation of fire safety.

Point systems constitute various processes of modeling and scoring causal and mitigating fire safety attributes to produce a rapid and simple fire safety evaluation. Point systems are useful and powerful tools that can provide valuable information on the risks associated with fire. Point systems have been applied to a variety of hazards and risk assessment projects to reduce fire safety costs, set priorities, and facilitate use of technical information. They provide an important link between the complex scientific principles of theoretical and empirical models, and the less than perfect circumstances found in real world applications compared with laboratory conditions.

Fire safety evaluation point systems have been referred to by various names such as risk ranking, index systems, and numerical grading. They originated as insurance rating schedules in the nineteenth century, but in the last few decades the basic concepts have appeared in a wide variety of formats.

In general, point systems assign values to selected variables based on professional judgment and experience. The selected variables represent both positive and negative fire safety features and the assigned values are then operated on by some combination of arithmetic functions to arrive at a single value or index. This value can be compared with other similar assessments or to a standard. The variables are referred to as attributes.

The nature of fire safety evaluation point systems is examined in this chapter and some significant examples that have gained widespread use are described. The chapter emphasizes how the principles of multiattribute evaluation from the field of decision analysis have been used

Evaluation of Fire Safety D. Rasbash, G. Ramachandran, B. Kandola, J. Watts and M. Law
 ISBN: 0-471-49382-1

in the development of robust models of fire safety evaluation. The processes described in this chapter include the identification of attributes and the methods of weighting them. The chapter also discusses rating within a scale and scoring methods.

13.2 Concepts

Point systems are popular because they are simplified models of complex systems. They assign values to important attributes of the problem and aggregate them into a score or index. The Glasgow Coma Scale is an excellent example of a point system in the field of medicine.

The Glasgow Coma Scale is widely used by neurologists for the evaluation and prognosis of head injuries (Yates, 1990). As shown in Table 13.1, the scale is based on three attributes, verbal response, motor response, and eye opening.

Each attribute is evaluated and the three selected values are added to produce a score. Patients receiving scores of 13 to 15 are considered to have minor head injuries with an excellent prognosis. Moderate head injury scores range from 9 to 12 and severe head injury or coma is defined as a scale value less than 9. This approach has been so successful that it is being proposed as a method to define death.

The Glasgow Coma Scale illustrates how powerful and useful a simple ordinal ranking system can be. In neurology as in fire safety, there is a great deal that we know and a great deal that we do not know. We do not need to let the limits of our knowledge deter us from the best use of what we do know. Point systems can be simple yet powerful ways to use our increasing body of knowledge in the evaluation and communication of fire safety.

If properly constructed, point systems offer a defensible combination of relevant attributes of fire safety. However, because they are heuristic models they are difficult to verify. The most valid point systems are those that follow the well-founded principals of multiattribute evaluation.

Table 13.1. Glasgow Coma Scale

Verbal response	
None	1
Incomprehensible sounds	2
Inappropriate words	3
Confused	4
Oriented	5
Motor response	
None	1
Abnormal extensor	2
Abnormal flexion	3
Withdraws	4
Localizes	5
Obeys	6
Eye opening	
None	1
To pain	2
To speech	3
Spontaneously	4

13.2.1 DEFENSIBILITY

Fire safety evaluation involves the analysis of many complex factors that are difficult to assess in a uniform and consistent way. Consequently, it is important that an evaluation can be defended and justified. Defensibility, both internally and externally, is one of the strongest assets of a scientifically constructed point system. Internal defensibility provides management with justification of fire safety policies and expenditures. It facilitates the allocation of limited resources among fire and other risks. External justification of priority setting is important in litigation and in dealing with regulatory agencies. Point systems consistently manage the multiple attributes involved in fire safety evaluation by providing a logical structure to the assessment of the options involved.

13.2.2 HEURISTIC MODELS

When formal optimizing algorithms do not exist or are inefficient, there still may be some sensible things to do. Intuition and experience can often provide good – but not necessarily optimal – solutions. Procedures that work but do not have a formal underlying theory are called *heuristics*.

Point systems are heuristic models of fire safety. They are processes of modeling and scoring fire hazard and exposure factors to produce a rapid and simple estimate of comparative evaluation. The processes heuristically relate known fire safety attributes that have varying degrees of accuracy in their measurement. The distinct advantage of a point system is a user-friendly comprehensive model that addresses all relevant aspects of fire safety with the capability of incorporating known relationships of behavioral and physical processes.

13.2.3 MULTIATTRIBUTE EVALUATION

By nature of the circumstances, fire safety decisions often have to be made under conditions in which the data are sparse and uncertain. The technical parameters of fire safety evaluation are very complex and normally involve a network of interacting components, the interactions generally being nonlinear and multidimensional. However, complexity and sparseness of data do not preclude useful and valid approaches. Such circumstances are not unusual in decision-making in business or other risk ventures and if such problems are not addressed, developments that could be useful to society may be inhibited. The space programme illustrates how success can be achieved when there is little relevant data. One applicable approach to fire safety evaluation is multiattribute evaluation.

As implied above, fire safety decisions require more than one attribute to capture all relevant aspects of the consequences. If the attributes for a decision problem are $x_1, x_2, x_3, \ldots, x_n$, then an evaluation function $E(x_1, x_2, x_3, \ldots, x_n)$ needs to be determined over these measures in order to conduct a performance assessment. However, determining an appropriate evaluation function over multiple performance measures is a complex problem. Subjective attribute values must be elicited by asking questions, and it seems difficult to answer questions that will directly determine an n-dimensional function.

Keeny and Raiffa (1976) have showed that if trade-offs among the attributes do not depend on the levels of the remaining attributes, then a single measure of the overall outcome of a system is given by

$$E(x_1, \ldots, x_i, \ldots, x_n) = \sum_{i=1}^{n} w_i R_i(x_i)$$

where the w_i are weighting constants greater than zero and the $R_i(x_i)$ are normalizing functions of the attributes. This and other multiattribute evaluation models are discussed later in this chapter.

13.2.4 APPLICATION

For many situations in which a quantitative fire safety evaluation is desirable, an in-depth theoretic analysis may not be cost-effective or appropriate. This could be the fundamental case where great sophistication is not required, where prioritization is the principal objective, or where it is necessary to institutionalize an approach to fire safety evaluation for a wide base of use.

Because of our limited phenomenological knowledge, the accuracy demanded for a fire safety evaluation is different from that for other engineering purposes. Often, establishing an order of magnitude will suffice. Time and resource expenditure increases as the depth of analysis is increased. In an age in which resources are scarce, and efficiency is prized, maximizing the utility of a point system approach to fire safety evaluation is clearly desirable for the many situations in which the evaluation of fire safety is fundamental.

Perhaps the most common implicit justification of point systems is the need for a simplistic process of fire safety evaluation. In most applications, the target of a point system is a broad class of products or facilities for which a detailed fire risk analysis of each individual case is not feasible. Point systems have an appeal to administrators charged with risk management decision-making responsibilities but who may be unfamiliar with the details and mechanics of the fire risk assessment process. Widespread implementation of a generalized approach to fire risk is contingent on its appeal to a broad class of users including architects, building officials, and property managers.

Point systems are not substitutes for detailed theoretic fire safety evaluation. They are planning tools useful for screening, ranking, and setting priorities.

13.3 Examples of established point systems

Point systems come in a large variety of formats and with a broad spectrum of purposes. For purposes of this section, existing point systems are divided into four categories. First is the Gretener system that is based on a multiplicative model. Second is the Dow Fire and Explosion Index that is an industry specific example. Next is the Fire Safety Evaluation System (FSES) and similar models that are additive but do not separate the attribute weights and values. Fourth are the methods with a theoretical basis in multiattribute evaluation models from the field of decision analysis. More in-depth reviews of some of these models may be found in the SFPE *Handbook of Fire Protection Engineering* (Watts, 2001).

13.3.1 GRETENER METHOD

One of the more standardized fire risk evaluation point systems is the Gretener method developed in Switzerland during the 1960s and 1970s, and now in use in many areas throughout Central Europe (Fontana, 1984). In the Gretener method, fire risk is measured as the ratio of negative features that increase risk to positive features that decrease risk. The basic relationship in the method is given by the equation

$$R = (P \cdot A)/(N \cdot S \cdot F)$$

where R = Risk, P = Potential hazard, A = Activation (ignition) hazard, N = Normal protection measures, S = Special protection measures, and F = Fire resistance of the structure.

In turn, each of these five factors comprising fire risk is the product of several components, for example, values of nine components are multiplied together to yield the value for Potential Hazard. The process is normalized so that a "standard" building has a computed fire risk value of 1.00. An acceptable risk is one for which the calculated ratio is less than or equal to 1.30.

The Gretener system has more than 30 years of application and acceptance in central Europe. However, the underlying logic for determining most of the component values is not apparent from the documentation. Gretener's method is insurance oriented with an emphasis on water supply and manual fire fighting that is greater than other fire safety evaluation point systems. Rasbash (1985) compared factors in the Gretener system with statistical estimates of savings or loss associated with the factors and found the expected annual financial loss was proportional to the square of the calculated risk, R.

A number of commercially available computerized models of the Gretener method have been developed in the last decade. These include FREM (Watts *et al.*, 1995), RiskPro (2000), F.R.A.M.E. (2000), and RISK DESIGN (FSD, 2000).

13.3.2 DOW FIRE AND EXPLOSION INDEX

A need for systematic identification of areas with significant loss potential motivated Dow Chemical Company to develop the Fire and Explosion Index and risk guide (Dow, 1966). The original edition issued in 1964 was a modified version of the "Chemical Occupancy Classification" rating system developed by Factory Mutual prior to 1957. It has been subsequently improved, enhanced, and simplified, and is now in its 7th edition (Dow, 1994).

Today there are many risk assessment methods available that can examine a chemical plant in great detail. The Fire and Explosion Index remains a valuable screening tool that serves to quantify the expected damage from potential fire, explosion, and reactivity incidents and to identify equipment that could likely contribute to the creation or escalation of an incident (Scheffler, 1994). Risks associated with operations in which a flammable, combustible, or reactive material is stored, handled, or processed can be evaluated with this system. The guide is intended to provide a direct and logical approach for determining the probable "risk exposure" of a process plant and to suggest approaches to fire protection and loss prevention design. An important application is to help decide when a more detailed quantitative risk analysis is warranted, as well as the appropriate depth of such a study.

Dow Indices is a software tool for implementing the Dow Fire and Explosion Index in an interactive, computer-based environment (Parikh and Crowl, 1998). The program includes a library of chemicals, on-line help, and a variety of visual tools to determine the dominant contributors to the overall hazard. The Dow Indices tool can be linked to existing chemical process simulators and can be coupled with economic evaluators, such as a cash flow analysis, using the maximum probable property damage and business interruptions loss predicted by the index.

13.3.3 FIRE SAFETY EVALUATION SYSTEM (FSES)

NFPA 101, *Life Safety Code* of the National Fire Protection Association (NFPA 101, 2000) is one of the most widely used voluntary codes for identifying a minimum level of fire safety. The Fire Safety Evaluation System (FSES) (Benjamin, 1979, Nelson and Shibe, 1980) is a point system approach to determining equivalency to the NFPA *Life Safety Code* for certain occupancies. The technique was developed in the late 1970s at the Center for Fire Research, National Bureau of Standards (presently the Building and Fire Research Laboratory, National Institute of Science and Technology). It has been adapted to new editions of the *Life Safety Code* and is presently published in NFPA 101A, *Alternate Approaches to Life Safety* (NFPA 101A, 2001).

The FSES was developed to provide a uniform method of evaluating fire safety to decide what measures would provide a level of safety equivalent to that provided by the *Life Safety Code*. The objective was to compile an efficient evaluation system that would present useful information with a minimum amount of effort by the user.

FSES for health care occupancies

Unlike the *Life Safety Code* itself, the FSES for health care occupancies begins with a determination of relative risk derived from basic occupant characteristics. Five occupancy risk parameters are used: patient mobility, patient density, fire zone location, ratio of patients to attendants, and average patient age. Values for these parameters, and all others in the FSES, were determined on the basis of the experienced judgment of a group of fire safety professionals and represent the opinions of that panel of experts. There is no documented process for validating or revising the values.

Safety features must offset the calculated occupancy risk. The FSES uses 13 fire safety parameters with up to seven levels of safety for each parameter. An important concept of the FSES is redundancy through simultaneous use of alternative safety strategies. This serves to ensure that failure of a single protection device or system will not result in a major life loss. Three fire safety strategies are identified: containment, extinguishment, and people movement.

The FSES determines if the measured level of fire safety is equivalent to that of the *Life Safety Code* by comparing the calculated level for each fire safety strategy to stated minimum values. Evaluating a range of code complying buildings with the evaluation system produces these minimum values.

A distinct advantage of point systems is that they lend themselves to computer programming and optimization techniques. ALARM 1.0 (Alternative Life safety Analysis for Retrofit Cost Minimization 1.0) is a personal computer software tool that helps decision makers in health-care facilities to achieve cost-effective compliance with NFPA 101, *Life Safety Code* (Webber and Lippiatt, 1994, 1996). The program is based on earlier work by Chapman and Hall (1982, 1983). It uses a mathematical optimization algorithm called *linear programming* to quickly evaluate all possible code compliance solutions and identify the least-cost means of achieving compliance. ALARM 1.0 generates a set of options from which the most appropriate code compliance strategy based on cost and design considerations can be selected. Also listed – for both individual zones and the entire building – are up to 20 alternative, low-cost compliance plans and the prescriptive solution for benchmarking purposes. The software includes the integrated code compliance optimizer, full-screen data editor, and file manager. ALARM 1.0 is available from the National Fire Protection Association (www.nfpa.org) through the One-Stop-Data-Shop.

Derivative applications

NFPA 101A now includes FSESs for health-care occupancies, correctional facilities, board and care homes, and business occupancies. One of the most widely used of these is Chapter 8, Fire Safety Evaluation System for Business Occupancies. NFPA 101 classifies the transaction of business other than mercantile, the keeping of accounts and records, or similar purposes, as a business occupancy. Typical examples are professional, financial, and governmental offices. The FSES for business occupancies was derived from a project to appraise the relative level of life safety from fire in existing office buildings and combination office–laboratory buildings of a US government agency (Nelson, 1986). It was based on the approach developed for health care occupancies and was subsequently incorporated into NFPA 101A. An analysis of the FSES for business occupancies, using the attribute value spread as a measure of importance to rank the fire safety attributes, found a difference between criteria for new and existing buildings of 6–10% (Watts, 1997).

Enhanced Fire Safety Evaluation System for Business Occupancies (Hughes Associates, Inc. (1999)) is a personal computer implementation of Chapter 8 in NFPA 101A and has been adopted as Chapter 9 of NFPA 101A. The software automates the calculation process and the generation of forms. Also, it provides the user with guidance and on-line help in making the decisions involved in completing the FSES. The help screens provide background information and reference material to assist the user in choosing attribute values. Another enhancement allows the user to interpolate between attribute values in the worksheet table. The program also allows "refinement" calculations that consider attributes in more depth. For example, the construction refinement calculation uses Law's fire severity calculation (Law, 1973) to estimate the fire duration in the worst-case space in the building. If this result is less than the fire resistance of the structural elements in the buildings, then the attribute value can be increased. The program is distributed through NFPA, and can be downloaded and installed from ftp://209.21.183.33/efsesinstall.exe where the user manual is also on-line.

Also derived from the original FSES is Section 3408, Compliance Alternatives, of the *BOCA National Building Code* (BOCA, 1996), an indexing system for fire safety in existing buildings. As stated in paragraph 3408.1, the purpose of this section is to maintain or increase safety in existing buildings without full compliance of other chapters of the Code. This system allows for older designs to be judged on their performance capabilities rather than forcing the buildings to comply with modern standards for new construction. Originally adopted in 1985, significant changes were effected between the 1993 and 1996 editions. Section 3408 is applicable to all occupancy use groups. For each use group, there are separate point values for each safety attribute and separate mandatory values to be considered as criteria for equivalency. A detailed comparison of both qualitative and quantitative aspects of the *BOCA National Building Code*, Compliance Alternatives and Chapter 8 of NFPA 101A shows some significant differences (Watts, 1998).

Chapter ILHR 70 of the Wisconsin Administrative (Code, 1995) is a building code for historic structures that is similar in many respects to the Compliance Alternatives of the *BOCA National Building Code*. Its purpose is to provide alternative building standards for preserving or restoring buildings or structures designated as historic. Subchapter IV of the Code is a point system called the *Building Evaluation Method*. It assesses life safety for a qualified historic building by comparing 17 building safety attributes with the requirements of the prevailing code. If an historic building has less of an attribute than is required by the prevailing code, a negative number is assigned. If an historic building has more of an attribute than is required by the prevailing code, a positive number is assigned. Thus, evaluation is directly related to the prevailing code. If the sum of all the attributes is greater than or equal to zero, the building is compliant. The same trade-offs previously would have been allowed under the variance petition process but are now codified. This adds a degree of certainty of approval that did not previously exist, often impeding development of historic buildings. Unlike other FSES applications, the Building Evaluation Method has no mandatory scores. If the total safety score is equal to or greater than zero, the building is considered code compliant. Also dissimilar to other FSES models, Wisconsin Subchapter IV does not vary, occupancy. A table for each attribute gives a set of numerical values, one of which is selected for each evaluation. Criteria for these values refer directly to the prevailing code. The same set of values applies for all applicable building uses and occupancies. This code can be accessed through the State of Wisconsin, Department of Commerce web site http://www.commerce.state.wi.us.

13.3.4 MULTIATTRIBUTE EVALUATION EXAMPLES

While the above models could also be considered as multiattribute evaluation models, none of them make the distinction between the intensity and importance of the risk attributes. That is, they do not directly convey the difference between a lot of a not-so-good attribute, a little bit of a very good attribute, and more importantly, all the other possible combinations in between. This issue was addressed in the 1980s at the University of Edinburgh and has been developed to incorporate the theoretical aspects of multiattribute evaluation in the construction of point systems.

Edinburgh model

Development of a hierarchical point system approach was initially undertaken at the University of Edinburgh, sponsored by the UK Department of Health and Social Services (Department of Fire Safety Engineering, 1982, Stollard, 1984, Marchant, 1988). The objective of this study was to improve the evaluation of fire safety in UK hospitals through a systematic method of appraisal. This approach was further developed at the University of Ulster for application to dwelling occupancies (Shields *et al.*, 1986, Donegan *et al.*, 1989).

Defining fire safety is difficult and often results in a listing of factors that together comprise the intent. These factors tend to be of different sorts. For example, fire safety may be defined in terms of goals and aims such as fire prevention, fire control, occupant protection, and so forth. These broad concepts are usually found in the introductory section of Building Codes and other fire safety legislation. Or, fire safety may be defined in terms of more specific hardware items such as combustibility of materials, heat sources, detectors, sprinklers, and so forth. These topics are more akin to items listed in the table of contents of Building Codes. A meaningful exercise is to construct a matrix of fire safety goals versus more specific fire safety features. This helps to identify the roles of these two concepts, in both theory and practice.

As a logical extension of this single fire safety matrix, consider that there are more than two categories of fire safety factors. This suggests a hierarchy of lists of things, or decision-making levels, that comprise fire safety. Such a hierarchy of fire safety decision-making levels is shown in Table 13.2.

This hierarchy of levels of detail of fire safety suggests that a series of matrices is appropriate to model the relationships among various fire safety factors, that is, a matrix of policy versus objectives would define a fire safety policy by identifying the specific objectives held most desirable. In turn, a matrix of objectives versus strategies would identify the relationship of these factors, and a matrix of strategies versus parameters would suggest where to use what. Thus, a matrix may be constructed to examine the association of any two adjacent levels in a hierarchy of fire safety factors.

An even more appealing aspect of this approach is that two or more matrices may be combined (multiplied) to produce information on the importance of fire safety factors to the overall fire safety policy at any level of management decision making. This process is used to produce a one-dimensional matrix or vector of parameter weights that specifies the relative importance of each fire safety parameter to the overall fire safety policy. Details of the method are discussed in a subsequent section of this chapter.

The resulting vector of parameter weights identifies the importance of a parameter to fire safety. To develop a fire safety evaluation of a specific building or space, it is also necessary to assess a parameter grade. This is the extent to which each parameter is present, or how much or

Table 13.2. Hierarchy of fire safety decision-making levels

Level	Name	Description
1	Policy	Course or general plan of action adopted by an organization to achieve security against fire and its effects
2	Objectives	Specific fire safety goals to be achieved
3	Strategies	Independent fire safety alternatives, each of which contributes wholly or partly to the fulfillment of fire safety objectives
4	Attributes	Components of fire risk that are determinable by direct or indirect measure or estimate
5	Survey items	Measurable features that serve as constituent parts of a fire safety parameter

what level of functionality of each parameter is available in the specific building or space; for example, the degree of fire resistance of structural members. These parameter grades are directly measurable or derived from various functions of items in a lower hierarchical level. The value of a parameter in a specific building or space is the product of its weight and its grade.

The sum of the parameter values, or the scalar product of the parameter weights and grades, yields a relative measure of fire safety. This may be used to rank facilities or it can be compared with a standard value.

Another important contribution of the Edinburgh model is the parameter interaction matrix. Construction of a square matrix of the parameters provides a systematic approach to the assessment of interdependence of each pair of parameters. This permits adjustment of results to reflect synergism and other associations of parameters in a consistent manner.

Central office fire risk assessment

Several incidents in the last decade have indicated the potential severity of a fire in telecommunications facilities. Interruption of a communications network in a telephone central office can result in serious impact on emergency services, health care facilities, financial institutions, and other organizations with intensive electronic telecommunications. Conformance with fire safety code requirements does not adequately address equipment susceptibility or service continuity. To deal with this problem, a point system identified as COFRA (Central Office Fire Risk Assessment) has been developed (Budnick *et al.*, 1997, Parks *et al.*, 1998).

Initial evaluation of the problem revealed that significant conflicts existed among demands for technical accuracy, ease of use, and implementation costs. From this, the Edinburgh model was chosen as the most efficient approach. However, the methodology was modified to evaluate the potential for fire damage to critical telecommunications equipment and service interruption. The problem was subsequently partitioned into separate components for life safety and for integrity of the communications network.

Extensive effort was directed to the development and documentation of parameter grades. To simplify the rating of parameters they were partitioned into measurable constituent parts. Usually, these parts were directly measurable survey items. Several parameters also had subparameters. Each parameter was analyzed with respect to the visible characteristics of a facility that would affect the contribution of the parameter to network integrity. The items were chosen for contributing significantly to the effectiveness of their respective parameters or subparameters and for being directly measurable. Each survey item was defined in sufficient detail to support these traits.

Decision tables were used to develop the logic for translating survey items into parameter grades (Watts *et al.*, 1995). Input to these tables included fire test results, fire hazard modeling, field experience from previous fire events, logic diagrams, and professional judgment. Subparameter weights were determined using the Analytic Hierarchy Process (AHP). The application has been coded and field-tested as a user-friendly software package for personal computers (Parks, 1996).

Other examples of multiattribute evaluation models

Recent applications of the multiattribute evaluation approach to point systems have developed in many forms and for many uses. In the US, the Historic Fire Risk Index has been developed to include an assessment of the cultural significance as a parameter of fire risk (Kaplan and Watts, 1999, Watts and Kaplan, 2001). In Hong Kong, aspects of fuzzy systems theory have been incorporated into a point system to evaluate existing high-rise buildings (Lo, 1999).

Although most point systems focus on life safety issues, FireSEPC (fire safety evaluation procedure for the property of parish churches) is insurance motivated and deals with the worth of the building (Copping, 2000). Using the hierarchical framework, the procedure rates the contribution of 18 fire safety components and compares the score to a "collated norm" developed from guidance documents.

In Sweden, there is a significant program for the development and verification of a point system for timber-frame, multistory, apartment buildings (Magnusson and Rantatalo, 1998, Hultquist and Karlsson, 2000). This work is of particular note due its comprehensive documentation and validation procedures.

13.4 Multiattribute evaluation

Evaluation of fire safety can be difficult. Many, sometimes conflicting, attributes must be juggled simultaneously. The field of management science has long dealt with this type of problem. They have developed a large body of knowledge on the subject of Multiattribute Evaluation, also known variously as Multiattribute Decision Analysis, Multicriteria Decision Making, and Multiattribute Utility Theory.

These methods apply to problems where a decision-maker must evaluate, rank, or classify alternatives characterized by two or more relevant attributes. The literature describing multiattribute evaluation theory, methods, and applications is vast. Summaries and descriptions of the principal methods are found in Yoon and Hwang (1995), and Norris and Marshall (1995). The five basic characteristics of multiattribute evaluation are applicable to fire safety problems:

1. *Multiple attributes* The nature of the decision is one of screening, prioritization assessment, or selection of an object from among alternative objects based on values of a set of attributes for each object or alternative. Thus, each problem has multiple decision criteria or performance attributes. These attributes must be generated for the specific problem setting. The number of attributes depends on the nature of the problem.
2. *Trade-offs among attributes* In the typical compensatory evaluation, good performance of one attribute can at least partially compensate for low performance of another attribute. This is also called *trade-off* or *equivalency*. Since most attributes have different measurement scales, accommodating trade-offs among them generally means that the method incorporates procedures for normalizing data that are not commensurate.
3. *Units that are not commensurate* The attributes of the problem are generally not all measurable in units that are directly proportional. In fact, some attributes may be impractical, impossible, or too costly to measure at all. This typically requires methods of subjective estimation.
4. *Attribute weights* The formal methods of analysis generally require information regarding the relative importance of each attribute, which is usually supplied by a cardinal scale. Weights can be directly supplied or developed by specific methods. In some simple cases the weights default to equality.
5. *Evaluation vector* The problem can be concisely expressed as a vector whose values correspond to the performance rating of each attribute for the specific object. If the attributes for a decision problem are $x_1, x_2, x_3, \ldots, x_n$, then an evaluation function $E(x_1, x_2, x_3, \ldots, x_n)$ needs to be determined over these measures to conduct a performance assessment.

Besides the information in the evaluation vector, multiattribute evaluation generally requires additional information, for example, information about the minimum acceptable, maximum acceptable, or target values of the attributes.

13.4.1 ATTRIBUTES

Multiattribute evaluation begins with the generation of attributes that provide a means of evaluating goal achievements. These attributes, also called parameters, elements, factors, variables, and so forth, identify the ingredients of fire safety.

Fire safety attributes are defined as components of fire risk that are quantitatively determinable by direct or indirect measurement or estimation. They are intended to represent factors that account for an acceptably large portion of the total fire risk. Usually they are not directly measurable. This is especially true for existing buildings where only limited information is readily available. Thus, attributes may be either quantitative or qualitative and both types of attributes are very important. Selection of attributes should result in a set that is nonconflicting, coherent, and logical.

Attribute generation

Fire safety is a complex system affected by many factors that may range from ignitability of personal clothing to availability of a heliport for evacuation. In practice, only a relatively small number of factors can be considered because of limits on computational effort and gaps in knowledge.

It is intuitively appealing to postulate that safety from fire is a Paretian phenomenon in that a relatively small number of attributes account for most of the problem. This is supported by general fire loss figures that suggest that a small number of factors are associated with a large proportion of fire deaths. It is necessary then to identify as attributes some defensible combination of factors that account for an acceptable portion of the fire risk. Pardee (1969) suggests that a desirable list of attributes should be as follows:

1. Complete and exhaustive. That is, all-important attributes should be represented.
2. Mutually exclusive. Independence of the attributes facilitates evaluation of trade-offs.
3. Restricted to highest degree of importance. Lower level criteria may be part of attribute rating discussed later in this section.

Keeny and Raiffa (1976) suggest the use of a literature survey or a panel of experts to identify the attributes of a particular problem.

Example list of attributes

A study of fire safety effectiveness statements conducted for the US Fire Administration focused on logical and reproducible means of identifying key life safety variables (Watts *et al.*, 1979). The study included an extensive survey of case histories, research and test data, logic diagrams, codes, fire models, reviews, inspection check lists, insurance rating schedules, and personal experience to identify a list of more than 100 life safety variables. This large number needed to be reduced to a more appropriately sized subset.

The reduction was conducted in two steps. The first pass screened the variables for redundancy, applicability, and for determining whether they were components of a well-defined fire safety system. This reduced the list to 66 variables. The second pass involved contingency analysis and functional analysis of each candidate variable to determine its independence and its importance.

Table 13.3 is the list of 19 key life safety variables or attributes resulting from this study. The attributes are placed into four group for convenience and clarity only.

Table 13.3. Key life safety variables (Watts *et al.*, 1979)

Fire development	
	Fire load
	Rate of heat release
	Toxicity of combustion products
	Obscuration by combustion products
Fire spread	
	Fire resistance of structural members
	Fire resistance of exit way enclosures
	Fire resistance of vertical shafts
	Fire resistance of hazardous area separation/compartmentation
Fire control	
	Automatic extinguishing system
	Automatic smoke control system
	System maintenance
	Suppression by municipal fire department
	Suppression by in-house staff
Exiting	
	Exit way dimensions
	Remoteness/independence of exits
	Height of building
	Automatic detection system
	Physiological/psychological condition

Such a detailed process is not typical of most point systems. Selection of attributes is usually more arbitrary, with correspondingly disparate results. Where the subject of attribute generation is addressed, approaches to the selection of attributes generally fall into one of three categories:

- Delphi, or some less formal consensus process that relies on expert judgment.
- Fire scenarios, ideally based on loss statistics, but usually employing subjective opinion.
- Cut set of a hierarchical success tree, providing an inclusive list.

The development of attribute lists in the applications described in the previous section of this chapter relied heavily on intuition and subjective judgment. It is most important that the evaluation vector include only those attributes that vary significantly among buildings and for which the variation is considered meaningful.

Delphi

Delphi is a noninteractive group judgment method for reaching consensus in decision making (Linstone and Turoff, 1975). The process involves a panel of experts who are asked to estimate otherwise unpredictable relationships of system variables. Traditionally the panel members do not meet. This is designed to eliminate elements of group dynamics that are personality dependent and may be undesirable in making a technical decision, such as dominant individuals, irrelevant communication, and pressures to conform.

The formal process is known as a Delphi exercise. Each individual of the respondent group is presented with a set of questions. A process monitor summarizes the results and presents them to each member as a stimulus to reevaluate their original answers and to elicit underlying reasons for differences. These steps are repeated until an acceptable level of consensus is achieved. It has been found that convergence tends to occur after two to five such rounds. When the process is conducted in real time with computer compilation of results, it is referred to as a Delphi conference.

The structured group communication process used in the FSES and the Edinburgh model is sometimes called a *modified Delphi exercise*. The panel is convened as a group but follows the iterative procedure of responding to a set of questions and reacting to feedback of aggregated results. This approach saves time and allows for discussion of complex issues that are difficult to fully describe in a written questionnaire. Delphi was used to both identify and weight attributes.

Operational considerations of Delphi applications to multiattribute evaluation in fire safety have been addressed in the literature (Shields *et al.*, 1987, Marchant, 1989). Dodd and Donegan (1995) discuss Delphi briefly among other approaches to subjective measurement in fire protection engineering.

13.4.2 ATTRIBUTE WEIGHTING

Not all fire safety attributes are equally important. The role of weight serves to express the importance of each attribute compared with the others. Hence the assignment of weights is a key component of multiattribute evaluation.

Although assigning weights by an ordinal scale is usually easier, most multiattribute evaluation methods require cardinal weights. The attribute weights are generally normalized to sum up to one, that is, if y_i is the raw weight of attribute I, then

$$w_i = \frac{y_i}{\sum_{i=1}^{n} y_i}$$

and

$$\sum_{i=1}^{n} w_i = 1$$

This produces a vector of n weights given by

$$W = (w_l, \ldots, w_j, \ldots, w_n)$$

where w_i is the resultant weight assigned to the ith attribute.

This relative importance of attributes is defined to be constant across building evaluations. There are many weight assessment techniques used in multiattribute evaluation. Eckenrode (1965), and Hwang and Yoon (1981) review some of these methods. Generally, hierarchical methods have been found effective in fire safety evaluation. In the Edinburgh model previously described, a weighting method was developed, which is outlined below.

Edinburgh method

A hierarchical matrix approach to developing fire safety attribute weights was derived in the Edinburgh study. The method uses a hierarchy of decision-making levels to generate weights that identify the importance of each fire safety attribute. The hierarchy generally consists of four levels: policy, objectives, strategies, and attributes (see Table 13.2).

The first step is to define corporate, organizational, or agency fire safety policy by the relative importance of each member of a set of fire safety objectives. No significant work has been done to identify just what it is that fire safety is trying to achieve (i.e. allocation of resources for fire safety is not generally directly associated with a specific corporate objective), so these objectives are a very subjective list. A list of fire safety objectives might include statements about life safety, property protection, continuity of operations, environmental protection, and heritage preservation.

In most applications, a modified Delphi exercise is used to define fire safety policy in terms of the specified list of objectives, that is, a group of experts is asked to rank fire safety objectives with respect to their importance to the policy. Each member of the Delphi group receives feedback in the form of response averages and the process iterates until an acceptable level of consensus is reached. The Delphi exercise yields a vector representing the relative importance of each objective to organizational policy. If there are l fire safety objectives then the policy vector, $\boldsymbol{P}$, is given by

$$\boldsymbol{P} = [o_1, \ldots, o_i, \ldots, o_l]$$

where o_i is the importance of the ith objective to the corporate, agency, or organizational policy.

The next decision-making level involves fire safety strategies. A list of strategies can be derived by taking a cut set of the NFPA Fire Safety Concepts Tree (NFPA 550, 1995). Examples of fire safety strategies are ignition prevention, limitation of combustibles, compartmentation, fire detection and alarm, fire suppression, and protection of exposed people or things.

A matrix of objectives versus strategies is constructed. Values of the cells are again supplied by Delphi or other subjective decision-making processes. Here the question to be answered is: How important is each strategy to the achievement of each objective?

Thus, we have a set of m strategies that define how the fire safety objectives are to be achieved and each strategy has a relative importance to each objective. This produces the objectives/strategies matrix, O, shown as

$$O = \begin{bmatrix} s_{1,1} & s_{1,2} \cdots & s_{1,m} \\ s_{2,1} & s_{2,2} \cdots & s_{2,m} \\ \vdots & \ddots & \vdots \\ s_{l,1} & s_{l,2} \cdots & s_{l,m} \end{bmatrix}$$

where $s_{i,j}$ is the importance of strategy j to objective i.

Continuing this procedure, the next level deals with the fire safety attributes. The list of n fire safety attributes is assessed as to their contribution to each of the m strategies. The resulting strategies/attributes matrix, $\boldsymbol{S}$, is then as follows

$$S = \begin{bmatrix} a_{1,1} & a_{1,2} \cdots & a_{1,n} \\ a_{2,1} & a_{2,2} \cdots & a_{2,n} \\ \vdots & \ddots & \vdots \\ a_{m,1} & a_{m,2} \cdots & a_{m,n} \end{bmatrix}$$

where $a_{i,j}$ is a value identifying the importance of the jth attribute to the ith strategy.

To simplify mathematical manipulation, the values of the matrices can be normalized. The three matrices are then multiplied together and the product is a vector that specifies the relative importance of each fire safety attribute to the overall fire safety policy.

$$[P] \cdot [O] \cdot [S] = [y_1, \ldots, y_j, \ldots, y_n]$$

The values in this vector are normalized to produce the attribute weights to be used in the evaluation.

$$W = [w_1, \ldots, w_i, \ldots, w_n]$$

A particular significance of this method is that the resulting vector is a transparent weighting of fire safety attributes that has an explicit link to declared fire safety goals and objectives.

Analytic hierarchy process (AHP)

The Analytic Hierarchy Process is a powerful multiattribute evaluation technique. It has been successfully used to generate attribute ratings as discussed in the next section. However, it may be awkward for establishing attribute weights. For practical purposes, the set of attributes used with AHP should be limited to six or seven. Above this number it is difficult to maintain an acceptable level of consistency in the method. Most fire safety evaluation point systems deal with 15 to 20 attributes. Theoretically, this constraint can be readily dealt with by partitioning the group of attributes into sets of seven or less. In fire safety evaluation, this partition may be difficult to construct without further compromising independence assumptions or the logic of the hierarchy.

13.4.3 ATTRIBUTE RATINGS

Each attribute weight represents a specific relative importance that is universal for all facilities within the scope of the evaluation method. Individual buildings will vary in the degree to which attributes exist or occur in a space. Attribute ratings or grades are a measure of the intensity level or degree of danger or security afforded by the attribute in a particular application.

The selected attributes may be either quantitative or qualitative. Because both types of attributes are very important, we need a method that accounts for attributes of a general nature. Qualitative attributes may be impractical, impossible, or too costly to measure directly. Likert scaling is most often used to empirically capture the essential meaning of the attribute and develop a scale upon which a surrogate measure or rating can be based.

Quantitative attributes are readily measured or quantified but may require judgment to convert to a compensatory measure. Each quantitative attribute typically has a different unit of measurement. Data transformation techniques become necessary since multiattribute evaluation scoring generally requires a homogenous type of data. Quantitative attribute ratings must be normalized to a scale that is common for all attributes.

Likert scaling

Scaling and scale construction are central to the measurement of any phenomenon. This includes objective conditions as well as subjective states. Scaling identifies each individual object so that valid and reliable differences among objects can be represented (Torgerson, 1958).

Likert scaling refers to a psychometric scale developed by Rensis Likert in which usually five choices are provided for each attribute, the alternatives being scored from one to five. Psychometric scaling methods are derived from psychophysical measurement scales such as loudness and optical density of smoke, but, while their purpose is to locate values on a linear (straight-line) scale, no direct quantitative physical values are involved. Likert scaling is commonly used to scale an individual's assessment of objects and various kinds of characteristics. No scaling model has more intuitive appeal than the Likert scale. A five-point Likert scale is used predominantly but a more detailed scale, such as a seven-point or nine-point scale, can also be used if its application does not stress the ability to distinguish differences in meaning or significance.

There are three underlying assumptions of Likert scaling (McIver and Carmines, 1981):

1. Each item is monotonically related to the underlying latent dimension continuum, that is, there is no obvious discontinuity or reversal of slope.
2. The sum of the item scores is monotonic (and approximately linear) with respect to the dimension measured.
3. The items as a group measure only the dimension sought. In other words, all items to be linearly combined should be related only to a single common factor. The sum of these items is expected to have all the important information contained in the individual items.

Only this last assumption tends to be somewhat problematic. It is difficult to decide conclusively that the items as a whole are measuring only a single phenomenon.

Scaling systems are classified as nominal, ordinal, interval or ratio. Nominal systems are like numbers on football jerseys that are simply used to distinguish and identify. An ordinal system is a ranking, indicating position in a series or order, such as first (1st), second (2nd), and third (3rd). An interval scale has meaningful differences between numbers, for example, the Celsius temperature scale. A ratio scale has a meaningful zero point such as the Kelvin temperature scale. Combinatorial calculations, addition and subtraction, are appropriate only on interval and ratio scales, for example, adding the numbers on football jerseys does not tell you anything. Multiplication and division can be performed only on ratio measurements; for example, you should not use Celsius temperature in the Stefan–Boltzman law.

In an experiment on Delphi methodology, the equivalency of three simple scaling techniques was examined and it was concluded that for practical purposes the results could be accepted as an interval scale (Scheibe *et al.*, 1975). The experiment showed that Likert scales used in Delphi exercises have the equal difference property and therefore combined calculations are appropriate. The intervals between point scores on a Likert scale are meaningful but ratios of scores are not interpretable.

Normalization of data

Typically each quantitative attribute has a different unit of measurement. In order to attain the compensatory trade-offs that are an essential characteristic of multiattribute evaluation, commensurable attribute units are necessary. Attribute ratings are therefore normalized to eliminate computational problems caused by differing measurement units in the evaluation vector. Thus, we need to construct or adopt a normalizing function $R_i(x_i)$ for each attribute i.

Normalization aims at obtaining comparable scales that allow comparison between attributes. Consequently, normalized ratings have dimensionless units and the larger the rating becomes, the more preference it has.

Fire safety attributes may be beneficial, detrimental, or nonmonotonic. Beneficial attributes offer monotonically increasing utility; the greater the attribute value, the more its preference, for example, fire resistance. Detrimental attributes are monotonically decreasing in utility; the greater the attribute value the less its preference, for example, rate of heat release. Nonmonotonic fire safety attributes are uncommon. One, perhaps unique, example is floor level, where, for life safety, ground level is preferred over stories above or below ground.

The most common form of normalization is linear. For beneficial attributes, the normalized rating of attribute i, r_i, is given by

$$r_i = \frac{x_i - x_i^{\vee}}{x_i^{\wedge} - x_i^{\vee}}$$

where x_i^ is the ceiling or maximum possible value of x_i, and x_i^v is the floor or smallest possible value of x_i. Thus, the expression x_i^ $- x_i^v$ is the range of all possible values of attribute x_i. If, as often happens, $x_i^v = 0$, then the normalized rating is given by the ratio of the attribute value to the maximum value, $r_i = x_i / x_i$^. The resultant ratings have the characteristic that $0 \leq r_i \leq 1$ and the attribute is more favorable as r_i approaches 1.

The linear normalized rating of a detrimental attribute i is given by

$$r_i = \frac{x_i{}^\wedge - x_i}{x_i{}^\wedge - x_i^\vee}$$

Again, the resultant ratings have the characteristic that $0 \leq r_i \leq 1$ and have been adjusted to be consistent with beneficial attributes so that the attribute is more favorable as r_i approaches 1.

There are statistical procedures for normalizing nonmonotonic attributes but they are typically not necessary in fire safety evaluation.

Where quantitative and qualitative attributes are mixed, the normalized ratings should be multiplied by the modulus of the Likert scale used for the qualitative attributes. For example, if a five-point Likert scale is used for qualitative data, then the normalized ratings should be multiplied by five. This is essential to maintain the compensatory capability of the scoring method.

Decision tables

Partitioning the attributes into measurable constituent parts can facilitate fire safety attribute rating. Usually, these parts will be directly measurable survey items. Sometimes there may also be intermediate subattributes. A survey item is a measurable feature of a building or building space that serves as a constituent part of one or more attributes or subattributes.

Decision tables are commonly used in decision analysis and documentation (CSA, 1970, Hurley, 1983). Their purpose is to provide orderly representation of information flow in elementary decisions. While such decisions can appear simple by relative comparison, their underlying logic may often be complex. The tabular approach is used to express decision logic in a way that encourages reduction of a problem to its simplest form by arranging and presenting logical alternatives under various conditions.

Many decision problems can be formulated as a set of attributes or conditions that can lead to certain conclusions or actions. The attributes and conclusions have an explicit or implied "if ... then" relationship. Each alternative combination of attributes that produces a conclusion is called a *decision rule*.

A decision table consists of four quadrants commonly separated by heavy or double lines. Attributes or conditions appear in the top half separated from conclusions or actions below, by a horizontal line. The right side of the table is vertically subdivided into columns called decision rules. Column numbers identify particular rules. At a time a single rule is examined, reading from top to bottom. A value for each applicable variable or survey item appears in each column or decision rule in the upper right quadrant. The outcomes of the decision rules are in the lower right quadrant. In the simplest form all the variables are binary, for example, Y and N mean yes and no. However, other indicators such as numeric data can also be used.

Decision Tables present a useful logic for using survey items to develop ratings for fire safety attributes (Watts *et al.*, 1995). In the COFRA model discussed in the previous section, 17 attributes were identified as the primary components of fire risk. One of these dealt with the expected fuel available for a fire. The attribute was called *Ordinary Combustibles.*

In developing the basis for grading this attribute, a logic tree (Figure 13.1) was derived from the NFPA Fire Safety Concepts Tree (NFPA 550, 1995). The attribute is broken down into two subattributes, *Ignition* and *Growth. Growth* is further divided into survey items *Fire Load* and *Fire Growth Rate. Ignition* has three parts, *Ignition Sources, Transfer Processes*, and *Fuel Ignitability. Transfer Processes* is also a subattribute, defined by survey items, *Equipment Maintenance* and *Housekeeping*. Thus, the attribute is evaluated in terms of both material properties and surrounding conditions.

From Figure 13.1 the subattribute *Growth* is a function of the survey items, *Fire Load* and *Fire Growth Rate*. Table 13.4 is the decision table representing this relationship with the survey items as attributes of modulus four. Note that one of the values for *Fire Load* (N = none) is dominant and therefore decision rule 1 represents 4 elementary decision rules.

As described above, a mathematically tractable approach is to assign attribute ratings as integers on a Likert scale of 0 to 5, where 0 is a theoretical optimum equivalent to zero risk and 5 is a worst feasible case. Thus, the range is defined phenomenologically and not by the state of the art. The ratings for the subattribute *Growth* are based on such a scale. For example, a moderate (M) *Fire Load* and a slow (S) *Fire Growth Rate* produce a rating of 2 (decision rule 3) for the subattribute *Growth*.

From Figure 13.1 it is seen that the *Ignition* subattribute is determined by three factors; two survey items, *Ignition Sources*, and *Fuel Ignitability*, and the subattribute, *Transfer Processes*. Table 13.5 is the decision table for the *Ignition* subattribute. With each of the three attributes having a modulus of three, there are 27 decision rules. As an example, if the three attributes all have values of M, the subattribute rating for *Ignition* would be 3 (decision rule 14).

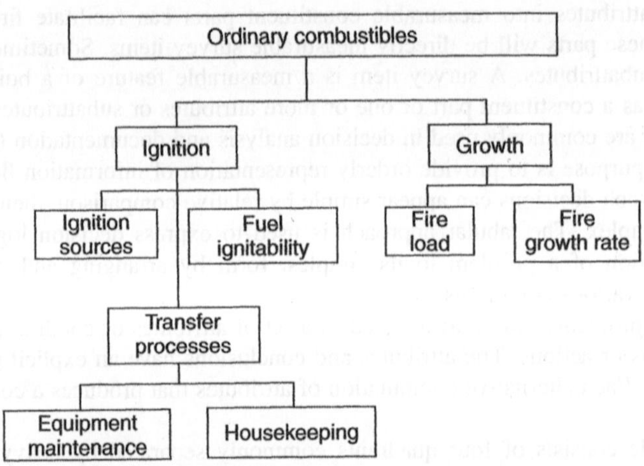

Figure 13.1. Logic tree for ordinary combustibles

Table 13.4. Decision table for subattribute growth

Survey items	Decision rules												
	1	2	3	4	5	6	7	8	9	10	11	12	13
Fire load (N, L, M, H)	N	L	M	H	L	M	H	L	M	H	L	M	H
Fire growth rate (S, M, F, V)	–	S	S	S	M	M	M	F	F	F	V	V	V
Growth	0	1	2	3	2	3	4	3	4	5	4	5	5

Table 13.5. Decision table for subattribute ignition

Survey items	Decision rule																										
	1	2	3	4	5	6	7	8	9	10	11	12	13	14	15	16	17	18	19	20	21	22	23	24	25	26	27
Ignition sources (L, M, H)	L	L	L	L	L	L	L	L	L	M	M	M	M	M	M	M	M	M	H	H	H	H	H	H	H	H	H
Transfer processes (L, M, H)	L	L	L	M	M	M	H	H	H	L	L	L	M	M	M	H	H	H	L	L	L	M	M	M	H	H	H
Fuel ignitability (L, M, H)	L	M	H	L	M	H	L	M	H	L	M	H	L	M	H	L	M	H	L	M	H	L	M	H	L	M	H
Ignition	1	1	2	1	2	3	2	3	4	1	2	3	2	3	4	3	4	5	2	3	4	3	4	5	4	5	5

There are three properties of decision tables that contribute to their effectiveness in point system fire safety evaluation.

1. Decision tables provide a disciplined way to rate fire safety attributes using data on survey items.
2. Decision tables provide a concise and standardized documentation of the detailed design of a point system.
3. Decision tables simplify transition to a computerized application.

Analytic hierarchy process (AHP)

Ratings for fire safety attributes can also be developed using pairwise comparisons and the Analytic Hierarchy Process. AHP has been widely reviewed and applied in the literature and its use is supported by several commercially available user-friendly software packages. In this approach, the relative importance of each survey item or subattribute is determined by setting up a square matrix, $\boldsymbol{A}$, and making pairwise comparisons.

$$A = \begin{bmatrix} 1 & a_{1,2} & \cdots & a_{1,n} \\ a_{2,1} & 1 & a_{2,n-1} & a_{2,n} \\ \vdots & & \ddots & \vdots \\ a_{n,1} & \cdots & a_{n,n-1} & 1 \end{bmatrix}$$

Each possible pair of items is examined and a subjective determination is made as to which is more important (preferred) and to what extent, so that a_{ij} represents how much item i is preferred over item j. Preference is assigned from a Likert scale of 1 to 9. For n items, there will be $n(n-1)/2$ such comparisons.

The diagonal of the matrix is by definition comprised of all 1s since each item has equal importance when compared with itself, that is, $a_{i,i} = 1$ for all i. Values on symmetrically opposite sides of the diagonal are reciprocals. That is, if an item A is x times as important as item B, then item B is $1/x$ times as important as item A or $a_{i,j} = a_{j,i}$ for all i and j. Relative importance of each item can then be calculated from the matrix using any of several methods. The best known and the most supported method by commercial software is the eigenvalue prioritization method (Saaty, 1980, 1990).

AHP also provides a heuristic check of inconsistency among paired comparisons. Perfect cardinal transitivity means that if A is twice as important as B and B three times as important as C, then A is exactly six times as important as C. Methods such as AHP that use pairwise comparisons, allow the decision maker to provide judgments that are not consistent in the sense that the resulting comparisons do not satisfy the property of perfect cardinal transitivity. AHP uses the principal eigenvector to calculate a measure of the degree of inconsistency (departure from perfect cardinal transitivity) exhibited by the set of pairwise comparisons. AHP software reports the results of this heuristic test of consistency.

When using pairwise comparisons and AHP, the number of subattributes or survey items for each attribute should be limited to around seven. This number is congruous with the theory that 7 ± 2 represents the greatest amount of information that an observer can give us about an object from an absolute judgment (Miller, 1956).

13.4.4 SCORING METHODS

A multiattribute evaluation may be viewed as a vector of attributes. The transformation of a vector to an appropriate scalar value is the purpose of the point system, that is, formulating an index to represent the effectiveness of the system. In fire safety, we do not yet have a thorough understanding of the functional relationships among components, so simple heuristic scoring techniques are used. The two most common are additive weighting and weighted product.

Additive weighting

Additive weighting is the most widely used method of fire safety evaluation point system scoring. The score is determined by adding the contribution from each attribute. Since two items with different measurement units cannot be added, a common numerical scaling system such as normalization, discussed in the previous section, is required to permit addition among attribute values. The total score for each evaluation can then be computed by multiplying the comparable rating for each attribute by the importance weight assigned to the attribute and then summing these products over all the attributes. Formally the evaluation score, S, in the additive weight method can be expressed as

$$S = \sum_{i=1}^{n} w_i r_i$$

where w_i is the weight of attribute i and r_i is the normalized rating of attribute I. Thus, the numeric evaluation is calculated as the scalar product of the weighting vector and rating vector of the attributes.

The underlying assumption of additive weighting is that attributes are preferentially independent. Less formally, this means that the contribution of an individual attribute to the total (multiattribute) score is independent of other attribute values. Therefore preferences regarding the value of one attribute are not influenced in any way by the values of the other attributes (Fishburn, 1976). Fortunately, studies (Edwards, 1977, Farmer, 1987) show that additive weighting yields extremely close approximations to "real" value functions even when independence among attributes does not exactly hold.

This model permits very lenient assumptions about individual components of a scale. Nunnally (1978) suggests that because each item may contain considerable measurement error or specificity, the importance of this additive model is that it does not take any particular item very seriously. Additive weighting also assumes that the characteristic weights are proportional to the relative value of a unit change in each attribute value function. This is what makes the method compensatory.

Additive weighting is the scoring model used in the FSES for Health Care Occupancies and the Edinburgh Model described earlier in this chapter.

Weighted product

In additive weighting, addition among attribute values is allowed only after different measurement units are transformed into a dimensionless scale by normalization. However, this transformation is not necessary if attributes are connected by multiplication. Attribute values are multiplied in the weighted product method. The weights become exponents associated with each attribute value, with a positive value for benefit attributes and a negative power for cost attributes. Formally the evaluation score, S, using the weighted product method can be expressed as

$$S = \prod_{i=1}^{n} x_i^{w_i}$$

where w_i is the weight of attribute i and x_i is the rating of attribute i.

Because of the exponent property, this method requires that all ratings be greater than one. When an attribute has fractional ratings, they should be consistently multiplied by some power of ten to meet this requirement.

The weighted product method may produce more variability in results. Since the attribute ratings are multiplied, a small measurement error in one attribute can generate a significant variation in the score. This makes the method less appropriate when attribute ratings are the result of subjective determinations with potentially large variances.

Weighted product is the scoring model used in the Gretener method described earlier in this chapter.

Analytic hierarchy process (AHP)

The Analytic Hierarchy Process discussed in the previous section on attribute rating is also widely used as a multiattribute evaluation scoring method. It has also been adapted to the Edinburgh point system model for a study of fire safety in dwellings (Shields and Silcock, 1986). However, there are limitations to this use of AHP in a fire safety evaluation point system.

The Analytic Hierarchy Process is not as intuitive or transparent as the arithmetic combining of attribute weights and ratings. Also, as has been previously discussed, AHP significantly restricts the number of attributes that can be considered. Judgment of pairwise comparisons quickly becomes cognitively onerous as the number of attributes increases. Seven attributes produce 21 pairwise comparisons, which is approaching the maximum reasonable effort for this process. Even AHP computer software limits the number of attributes to nine. Finally, although this is seldom the case in fire safety evaluation, if the practical range of the attribute ratings is not known, AHP can be subject to distortion and rank-reversal.

Although AHP has the advantage of measurable consistency in pairwise comparisons, it does not necessarily produce a more accurate score. Karni *et al.* (1990) compared AHP and simple additive weighting in real life cases and found that the resultant rankings did not differ significantly.

13.5 Criteria

The fire protection community has seen a proliferation of point systems in the last two decades. Some of these appear from their documentation to be quite robust while others have no rational substantiation for their existence. Like any analytical techniques, point systems have their limitations and should not be used uncritically.

The purpose of a point system is to provide a useful aid to decision making. Usefulness requires the methodology to be simple yet credible. Applying it must be not only easy but also sophisticated enough to provide a minimum of technical validity. Credibility can also be improved through consistency and transparency. Development should be systematic and it should be clearly discernible to all interested parties that the relevant technical issues have been appropriately covered. On the basis of a review of many existing point systems, ten criteria have been proposed as an aid in future development and assessment (Watts, 1991).

1. *Development and application of the method should be thoroughly documented according to standard procedures.* One hallmark of professionalism is that as a study proceeds, a record is made of assumptions, data, attribute estimates and why they were chosen, model structure and details, steps in the analysis, relevant constraints, results, sensitivity tests, validation, and so on. Little of this information is available for most point systems.

Beyond facilitation of review, there are other practical reasons not to slight the documentation: (1) If external validation is to be conducted, adequate documentation will be a prerequisite; (2) During the life cycle of a point system, the inescapable changes and adjustments will require appropriate documentation; (3) Clear and complete documentation enhances confidence in the method, its absence inevitably carries with it the opposite effect.

The value of the documentation will be improved if it follows established guidelines. Standard formats for documentation are primarily directed at large-scale computer models (e.g. Gass, 1984, ASTM, 1992) but can be readily adapted in principle to more general applications.

2. *Partition the universe rather than select from it.* One of the least well-established procedures in point systems is the choice of attributes. In following a systemic approach, being comprehensive is best. In the Edinburgh model, this is achieved by using the NFPA Fire Safety Concepts Tree (NFPA 550, 1995). The Tree branches out from the holistic concept of fire safety objectives. A cut set on the Tree will then identify a group of attributes that encompasses all possible fire safety features.
3. *Attributes should represent the most frequent fire scenarios.* In determining the level of detail of the attributes, it is necessary to look at those factors that are most significant, statistically, or by experienced judgment. This criterion may also be used as an alternative to Criterion 2, provided the need for systemic comprehensiveness is satisfied.
4. *Provide operational definitions of attributes.* If the methodology is to be used by more than a single individual, it is necessary to ensure precise communication of the intent of key terms. Many fire risk attributes are esoteric concepts that have a wide variety of interpretations even within the fire community.
5. *Elicit subjective values systematically.* Most point system methods rely heavily on experienced judgment. The use of formalized, documented procedures significantly increases credibility of the system. Similarly, the use of recognizable scaling techniques will enhance credibility.
6. *Attribute values should be maintainable.* One variable that is not explicitly included in point systems is time. Yet the influence of time is ubiquitous. It influences the fire risk both internally (e.g. deterioration) and externally (e.g. technological developments). In order for a method to have a reasonable useful lifetime, it must be amenable to updating. This implies that procedures for generating attribute weights and ratings must be repeatable. Changes over time and new information dictate that the system can accommodate revisions.
7. *Treat attribute interaction consistently.* Most often this will consist of an explicitly stated assumption of no interactive effect among attributes. Where interactions are considered, it is important that they be dealt with systematically to avoid bias. The Edinburgh model interaction matrix is one approach to this assessment.
8. *State the linearity assumption.* While this assumption is universal in point systems, it is also well known that fire risk variables do not necessarily behave in a linear fashion. It is important to the acceptance of fire safety evaluation point systems and their limitations that such assumptions are understood.
9. *Describe fire risk by a single indicator.* The objective of most point system methods is to sacrifice details and individual features for the sake of making the assessment easier. Information should be reduced to a single score even in the most complex applications. Techniques have been espoused to combine technical, economic, and sociopolitical factors (Chicken and Hayns, 1989). The results should be presented in a manner that makes their significance clear in a simple and unambiguous way. Unless all those involved can understand and discuss the meaning of the evaluation there will not be general confidence in its adequacy.

10. *Validate results.* Some attempt should be made to verify that the method does in fact differentiate between lesser and greater fire risks with sufficient precision. The accuracy demanded here is not the same as for other engineering purposes. Establishing an order of magnitude will generally suffice.

13.6 Summary

Point systems have proliferated because of their high utility and relative ease of application. Fire safety evaluation involves a large number of multifarious factors that are hard to assess in a uniform and consistent way. Analysis of such a complex system is difficult but not impossible as evidenced by activities in the fields of nuclear safety and environmental protection. Detailed risk assessment can be an expensive and labor-intensive process and there is considerable scope for improving the presentation of results. Point systems can provide a cost-effective means of fire safety evaluation that is sufficient in both utility and validity.

For analytical and procedural simplicity, common practice neglects both uncertainties and imprecision inherent in the evaluation vector data and in the additional elicited information about the attributes and objects. Neglect of uncertainty occurs when uncertain values are represented by their expected values rather than by probability distributions. Neglect of imprecision occurs when ratings such as "good" and "bad" are converted to scalar numbers rather than ranges. All applications to fire safety follow this practice. Chance or random error is involved in any type of measurement. However, as Nunnally (1978, p. 67) observes, "this unreliability averages out when scores on numerous items are summed to obtain a total score, which then frequently is highly reliable."

Many formal methods of multiattribute evaluation are available in the field of decision analysis. Choosing among them is not as important as being logical and consistent in application of the chosen method. In one study of four compensatory methods, no significant difference was found in the appropriateness of the method or its ease of use (Hobbs *et al.*, 1992). The advice generally given is that whichever method users feel comfortable with should be used.

Many multiattribute evaluation algorithms can be easily executed on personal computers. The electronic spreadsheet is an especially powerful tool for multiattribute evaluation analysis. It can readily store and manipulate evaluation vectors.

A large number of commercial software programs are available for additive weighting and AHP. Several computer programs specific to fire safety evaluation have also been developed as previously discussed in this chapter.

Point Schemes try to obtain a meaningful index from multidimensional data to evaluate fire safety. They are relatively simple models that do not rely exclusively on demonstrated principles of physical or management science. But their credibility is enhanced to the extent that they do employ such principles.

Glossary

Some terms used in this chapter have origin in other fields and disciplines such as decision analysis, economics, finite mathematics, linear algebra, management science, operational research, psychology, and set theory. They may be unfamiliar or have different connotations than in common use. To assist the reader, a brief glossary is included here with definitions of such terms as may be unclear in the context of the chapter. Some repeated acronyms are also included here.

AHP – Analytic Hierarchy Process. A decision analysis tool.

Attribute – A characteristic that can be measured to indicate the degree to which an objective is achieved or fulfilled. In some point systems, attributes are also referred to as parameters or factors.

Cardinal number – A number, such as 3 or 11 or 412, used in counting to indicate quantity but not order.
Commensurate – Having a common measure.
Continuum – A coherent whole characterized as a collection, sequence, or progression of values or elements varying by minute degrees, for example "good" and "bad" stand at opposite ends of a continuum instead of describing the two halves of a line.
Cut set – A set of nodes or events in a network whose removal disconnects the network.
FSES – Fire Safety Evaluation System. Class of point systems published in NFPA 101A.
Heuristic – A rule or algorithm for producing a solution that may not be optimal. Having no formal proof but based on reasoned processes.
Likert scale – measurement scale that rates attributes as 1, 2, 3, 4, or 5, reading from unfavorable to favorable.
Matrix – A rectangular array of numbers.
Monotonic – Having the property either of never increasing or of never decreasing as the values of the independent variable increase.
Ordinal number – A number indicating position in a series or order. The ordinal numbers are first (1st), second (2nd), third (3rd), and so on.
Paretian – After Vilfredo Pareto (1848–1923), an Italian economist and engineer (University of Turin 1869) known, among other things, for his application of mathematics to economic analysis. His law of income distribution characterizes the accumulation of wealth in society as a consistent pattern fitting a Pareto distribution. The theory contends that most of the money is held by very few people. The term is used here to refer to the characteristic that most of fire safety can be explained with a small set of attributes.
Scalar product – A real number that is the result of multiplication of two vectors. It is equal to the sum of the products of the corresponding values in each vector. It is also referred to as dot product and inner product.
Surrogate measure – Determination that indirectly gauges the fulfilment of the objective based on related factors. It is also referred to as a proxy measure.
Vector – A single row or column matrix, that is, a linear array of numbers.

Nomenclature

$a_{i,j}$	importance of jth attribute to ith strategy
$a_{m,n}$	how much item m is preferred over item n
A	activation (ignition) hazard; square matrix
E	evaluation function
F	fire resistance of the structure
N	normal protection measures
o_i	importance of ith objective
O	objectives/strategies matrix
P	potential hazard; policy vector
r_i	normalized rating of attribute I
R	risk
R_i	normalizing function
$s_{i,j}$	importance of strategy j to objective I
S	special protection measures; strategies/attributes matrix; evaluation score
w_i	weighting constant; resultant weight assigned to the ith attribute
W	vector of n weights
x_i	attribute

$x_i^\wedge$	maximum possible value of x_i
$x_i^\vee$	smallest possible value of x_i
y_i	raw weight of attribute *I*

References

ASTM E-1472 (1992). *Documenting Computer Software for Fire Models*, American Society for Testing and Materials, Philadelphia.

Benjamin, I A (1979). A firesafety evaluation system for health care facilities. *Fire Journal*, **73**(2).

BOCA (1996). *The BOCA National Building Code/1996*, Building Officials and Code Administrators International, Country Club Hills, IL.

Budnick, E K, McKenna, L A Jr and Watts, J M Jr (1997). Quantifying fire risk for telecommunications network integrity. *Fire Safety Science – Proceedings of the Fifth International Symposium*, International Association for Fire Safety Science, pp. 691–700.

Chapman, R E and Hall, W G (1982). Code compliance at lower costs: a mathematical programming approach. *Fire Technology*, **18**(1), 77–89.

Chapman, R E and Hall, W G (1983). *User's Manual for the Fire Safety Evaluation System Cost Minimizer Computer Program*, NBSIR 83-2796, National Bureau of Standards, Washington.

Chicken, J C and Hayns, M R (1989). *The Risk Ranking Technique in Decision Making*, Pergamon Press, Oxford.

Copping, A G (2000). Fire safety evaluation procedure for the property protection of English Parish churches: a tool to aid decision making. *Proceedings, International Conference on Fire Protection of Cultural Heritage, Aristotle University of Thessaloniki*, Greece, pp. 255–268.

CSA (1970). *Decision Tables*, Standard Z243.1-1970, Canadian Standards Association.

Department of Fire Safety Engineering (1982). *Fire Safety Evaluation (Points) Scheme for Patient Areas Within Hospitals*, University of Edinburgh.

Dodd, F J and Donegan, H A (1995). Subjective measurement in fire protection engineering. *SFPE Handbook of Fire Protection Engineering*, P J DiNenno *et al.* (Eds.), 2nd Edition, National Fire Protection Association, Quincy, MA, Section 5, Chapter 5, pp. 5-46–5-51.

Donegan, H A, Shields, T J and Silcock, G W (1989). A mathematical strategy to relate fire safety evaluation and fire safety policy formulation for buildings. *Fire Safety Science – Proceedings of the Second International Symposium*, T Wakamatsu *et al.* (Eds.), Hemisphere, New York, pp. 433–441.

Dow (1966). Process safety manual. *Chemical Engineering Progress*, **62**(6).

Dow (1994). *Dow's Fire and Explosion Index Hazard Classification Guide*, 7th Edition, AICHE Technical Manual, The Dow Chemical Company, American Institute of Chemical Engineers, New York.

Eckenrode, R T (1965). Weighting multiple criteria. *Management Science*, **12**, 180–192.

Edwards, W (1977). Use of multi-attribute utility measurement for social decision making. *Conflicting Objectives in Decisions*, D E Bell, R L Keeny and H Raiffa (Eds.), John Wiley, New York, pp. 247–276.

Farmer, T A (1987). Testing the robustness of multi-attribute utility theory in an applied setting. *Decision Sciences*, **18**, 178–193.

Fishburn, P C (1976). Utility independence on subsets of product sets. *Operations Research*, **24**, 245–255.

Fontana, M (1984). *SIA 81, Swiss Rapid Risk Assessment Method*, Institute of Structural Engineering, ETH, Zurich.

F.R.A.M.E. (2000). Fire Risk Assessment Method for Engineering, http://user.online.be/~otr034926/webengels.doc.htm, 7 September.

FSD (2000). Risk Design, http://www.fsd.se/eng/index.html, 7 September.

Gass, S I (1984). Documenting a computer based model. *Interfaces*, **14**(3), 84–93.

Hobbs, B F Chakong, V Hamadeh, W and Stakhiv, E Z (1992). Does choice of multicriteria method matter. *Water Resources Research*, **28**, 1767–1780.

Hughes Associates, Inc. (1999). *Enhanced Fire Safety Evaluation System for Business Occupancies Software*, Version 1.2, Hughes Associates, Inc, Baltimore.

Hultquist, Hans and Karlsson Bjorn (2000). *Evaluation of a Fire Risk Index Method for Timber-frame, Multi-storey Apartment Buildings, Report 31XX, Department of Fire Safety Engineering*, Lund University, Sweden.

Hurley, R B (1983). *Decision Tables in Software Engineering*, Van Nostrand Reinhold, New York.

Hwang, C L and Yoon K (1981). *Multiple Attribute Decision Making: Methods and Applications*, Springer-Verlag, New York.

Kaplan, M E and Watts, J M Jr (1999). A prototypical fire risk index to evaluate fire safety in historic buildings. *APT Bulletin*, **30**(2–3), 49–55.

Karni, R Sanchez, P and Tummala, V M R (1990). A comparative study of multiattribute decision making methodologies. *Theory and Decision*, **29**, 203–222.

Keeny, R L and Raiffa, H (1976). *Decisions With Multiple Objectives*, John Wiley, New York.

Law, M (1973). Prediction of fire resistance. *Proceedings, Symposium No. 5, Fire-Resistance Requirements for Buildings*, A New Approach, Joint Fire Research Organization, Her Majesty's Stationery Office, London.

Linstone, H A and Turoff, M (Eds.) (1975). *The Delphi Method: Techniques and Applications*, Addison-Wesley, London.

Lo, S M (1999). A fire safety assessment system for existing buildings. *Fire Technology*, **35**(2), 131–152.

Magnusson, S E and Rantatalo, T (1998). *Risk Assessment of Timberframe Multistorey Apartment Buildings: Proposal for a Comprehensive Fire Safety Evaluation Procedure*, IR 7004, Department of Fire Safety Engineering, Lund University, Sweden.

Marchant, E W (1988). Fire safety engineering – a quantified analysis. *Fire Prevention*, (210), 34–38.

Marchant, E W (1989). Problems associated with the use of the Delphi technique – some comments. *Fire Technology*, **24**(1), 59–62.

McIver, J P and Carmines, E G (1981). *Unidimensional Scaling*, Sage Publications, London.

Miller, G A (1956). The magic number seven, plus or minus two. *Psychological Review*, **63**, 81–97.

NFPA 101 (2000). *Life Safety Code*, National Fire Protection Association, Quincy, MA.

NFPA 101A (2001). *Alternative Approaches to Life Safety*, National Fire Protection Association, Quincy, MA.

NFPA 550 (1995). *Guide to the Fire Safety Concepts Tree*, National Fire Protection Association, Quincy, MA.

Nelson, H E (1986). *A Fire Safety Evaluation System for NASA Office/Laboratory Buildings@*, NBSIR 86-3404, National Bureau of Standards, Gaithersburg, MD.

Nelson, H E and Shibe, A J (1980). *A System for Fire Safety Evaluation of Health Care Facilities*, NBSIR 78-1555, Center for Fire Research, National Bureau of Standards, Washington, DC.

Norris, G A and Marshall, H E (1995). *Multiattribute Decision Analysis Method for Evaluating Buildings and Building Systems*, NISTIR 5663, National Institute of Standards and Technology, Gaithersburg, MD.

Nunnally, J C (1978). *Psychometric Theory*, McGraw-Hill, New York.

Pardee, E S (1969). *Measurement and Evaluation of Transportation System Effectiveness*, RAND Memorandum, RM-5869-DOT, Santa Monica.

Parikh, P B, and Crowl, D A (1998). Implementation and application of the Dow hazard evaluation indices in a computer-based environment. *Proceedings: International Conference and Workshop on Reliability and Risk Management*, San Antonio, TX, American Institute of Chemical Engineers, New York, pp. 65–83.

Parks, L (1996). COFRA-2: A tool to aid in telecommunications central office fire risk assessment. *Proceedings, Fire risk and Hazard Assessment Symposium*, National Fire Protection Research Foundation, Quincy, MA, pp. 523–540.

Parks, L L Kushler, B D Serapiglia, M J McKenna, L A Jr, Budnick, E K and Watts, J M Jr (1998). Fire risk assessment for telecommunications central offices. *Fire Technology*, **34**(2), 156–176.

Rasbash, D J (1985). Criteria for acceptability for use with quantitative approaches to fire safety. *Fire Safety Journal*, **8**(2), 141–158.

RiskPro (2000). SimCo Consulting, Wantirna South, Victoria, Australia.

Saaty, T L (1980). *The Analytic Hierarchy Process*, Wiley, New York.

Saaty, T L (1990). *Multicriteria Decision Making: The Analytic Hierarchy Process*, RWS Publications, Pittsburgh.

Scheffler, N E (1994). *Improved fire and explosion index hazard classification*. AIChE Spring National Meeting.

Scheibe, M, Skutsch, M and Schofer, J (1975). Experiments in delphi methodology. *The Delphi Method: Techniques and Applications*, A L Harold and M Turoff (Eds.), Addison-Wesley, London, pp. 262–282.

Shields, T J Silcock, G W (1986). An application of the hierarchical approach to fire safety. *Fire Safety Journal*, **11**(3), 235–242.

Shields, T J Silcock, G W and Bell, Y (1986). Fire safety evaluation of dwellings. *Fire Safety Journal*, **10**(1), 29–36.

Shields, T J Silcock, G W H Donegan, H A and Bell, Y A (1987). Methodological problems associated with the use of the Delphi technique. *Fire Technology*, **23**(2), 175–185.

Stollard, P (1984). The development of a points scheme to assess fire safety in hospitals. *Fire Safety Journal*, **7**(2), 145–153.

Torgerson, W S (1958). *Theory and Methods of Scaling*, John Wiley, New York.

Watts, J M Jr (1991). Criteria for fire risk ranking. *Fire Safety Science – Proceedings of the Third International Symposium*, G Cox and B Langford (Eds.), Elsevier, London, pp. 457–466.

Watts, J M Jr (1995). Software review: fire risk evaluation model. *Fire Technology*, **31**(4), 369–371.

Watts, J M Jr (1997). Analysis of the NFPA fire safety evaluation system for business occupancies. *Fire Technology*, **33**(3), 276–282.

Watts, J M Jr (1998). Fire risk evaluation in the codes: a comparative analysis. *Proceedings: Second International Conference on Fire Research and Engineering*, Society of Fire Protection Engineers, Bethesda, MD, pp. 226–237.

Watts, J M Jr (2001). Fire risk indexing. *SFPE Handbook of Fire Protection Engineering*, P J DiNenno *et al.* (Eds.), 3rd Edition, National Fire Protection Association, Quincy, MA, Section 5, Chapter 10.

Watts, J M Jr and Kaplan, M E (2001). *Fire risk index for heritage buildings, Fire Technology*, submitted.

Watts, J M Jr, Budnick, E K and Kushler, B D (1995). Using decision tables to quantify fire risk parameters. *Proceedings – International Conference on Fire Research and Engineering*, Society of Fire Protection Engineers, Boston.

Watts, J, Milke, J A Bryan, J L, Dardis, R and Branigan, V (1979). *A Study of Fire Safety Effectiveness Statements*, Department of Fire Protection Engineering, University of Maryland, College Park, NTIS, PB-299169.

Webber, S F and Lippiatt, B C (1994). *ALARM 1.0: Decision Support Software for Cost-Effective Compliance with Fire Safety Codes*, NISTIR 5554, National Institute of Science and Technology, Gaithersburg, MD.

Webber, S F and Lippiatt, B C (1996). Cost-effective compliance with fire safety codes. *Fire Technology*, **32**(4).

Wisconsin Administrative Code (1995). Chapter ILHR 70, Historic Buildings, Department of Industry, Labor, and Human Relations, Madison, WI.

Yates, D W (1990). ABC of major trauma: scoring systems for trauma. *British Medical Journal*, **301**(6760), 1090–1094.

Yoon, K P and Hwang, C-L (1995), *Multiple Attribute Decision Making*, Sage Publications, London.

14 LOGIC TREES

14.1 Introduction

In practice, it is impossible to specify and control all the factors affecting the occurrence and spread of fire in a building, although the relative influence of some of the factors in enhancing or reducing fire risk can be assessed quantitatively by modeling techniques as described in Chapter 11. Hence, it would be realistic to treat fire as a random phenomenon and adopt a probabilistic approach to the evaluation of fire risk and determination of fire protection requirements. In this approach, as described in Chapter 7, fire risk is expressed as the product of two components:

(i) the probability of fire occurring within a specified period of time, say a year (the frequency),
(ii) the probable damage to life, property, and the environment in the event of a fire occurring (the consequences).

In regard to the second component, there are, essentially, three types of models in which probabilities enter the calculations, explicitly. These are

(a) Statistical methods
(b) Stochastic models
(c) Logic trees.

The basic characteristics of the first kind of probabilistic models have been discussed in Chapter 7. More advanced models of this kind have been described in Chapter 8 dealing with risk to life and in Chapter 9 concerned with property damage. These models include statistical methods such as probability distributions, which provide estimates of fire risk for a group or type of buildings. Although these estimates can be modified and used for assessing the risk of fire in a particular building, more accurate predictions for fire spread in a building of given features are provided by stochastic models discussed in the next chapter (Chapter 15).

However, for most practical problems in fire protection, it would be sufficient to carry out an analysis on the basis of a Logic Tree, particularly the Fault Tree, which can provide an estimate of the probability of occurrence of an undesirable (top) event. This chapter is mainly concerned with

Evaluation of Fire Safety D. Rasbash, G. Ramachandran, B. Kandola, J. Watts and M. Law
 ISBN: 0-471-49382-1

Event Trees and Fault Trees and their application to fire safety problems. The chapter includes a brief discussion about Decision Trees, which are generally used in economic problems for identifying a cost-effective fire protection strategy. These logic trees and their application in fire protection economics are discussed in detail by Ramachandran (1998).

14.2 Management of fire safety

Before we discuss the structure of logic trees and their application to fire protection, it is necessary to explain briefly the role played by these trees in the implementation of an effective management programme to reduce the risk of fire occurrence and its consequences. As described by Ramachandran (1987), management of fire risk involves four main stages – risk identification, risk evaluation, risk reduction, and risk transfer.

The first stage is concerned with identifying all possible causes and contributing factors responsible for the occurrence and spread of fires and establishing the sequence of events, which could lead particularly to a fire of undesirable magnitude. This process enables appropriate risk-reducing measures to be put in place to reduce the frequency of serious fires and to limit their consequences.

The second stage of the management process is concerned with a quantitative evaluation of current level of risk in order to judge whether this level is acceptable in terms of the damage to life, property, and the environment. This is done, first by formulating an Acceptance Criterion and then assessing the results against it. If the calculated level of risk is found to be not acceptable, then appropriate fire prevention, detection, control, and mitigating measures are adopted to reduce the risk to an acceptable level; this forms the third stage in fire risk management programme.

In spite of all fire safety measures, there may still be a "residual risk," which although small, could lead to a large fire resulting in serious consequences. In order to mitigate the adverse effects of financial damage likely to occur from such a fire, a property owner can insure his/her building and its contents. This is generally referred to as *risk transfer*, which is the fourth stage in a risk management scheme.

The second and third stages of risk management constitute what is generally known as Quantitative Risk Assessment (QRA). QRA has the following four main functions:

(i) It provides a numerical measure for fire risk,
(ii) It assists in a quantitative evaluation of the effectiveness of fire safety measures in reducing the fire risk,
(iii) It enables a comparison of the effectiveness of different fire safety strategies,
(iv) It demonstrates to the Regulatory Authority that specific fire safety and risk targets are being achieved and maintained.

As mentioned in the previous section, mathematical models available for carrying out QRA are of two types – deterministic (Chapter 11) and nondeterministic (Ramachandran, 1991). The latter type can be further classified into probabilistic (Chapters 7, 8, and 9) and stochastic (Chapter 15) models. Logic trees come under the first type of nondeterministic models but are semiprobabilistic since they only evaluate constant discrete values for the probabilities of occurrence of events. Probability distributions are generally considered in these models. Logic trees provide simple analytical tools for a QRA.

14.3 Logic trees

The primary objective of a QRA is to estimate the likelihood of the accidental scenarios identified at the risk or hazard identification stage being realized. This is done by calculating the probability

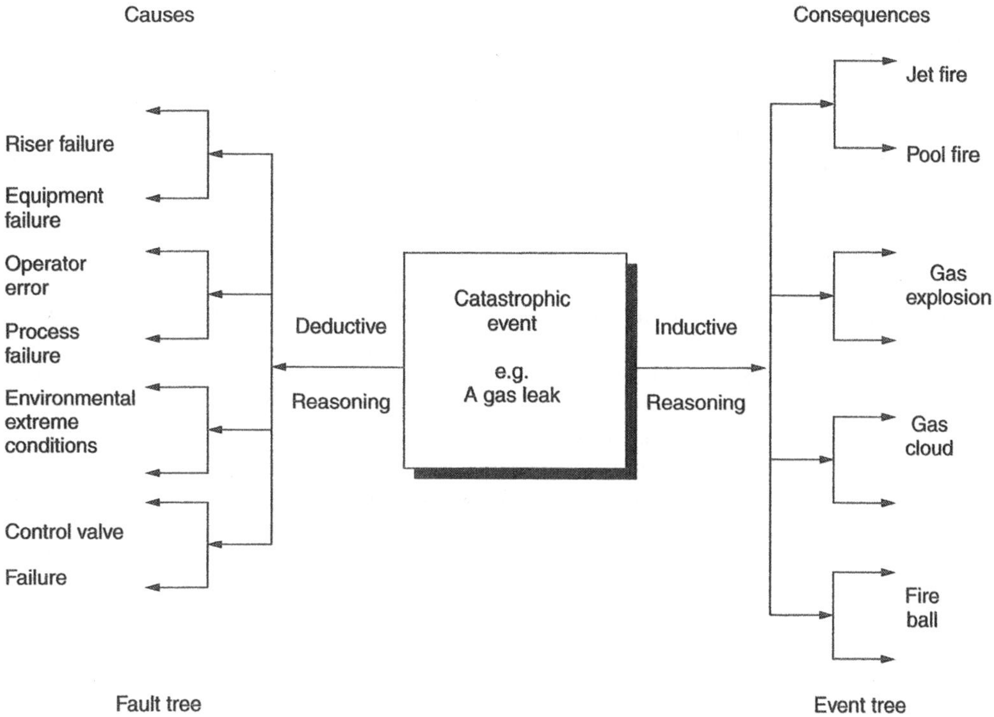

Figure 14.1. Fault tree and event tree

or frequency of occurrence of undesirable events. The probability of occurrence of an undesirable top event is estimated by placing in correct sequential order the subevents leading to the top event and specifying the probabilities of occurrence of the subevents. Probabilities associated with the subevents are combined in a suitable logical manner to derive the probability of occurrence of the top event. This calculation is facilitated by the use of logic diagrams, which form graphical representations of sequence of events. This way, logic trees constitute invaluable techniques of risk management. The commonly used diagrams are called *Event Trees* and *Fault Trees.*

Fault trees attempt to trace the root causes of a given final event by working backward using deductive reasoning. The event trees, on the other hand, use inductive reasoning and work forward to define the consequent events and paths, which result from a given initiating or primary event. The distinction between a fault tree and an event tree is explained in Figure 14.1 using a gas leak as an example of a catastrophic event.

14.4 Decision tree

As mentioned in Section 14.1, decision trees are logic diagrams used to compare the outcome of different safety strategies or courses of action in order to identify the most effective strategy or course of action. Decision tree is the appropriate logic diagram to use if the object is to identify a strategy or course of action optimizing a prescribed economic, regulatory, or technological criterion. The possible choices for strategies or courses of action can be conveniently represented in a Decision Tree, which is a mechanism for systematically organizing complex alternatives. The various alternatives are represented by the branches of such a tree emanating from decision forks.

The effects or consequences of decisions are represented by branches coming off the probability forks if they vary significantly, and create uncertainty in the selection of the "best" alternative.

In the field of fire safety, an economic objective may be to identify the most cost effective fire protection strategy that minimizes the total cost of fire protection and insurance. Figure 14.2 is an example showing eight options available to a property owner for selecting the most cost effective option. The options arise from four fire protection choices and two fire insurance levels. The fire protection alternatives are no sprinklers or detectors, only sprinklers, only detectors, and sprinklers and detectors. The insurance options are full insurance coverage (no self-insurance) and no insurance (full self-insurance).

After constructing a decision tree such as that shown in Figure 14.2, the costs associated with each option are enumerated and estimated. The costs are mostly costs of installing and maintaining fire protection devices and fire insurance premiums. If full insurance cover is not obtained and a self-insurance deductible is accepted, a property owner has to bear the entire amount of damage if a fire with loss less than the deductible amount occurs. If the loss in a fire exceeds the deductible, the owner has to bear a loss equivalent to the deductible. In this case, the owner can get compensation from the insurance company only for an amount equivalent to loss minus the deductible. The cost toward fire damage, which a property owner has to bear is, hence, a random variable depending on the probability of fire occurring and the probable damage if a fire occurs. In addition to this uncertain cost due to fire damage, the owner has to pay an appropriate fire insurance premium for obtaining cover for a loss exceeding the deductible.

If full insurance cover is obtained, the insurance company will reimburse the property owner almost the entire amount of damage when a fire occurs. In this case, the owner need not make any provision toward fire damage in his/her financial planning. If the property is fully self-insured with no insurance cover, the owner has to make a provision for the entire amount of damage likely to occur if a fire breaks out.

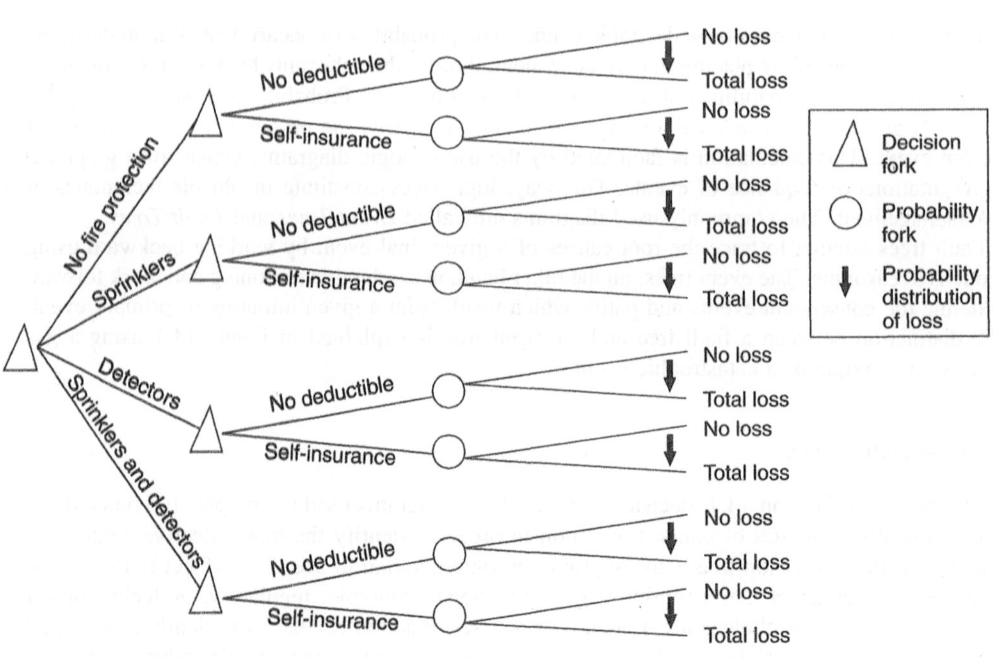

Figure 14.2. Decision tree for investment in fire protection and/or insurance

The next step in the above analysis is to express all the costs on an annual basis, and calculate the total annual cost for each option. The annual cost toward fire damage is estimated by multiplying the annual probability of fire occurrence by the expected value of the amount the property owner has to bear if a fire occurs. To facilitate a comparison of different options, the total annual cost for each option can be entered at the end of the corresponding branch of the decision tree (Figure 14.2). The property owner can then select for adoption the option with the least total annual cost.

An analysis of the decision tree type has been carried out by Helzer *et al.* (1979) to evaluate different strategies for reducing residential upholstered furniture fire losses. Three options were evaluated: no action, mandatory smoke detector installations, and a proposed upholstered furniture standard under consideration by the Consumer Product Safety Commission. The options were evaluated on the basis of minimizing the sum of the total cost plus the loss to society over time. Subject to some assumptions used in the report, the analysis showed that the detector option and the proposed standard were essentially equivalent and preferred to the no-action option. The proposed standard was judged to be more effective in saving lives, whereas the detector choice was less costly to implement. A sensitivity analysis showed that the results were particularly sensitive to the cost of the proposed standard, the value assigned to loss of life, and the standard of the upholstered furniture replacement.

14.5 Event tree

The construction of an event tree starts by defining an initiating event leading to the final outcome (a gas release, fire, or explosion) following a series of branches each denoting a possible outcome of a chain of events. The main elements are therefore the event definitions and logic vertices. Figure 14.3 is an example of an event tree representing a range of outcomes resulting from a release of flammable liquid as an initiating event. The fluid may be in the form of a liquid or gas or a mixture of the two. Depending on the intermediate events, the outcome may be a jet fire, a pool fire, or a vapor cloud explosion. The probabilities associated with each outcome are given in this figure. The probability of any outcome is the product of the probabilities of events leading to the outcome.

Figure 14.4 is another example of an event tree for possible events that might follow release of gas from a domestic supply. In this example, a gas leak is the initiating event and an explosion is the final hazard. The events or conditions for each branch are listed across the top. These are known as *nodal events*. At each node, taking an upward branch implies that the event happens, and taking a downward branch implies that the event does not happen. Probabilities are assigned to each event and these for any branch are multiplied to estimate the final probability for explosion. The figure of 3.6×10^{-3} at the top, for example, is the product of five factors, 0.1, 0.5, 0.9, 0.1, and 0.8, and gives the probability of explosion due to the chain of events indicated by the branch. There are two other such chains with explosion probabilities of 4×10^{-3} and 4×10^{-2}. The sum of all the three probabilities is 4.76×10^{-2} (as indicated in Figure 14.4), which is the overall probability of explosion, given a gas leak has occurred.

Event tree can be used to design a protection measure to reduce the probability of occurrence of the final hazard. Considering Figure 14.4, for example, if the probability of 4.76×10^{-2} for the occurrence of explosion is unacceptable, one might install a gas-leak detector. If this was a reliable device, it could effectively remove the "No" branch to "smell noticed." This would lower the overall probability of 0.0476 to 0.0076 ($= 0.0036 + 0.004$).

Event trees give a good pictorial display of event sequences, but it is important to remember that the outcomes are only relative to the specific cause being analyzed. If the same outcome could also arise from other causes, these would not be shown. Event trees can also accommodate

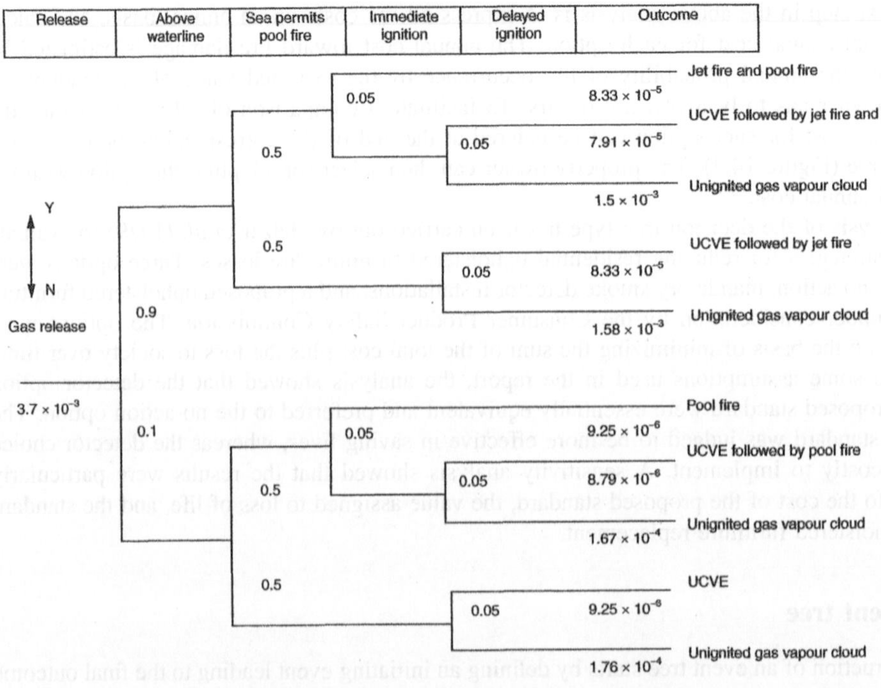

Figure 14.3. Event tree analysis for a gas release

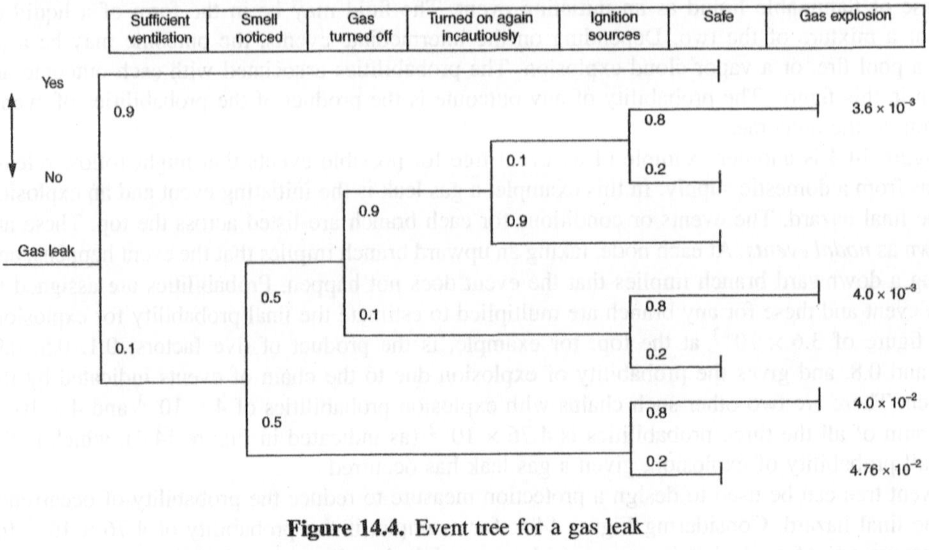

Figure 14.4. Event tree for a gas leak

a whole range of outcome on one tree. A significant disadvantage with event tree is that the output is diffused and not focused on to a particular undesirable event. Consequently, event trees can quickly become extremely large and unwieldy. Overall, event trees find greater use in analyzing the response of an equipment to various forms of accidents and malfunctions.

14.6 Fault tree

Fault trees are the most widely used techniques for developing failure logic and hazard assessment. A fault tree is a graphical representation of logical relations between an undesirable top event and primary cause events. It uses deductive arguments to arrive at the root causes.

The construction of a fault tree starts with the definition of the top event (the undesired event), which may have been identified at the Hazard Identification stage. It could be a fire or explosion or a gas release. The tree is constructed by placing in correct sequential order various cause events. This is generally done by working backward from the top event and specifying the events, causes, faults, or conditions that could lead to the occurrence of the top event, working backward from each of these, which in effect become secondary top events and so on. This process is continued and terminated when a final set of basic events, faults, or conditions are identified. There is no logical reason to continue beyond basic events because their contribution is negligibly small. A diagrammatic representation of the process would then generate the branches of a tree.

The events in a fault tree are connected by logic gates that show what combination of the constituent events could cause a particular top event. These are mainly AND gates in which all the constituent events have to occur simultaneously, and OR gates in which only one of the constituent events needs to occur to cause the occurrence of the specific top event.

Figure 14.5 illustrates the logic underlying the AND and OR gates. This example is concerned with the undesirable event of flames generated by a fire in a room reaching the ceiling. For this top event to occur, all of the following three factors should be present:

(i) An ignition source (A) – source A_1 or A_2
(ii) Heat transfer condition (B) – Conditions B_1 or B_2
(iii) A material (C) – material C_1 or C_2 or C_3.

The events (factors) A, B, and C are connected by an AND gate, whereas the subevents or factors for each of these events are connected by an OR gate. For purposes of illustration, hypothetical probabilities are specified in Figure 14.5 for the basic events. Since A_1 and A_2 are connected by an OR gate, their probabilities are added to provide an estimate for the probability associated with the factor A. Likewise, the probabilities of B_1 and B_2 are added to estimate the probability of B and the probabilities of C_1, C_2, and C_3, are added to estimate the probability of C. Since A, B, and C are connected by an AND gate, their probabilities are multiplied to provide an estimate of the probability (0.0214) of occurrence of the top event, flames reaching the ceiling.

From this it can be seen that the fault tree analysis approach is particularly useful in

- highlighting the weaknesses of a system from a safety point of view,
- identifying failure modes deductively,
- providing a graphical representation to help in safety management,
- providing options for qualitative or quantitative system reliability analysis,
- providing a methodology for analyzing one particular system failure at a time.

Despite their usefulness, fault trees are not free from their problems and limitations. These include the following:

- Fault trees do not show sequence of events.
- The use of binary logic approach means that only two states can be shown.
- Time and rate dependence cannot be represented easily.

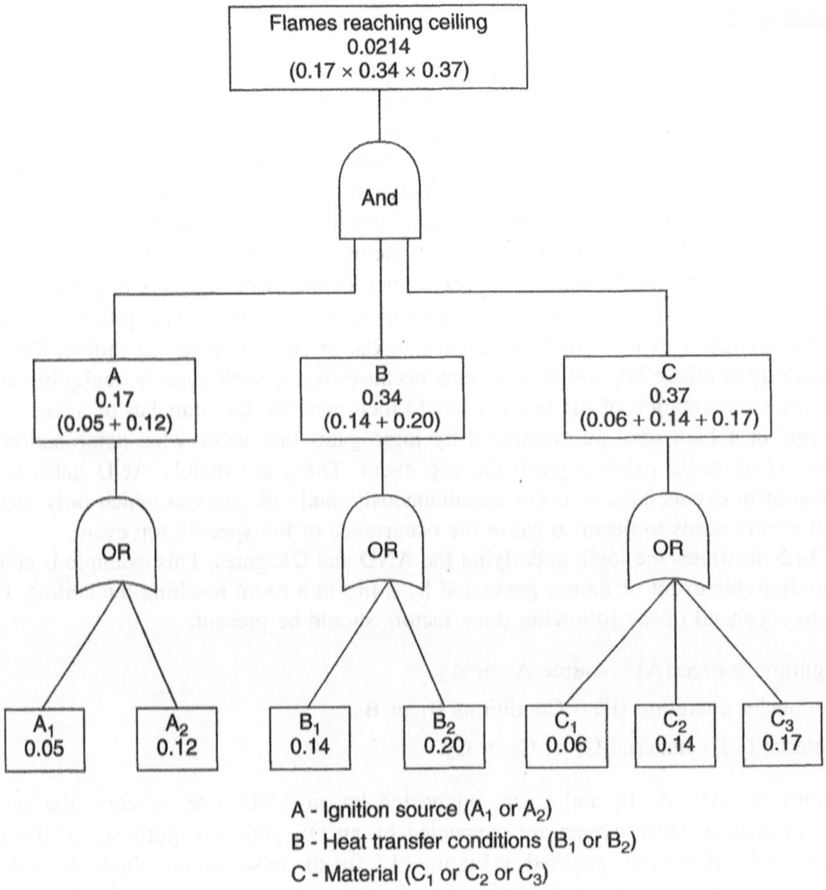

Figure 14.5. Fault tree for flames reaching the ceiling

The method described above, based on a multiplication theorem for AND gates and an addition theorem for OR gates, only provides an approximate value for the probability of occurrence of a top event. A more accurate value for this probability can be obtained by applying complex calculation techniques (Boolean Algebra) relating to cut sets or path sets. A discussion of these techniques is beyond the scope of this book.

References

Helzer, S G, Buchbinder, B and Offensend, F L (1979). *Decision Analysis of Strategies for Reducing Upholstered Furniture Fire Losses*, Technical Note 1101, National Bureau of Standards, USA.

Ramachandran, G (1987). Management of fire risk, risk assessment and management. *Advances in Risk Analysis*, L B Lave (Ed.), Plenum Press, New York, pp. 289–308.

Ramachandran, G (1991). Non-deterministic modelling of fire spread. *Journal of Fire Protection Engineering*, **3**(2), 37–48.

Ramachandran, G (1998). *The Economics of Fire Protection*, E. & F. N. Spon, London.

15 STOCHASTIC FIRE RISK MODELING

15.1 Introduction

Apart from changes in environmental conditions, the spread of fire in a building is governed by physical and chemical processes evolved by a variety of burning materials arranged in different ways. Multiple interactions among these processes at different times cause uncertainties in the pattern of fire development and performance of passive and active fire protection systems. Uncertainties are also involved in the behavior and movement of occupants of a building and in the spread of combustion products that block the escape routes and cause damage to life. A deterministic model (Chapter 11) cannot provide any evaluation of these uncertainties (errors) although it can simulate different patterns of fire growth or occupant behavior by varying the input values to the parameters of the model. The uncertainties can be quantified in terms of probabilities by applying nondeterministic models (Ramachandran, 1991).

Nondeterministic models estimate and predict the movement of fire, smoke, and occupants within limits of confidence expressed in probabilistic terms. Depending upon the constancy or transience of the lack of certainty (i.e. probability), a nondeterministic model may be either probabilistic or stochastic. The former type includes probability distributions (Chapters 8 and 9), which are generally useful in assessing fire risk in a group of buildings and semiprobabilistic techniques such as logic trees (Chapter 14). These models deal with the final outcomes (e.g. area damaged, spread beyond room), which may be sufficient in fire protection and insurance problems not requiring a detailed knowledge of the processes governing fire spread and occupant evacuation.

When evaluating fire risk in a particular building, a more sophisticated approach is provided by stochastic models, which constitute the subject matter of this chapter. These models consider the chain of critical events occurring in space and time and the probabilities connecting these events. Two types of stochastic models are discussed in detail – Markov Chains and Networks. Attention is also drawn to the application of other stochastic models, which include Random Walk, Percolation Theory, and Epidemic Models. (More detailed discussions of these models with reference to fire spread are contained in the Second Edition (1995) of the SFPE Handbook of Fire Protection Engineering.)

Evaluation of Fire Safety D. Rasbash, G. Ramachandran, B. Kandola, J. Watts and M. Law
 ISBN: 0-471-49382-1

15.2 General model

In a simple model, the random (stochastic) process associated with the burning of a particular object in a room may be assumed to be Poisson such that the duration of burning follows an exponential probability distribution (Ramachandran, 1985). In this case, a fire can only be in one of two states – extinguished or spreading. The probability distribution of the duration of burning can also have other forms such as uniform and lognormal. A third state can be added to denote fire burning without extinguishment or spread.

A fire involving the first object ignited becomes capable of spreading to another object after some time if it survives the "incubation" or "latent" period. This time and the spread probability also depend upon the distance between the two objects. Fire spread may not occur if the second (unignited) object is located beyond a "critical distance" from the first (burning) object.

In the language of stochastic modeling, the spread probability λ_1 (t) is the "transition probability" at time t for spread beyond the first object. It may be redefined as λ_{12} (t) to denote the spread from the first to the second object. Other objects in a room may be considered as the first or second object such that, in the general case, λ_{ij} (t) is the transition probability for spread from the ith to the jth object at time t. On the basis of the distances between them, the objects can be arranged in order to represent the sequential spread of fire from object to object at any time t. This simple analysis may be sufficient for all practical purposes, although a fire from one object can spread to any other object directly or indirectly through the ignition of another object.

Conceptually, therefore, as a fire starting in a compartment spreads from the object first ignited to other objects, there is a chain of ignitions that could lead to flashover and fully developed conditions. Depending upon the fire resistance of structural boundaries, the fire could spread beyond the compartment, floor to floor, and finally beyond the building of origin. There is, however, a chance (probability) that this chain could break at some stage because of the arrangement of objects, environmental factors, fire fighting, and other reasons. Hence, it is probable that the fire could burn out itself or be extinguished before involving the entire compartment or building. Statistics of actual (not experimental) fires support this theory.

It is, therefore, apparent that the fire chain contains different objects with the spread between them characterized by critical events (transitions) occurring sequentially in space and time and connected by probabilities. The fire stays in each object governed by a "temporal" probability distribution and moves from object to object at any time t according to the "transition" probability λ_{ij} (t). This stochastic process ends when the fire reaches the "absorbing state" denoted by burning out or extinguishment. The fire has to remain in an absorbing state once it enters this state.

Equations can be derived for the general stochastic model described above and used to evaluate the probabilities λ_{ij} (t) for any particular building with known design features, distribution and combustible nature of materials, and installed fire protection devices. However, this method requires a detailed survey of the building, a considerable amount of statistical and experimental data, and complex calculations. The problem can be simplified by adopting a Markov model for fire spread within a room or compartment and a Network model for spread throughout a building.

15.3 Markov model

15.3.1 MATHEMATICAL REPRESENTATION

For practical reasons, it may be sufficient to consider fire spreading through a number of spatial modules (Watts, 1986), phases (Morishita, 1977), or realms (Berlin, 1980). These stages of fire growth may generally be defined as *states* such that a fire spreads, moves, or makes a transition from state to state. As discussed in the previous section, the movement of the fire from state to state is governed by a "transition probability," which is a function of time since the start of the fire. The fire also spends a random length of time in each state before making the transition; this duration follows a "temporal" probability distribution. Representing mathematically, if the fire is in state a_i at the nth minute, it can be in state a_j at the $(n+1)$th minute according to the transition probability λ_{ij} (n). The transition probabilities are most conveniently handled in matrix form. We may write, dropping (n) for convenience, with m states;

$$P = \begin{vmatrix} \lambda_{11} & \lambda_{12} & \cdot & \cdot & \lambda_{1m} \\ \lambda_{21} & \lambda_{22} & \cdot & \cdot & \lambda_{2m} \\ \cdot & \cdot & \cdot & \cdot & \cdot \\ \cdot & \cdot & \cdot & \cdot & \cdot \\ \lambda_{m1} & \lambda_{m2} & \cdot & \cdot & \lambda_{mm} \end{vmatrix}$$

where

$$\sum_{j=1}^{m} \lambda_{ij} = 1, \quad i = 1, 2, \ldots\ldots\ldots\ldots, m$$

The probability distribution of the system at time n can be expressed as the vector

$$p = (q_1 q_2 q_3 \ldots\ldots\ldots\ldots q_m)$$

where q_i is the probability of the fire burning in the ith state at time n. Since a fire can be in one of the m states at a given time

$$\sum_{i=1}^{m} q_i = 1.$$

The mth state may denote the state of fire having been extinguished if such a state is included in the model considered. The vector is given by the product **p**. **P** expresses the probabilities of burning in different states one transition (minute) later.

As an example, consider a model of fire growth in a room where the ith state represents i objects burning. Suppose, with $m = 4$ and no extinguishment, the process stops with the occurrence of flashover when all the four objects are ignited. There is no recession in growth and hence there is no transition to a lower state from a higher state. With these assumptions, let the transition matrix be

$$P = \begin{vmatrix} 0.4 & 0.3 & 0.2 & 0.1 \\ 0 & 0.5 & 0.3 & 0.2 \\ 0 & 0 & 0.6 & 0.4 \\ 0 & 0 & 0 & 1 \end{vmatrix}$$

If, at time n, the probabilities of the fire burning in different states is given by

$$p_n = (0.1 \quad 0.2 \quad 0.3 \quad 0.4)$$

it may be seen by performing the matrix multiplication, that the probability of fire burning in different states at time $(n+1)$ is given by

$$P_{n+1} = (0.04 \quad 0.13 \quad 0.26 \quad 0.57)$$

Hence, at time $(n+1)$, the probability of the fire being in the third state, for example, is 0.26 and the probability of flashover (fourth state) is 0.57. Berlin (1988) and Watts (1986) have described similar examples on the use of matrix **P** by modeling a "random walk" among five adjacent spaces.

15.3.2 MARKOV CHAINS

Markov chains are used for repetitive situations in which there is a set of probabilities that define the likelihood of transition from one state to another. A chain comprises a sequence of such transitions. In a Markov chain, the transition probabilities satisfy the following properties: (Berlin, 1988)

1. Each state belongs to a finite set of all possible states.
2. The characteristics of any state do not depend upon any other previous state.
3. For each pair of states $[i, j]$ there is a probability λ_{ij} that state j occurs immediately after state i occurs.

The transition probabilities can be specified in a matrix form **P** as discussed in Section 15.3.1. State i is an absorbing state if row i of **P** has a value of $\lambda_{ij} = 1$ and all other values in the row are zero.

15.3.3 MARKOV PROCESS

The next step is to consider a slightly more complex model called the *Markov process*, a stochastic or random process in which the probability of occurrence of some future state of the system, given its present state, is not altered by information concerning past states, that is, the history of the process has no influence on its future. This lack of a historical influence is often referred to as a *memoryless* or *Markovian* property of a process.

In a Markov process with stationary transition probabilities, the value of λ_{ij} (n) is a constant independent of the time variable n. Following this process, Berlin (1980) estimated stationary transition probabilities for six realms (states) for residential occupancies as follows:

(i) the nonfire state
(ii) sustained burning
(iii) vigorous burning
(iv) interactive burning
(v) remote burning
(vi) full room involvement.

These realms were defined by critical events such as heat release rate, flame height, and upper room gas temperature. Development of fire over time was considered as a "random walk" through these realms.

On the basis of data from over a hundred full-scale fire tests, Berlin (1980) calculated transition probabilities as in Table 15.1. The information in this Table indicates that, when a fire is in Realm III, there is a 75% chance of growth to Realm IV and a 25% chance of recession to Realm II. Figure 15.1 developed by Ramachandran (1995), is the transition diagram defined by the transition probabilities in Table 15.1. Realm I, no fire, is an absorbing state, since all fires eventually terminate in this state. The process also ends when Realm VI (full room involvement) is reached; for this reason, this state also is an absorbing state. Berlin used uniform, normal, and lognormal distributions to describe temporal probability distributions for the different states.

Among many questions asked about fire development using the Markov model, the maximum extent of fire growth represents the most extreme condition. The portion of fires that do not grow

Table 15.1. Transition descriptors for a typical room in a residential occupancy fire type: smoldering fire in a couch with cotton cushions

Realm transition		Transition probability	Temporal distribution		
From	To			Mean	Std.dev
II	I	0.33	Uniform	2	5
II	III	0.67	Lognormal	8.45	0.78
III	II	0.25	Uniform	1	2
III	IV	0.75	Normal	5.55	3.22
IV	III	0.25	Uniform	1.5	9
IV	V	0.75	Uniform	0.5	3.5
V	IV	0.08	Uniform	0.6	6.0
V	VI	0.92	Lognormal	5.18	4.18

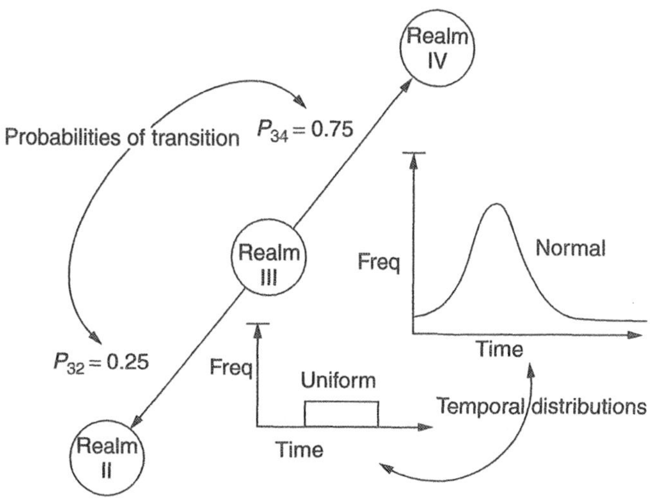

Figure 15.1. Realm transition descriptors

beyond Realm II is the probability (0.33) of transition from Realm II to I. If M_3 is the long-run (limiting) probability of fire reaching Realm III, but not growing beyond, then (Berlin, 1980)

$$M_3 = \frac{\lambda_{21} + \lambda_{23} \cdot \lambda_{31}}{1 - \lambda_{23} \cdot \lambda_{32}} - \lambda_{21} \qquad [15.1]$$

Using the figures in Table 15.1 and noting that $\lambda_{31} = 0$, it may be seen that $M_3 = 0.07$. Beyond Realm III is more difficult, as described by Berlin. Probabilities of maximum extent of flame development as estimated by Berlin are given in Table 15.2. Berlin has also discussed other fire effects such as probability of self-termination and distribution of fire intensity.

Berlin has estimated that 99% of all fires will terminate within 12 transitions. This result is based on the assumption of stationary transition probabilities that may be nearly true for a few fluctuations between the same realms where different materials would be contributing to the burning process. However, the fire will eventually consume all fuels in which case the

Table 15.2. Maximum extent of flame development

Maximum flame extent	Probability of flame extent
Realm II	0.33
Realm III	0.07
Realm IV	0.02
Realm V	0.58

probabilities of termination from all realms will be equal to 1. Therefore, Berlin's approach represents a worst-case analysis.

One of the major weaknesses of the Markov model regards the "stationary" nature of the transition probabilities. It is assumed that these probabilities remain unchanged regardless of the number of transitions representing the passage of time. The length of time a fire burns in a given state affects future fire spread. For example, the probability of a wall burn-through increases with fire severity, which is a function of time. The time spent by fire in a particular state may also depend on how that state was reached, that is, whether the fire was growing or receding. Some fires may grow quickly and some may grow slowly depending on high or low heat release. In a Markov model with stationary transition probabilities, no distinction is made between a growing fire and a dying fire.

15.4 State transition model – spread within a room

According to Berlin's Markov model (Section 15.3.3), a fire in a particular realm can either grow to a higher realm or recede to a lower realm. There is no transition to the nonfire (absorbing) state (Realm I) from any realm higher than Realm II except Realm VI (full room involvement), which is also an absorbing state. Receding to a lower state may be true to some extent when describing fire growth in terms of flamespread, but such an assumption is not possible in the case of spatial spread of fire in which fire spreads sequentially from one object to another. According to this model, if fire spreads to an object it cannot spread backward to the object from which it spread. The fire involving an object either spreads forward to other objects or gets extinguished or stays with the object without spreading.

Consistent with the fire statistics available, particularly in the United Kingdom and the United States, a simplified model based on the following three main states can be considered for fire development in a room.

S_1 – fire confined to the object first ignited,
S_2 – fire spreading beyond the object first ignited but confined to the contents of the room, and
S_3 – fire spreading beyond the room of origin but confined to the building.

A fourth state may be added to denote extinguishment or burning out (self-termination) of fire; this is an "absorbing state" since a fire process cannot leave this state after entering it. The third state, S_3, is also an absorbing state, since a spreading fire will eventually terminate within the building of origin; spreading beyond the building is not considered.

The three states ($i = 1, 2, 3$) mentioned above generate a State Transition Model, which is a special case of a Markov model. This model was used by Ramachandran (1985) to evaluate the transition (spread) probabilities $\lambda_i(t)$ and the probabilities $\mu_i(t)$ for extinguishment or transition to the fourth state. The parameters $\mu_i(t)$ and $\lambda_i(t)$ are the probabilities of extinguishment and spread for the ith state at time t. They are conditional probabilities given that the fire has spread to the ith state. The value of $\lambda_3(t)$ was taken as zero since fire spread beyond the building was not considered. The probability of burning in a state without spreading was also considered with the aid of the parameter $w_i(t)$ $[= 1 - \lambda_i(t) - \mu_i(t)]$. The duration of burning was divided into subperiods, each of a fixed length of 5 min.

Statistics furnished by fire brigades in the United Kingdom related to fires that were extinguished during each time period since ignition. Hence, Ramachandran used the extreme value technique, with some assumptions, to estimate the number of fires that were burning in a particular stage at the beginning of each subperiod. With the aid of these estimates and the actual numbers that were extinguished, approximate values were obtained for the extinguishment and spread probabilities (as functions of time) and probability distributions of duration of burning in each state. Four materials ignited first in the bedroom of a dwelling were considered for illustrating the application. Aoki (1978) described fire growth with similar states based on the spatial extent of spread; his analysis was similar to that of Ramachandran (1985). Morishita (1977) considered eight phases of spatial spread of fire, which included spread to the ceiling.

In a later study, Ramachandran (1988) added another state between S_2 and S_3 to denote the event of fire involving the structural barriers of a room assumed to occur after the fire had spread beyond S_1 and S_2 but was still confined to the room. This intermediate state was considered as generally consecutive to S_2, although a fire can spread directly from S_1 and involve the structural boundaries. Fire statistics available in the United Kingdom permit the incorporation of this additional state into a state transition model. As shown in Figure 15.2, only the limiting probabilities λ_i and μ_i were estimated from the following equations:

$$\begin{aligned} E_1 &= \mu_1;\ \lambda_1 = 1 - \mu_1 \\ E_2 &= \lambda_1 \cdot \mu_2;\ \lambda_2 = 1 - \mu_2 \\ E_3 &= \lambda_1 \cdot \lambda_2 \cdot \mu_3;\ \lambda_3 = 1 - \mu_3 \\ E_4 &= \lambda_1 \cdot \lambda_2 \cdot \lambda_3 \cdot \mu_4 \end{aligned} \qquad [15.2]$$

Fire statistics (Table 7.3) provided estimates for E_i ($i = 1$ to 4), the proportion of fires extinguished in the ith state. The condition that

$$\sum_{j=1}^{4} E_j = 1$$

follows from the assumption that $\mu_4 = 1$ and hence $\lambda_4 = 0$; fire spread beyond the building of origin was not considered.

In Figure 15.2, the product $\lambda_1 \cdot \lambda_2$ $(= E_3 + E_4)$ may be regarded as the probability of structural involvement with a high level of fire severity and λ_3 as the probability of failure of the structural boundaries of the room. For the following reasons, fire statistics do not provide a valid estimate of λ_3. Figures for the number of fires that spread beyond the room of origin include fires that spread by destruction of barrier elements (wall, floor, ceiling) as well as those that spread by convection through a door or window left open or through some other opening. In the latter case, the barrier elements would still be structurally sound. A "room" as recorded in fire brigade reports is not necessarily a "fire compartment." Using a probabilistic model (Ramachandran, 1990) or other methods, the value of λ_3 for any compartment of given fire resistance can be estimated and multiplied by the probability of structural involvement to provide an estimate of probability of spread beyond the compartment of origin by the destruction of barrier elements. The probabilities provided by a stochastic model can be regarded as "noise" terms superimposed over a deterministic trend in fire growth over space and time predicted by a model such as exponential model. The estimates of time for the different states in Figure 15.2 are given in Table 7.3.

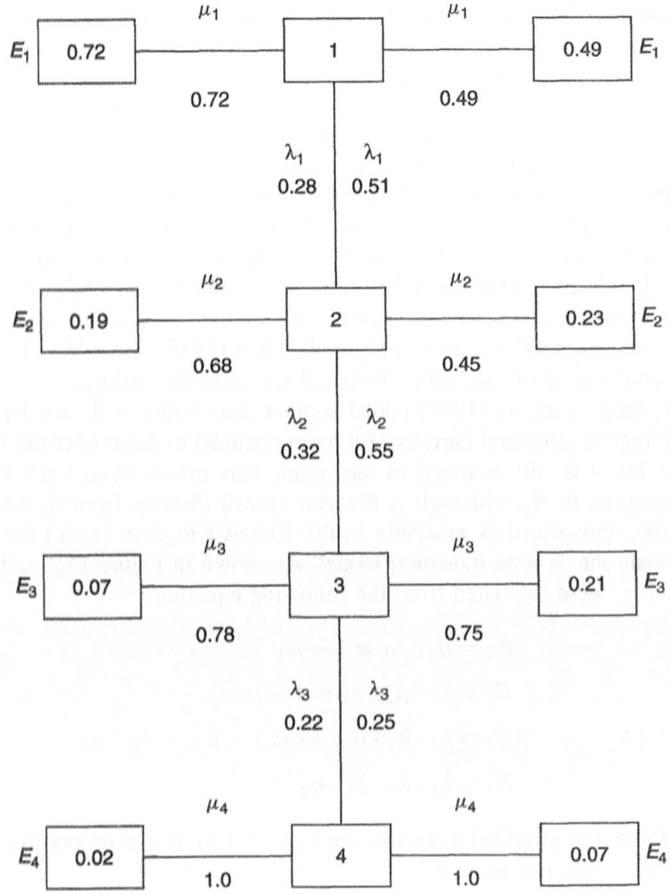

Figure 15.2. Probability tree (textile industry)
E_1 = Probability of confinement to item first ignited.
E_2 = Probability of spreading beyond item first ignited but confinement to contents of room of fire origin.
E_3 = Probability of spreading beyond item first ignited and other contents, but confinement to room of fire origin and involvement of structure.
E_4 = Probability of spreading beyond room of fire origin, but confinement to building.

To represent the interaction between human behavior and fire dynamics, Beck (1987) developed a series of stochastic state transition models and interrelated deterministic models. His sequential fire-growth model was based on the six realms defined by Berlin (1980) with the remote burning state denoting flashover. His results reproduced in Table 15.3 are applicable to office buildings. P_i for this Table is the same as E_i for Equation [15.2]. Adopting a different notation and starting with $P_1 = \mu_1$, the conditional probabilities of extinguishment, μ_i, and conditional probabilities of spread, λ_i, were calculated according to Equation [15.2]. The probability of a fully developed fire, given a fire, defined by P_{FDF}/F is given by the product $\lambda_1 \cdot \lambda_2 \cdot \lambda_3 \cdot \lambda_4$. The probability of spread beyond the compartment of fire origin, P_{VI}, is given by the product of P_{FDF}/F and λ_5 or by $\lambda_1 \cdot \lambda_2 \cdot \lambda_3 \cdot \lambda_4 \cdot \lambda_5$.

Beard (1981/82) proposed a state transition model by considering a number of "critical events" with directional characteristics that a fire may pass through and the times between critical events.

Table 15.3. Probabilities of extinguishment: fire-growth and suppression model (offices)

System configuration	P_{I}	P_{II}	P_{III}	P_{IV}	P_{V}	P_{VI}	P_{FDF}/F
No sprinkler	0.5673	0.0038	0.0017	0.3282	0.0666	0.0324	0.0990
Sprinkler	0.5673	0.3827	0.0201	0.0232	0.0045	0.0022	0.0067

For example, critical heat event CHE2U referred to *fire passing through 2 kW on the way up* whereas CHE2D referred to *fire passing through 2 kW on the way down.* The time between two critical events was assumed to have a "temporal" probability distribution independent of the time between earlier critical events. A particular succession of critical events formed a "chain"; specific times between critical events were referred to as a *sequence* within the chain. On the basis of assumed forms for "transition probabilities" and "temporal" probability distributions, Monte Carlo simulation was employed to generate randomly particular chains and sequences. A general sequence to smoke and toxic gases was related to the corresponding sequence for the burning rate. On the basis of the concentration of carbon monoxide, Beard employed the concept of "fraction of fatality" with fatality resulting at a unit value for this fraction. He applied the model to a particular case involving flaming ignition on a bed in a hospital ward. He concluded that there would be a very large (greater than 80%) likelihood of having multiple fatalities if a fire goes above 50 kW. One of the several assumptions used by him was that the fire did not spread beyond the ward.

15.5 State transition model – spread from room to room

As discussed earlier, there is a probability p_f for flashover occurring in a room or compartment, which depends on the objects in the room and their spatial arrangement apart from ventilation and other factors. Given flashover, the fire can breach the structural boundaries of the room with a probability p_b and spread beyond the room with a probability p_s $(= p_f \cdot p_b)$. The value of p_b depends on the level of fire severity attained after flashover and the fire resistance of the structural elements such as walls, ceilings, and floors. The probability of failure of a room or compartment of given fire resistance, p_b, can be estimated from the joint probability distribution of fire severity and fire resistance expressed in units of time (Ramachandran, 1990). Fire resistance of a compartment will be reduced and the failure probability p_b increased by weakness caused by penetrations such as piping or cables through walls, doors, windows, or other openings in the structural barriers.

Each room or corridor in a building has, therefore, an independent probability p_s of fire spreading beyond its boundaries. Using these probabilities for different rooms and corridors, fire spread in a building can be considered as a discrete propagation process of burning among points, which abstractly express the rooms, spaces, or elements of a building. In a simple analysis, states classified by burning situation of individual points can be incorporated in a state transition model (Morishita, 1985).

Consider, for example, three adjoining rooms R_1, R_2 and R_3, which provide the following four states with the fire commencing with the ignition of objects in R_1.

1. Only R_1 is burning.
2. R_1 and R_2 are burning (and not R_3).
3. R_1 and R_3 are burning (and not R_2).
4. All the three rooms are burning.

There is no transition from the

(a) first to the fourth state
(b) second to the third state
(c) third to the second state
(d) second or third or fourth to the first state (recession of fire growth).

A transition from the second to the fourth state involves the spread of fire to R_3 from R_1 or R_2. The probability for this transition is, therefore, the sum of probabilities for spread from R_1 to R_3 and from R_2 to R_3. Likewise, the probability of transition from the third to the fourth state is the sum of probabilities for spread from R_1 to R_2 and from R_3 to R_2. A fire can burn in the same state without transition to another state. The process terminates when the fourth state is reached.

With the assumptions mentioned above, a transition matrix (**P**) can be formed specifying the probability of fire spread from one room to another. The unit of time may be longer than 1 min, say, 5 min, since we are considering spread from room to room after the occurrence of flashover. The values of transition probabilities may be considered as constants in a state transition model with stationary transition probabilities. Starting with the initial period when only R_1 is burning, the probability distributions of the system for later periods can be obtained by repeating the matrix multiplication as described in Section 15.3.1. This process will generate for each state, a probability distribution for burning in that state as a function of time, which will provide an estimate of average transition time to that state. Morishita (1985) proposed a method of estimating the average time for transition to the fourth state denoting the burning of all the three rooms, on the basis of partitioning of the matrix P. He has also discussed the stochastic process for a system in which extinguishment is attempted. For purposes of illustration, he has applied the model to a hypothetical small house.

15.6 Network models

15.6.1 ELMS AND BUCHANAN

The model described in the previous section can be extended to provide cumulative probabilities, at time n or limiting, for the burning of more than three rooms, but this will involve tedious and complex calculations. It would be simpler to consider fire spread between two given rooms through intermediate rooms and corridors in terms of discrete values attached to the probability p_s of spread beyond a room. This probability may be the limiting value for the cumulative probability given by, say, E_4 ($= \lambda_1 \cdot \lambda_2 \cdot \lambda_3$) in Figure 15.2. Alternatively, the time taken by fire to breach the boundaries of a room may be ascertained from deterministic (scientific) models and a probability assigned to this time (t_s) used for p_s in the stochastic model. The duration t_s is the sum of t_f representing the time to the occurrence of flashover after the start of established burning and t_b representing the time for which the barriers of the room can withstand fire severity after flashover. The latter time may be the endurance of barrier elements as measured by a standard fire resistance test such as ISO834. As mentioned earlier, $p_s = p_f \cdot p_b$.

Consider, as an example, the simple layout of Figure 15.3a relating to four rooms and the corresponding graph shown in Figure 15.3b, which also shows the probability (p_{ij}) of fire spread between each pair of rooms (i, j). The probability p_{ij} refers to p_s as defined in this paper, whereas Dusing *et al.* (1979) and Elms and Buchanan (1981) have considered only the barrier failure probabilities denoted by p_b ignoring the probability of flashover denoted by p_f. The

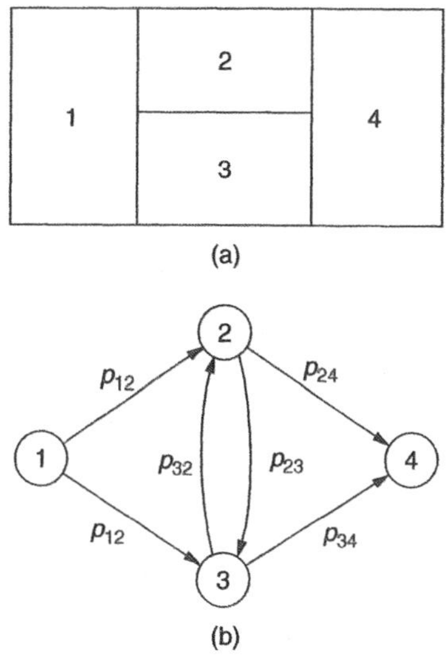

Figure 15.3. (a) Room layout; (b) Corresponding graph

specific problem considered by these authors was to compute the probability of fire spread from Room 1 to 4, which might follow any of the four paths –

$$(1) \rightarrow (2) \rightarrow (4); (1) \rightarrow (3) \rightarrow (4); (1) \rightarrow (2) \rightarrow (3) \rightarrow (4);\ \text{and}\ (1) \rightarrow (3) \rightarrow (2) \rightarrow (4).$$

Using the event space method, Elms and Buchanan have considered first all possible "events" or combinations of fire spreading or not spreading along various links. If a_{ij} represents spread of fire along link ij, and $\overline{a}_{ij}$ represents fire not spreading along the link, then one event might be

$$[a_{12}, \overline{a}_{13}, a_{23}, \overline{a}_{32}, \overline{a}_{24}, a_{34}]$$

There will be $2^6 = 64$ events, which will all be exclusive as any pair of events will contain at least one link for which fire spreads in one event and does not spread in the other. The probability of each event occurring is the product of the probabilities of its elements, assuming that the elements are independent. Thus, for the example given above, the event probability will be

$$p_{12}(1 - p_{13})p_{23}(1 - p_{32})(1 - p_{24})p_{34}$$

and the overall probability is the sum of all 64 event probabilities.

The complete event space can be represented as a tree with 64 branches. The probability of fire spread for each event (branch) is obtained by multiplying all the link probabilities in a branch. However, not all branches have to be computed in full. The computation can be curtailed, while still allowing for all cases. For this purpose, Elms and Buchanan (1981) have described a method of constructing the tree and its ordering for identification or search of possible paths of links leading to Node (room) 4 from Node 1. This procedure is known as a *depth-first* search of a graph. In this algorithm, a path is a series of nodes or rooms in a building and the construction of a branch, which is part of a particular event, is based on an underlying path. Each branch allowing fire spread must contain at least one path. Figure 15.4 shows the actual tree as it would be computed by the algorithm. The total probability of spread from Node 1 to Node 4 is given by the sum of all the branch probabilities. The calculation is carried out for each pair of rooms and

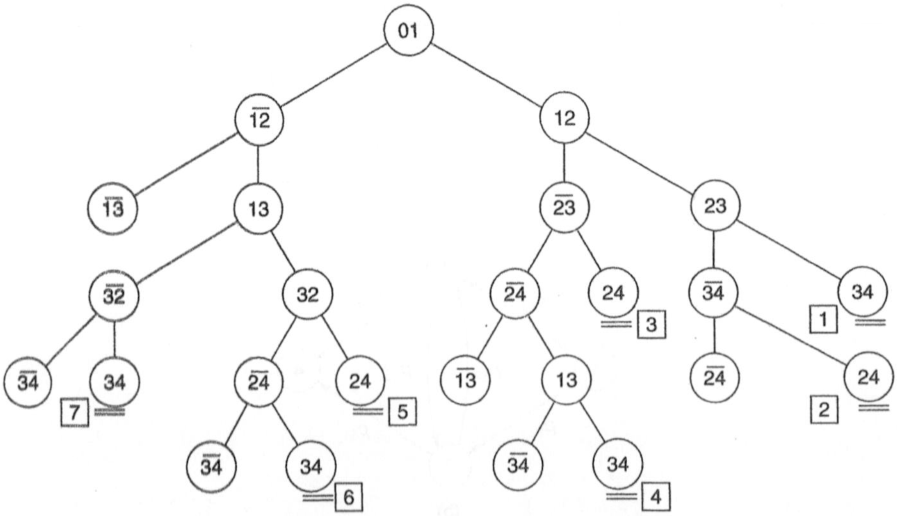

Figure 15.4. Modified event space tree

the results assembled in a "fire spread matrix." The diagonal elements of the matrix are unity. Various means have been employed to curtail the algorithm to prevent the computer developing excessively lengthy branches, which would, as the branch probability decreases with branch length, have little effect on the result.

In the computer-based technique of Elms and Buchanan as described above, a building is represented as a network by defining compartments as nodes and the links between these nodes as possible paths for fire spread from compartment to compartment in a multicompartment building. The core of this model is a probabilistic network analysis to compute the probability of fire spreading to any compartment within the building. The authors (1988) added a series of further refinements when they applied the model to analyze the effects of fire resistance ratings on the likely fire damage to buildings.

15.6.2 PLATT

Elms and Buchanan (1981) did not consider the dimension of time explicitly, although it was implicit in many of their functions. The probability of fire spreading from one compartment to another was considered irrespective of how long it might take. As a result, the analysis did not take into account any intervention, for example, the fire service. In this respect, the model represented a worst-case scenario involving the probable effects of a fully developed fire.

Platt (1989) has proposed a new network model in which fire resistance and severity are related to real time instead of equating these two parameters to the time based on ISO standard fire, which is not necessarily representative of the real time. The model computes the probability that fire will have spread to any part of a building after an elapsed time t. The essential features of Platt's model are described in the following paragraphs.

The spread of fire to an adjacent compartment may be via the following paths:

(a) Through an open doorway
(b) Through vertical spread via windows
(c) Through a barrier such as a wall, closed door, and ceiling.

Two models are considered for an estimation of fire growth as a function of time. The first model is based on the exponential relationship between fire area and growth time as suggested by Ramachandran (1986). The second model uses the parabolic relationship between rate of heat release and growth time as proposed by Heskestad (1982). This model is used in conjunction with the relationship between compartment temperature and rate of heat flow to provide an estimate of flashover time that is taken as the time when ceiling temperature reaches 600 °C.

A figure of 49% is used to represent the probability that the initial fire will not result in flashover. Subsequent ignitions caused by fire spread are given a 100% probability of reaching flashover. This assumption may overestimate the spread of fire since a barrier may "fail" in the latter stages of a fully developed fire, which may not then have the "momentum" to initiate further ignition.

The real time of the fire duration, t, representing fire severity, S, is estimated by the ratio of fuel load to rate of combustion, which is a function of ventilation and dimensional characteristics of the compartment. A formula suggested by CIB (1986) is used to estimate the "equivalent time," t_e, involving the "real" parameters of the compartment. The approved fire resistance rating (FRR) of a barrier element, modified for "weakness" and another factor, is multiplied by the ratio (t/t_e) to yield an "equivalent FRR" denoted by R. As defined above, R and fire severity, S, are not independent but, quite rightly, they have been assumed to be independent random variables with lognormal probability distributions. Under this assumption, the probability of fire spread through a barrier is estimated through the safety factor (R/S), which is also a lognormal variate. The probability of fire spread via an open door is assumed to be 100%. The probability of fire spreading vertically up the facade of a building via windows is equated to the probability that the height of the external flames is greater than or equal to the height of the spandrel.

A comparison is then made between these values and the design values of the barrier and door fire resistance and the spandrel heights. The output from these comparisons is a series of probabilities that fire will spread via each of the three possible paths described earlier. Combining these individual probabilities gives an overall probability of fire spreading to an adjacent compartment. Repeated for each compartment within the building, these values collectively form the adjacency fire spread matrix, whose values represent the probability that fire will spread from compartment i to an adjacent compartment j. The expected time for fire to spread to an adjacent compartment, given that fire does spread, provides values for the adjacency fire spread time matrix.

By combining the two matrices, the analysis computes the probability of fire spreading from an initial compartment i to any compartment j. The fire may spread along any path, but is conditional on having arrived at compartment j in a given time. The resulting matrix, Global Fire Spread Matrix, may be considered as a *three-dimensional matrix*, with each layer being evaluated at a different time. Once the fire spread matrix has been formed, Platt's model (1989) is very similar to that of Elms and Buchanan (1988) except that in the former model, the probability of spread is dependent on time, whereas in the latter model, the probability of fire spreading is irrespective of the time taken.

15.6.3 LING AND WILLIAMSON

Ling and Williamson (1986) have proposed a model in which a floor plan is first transformed into a network similar to the process described by Dusing *et al.* (1979) and Elms and Buchanan (1981). Each link in their network represents a possible route of fire spread and those links between nodes corresponding to spaces separated by walls with doors are possible exit paths similar to those developed by Berlin *et al.* (1980). The space network is then transformed into a probabilistic fire spread network as in the example in Figure 15.5 with four rooms, Rm 1 to Rm 4, and two corridor segments c_1 and c_2. In this figure, Rm 1 has been assumed as the room of

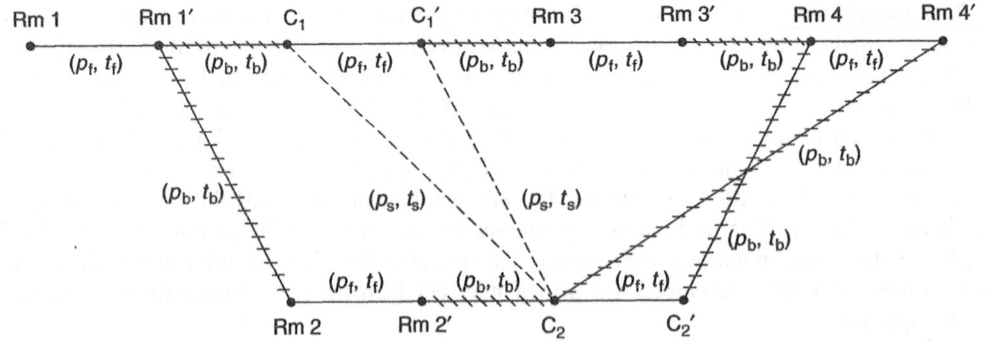

Figure 15.5. Probabilistic network of fire spread of Room 1 to C_2. ——: Fire growth within compartment; -\\-\\-: Fire breaches barrier elements; - - - : Fire spread along corridor

fire origin but it would be a simple modification to reformulate the problem for another room of origin. With Rm 1 and Rm 1′ with a "prime" denoting the preflashover and postflashover stages, the first link is represented by

$$\text{Rm } 1 \xrightarrow{(p_f,\, t_f)} \text{Rm } 1'$$

where p_f represents the probability of flashover and t_f represents the time to flashover. The nodes denoted by a prime represent a fully developed (i.e. postflashover) fire in the compartment.

In Figure 15.5, three different types of links are identified. The first corresponds to the fire growth in a compartment, the second to the fire breaching a barrier element, and the third to fire spread along the corridors. To each link i, a pair of numbers (p_i, t_i) is assigned with p_i representing the distributed probability that a fire will go through link i and t_i representing the time distribution that it will take for such a fire to go through link i. The section of the corridor, c_1, opposite Room 1 is treated as a separate fire compartment and is assigned a (p_f, t_f) for the link from c_1 to c_1'. The number pair (p_s, t_s) represents the probability and time for the preflashover spread of fire along the corridor from c_1 to c_2. As a first approximation, p_s may be considered to be governed by the flamespread classification of the corridor's finish materials on the walls and ceiling, as measured by a test method such as the ASTM E-84, Tunnel Test.

Once full involvement occurs in the section c_1 of the corridor outside Rm 1 (i.e. node c_1' is reached), the fire spread in the corridor is influenced more by the ventilation in the corridor and by the contribution of Rm 1 than by the materials properties of the corridor itself. Thus, there is a separate link, c_1' to c_2, which has its own (p_s, t_s). The number pair (p_b, t_b) represents the probability of failure of the barrier element with t_b representing the endurance of the barrier element.

Once one has constructed the probabilistic network, the next step is to "solve" it by obtaining a listing of possible paths of fire spread with quantitative probabilities and times associated with each path. For this purpose, Ling and Williamson have adopted a method based on "emergency equivalent network" developed by Mirchandani (1976) to compute the expected shortest distance through a network. (The word "shortest" has been used instead of "fastest" to be consistent with the literature.) This new "equivalent" network would yield the same probability of connectivity and the same expected shortest time as the original probabilistic network. In this method, each link has a Bernoulli probability of success and the link delay time is deterministic.

It must be noted that there are multiple links between nodes in the equivalent fire spread network. For example, the door between Rm 1 and the corridor could be either open or closed at the time the fire flashed over in Rm 1. Ling and Williamson assumed as an example that there is a 50% chance of the door being open and that an open door has zero fire resistance. Furthermore, they assumed that the door if closed would have a 5-min fire rating. With further assumptions, they constructed the equivalent fire spread network (Figure 15.6) with 12 possible paths for the example in Figure 15.5 to find the expected shortest time for the fire in Rm 1 to spread to the portion of corridor c_2. This network changes to Figure 15.7 with 10 possible paths if self-closing, 20-min, fire-rated doors had been installed in the corridor, assuming that the reliability of the self closures is perfect and that door stops had not been allowed. Note that the links have been renumbered for Figure 15.7.

For the two equivalent networks shown in Figures 15.6 and 15.7, all of the possible paths are listed in Tables 15.4 and 15.5 with increasing time and with all the component links identified. Each of these paths can be described by a fire scenario in words; for instance, Path 1 in Table 15.4 consisting of links l_1, l_2, and l_4 would be

"The fire flashes over, escapes from Rm 1 through an open door into the corridor c_1, and spreads along the corridor to c_2."

The probability of that scenario (0.13) is strongly dependent on the probability (0.5) for the occurrence of flashover in Rm 1 and of the probability (0.5) that the door will be open. The time

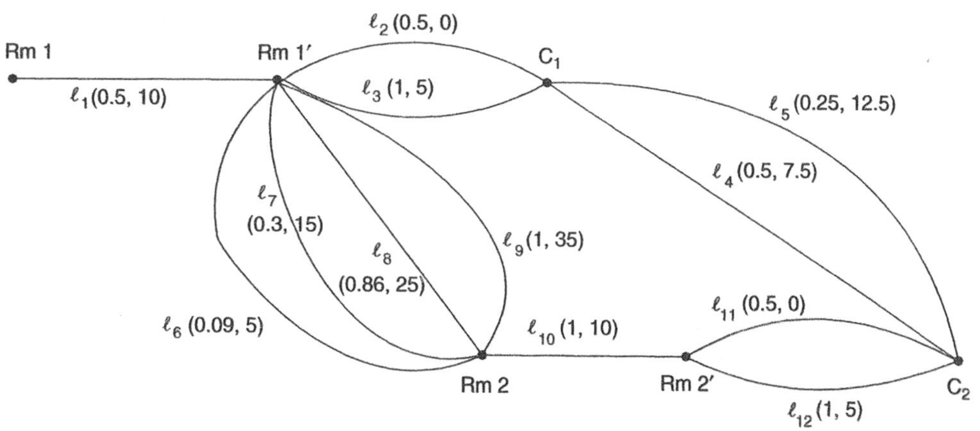

Figure 15.6. Equivalent fire spread network with 5-min unrated doors

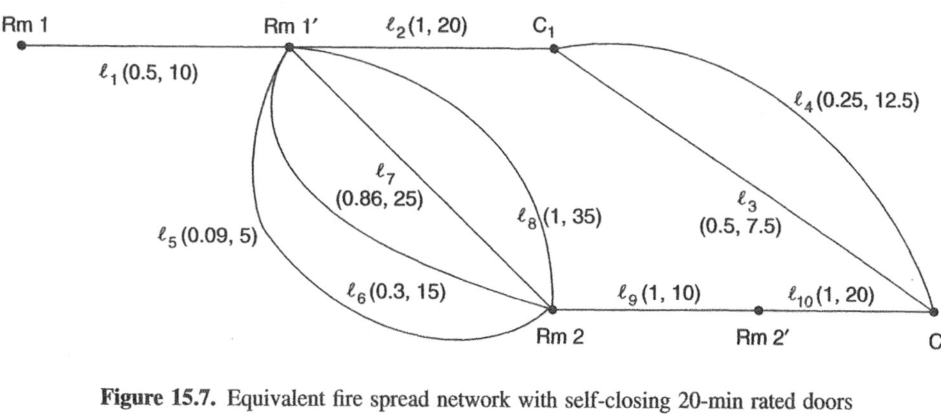

Figure 15.7. Equivalent fire spread network with self-closing 20-min rated doors

Table 15.4. Pathways through the example fire spread equivalent network assuming 5-min unrated corridor doors as shown in Figure 15.6

Paths	Component links	Probability p_i	Time t_i (minutes)
1	1–2–4	1/8 = 0.13	17.5
2	1–2–5	1/16 = 0.06	22.5
3	1–3–4	1/4 = 0.25	22.5
4	1–6–10–11	1/44 = 0.02	25.0
5	1–3–5	1/8 = 0.13	27.5
6	1–6–10–12	1/22 = 0.05	30.0
7	1–7–10–11	3/40 = 0.08	35.0
8	1–7–10–12	3/20 = 0.15	40.0
9	1–8–10–11	3/14 = 0.21	45.0
10	1–8–10–12	3/7 = 0.43	50.0
11	1–9–10–11	1/4 = 0.25	55.0
12	1–9–10–12	1/2 = 0.50	60.0

Table 15.5. Pathways through the example fire spread equivalent network assuming self-closing 20-min rated corridor doors as shown in Figure 15.7

Paths	Component links	Probability p_i	Time t_i (minutes)
1	1–2–3	1/4 = 0.25	37.5
2	1–2–4	1/8 = 0.13	42.5
3	1–5–9–10	1/22 = 0.5	45
4	1–6–9–10	3/20 = 0.15	55
5	1–7–9–10	3/7 = 0.43	65
6	1–8–9–10	1/2 = 0.50	75

of 17.5 min is composed of the times of 10 min for flashover and 7.5 min for fire to spread in the corridor from c_1 to c_2.

Ling and Williamson have derived a formula for calculating from the figures in Table 15.4 and Table 15.5, the probability of connectivity R, which is 0.5 for both the networks (Figures 15.6 and 15.7). This probability is a direct result of the assumed probability of 0.50 for flashover in the room of fire origin and the occurrence of unity probabilities in the remaining links, which make up certain paths through the network. According to another formula, the expected shortest time is 29.6 min for Figure 15.6, which increases to 47.1 min for Figure 15.7 because of the presence of the 20-min, fire-rated door. The equivalent fire spread network thus facilitates an evaluation of design changes and affords ready comparison of different strategies to effect such changes.

15.7 Random walk

In a simple stochastic representation, the fire process involving any single material or number of materials can be regarded as a random walk. The fire takes a random step every short period either to spread with a probability λ or to get extinguished (or burn out) with a probability μ ($= 1 - \lambda$). The parameter λ denotes the success probability of the fire, whereas μ denotes the success probability of an extinguishing agent. The problem is similar to two gamblers, A (fire) and B (extinguishing agent), playing a sequence of games, the probability of A winning any particular game being λ. If A wins a game, he acquires a unit stake by destroying, say, a unit of the floor area; if he loses the game, he does not gain any stake. In the latter case, A does not lose his own stake to B; an already burnt out area is a loss that cannot be regained. Extinguishment can also be considered as an "absorbing boundary" to the random walk just as an "absorbing state" in a state transition model discussed in Section 15.4.

A random walk as described above will lead to the following exponential model, where $Q(t)$ is the probability of duration of burning exceeding t and $c = \mu - \lambda$:

$$Q(t) = \exp(-\mu t) = \exp[-(1 + c)t/2] \quad [15.3]$$

The fire-fighting effort is adequate if c is positive with μ greater than λ and hence greater than half; it is inadequate if c is negative with μ less than λ and hence less than half. If $c = 0$ such that $\mu = \lambda = 1/2$, there is an equal balance between fire-fighting efforts and the propensity of fire to spread.

Associated with the random variable t denoting time, there is another random variable x denoting damage, which may be expressed in terms of, say, area destroyed. Damage in fire has an exponential relationship with duration of burning (Ramachandran, 1986) such that the logarithm of x is directly proportional to t as a first approximation. This assumption would lead to Pareto distribution

$$\phi(x) = x^{-w}, x \geq 1 \quad [15.4]$$

denoting the probability of damage exceeding the value x. This distribution is used in economic problems concerned with, for example, income distribution to describe the fact that there are a large number of people with low incomes and a small number of people with high incomes. The damage is small in most of the fires, with high levels of damage occurring only in a small number of fires.

The use of Pareto distribution for fire damage originally proposed by Benckert and Sternberg (1957) was later supported by Mandelbrot (1964), who derived this distribution following a random walk process. For all classes of Swedish houses outside Stockholm, the value of the exponent w was found to vary between 0.45 and 0.55. A value of $w = 0.5$ in equation [15.4] would imply, as discussed with reference to equation [15.3], an equal balance between fire-fighting efforts and the propensity of fire to spread and cause damage.

The parameter μ in equation [15.3] is known as the *hazard* or *failure rate* given by the ratio

$$f(t)/Q(t)$$

where $f(t)$ is the probability density function obtained as the derivative of $F(t) = 1 - Q(t)$. A constant value for μ would denote a "random failure." For a Pareto distribution in equation [15.4], the failure rate is (w/x) such that, with a constant value for w, the failure rate would decrease as x increases indicating that, in terms of damage, a fire can burn forever without getting extinguished. A constant value of μ or w is somewhat unrealistic particularly for a fire that is fought and extinguished at some stage. A fire will also burn out itself when all the available fuel is consumed or when it stops spreading because of the arrangement of objects in a room or building. For the reasons mentioned above, although the failure rate can be decreasing in the early stages of fire development denoting a success for fire in spreading, it would eventually increase ("wear out failure") since fire-extinguishing efforts would succeed ultimately (Ramachandran, 1969). There may be an intermediate stage with μ remaining as a constant such that the failure rate as a function of time would resemble a "bath tub."

In the context of fire spread, random walk as described above is a one-dimensional process describing the damage by random functions of time rather than by a random function of time and space. The random walk indicates the position of the fire, that is, damage at any time. Every unit of time, there is a change in position indicated by an increment to the damage or no change due to absorption (extinguishment or burning out). Generally, the walk is considered in discrete time. If the walk is continuous in time such that the increments are Gaussian, this leads to a Diffusion Process (Karlin, 1966). (A Diffusion Process is an approximation of the Brownian Motion – a phenomenon well known in many branches of science and technology.) The normal or diffusion term is one of two possible components of a general additive stochastic process, the other component being a discontinuous or transition term arising from occurrences of events at random times. The Markov chains discussed earlier belong to the second type of component. A linear superposition of the two components provides a solution to an equation governing a general additive process.

15.8 Percolation process

In random walk and diffusion models, randomness is a property of the moving object, whereas in a percolation process, randomness is a property of the space in which the object moves (Hammersley and Handscomb, 1964). Thus, the transition that the object suffers when at a particular point is random, but if the object ever returns to this point, it would suffer the same transition as before. The process is described by a stochastic field on the space, a vector field of transition numbers. Percolation process deals with deterministic flow in a random medium in contrast with random walk and diffusion models, which are concerned with random flow in a deterministic medium.

Broadbent and Hammersley (1957) considered the walk as taking place on a graph consisting of a number of sites, connected by directed "bonds," the passage being possible only along such a "bond." If such a graph obeys certain connectivity requirements, it is termed a *crystal*. In a randomized version, each bond of the crystal has an independent probability of being blocked and it is desired to know what effect this has on the probability of communication from one site A to another site B; this is not the same as from B to A since communication has a direction.

If fire is considered as the moving object, the movement takes place in a space or medium, which has a certain random property, although the object (fire) itself has some randomness associated with it. Buildings in an area, for example, are somewhat randomly distributed. Buildings are also connected by directed bonds, spread (flow) of fire being possible only along the bonds. Each bond has an independent probability of blocking or preventing fire spread; this depends on the nature of a building and its contents, wind conditions, and the distances between buildings. A percolation problem also arises when one considers a network, some of whose links, chosen at random, may be blocked and one wishes to know the effect of this random blockage on flow through the network. Such a problem would be encountered in predicting fire spread, particularly in a forest or from building to building in an urban area.

Apparently for the reasons mentioned above, Hori (1972) considered percolation process to modeling of fire spread from building to building. Sasaki and Jin (1979) were concerned with the actual application of this model and estimation of probabilities of fire spread. By using the data contained in the fire incidence reports for Tokyo, urban fires were simulated and the average number of burnt buildings per fire estimated. Apart from distances between buildings and wind velocity, the following factors were also regarded as having some effect on the probability of fire spread – building construction, building size and shape, window area, number of windows, indoor construction material, furniture, wall, fence, garden, and tree.

In the model mentioned above, the first factor was classified into three groups: wooden construction, mortar (slow burning) construction, and concrete (fireproof) construction. The wind velocity was classified into two groups – 0 to 2.5 m/s, 2.5 to 5.0 m/s. In the former case, fire spread was assumed to be undirectional (isotropic) whereas, in the latter case, the data were subdivided into smaller groups according to the directions of fire spread: the windward direction, the leeward direction, direction perpendicular to that of the wind (the sideward direction). Fire incidents in which wind velocity was larger than 5.1 m/s were excluded owing to their small number. If the number of burnt buildings was i and that of unburnt ones was j the probability of fire spread was expressed as $i/(i + j)$.

The data were divided by every meter of the distance between buildings or by every 2 m or more in case the data were few. A negative exponential function was used for estimating the probability of spread between two buildings that would decrease with increasing distance between the buildings. The analysis revealed that building construction was the main factor responsible for fire spread. The simulations did not evaluate the changes in the pattern of spread according to time.

Nahmias *et al.* (1989) have examined the feasibility of applying percolation theory to the spread of fires in forests. The authors studied the effect of randomness on the propagation of fire using a square network model containing combustible and noncombustible blocks randomly distributed, with a variable concentration, the parameter q denoting the fraction of noncombustible elements. In the absence of wind, the propagation was found to be consistent with a model of invasion percolation on a square site lattice with nearest neighbor interaction leading to a threshold not far from the theoretical value $q = 0.39$. The threshold was larger with wind blowing on the model. The largest threshold value obtained was $q = 0.65$. The final state of the model after combustion was represented for different values of wind velocity and fraction values q. The observation of this state can bring out the directed, nonlocal, and correlated characters of the contagion.

15.9 Epidemic theory

Albini and Rand (1964) predicted fire spread in a large urban area, using a model that has some similarity with the chain-binomial models of Reed and Frost (Bailey, 1964) for the spread of an epidemic. The authors envisaged fires in "locales" that may be single buildings or blocks of buildings. A number of these are presumed to be alight initially and randomly distributed and to stay alight for a time T in the absence of fire fighting. At time T, this "generation" of fires can spread fire and then die out, leaving a second "generation" to burn for a second period T and so on.

Fire spread is assumed to take place only at the end of each fire interval. For the $(n + 1)$th interval, the a priori probability that any locale is burning is P_n and that it has not yet been burnt is A_n. It follows that

$$A_n = (1 - P_0)(1 - P_1)\dots(1 - P_n)$$

$$P_{(n+1)} = A_n.B_n$$

where B_n is the probability that during the $(n + 1)$th interval, fires spread into a "locale" previously unburnt. To obtain B_n, Albini and Rand introduced parameters defining the following three probabilities:

1. Probability that during the $(n + 1)$th interval, there are just m locales burning out of N possible locales adjacent to a given locale.
2. Probability that at least one of the m burning locales spreads fire.
3. "A priori" probability that fire will spread during any interval of duration T from a burning locale to an unburnt neighbor.

On the basis of the parameters mentioned above, the authors obtained an upper and lower approximation for B_n and narrow limits for $(1 - A_n)$, the probability of a locale being burnt.

The Albini and Rand model allowing for fire fighting was based on a number of idealizations. Firstly, fire-fighting effort was assumed to be constant. The authors introduced a parameter M

for the fraction of burning locales that all firemen in a city could extinguish during the given time interval out of all possible burning locales. Fire fighting was assumed to go on throughout the time interval. A fire not extinguished may or may not spread; if extinguished, it cannot spread. Under the assumptions mentioned above, the authors have derived an expression for the probability $P_{(n+1)}$ defined earlier. Albini and Rand considered directional spread of fire assuming that, from an isolated locale, the probability of spread forward and backward was the same, and the directional element in the spread arose only from the initial condition. Spatial variation was included in the model by connecting the probability of spread to the probability that any building was itself burning and separated from any of its neighbors not yet burning by less than the appropriate "safe" distance for radiation or brand transfer.

Thomas (1965) drew attention to the possible relevance of the epidemic theory to fire spread in a building and compared the model of Albini and Rand with a deterministic epidemic model based on a continuous propensity to spread fire. He found the results of both the models to be in reasonable agreement as to their basic features, but concluded that neither would be appropriate for dealing with spread in a single building where the number of "locales" is not large. For such a situation, a stochastic treatment would be necessary to allow for the finite chance that the initiating fires can burn out before spreading, a chance that is negligible when the number of initial fires is large.

Acknowledgment

This chapter is a reduced version of Chapter 15, Section 3 on Stochastic Models of Fire Growth contributed by G. Ramachandran to the Second Edition of the SFPE Handbook of Fire Protection Engineering published by the National Fire Protection Association, Quincy, MA, USA in 1995. The author wishes to thank the Society of Fire Protection Engineers, USA for its permission to reproduce this chapter.

Symbols

A_n	probability that a locale has not been burnt
a_i	state i
aj	state j
a_{ij}	fire spread along link ij
$\bar{a}_{ij}$	no fire spread along link ij
B_n	probability that during $(n+1)$th interval fire spreads into a locale previously unburnt
c	$= (\mu - \lambda)$
E_i	proportion of fires extinguished in the ith state
$F(t)$	probability of duration of burning being less than or equal to t
$f(t)$	density function of t (derivative of $F(t)$)
M	fraction of burning locales in a city, which firemen can extinguish
M_3	long run probability of fire reaching Realm III but not beyond
m	number of states *or* number of burning locales
N	number of possible locales
n	time
P	transition matrix
P_{FDF}/F	probability of a fully developed fire, given a fire
P_n	probability that a locale is burning
p_b	probability of breach of boundary of a room

p_f	probability of flashover
p_{ij}	probability of spread from room i to room j
p_i	distributed probability that a fire will go through link i
p_n	probability distribution of fire burning in different states at time n
p_s	probability of spread beyond a room
(p_b, t_b)	probability and time for failure of barrier
(p_s, t_s)	probability and time for preflashover spread along corridor
$Q(t)$	probability of duration of burning exceeding t
q	fraction of noncombustible elements
q_i	probability of fire burning in the ith state at time n
R	equivalent fire resistance
S	fire severity
S_1, S_2, S_3	states of fire spread
T	time period in epidemic theory
t	real time of fire duration in representing fire severity
t_b	time for which barriers of room withstand fire after flashover
t_e	equivalent time
t_f	time for occurrence of flashover after start of established burning
t_i	time distribution that fire will take to go through link i
t_s	$= (t_f + t_b)$
w	exponent in Pareto distribution
$w_i(t)$	$= [1 - \lambda_i(t) - \mu_i(t)]$
x	damage
λ	success probability of fire spread
$\lambda_1(t)$	transition probability at time t for spread beyond first object
$\lambda_{12}(t)$	transition probability for spread from first to second object at time t
$\lambda_i(t)$	conditional probability of spread for ith state at time t given that the fire has spread to the ith state
$\lambda_{ij}(n)$	transition probability for state a_i at time n changing to state a_j at time $(n+1)$
$\lambda_{ij}(t)$	transition probability for spread from ith object to jth object at time t
μ	success probability of extinction
$\mu_i(t)$	conditional probability of extinction for ith state at time t given that fire has spread to the ith state

References

Albini, F A and Rand, S (1964). *Statistical Considerations on the Spread of Fire*, IDA Research and Engineering Support Division, Washington, DC.

Aoki, Y (1978). Studies on Probabilistic Spread of Fire, Research Paper No. 80, Building Research Institute, Tokyo, Japan.

Bailey, N J T (1964). Reed and frost model. *The Elements of Stochastic Processes*, John Wiley, New York, Section 5, Chapter 12.

Beard, A N (1981/82). A stochastic model for the number of deaths in a fire. *Fire Safety Journal*, **4**, 164–169.

Beck, V R (1987). A cost-effective decision making model for building fire safety and protection. *Fire Safety Journal*, **12**, 121–138.

Benkert, L G and Sternberg, I (1957). An attempt to find an expression for the distribution of fire damage amount. *Transactions of the Fifteenth International Congress of Actuaries*, **11**, 288–294.

Berlin, G N (1980). Managing the variability of fire behaviour. *Fire Technology*, **16**, 287–302.

Berlin, G N (1988). Probability models in fire protection engineering. *SFPE Handbook of Fire Protection Engineering*, National Fire Protection Association, Quincy, MA, USA, Chapter 4– 6, pp. 43–52.

Berlin, G N, Dutt, A and Gupta, S M (1980). Modelling emergency evacuation from group homes. *Annual Conference on Fire Research*, National Bureau of Standards.

Broadbent, S R and Hammersley, J M (1957). Percolation processes, 1, crystals and mazes. *Proceedings of the Cambridge Philosophical Society*, **53**, 629–641.

CIB W14 Workshop (1986). Design guide: structural fire safety. *Fire Safety Journal*, **10**(2), 81–138.

Dusing, J W A, Buchanan, A H and Elms, D G (1979). *Fire Spread Analysis of Multicompartment Buildings*, Department of Civil Engineering, Research Report 79/12, University of Canterbury, New Zealand.

Elms, D G and Buchanan, A H (1981). *Fire Spread Analysis of Buildings*, Research Report R35, Building Research Association of New Zealand.

Elms, D G and Buchanan, A H (1988). *The Effects of Fire Resistance Ratings on likely Fire Damage in Buildings*, Department of Civil Engineering, Research Report 88/4, University of Canterbury, New Zealand.

Hammersley, J M and Handscomb, D C (1964). Percolation processes. *Monte Carlo Methods*, Methuen & Co Ltd, London, Chapter 11.

Heskestad, G (1982). *Engineering Relations for Fire Plumes*, Technology Report 82–8, Society of Fire Protection Engineers, USA.

Hori, M (1972). Theory of percolation and its applications. *Nippon Tokeigakkaishe*, **3**, 19.

Karlin, S (1966). *A First Course in Stochastic Processes*, Academic Press, New York.

Ling, W T C and Williamson, R B (1986). The modelling of fire spread through probabilistic networks. *Fire Safety Journal*, **9**.

Mandelbrot, B (1964). Random walks, fire damage amount and other Paretian risk phenomena. *Operations Research*, **12**, 582–585.

Mirchandani, P B (1976). *Computations and Operations Research*, Vol. 3, Pergamon Press, pp. 347–355.

Morishita, Y (1977). *Establishment of Evaluating Method for Fire Safety Performance*, Report, Research Project on Total Evaluating System on Housing Performances, Building Research Institute, Tokyo, Japan.

Morishita, Y (1985). A stochastic model of fire spread. *Fire Science and Technology*, **5**(1), 1–10.

Nahmias, J, Tephany, H and Guyon, E (1989). Propagation de la Combustion sur un Reseau, Heterogene Bidimensionnel. *Revue Phys. Appl.*, **24**, 773–777.

Platt, D G (1989). *Modelling Fire Spread: A Time-Based Probability Approach*, Department of Civil Engineering, Research Report 89/7, University of Canterbury, New Zealand.

Ramachandran, G (1969). The Poisson process and fire loss distribution. *Bulletin of the International Statistical Institute*, **43**(2), 234–236.

Ramachandran, G (1985). Stochastic modelling of fire growth. *Fire Safety: Science and Engineering*, ASTM STP 882, American Society for Testing and Materials, Philadelphia, PA, USA, pp. 122–144.

Ramachandran, G (1986). Exponential model of fire growth. *Fire Safety Science: Proceedings of the First International Symposium*, Hemisphere Publishing Corporation, New York, pp. 657–666.

Ramachandran, G (1988). Probabilistic approach to fire risk evaluation. *Fire Technology*, **24**(3), 204–226.

Ramachandran, G (1990). Probability-based fire safety code. *Journal of Fire Protection Engineering*, **2**(3), 75–91.

Ramachandran, G (1991). Non-deterministic modelling of fire spread. *Journal of Fire Protection Engineering*, **3**(2), 37–48.

Sasaki, H and Jin, T (1979). *Probability of Fire Spread in Urban Fires and their Simulations*, Report No 47, Fire Research Institute, Tokyo, Japan.

Thomas, P H (1965). *Some Possible Applications of the Theory of Stochastic Processes to the Spread of Fire*, Internal Note No. 223, Fire Research Station, Borehamwood, Herts, UK.

Watts, J M Jr (1986). Dealing with uncertainty: some applications in fire protection engineering. *Fire Safety Journal*, **11**, 127–134.

16 FIRE SAFETY CONCEPTS TREE AND DERIVATIVE APPROACHES

16.1 Introduction

In the 1960s and 1970s, the potential application of a "systems approach" to fire safety was recognized by several organizations. The National Fire Protection Association (NFPA) created a new technical committee with the mission of developing systems concepts and criteria for fire protection in structures. The product of this committee was a comprehensive logic diagram first published in 1974 and revised in 1980.

Subsequent attempts to quantify portions of this tree have led to several quantitative approaches to the evaluation of fire safety. The probabilistic Goal-Oriented Systems Approach developed by the US General Services Administration evolved from a similar logic tree. It has, in turn, produced two additional approaches – a modification of the mathematics of the original, referred to in this chapter as the modified approach and the WPI (Worcester Polytechnic Institute) engineering method, a more comprehensive approach under continuing development.

16.2 Fire Safety Concepts Tree

The Fire Safety Concepts Tree is a logic diagram intended to cover all possible mechanisms for achieving fire safety objectives. It was published by the NFPA as a wall chart (NFPA, 1980) and documented in NFPA 550, *Guide to the Fire Safety Concepts Tree* (NFPA, 1986).

16.2.1 CONFIGURATION OF THE TREE

The Fire Safety Concepts Tree is comprised of elements of building fire safety and their relationships. Focus is on achievement of stated fire safety objectives, the top element of the tree (see Figure 16.1). The tree helps in evaluating or designing a building by progressively moving through various levels of fire safety elements in a logical manner. The tree provides the logic required to achieve fire safety, that is, it provides conditions whereby the fire safety objectives can be satisfied but does not identify the minimum conditions required to achieve those objectives. Success depends upon the completeness by which each level is satisfied. Lower levels on the tree

Evaluation of Fire Safety D. Rasbash, G. Ramachandran, B. Kandola, J. Watts and M. Law
 ISBN: 0-471-49382-1

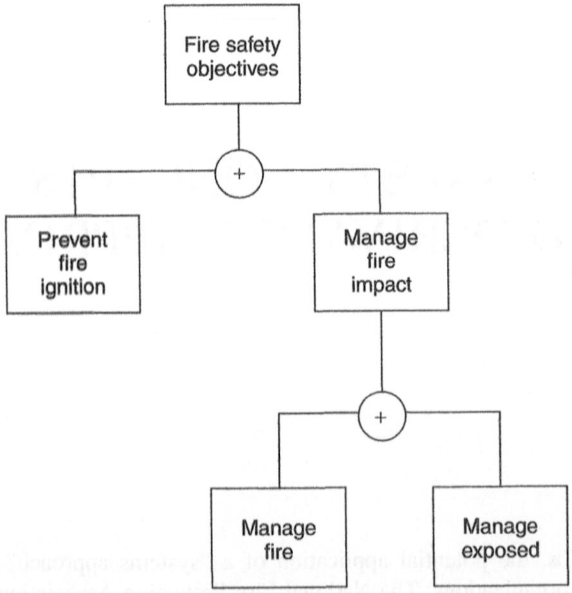

Figure 16.1. Major branches of the Fire Safety Concepts Tree

do not represent a lower level of importance or performance. They represent means for achieving the next higher level.

Logic gates

Elements are related by logical associations called AND gates and OR gates. These are standard operations from Boolean algebra.

The AND gate is the logic operation that says all of the inputs must coexist simultaneously to produce the output. All of the inputs are necessary and there is no redundancy. The printed symbol for an AND gate is a circle with a dot in the middle (⊙).

A circle with a plus in it (⊕) is the symbol used to designate an OR gate. The OR gate is a logic operation whereby any of several possible inputs will produce the indicated output. It is the "inclusive or," which means all the concepts below the gate may be included but only any one of them is necessary. In theory, this implies that all but one input can be ignored. However, theoretically perfect fire safety inputs can never be achieved.

The likelihood of realizing fire safety objectives is increased by the presence of more than one input. This is an example of reliability through redundancy. Thus, OR gates in the Fire Safety Concepts Tree suggest where the reliability of achieving an objective is improved by the use of more than one strategy. It is also important to note that the inputs to an OR gate are intended to be exhaustive. This means they encompass every possible way of achieving the respective output.

Major tree branches

Like its botanical counterpart, the Fire Safety Concepts Tree has many branches. The three major branches represent the concepts of "Prevent Fire Ignition," "Manage Fire," and "Manage Exposed."

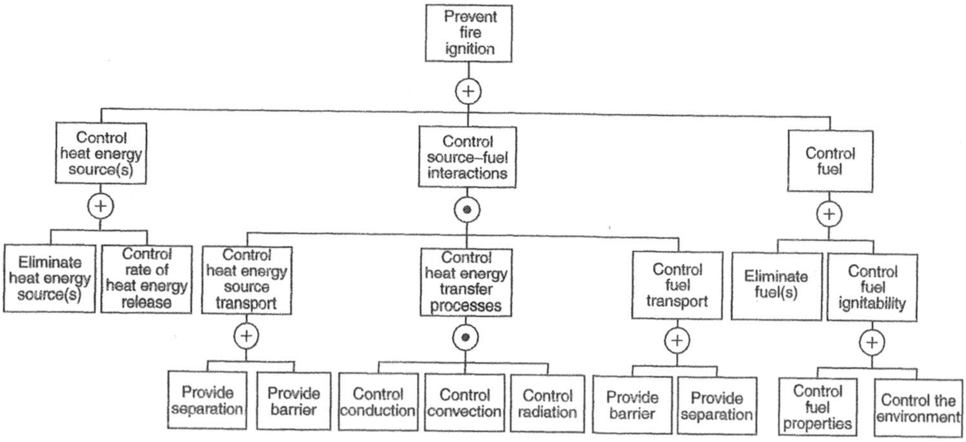

Figure 16.2. Prevent Fire Ignition branch of the Fire Safety Concepts Tree

The "Prevent Fire Ignition" branch of the tree (Figure 16.2) is essentially a form of fire prevention code. Most of the elements described in this branch require continuous monitoring for success. Consequently, the responsibility for satisfactorily achieving the goal of fire prevention is essentially an owner/occupant responsibility. The designer, however, could incorporate certain features into the building that may help the owner/occupant in preventing fires. It is impossible to prevent the start of fires completely in a building. Therefore, to realize the overall fire safety objective, from a building design viewpoint, a high degree of success in the "Manage Fire Impact" branch assumes a significant role. According to the logic of the Tree, the impact of the fire can be managed either through the "Manage Fire" or "Manage Exposed" branches.

The "Manage Fire" branch (Figure 16.3) shows how fire safety objectives can be achieved by managing the fire itself. This is accomplished by (1) controlling the combustion process, (2) suppressing the fire, or (3) controlling the fire by construction. Here again, any one of these branches of the Tree will satisfy the "Manage Fire" element. Thus, for instance, in some fires, success is achieved where the building construction controls the fire. In other fires, success is achieved by controlling the combustion process, either by controlling the fuel or the environment. In practice, a combination of these elements is used to meet fire safety objectives.

The "Manage Exposed" branch (Figure 16.4) indicates that fire impact can be managed by coordinating fire safety measures directly involving that which is "exposed" – people, property, or functions, depending upon the fire safety objectives being considered. Success is achieved either by limiting the amount exposed or by safeguarding the exposed. For example, the number of people as well as the amount or type of property in a space may be restricted. Where this is impractical, objectives may still be met by incorporating design features to safeguard the exposed.

The exposed people or property may be safeguarded either by moving them to a safe area of refuge or by defending them in place. For example, people in institutionalized occupancies such as hospitals, nursing homes, or prisons must generally be defended in place. On the other hand, alert, mobile individuals, such as those expected in offices or schools, could be moved to safeguard them from fire exposure.

16.2.2 APPLICATIONS

Among the many applications of the NFPA Fire Safety Concepts Tree is its ability to quickly and easily communicate a total picture of building fire safety. As a pictorial description of the concept

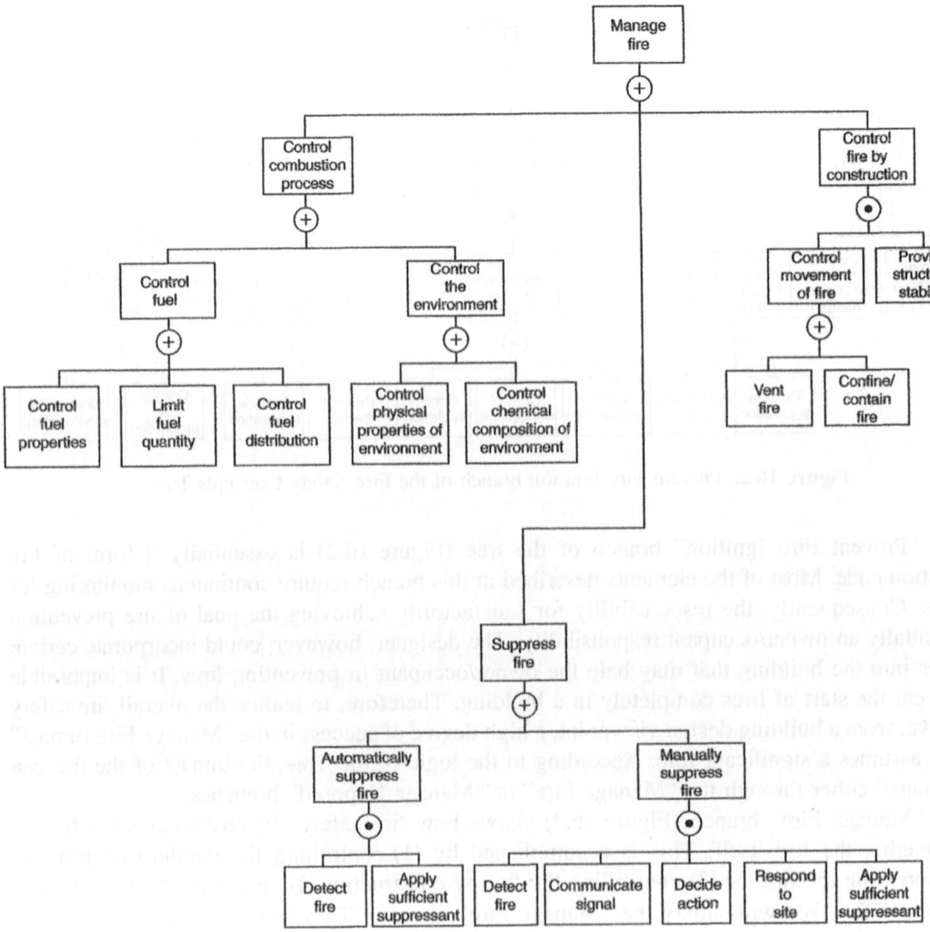

Figure 16.3. Manage Fire branch of the Fire Safety Concepts Tree

of fire safety, the Tree identifies the role of specific code requirements and suggests alternative approaches to achieving equivalent levels of acceptable risk.

Communication

The Fire Safety Concepts Tree is a simple pictorial description of the total concept of fire safety incorporated in codes and standards. It is convenient for communication between fire safety specialists and others to help identify the role that specific requirements play. In building design, once the basic fire safety objectives for a building have been identified, the designer can analyze the paths through the tree by which these objectives can be met. Alternative strategies and means to improve fire safety reliability can be identified. The tree can then be used to communicate the fire safety concepts of the design to clients, management, and code officials.

Checklist

The Fire Safety Concepts Tree can be used as a checklist for fire safety evaluation. Inputs to AND gates in the Tree are comparable to lists of items required to achieve the output objective

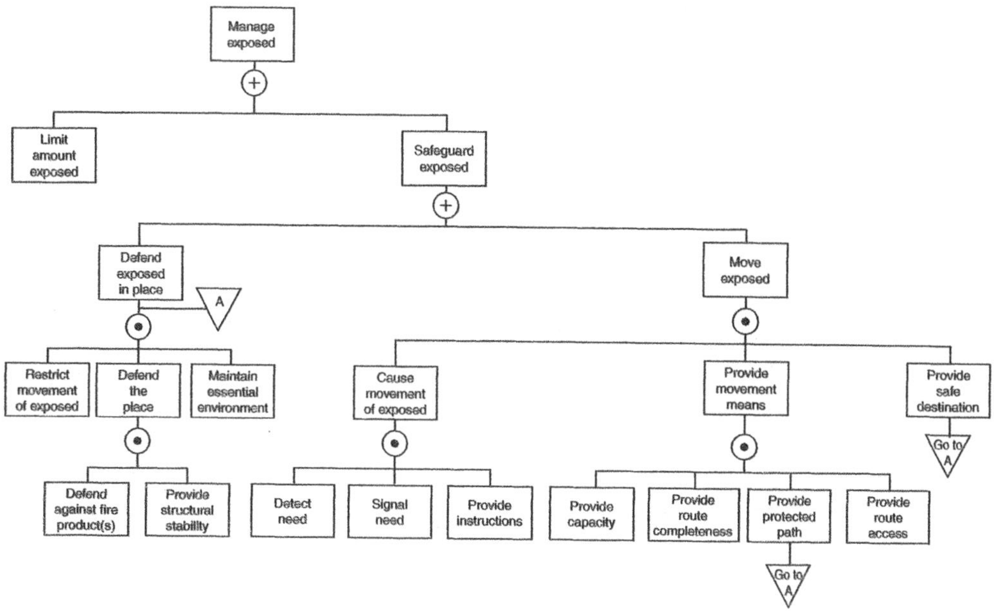

Figure 16.4. Manage Exposed branch of the Fire Safety Concepts Tree

or strategy. Each must be present to an adequate degree. Inputs to OR gates comprise a list of options. Generally, they are examined to see if, taken together, they are adequate to successfully achieve the output element. Using the Tree as a checklist, fire safety redundancies and deficiencies can be specifically identified and evaluated.

Trade-offs

An important feature of fire-safe design is the subject of alternatives or "trade-offs." The logic structure of the Fire Safety Concepts Tree identifies those areas of fire-safe design that are amenable to such trade-offs. The only legitimate areas in which alternatives can be established are among those factors below an "OR" gate in the Fire Safety Concepts Tree. For example, decisions can be made regarding evacuation versus temporary refuge and their implications on the functions of the building. Factors below an "AND" gate are necessary and thus cannot be considered as alternatives.

The Fire Safety Concepts Tree does not, in its present form, provide a numerical fire safety score for a building. Other approaches, described in the following sections, have been developed to provide such scores for portions of the Tree.

16.3 The GSA Goal-Oriented Systems Approach

Any effective system must be responsive to management objectives. In the development of the Goal-Oriented Systems Approach, the GSA policy statement on safety was reformulated and a probabilistic criterion was developed for mission-focused goals. This goal criterion is expressed in terms of the probability of limiting fire involvement in each of successive spatial or structural modules within a building. Figure 16.5 represents the GSA mission continuity goals for general level and critical operations.

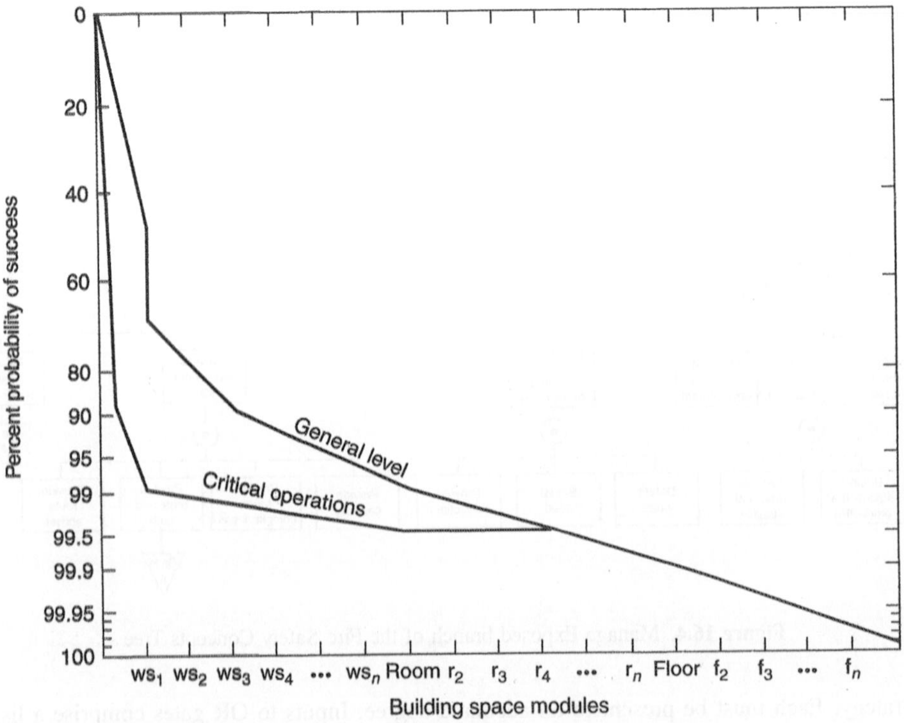

Figure 16.5. GSA mission focused goals. ws = workstation, r = room, f = floor

Quantitative application of the Goal-Oriented Systems Approach involves a probability calculation for each workstation, room, and floor of a specific building. Where calculated probabilities fall within the area under the goal curve of Figure 16.5, the required objectives have been met. The methodology for these probabilistic determinations is the principal focus of this paper.

The GSA Goal-Oriented Systems Approach was primarily developed by Harold E. Nelson and used by GSA as an adjunct to their prescriptive Building Fire Safety Criteria. It was revised in 1975 and published as the appendix to an in-depth review (Watts, 1977).

16.3.1 COMPONENTS OF THE GOAL-ORIENTED SYSTEMS APPROACH

There are two basic components of the Goal-Oriented Systems Approach as it is presently practiced. The qualitative component covers all aspects of fire protection, while the quantitative component addresses itself to that aspect of fire protection about which we have the most specific knowledge.

The underlying structure of the Goal-Oriented Systems Approach is that of a logic tree. The nature of this tree evolved from the fault trees developed in the field of systems safety as primarily practiced in the aerospace industry. The GSA fire safety tree is intended to represent every conceivable means of providing fire safety. Thus, the elements of the tree represent a collectively exhaustive set of fire protection measures, and the tree provides a qualitative tool for examining all possibilities for fire safety design. The GSA fire safety tree is very similar to the tree promulgated by the NFPA.

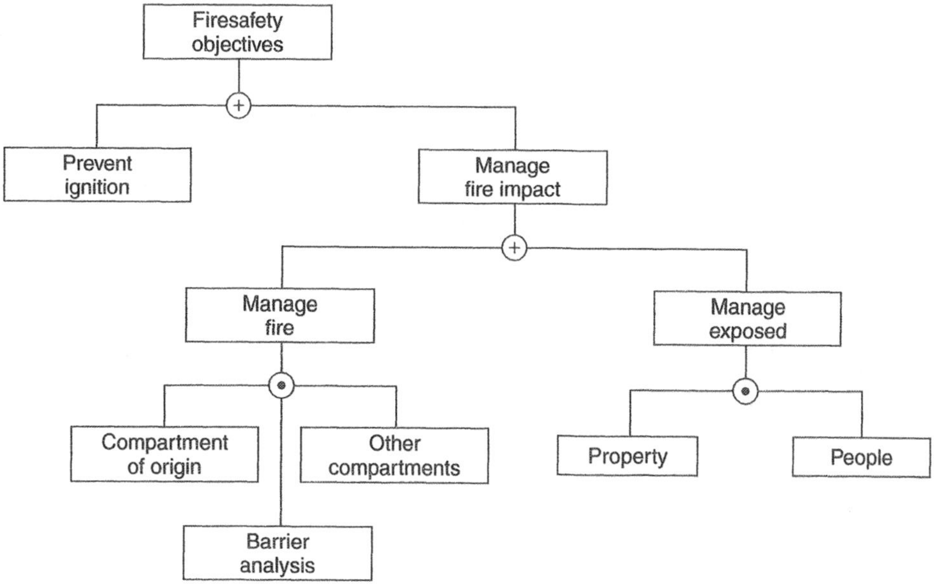

Figure 16.6. Major branches of GSA fire safety tree

Fault trees are often used as a framework for the quantitative analysis of system safety. A branch of the GSA fire safety tree, which is particularly amenable to this type of analysis, is concerned with the management of fire, as opposed to the prevention of fire or the management of persons or property exposed to the effects of fire (Figure 16.6). For this branch, knowledge and data appear adequate to support a probabilistic measure of the level of fire safety.

16.3.2 QUANTITATIVE ASPECTS

The focus here is on the portion of the Goal-Oriented Systems Approach that is quantitative in nature. This is the aspect of the approach that derives probabilities of success of limiting fire spread.

Probability curves

Presentation of derived probabilities is made in the form of "curves," such as the GSA objectives curve portrayed in Figure 16.5. Most of the present applications of the Goal-Oriented Systems Approach consist of a defined sequence of such curves.

The common abscissa of these curves is a nonlinear, discontinuous scale representing floor area. Specific points on the scale identify spatial modules of increasing size within a building. The initial modules represent workstations or fuel packages that are semicontiguous combustible materials in which a fire may originate or among which a fire may spread. An example would be a desk, chair, and wastebasket in close proximity to one another. Thus, a fire starting in the wastebasket would ignite the desk and chair by direct flame impingement, whereas spread to an adjacent workstation would most likely be by radiative heat transfer from the first workstation. Total room involvement is defined as fire spread among n workstations where $n \geq 1$. In most compartments, n will take a value of 3 or 4, that is, the entire room will be involved simultaneously with the involvement of the third or fourth fuel package. The sequence of fire spread is then considered from room to room where n rooms represent an entire floor. Similarly, the building is considered to be composed of n floors. Thus, n is an arbitrary variable used to indicate a terminal number of workstations, rooms, or floors.

The ordinate of the curves is the cumulative probability of success of limiting fire spread. For unknown reasons, the scale is inverted, placing the origin at the upper left. The scale used is primarily a linearized cumulative normal probability distribution apparently selected for convenience and availability. The extreme portions have been altered or adapted in various ways by GSA and other users of the approach. Since the abscissa is not continuous, there can be no mathematical significance of the normal distribution to the curves. Thus, the curves of the Goal-Oriented Systems Approach are in fact discrete points, which are not truly related in a continuous manner; however, they are connected to facilitate the effectiveness of their graphical presentation.

This appears to be a strong point of the approach. Presentation of fire-spread probabilities on a single graph provides the designer with an effective tool for communication with owners, occupants, and regulatory officials.

The compartment of origin

The first probabilistic evaluations are made for the workstations or fuel packages within the compartment of origin. These evaluations are based on the relevant portion of the logic tree (Figure 16.7), which dictates that the limitation of fire spread to a work space is achieved by self-termination of the fire (i.e. it just goes out), by manual suppression (e.g. fire department), or automatic suppression (e.g. automatic sprinkler system).

GSA has developed, from staff experience and available technical data, a series of plots of the probability of self-termination for various types of office occupancies, which are referred to as "I-curves" (Figure 16.8). The designer or fire protection engineer, must make similar judgment decisions for the suppression probabilities ("M-curves" for manual suppression and "A-curves" for automatic suppression). Then, following the indicated logic of the tree, that is, an OR gate, the probability of limiting the spread of fire to workstation i is given by

$$P\{L_i\} = P\{I_i + A_i + M_i\} \qquad [16.1]$$

which by Boolean algebra is readily calculable from

$$P\{L_i\} = 1.0 - P\{\tilde{I}_i\}P\{\tilde{A}_i\}P\{\tilde{M}_i\} \qquad [16.2]$$

where $\sim$ indicates the complement of the respective event (i.e. $P\{I_i\} = 1 - P\{\tilde{I}_i\}$).

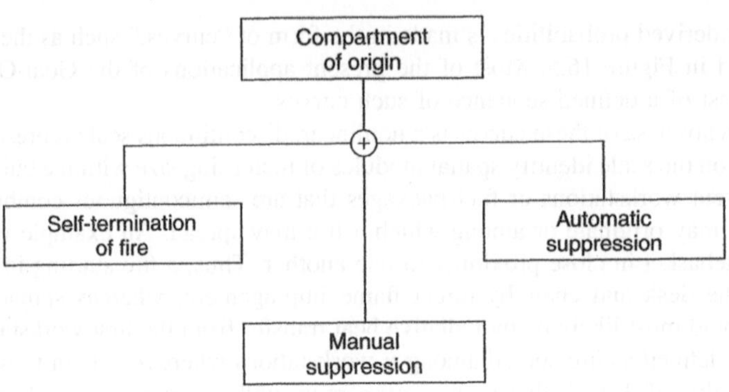

Figure 16.7. Compartment of origin branch of GSA fire safety tree

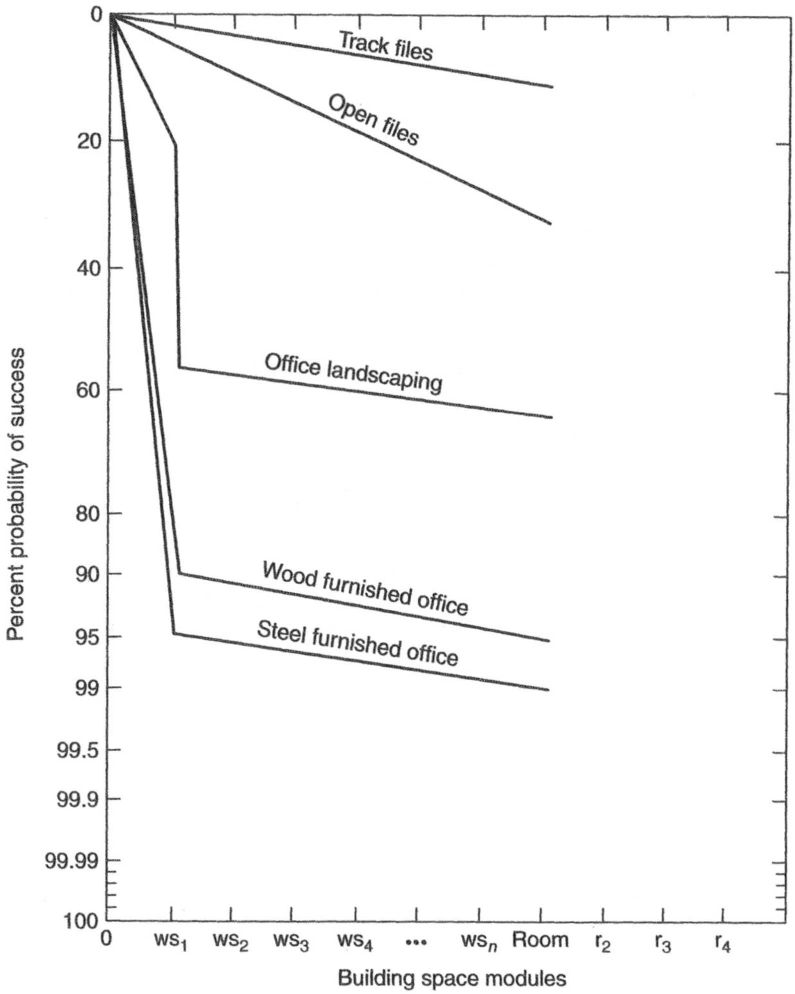

Figure 16.8. GSA I-curves. ws = workstation, r = room

The probabilities of limitation of fire spread for workstations 1, 2, . . . , *n*, when connected together, are referred to as the "L-curve" for the compartment of origin.

Barrier analysis

When a fire reaches the physical boundaries of the compartment, it encounters its first material barrier to further spread. Determination of the capability of a structural barrier to retard fire spread is the most sophisticated aspect of the Goal-Oriented Systems Approach.

Three failure modes of fire barriers are considered in the traditional fire testing procedure: passage of flame or hot gases, transmission of heat, and inability to sustain the applied load (ASTM, 1988). The first of these failure modes is handled directly by an assessment of the percentage of openings, orifices, holes, or other means by which the passage of flame or hot gases may take place. GSA uses a judgment analysis to graphically define the probability of limiting fire spread versus the percentage of openings for several types of barriers. This probability is designated by GSA as $P\{0\}$.

The other failure modes are dependent on the severity of the fire. Traditionally, fire severity was estimated by a relationship of the amount of combustibles to the standard ASTM fire test (Ingberg, 1928). On this basis, GSA estimated cumulative probability distributions of probable fire severities for several furnishing conditions. Response of different types of barriers to fires of varying severity for thermal resistance (T) and structural integrity (D), respectively were also estimated. The total probability for each of these is then found by conditioning on the severity probability. Thus, if the probability of a fire of severity i is given by $P\{H_i\}$ and the conditional probability of thermal resistance is given by $P\{T|H_i\}$, then the total probability of thermal resistance over the range of fire severities is

$$P\{T\} = \sum_{i=1}^{n} P\{T|H_i\}P\{H_i\} \qquad [16.3]$$

Similarly, when the conditional probability of sustaining the applied load is designated as $P\{D|H_i\}$, the total probability is

$$P\{D\} = \sum_{i=1}^{n} P\{D|H_i\}P\{H_i\} \qquad [16.4]$$

The discrete representation of total probability was used in equations [16.3] and [16.4], since the method designated by GSA involves only empirical distributions. As will be shown in the next section, the procedure may be streamlined by the use of theoretical distributions.

The determination of the probability of the success of a barrier in limiting the spread of fire now follows the Boolean logic of that portion of the fire safety tree indicated by Figure 16.9. Thus, the probability of the success of barrier j is given by

$$P\{Fj\} = P\{Oj \cdot Tj \cdot Dj\} \qquad [16.5]$$

which is calculated by

$$P\{Fj\} = P\{Oj\}P\{Tj\}P\{Dj\} \qquad [16.6]$$

Construction of the L-curve

The "L-curve" of a building is the fire safety evaluation measure of the Goal-Oriented Systems Approach. It represents the cumulative probability of limiting fire spread at each of the spatial modules considered. The "L-curve" is derived in a step-by-step process of calculation at each

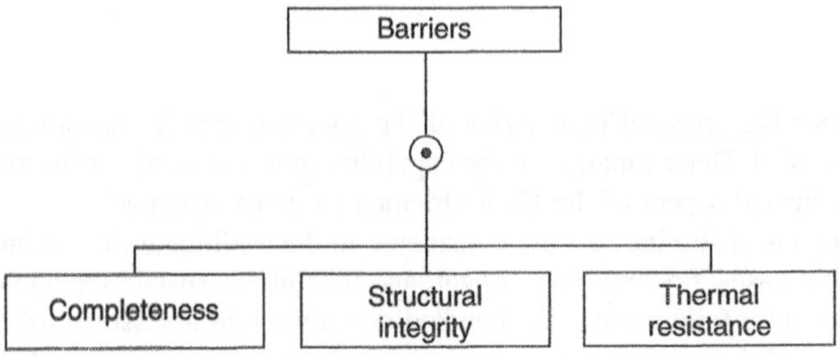

Figure 16.9. Barrier branch of GSA fire safety tree

module and at each barrier. The residual probability of failure, $P\{L\}$, at each step is reduced by the probability of success of the specific module or barrier, for example:

$$P\{L_{i+1}\} = P\{L_i\} + P\{\tilde{L}_i\}P\{\lambda_i\} \qquad [16.7]$$

where $P\{\lambda_i\}$ is given by equation [16.2] for a space and by equation [16.6] for a barrier, that is, the probability of success of limiting fire spread at a point on the L-curve, designated by L_{i+1}, is equal to the probability of success at the previous point, L_i, plus the residual probability of failure reduced by the probability of success of the ith barrier $P\{F_i\}$. The L-curve is then found by connecting these points as, for example, the points, "a" through "q" on Figure 16.10.

The resultant L-curve is compared to the identified goals of the owner or occupant of the building. In Figure 16.10, the fire protection does not meet the general level goal criteria of the GSA.

16.3.3 LIMITATIONS OF THE GSA APPROACH

The Goal-Oriented Systems Approach is an early attempt at the application of systems concepts to fire safety. As such, there are several limitations to its applicability.

The approach is highly dependent on the selection of the scenario of fire spread, that is, the abscissa of the "L-curve." There is no formal methodology for the determination of the most likely path of fire development. The process is basically one of applying the professional judgment of experienced fire protection engineers. Alternatively, all possible scenarios would have to be evaluated and weighed by their likelihood of occurrence.

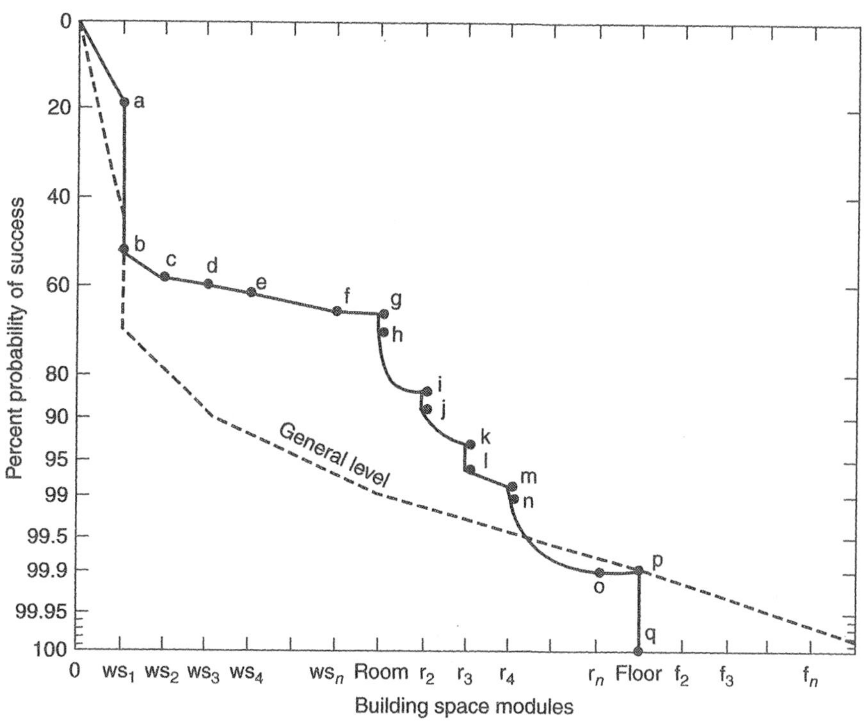

Figure 16.10. GSA L-curve. ws = workstation, r = room, f = floor

The concept of time is ignored throughout the approach. This may be acceptable for the barrier analysis but is a questionable assumption with respect to suppression. The capability of a suppression system will generally be more dependent on how fast the fire spreads from one module to another, rather than on the likelihood of fire spread. This may be handled by introducing the concept of "suppressibility" of fires as a probability distribution, which can then be matched by a "suppressibility" distribution for a specific automatic or manual extinguishing system. This technique is illustrated in the next section. If the Goal-Oriented Systems Approach is to be extended to cover aspects of life safety, then the introduction of a temporal factor becomes paramount.

A further limitation is in the rather burdensome treatment of barrier analysis by the use of discrete probability distributions. The continuous case is a better model of the real-world physical process and is easier to use. A significant amount of work has been done on the application of the convolution integral to stress–strength models, and this body of knowledge can be advantageously used in the barrier analysis as well as in the aforementioned concept of "suppressibility."

Sensitivity analyses have shown that the results of the application of the Goal-Oriented Systems Approach are not very sensitive to change in the input parameters (Watts, 1979). This bodes well for the use of professional judgment in the determination of the input distributions. However, it also suggests that greater resolution would be ineffective. That is, any one particular detail of fire protection may have no perceptible effect. This may suggest either an inadequate measure of fire safety or an improper manipulation of the variables.

While the "L-curve" has the distinct advantage of graphical representation, it suffers greatly from the lack of rationality in its axes. Mathematically, the curve is a connected series of points with intuitively defined relationships. Alternatively, the abscissa could be represented by a logarithmic scale of floor area. In either case, a defined, conventional scale should replace the current ordinate.

A final criticism related to the previous two has to do with the measure of fire safety. Very small or very large probabilities when expressed as decimals or percentages are misleading. While mathematically correct, this formulation lends itself readily to false interpretations of implications. For example, the distinction between 1/7000 and 1/8000 is lost when they are portrayed as the compliments of their decimal equivalents: 0.9998371 and 0.999875.

16.3.4 SUMMARY OF THE GSA APPROACH

The distinct contribution of the Goal-Oriented Systems Approach is in the systematic consideration of fire safety. At the time of its development, it was the most inclusive systematic approach to building fire safety ever issued in the United States. It served to spur interest in systems approaches, and is recognized in the United States as the principal motivational effort in the development of the application of systems concepts to fire protection engineering.

16.4 Modified approach

The Goal-Oriented Systems Approach developed by GSA is an unusually complete and rational fire safety design method. However, the method as developed is substantially intuitive. Watts (1979) undertook a study to establish a scientific basis for the approach, addressing the following issues:

- Are there underlying theoretical concepts in the GSA approach that may be used to develop a more direct method?
- Can the GSA approach be improved with respect to flexibility of scope, simplicity of application, and validity of concepts?
- How sensitive is the approach to the limited availability of probabilistic data?

This study produced a modified systems approach to building fire safety.

16.4.1 POSTULATES OF FIRE SPREAD

Three postulates of fire spread in structures may be induced from the quantitative component of the Goal-Oriented Systems Approach. These postulates are as follows:

1. *Limitation of fire spread may be achieved by containment or by termination.*
Limitation of fire spread represents an event or condition whereby fire will not spread from one module to the next, and therefore, implies that the next module is secure for mission continuity. Limitation of fire spread is thus equivalent to the event L in the GSA approach. Containment is the event or condition by which heat transfer between modules is physically prevented. This will usually be effected by spatial separation or by a thermal barrier. Termination is the event or condition of cessation of the combustion reaction prior to the normal consumption of available fuel. Termination may be due solely to the physicochemical characteristics of the involved module, or it may be abetted by a suppression methodology. Therefore, containment and termination are equivalent to GSA events F and G, respectively.

2. *Termination will not occur if ignition is by massive energy transfer.*
Massive energy transfer is an event or condition of modular fire spread, which results in extensive fire involvement. Between compartments, a massive energy transfer may be effected by the disintegration or collapse of a physical barrier. Thus, massive energy transfer is equivalent to the complement of the event D in the GSA approach.

A third postulate applies to the sequential fire spread among modules.

3. *Limitation of fire spread to a sequential module is achieved if the fire is limited to any previous module.*
This postulate is the essence of the "L-Curve" in the Goal-Oriented Systems Approach, and is similar in principle to other probabilistic models, which produce a geometric distribution of the number of rooms burned.

The first two postulates of fire spread may be combined as a Boolean statement

$$L = F \cup (G \cap D)$$

where
$A \cup B$ = the union of A and B,
$A \cap B$ = the intersection of A and B.

This statement says that termination is the intersection of event G and the absence (complement) of massive energy transfer, and the limitation of fire spread results from the union of containment and termination. This expression holds, in general, within any module i.

Thus:

$$L_i = F_i \cup (G_i \cap D_i)$$

By the third postulate, fire spreads sequentially through spatial modules. Thus, the limitation at the nth module is the union of the limitation within all modules 1 through n.

$$\begin{aligned} L_n &= L_1 \cup L_2 \cup \ldots \cup L_n \\ &= \bigcup_{i=1}^{n} L_i \\ &= \bigcup_{i=1}^{n} [F_i \cup (G_i \cap D_i)] \end{aligned}$$

We now have a Boolean statement as to the means by which the limitation of fire spread is achieved in any module of a structure. This statement may also be written in terms of probabilities:

$$P\{L_n\} = P\left\{\bigcup_{i=1}^{n} [F_i \cup (G_i \cap D_i)]\right\}$$

Assuming independent, mutually exclusive events, the equation can be written as

$$P\{L_n\} = P\left\{\sum_{i=1}^{n} (F_i + G_i D_i)\right\}$$

It is assumed that there is no barrier to the ignition of the first module. Hence, F_i and D_i do not exist and the equation becomes

$$P\{L_n\} = P\left\{G_1 + \sum_{i=2}^{n} (F_i + G_i D_i)\right\}$$

where

$P\{L_i\}$ = Probability of success in limiting fire involvement to the ith room,
$P\{F_i\}$ = Probability of success of the compartmentation barrier between room i and room $(i + 1)$,
$P\{D_i\}$ = Probability of structural integrity of the ith barrier, and
$P\{G_i\}$ = Probability of success in limiting the fire involvement in room i as if it were the room of origin, that is, limitation due to fuel, environmental, and control factors within the room.

16.4.2 STRESS–STRENGTH MODELS

Reliability theory is a body of mathematical models and methods, which deals with problems in predicting, estimating, or optimizing the probability of the proper functioning of a system. Among the more common models in reliability theory are those depicting a stress-strength relationship, where the reliability of a component in successfully completing its mission is defined as the probability that its strength exceeds the stress encountered during its operation. Watts (1983) has

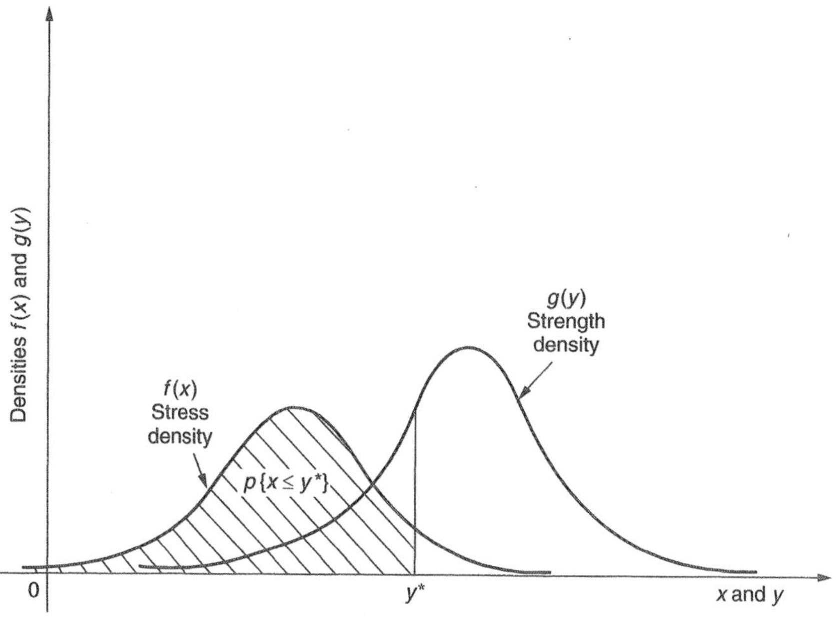

Figure 16.11. Stress–strength model

illustrated how the convolution integral may be used as a stress–strength model fire safety tree elements.

Let X be a random variable denoting the maximum stress encountered, and let Y be a random variable denoting the effecting strength. Since the units of stress and strength are the same, their probability density functions may be plotted on the same axes as shown in Figure 16.11. When strength of the system is y^*, then the reliability of the system (i.e. the probability that the stress will be less than the strength) is the area under the stress curve to the left of y^*

$$P\{X \leq y^*\} = \int_{-\infty}^{y^*} f(x)\,\mathrm{d}x$$

If the exact strength y^* is unknown, the reliability is also a function of the strength distribution $g(y)$:

$$\begin{aligned} P\{X \leq Y\} &= \int_{-\infty}^{\infty} \int_{-\infty}^{y^*} f(x) g(y)\,\mathrm{d}x\,\mathrm{d}y \\ &= \int_{-\infty}^{\infty} F_x(y) g(y)\,\mathrm{d}y \end{aligned}$$

which is the usual form of the general stress–strength model.

Stress–strength model of a fire barrier

Let R be a random variable that represents the fire resistance of the barrier, and let S represent the severity of the fire to which the barrier is exposed. Then, the characteristic of interest is the probability that the fire resistance is greater than the fire severity:

$$P\{R > S\} = P\{R/S > 1\}$$
$$= P\{X > 1\}$$

where $X = R/S$ and:

$$\ln X = \ln R - \ln S$$

by the properties of logarithms.

If R and S are lognormal random variables, then $\ln R$ and $\ln S$ are normally distributed, and a linear combination of independent, normally distributed random variables is also normally distributed. Assuming then, that the fire severity and the fire barrier are independent,

$$Y = \ln X = \ln R - \ln S$$

is a normally distributed random variable with mean $\mu = \mu_{\ln R} - \mu_{\ln S}$ and variance $\sigma^2 = \sigma^2_{\ln R} + \sigma^2_{\ln S}$. Now, the probability of interest may be expressed in terms of the normal random variable Y:

$$P\{X > 1\} = P\{Y > \ln 1\}$$
$$= P\{Y > 0\}$$

The standard normal variate is a normally distributed random variable with a zero mean and unit standard deviation. Any normal variate (x) may be represented as a standard normal (z) by the following transformation

$$z = (x - \mu)/\sigma.$$

Thus:

$$P\{Y > 0\} = P\{Z > -(\mu/\sigma)\}.$$

Values of the standard normal distribution are tabulated in most texts on probability and statistics. For any standard normal variable

$$P\{X > x\} = P\{X < (-x)\}.$$

Therefore, the probability may be written in the more usual form

$$P\{R > S\} = P\{Z < (\mu/\sigma)\}.$$

Thus, the probability of a given barrier withstanding a given fire may be represented as a standard normal random variable.

In the revised GSA version of the Goal-Oriented Systems Approach, use of the "total probability theorem" to calculate the thermal resistance and structural integrity of a barrier is a discrete form of a stress–strength model.

This application of the stress-strength model can be extended to evaluate suppressibility of a preflashover fire by automatic sprinklers. The stress of preflashover fire severity can be represented by a probability distribution of the rate of heat release. Similarly, strength of a suppression system can be modeled as a probability distribution of heat absorption capacity, including the wetting of unburned fuel.

16.4.3 PROBABILITY DISTRIBUTIONS

The primary inputs to the modified approach are continuous probability distributions for the major components of fire safety: the preflashover fire, automatic sprinklers, the postflashover fire, and barriers. Selection of an appropriate probability distribution has been identified as the essence of statistical modeling. Two steps comprise this process: an a priori analysis of the physical processes being described and a verification of the model with observed data.

A priori analysis

The normal distribution is representative of so many randomly fluctuating phenomena that it is usually a first choice where there is little information on which to base a selection. The normal distribution was chosen by Lie (1972) as his model of fire severity. Burros (1975), in his refinement of Lie's work, notes that negative fire severity is nonexistent and suggests a truncated distribution (range: zero to $+\infty$ rather than $-\infty$ to $+\infty$) such as the lognormal. Ramachandran (1972) also assumed a lognormal distribution of fire severity in his work on fire resistance while Rennie (1961) and Benckert (1962) used the lognormal distribution as a model of fire damage based on insurance claims. Thus, there are a priori indications of the suitability of the lognormal distribution to be found in previous work and in the related literature.

Model verification

Fire load, the weight of combustibles per unit floor area, has long been used as a surrogate measure or parameter of fire severity. The National Bureau of Standards (Culver, 1976) conducted a survey of fire load in 1044 offices in 23 federal and private office buildings throughout the US from 2 to 49 stories high. These data have been plotted as an exponential distribution, a normal distribution, and a lognormal distribution. The lognormal was the closest to a straight-line fit (Watts, 1979, appendix A3).

The characteristic of interest in the preflashover fire is the rate at which heat is released by the burning fuel. Pape *et al.* (1976) assembled a significant amount of data on the heat release rates of various furniture items. Data on the burning rate of cotton upholstered chairs were fit to a lognormal distribution. A Kolmogorov–Smirnoff goodness of fit test showed the null hypothesis that the distribution is lognormal could not be rejected at the 0.01 level of significance (Watts, 1979, Appendix A4.3).

Thus, the identification of the appropriate probability distribution consists of selecting the parameters of a lognormal distribution. The two parameters of the lognormal distribution are the mean μ and the standard deviation σ. There are several parameter estimation techniques in the literature on probability and statistics that can aid in the appropriate selection. Mathematical operation using these parameters is the essence of the modified approach.

16.4.4 MODIFIED APPROACH

The modified approach is a synthesis of the basic concepts of the Goal-Oriented Systems Approach with more formalized inherent theoretical models. The objective is to develop a meaningful framework whereby intuition, experience, and existing data may be utilized with theoretically sound analytical techniques to produce a probabilistic measure of fire safety. The resulting eclectic model represents a significant departure from the methodology of the original Goal-Oriented

Systems Approach, but retains the underlying concepts. The modified approach is theoretically based, intuitively acceptable, and easier to use.

Overview

The essence of the modified approach is in the concepts of containability and suppressibility of a fire in a compartment. Each of these is estimated by a stress–strength relationship of the severity of a fire versus the resistance of a barrier or of an automatic extinguishing system.

For the case of the barrier, the severity of the fire and resistance of the barrier are modeled as lognormal distributions of the same dimension, for example, hours of duration. The stress-strength relationship then identifies the adequacy of the barrier to contain the fire. The ultimate effectiveness of the barrier also includes a factor of reliability, estimated as the expectation that the barrier is not immediately penetrable via openings or defective assembly.

The suppressibility of both the fire and the automatic extinguishing system are similarly modeled as lognormal distributions with a consistent dimension, for example, heat release or absorption. The stress-strength relationship predicts the adequacy of the suppression system and the expected reliability is estimated. The product of adequacy and reliability yields a measure of the effectiveness of the automatic extinguishing system.

A third concept, self-termination of the fire, is also estimated as an expected value.

The probability of limiting the extent of a fire to the room of origin is then the Boolean sum of these three factors: the effectiveness of the barrier, the effectiveness of suppression, and the expected value of the self-termination.

The probability of limiting the fire to within successive barriers is found by assuming a simple Markov process of fire spread whereby the probability of success of fire limitation at a given barrier is the intersection of the probability of failure of the previous barrier, and the probability of the effectiveness of the present barrier.

Procedures

The overall process for calculating probabilities of limiting fire spread is illustrated in Figure 16.12. There are four parts to the process identified as procedures A to D. The four steps of procedure A are included in Figure 16.12. The remaining procedures are represented symbolically in their totality.

The steps in procedure A are also listed in Table 16.1.

Step A-1 produces a probability of limiting fire spread to the compartment of origin. The procedure for this calculation is outlined in procedure B as listed in Table 16.2.

Table 16.1. Procedure A: Calculating the probability of limiting fire spread to successive building compartments

Step A-1	Calculate the probability of limiting fire spread to the compartment of origin (ρ_1) as indicated in Procedure B (Table 16.2).
Step A-2	For each successive compartment, $P(E_s) = 0$.
Step A-3	Calculate the probability of limiting fire spread to each successive compartment as if it were the compartment of origin ($\rho_i, i = 2, 3, \ldots, n$) again using Procedure B (Table 16.2).
Step A-4	Calculate the probability of limiting the spread of fire to any successive compartment (n): $$P_n = 1 - \prod(1 - \rho_i)$$

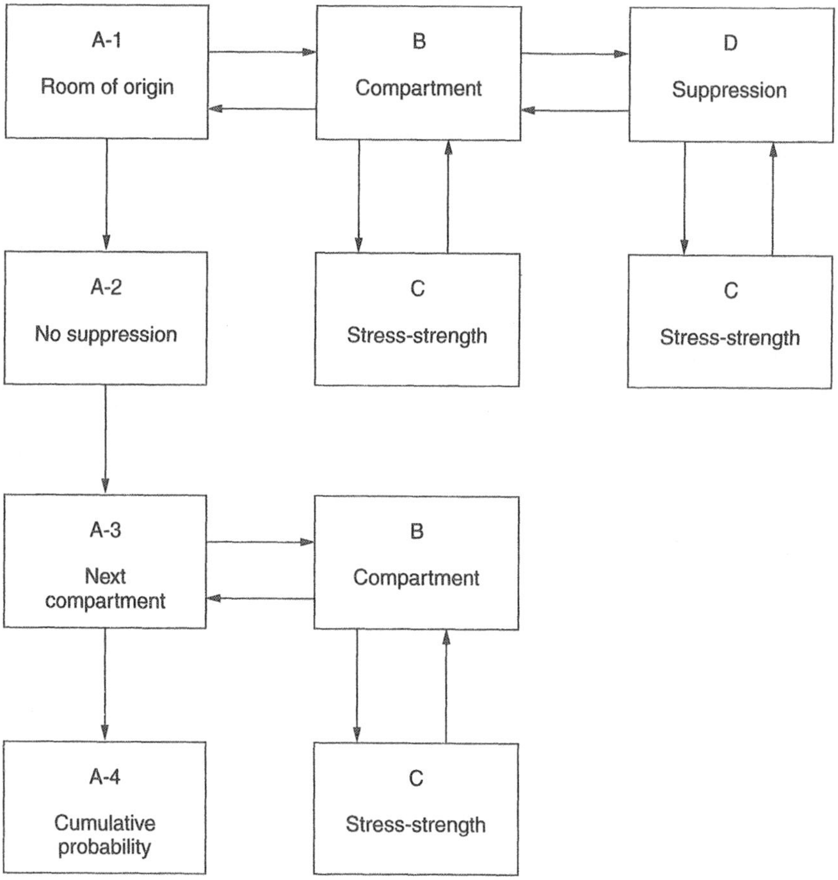

Figure 16.12. Process for calculating probabilities of limiting fire spread in buildings by the modified approach

Table 16.2. Procedure B: Calculating the probability of limiting fire spread within a compartment

Step B-1	Input mean and standard deviation of lognormal distribution of postflashover fire severity: $LN(\mu_{post}, \sigma_{post})$.
Step B-2	Input mean and standard deviation of lognormal distribution of barrier capacity: $LN(\mu_b, \sigma_b)$.
Step B-3	Calculate the probability that the barrier is adequate, $P(A_b)$, by the stress–strength relationship of Procedure C (Table 16.3).
Step B-4	Input the barrier reliability: $P(R_b)$.
Step B-5	Calculate the effectiveness of the barrier: $$P(E_b) = P(A_b) \cdot P(R_b)$$
Step B-6	If there is a suppression system $P(E_s) > 0$, go to Procedure D (Table 16.4).
Step B-7	Input the probability of self-termination of the fire: $P(T)$.
Step B-8	Calculate the probability of limiting fire spread to compartment i: $$P_i = 1 - [1 - P(T)] \cdot [1 - P(E_s)] \cdot [1 - P(E_b)]$$

Procedure B has two basic parts – estimating barrier effectiveness and combining this with suppression effectiveness, and self-termination. Estimating the probability of barrier effectiveness involves use of the stress–strength relationship applied to barrier adequacy. The stress–strength calculation is outlined in procedure C which is listed in Table 16.3.

The stress–strength relationship produces a probability that the particular fire control measure in question is adequate. This is adjusted for the reliability of the assembly or system in question.

The other basic part of procedure B involves the effectiveness of fire suppression as indicated in step B-6 of Table 16.2. The estimation of suppression effectiveness is a separate procedure, D, listed in Table 16.4.

The effectiveness of suppression also requires the stress–strength calculation. Therefore, as in procedure B, procedure D (step D-3) refers to the calculations in procedure C (Table 16.3).

These relationships are also illustrated in Figure 16.12. Step A-1 refers to procedure B which in turn refers to procedures C and D. Procedure D also refers to procedure C for the stress–strength calculation.

Table 16.3. Procedure C: Calculating probabilities of the adequacy of fire control alternatives (barriers and suppression) by the lognormal stress–strength relationship

Step C-1	Transform the parameters (mean and standard deviation) of the lognormal distributions of stress and strength to parameters of the normal distributions, $Y_i = \ln X_i$: $$\mu_y = \ln\,[\mu_x - (\sigma_y^2/2)]$$ $$\sigma_y = \ln\,[(\sigma_x/\mu_x)^2 + 1]$$
Step C-2	Calculate the parameters of the normally distributed difference between the stress and strength distributions, $W = \ln X_1 - \ln X_2$: $$\mu_w = \mu_{y1} - \mu_{y2}$$ $$\sigma_w = \sqrt{(\sigma_{y1}^2 + \sigma_{y2}^2)}$$
Step C-3	Find the standard normal variate corresponding to the condition of adequacy, $P\{W \geq 0\}$: $$z = \mu_w/\sigma_w$$
Step C-4	Identify the probability $P\{X > z\}$ from standard tables or by numerical methods.

Table 16.4. Procedure D: Calculating the probability of fire control in a compartment by suppression

Step D-1	Input mean and standard deviation of lognormal distribution of preflashover fire severity: $LN(\mu_{pre}, \sigma_{pre})$.
Step D-2	Input mean and standard deviation of lognormal distribution of capacity of the suppression system: $LN(\mu_s, \sigma_s)$.
Step D-3	Calculate the probability that the suppression system is adequate, $P(A_s)$, by the stress–strength relationship of Procedure C (Table 16.3).
Step D-4	Input the suppression system reliability: $P(R_s)$.
Step D-5	Calculate the effectiveness of the suppression system: $$P(E_s) = P(A_s) \cdot P(R_s)$$

Step A-2 in Figure 16.12 and Table 16.1 assumes that to be effective, a suppression system will control the fire in the compartment of origin. Therefore, the effectiveness of suppression is automatically set to zero for compartments other than the room of origin.

Step A-3 is an iterative step that repeats itself for all remaining compartments of interest. For each such compartment, the probability of limiting fire spread is computed by procedure B, which in turn utilizes procedure C.

Step A-4 calculates the cumulative probability of success in limiting fire spread at each compartment by the following equation:

$$P_n = 1 - \prod_{i=1}^{n}(1 - \rho_i)$$

The values produced by this computation are comparable to the points on the L-curve of the GSA approach.

16.4.5 EXAMPLE

One of the first applications of the GSA Goal-Oriented Systems Approach was the Richard B. Russell Federal Courthouse and Office Building in Atlanta, Georgia. This structure, referred to as the Atlanta Federal Building, is 24 stories high and contains over one million square feet of floor area. The lobby floor of the building is essentially unoccupied, the 2nd through 14th floors are office space, the 15th and 24th floors are mechanical equipment spaces, the 16th houses the US Marshal's Offices, the 17th through 23rd floors contain two-level courtrooms and auxiliary activities. Two below grade levels contain parking, maintenance shops, storage, and similar support functions.

The entire building is fitted with a hydraulically calculated, fully supervised automatic sprinkler system. On floors 2 through 14, the general office space, there is a central core area, which is separated from the remainder of the building as an area of refuge from fire. The separating walls are nonbearing, concrete masonry unit partitions.

Two critical events were considered – the limitation of fire spread within the general office space and the prevention of fire spread to the central core area of the structure. Data necessary for the application of the modified approach was gleaned from documents describing the application of the GSA approach. Input parameters are summarized in Table 16.5.

The probabilities of the critical events for both the original GSA approach and the modified approach are presented in Table 16.6.

The "L-Curve" for the office floors, developed by the designing fire protection engineers using the GSA Goal-Oriented Systems Approach, is shown in Figure 16.13. Also plotted in this figure are the results from the modified approach. The modified approach produces a more conservative

Table 16.5. Input data for application of the modified approach to Atlanta Federal Building

Distribution of postflashover severity: LN (28.0, 24.7)
Distribution of barrier resistance: LN (82.6, 28.8)
Probability of barrier reliability: $P(R_b) = 0.9995$
Distribution of preflashover severity: LN (0.054, 0.049)
Distribution of suppression resistance: LN (0.152, 0.048)
Probability of suppression reliability: $P(R_s) = 0.99$
Probability of self-termination: $P(\mathrm{T}) = 0.983$

Table 16.6. Probabilities of fire limitation in the Atlanta Federal Building by GSA and modified approaches

	GSA	Modified
Limit to office area	0.9996	0.9988
Prevent spread to core	0.99999	0.99993

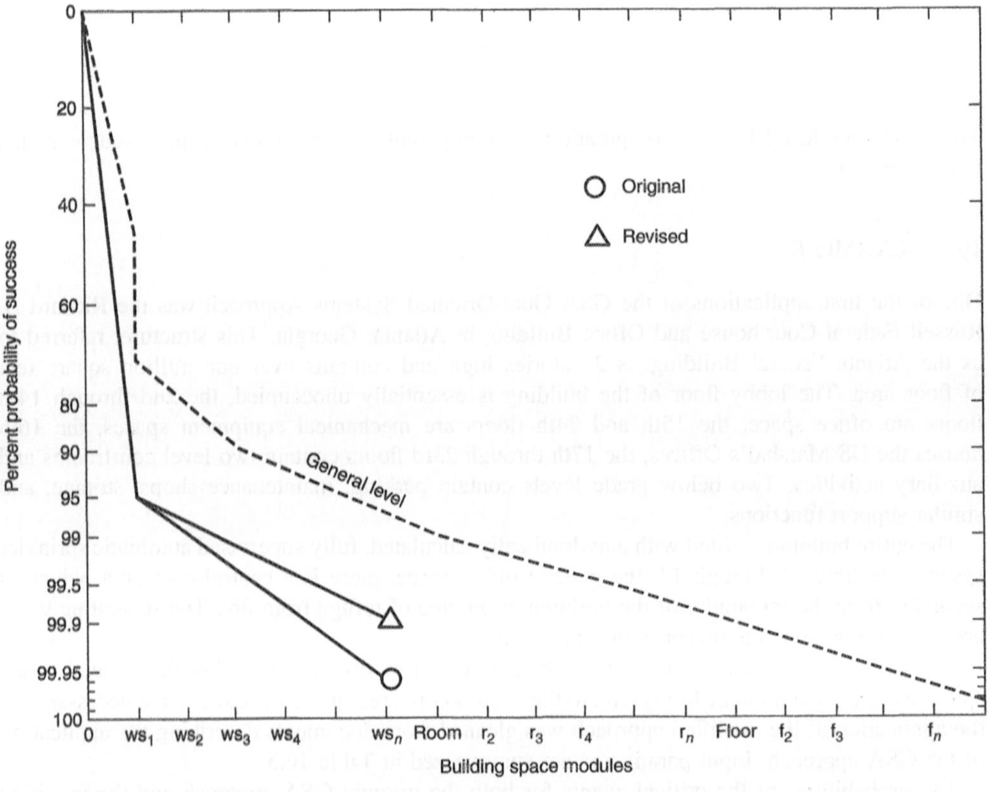

Figure 16.13. "L-curves" for Atlanta Federal Building using GSA and modified approaches

value, which is still within the goal level set by GSA. The probabilities of preventing fire spread to the central core for the two approaches are essentially coincident near the abscissa of Figure 16.13.

16.4.6 LIMITATIONS OF THE MODIFIED APPROACH

The modified approach is not without significant limitations to its application. The limitations can be generally described as imperfections in dimensionality, comprehensiveness, and interpretation.

This approach addresses only inanimate objectives, thus it is sufficient for evaluation of fire safety only in relation to spatial development. If animate goals are to be considered, namely, life safety, then it can be argued as necessary to introduce a temporal factor to model the mobility of the exposed in relation to the progress of the fire; that is, the undesirable event is the simultaneous exposure to fire in the dimensions of both space and time.

The modified approach does not represent an entire fire safety system. The approach is only one element of the system and is highly dependent for its appropriateness on the selection of scenarios. No formal methodology is proposed for determination of these likely paths of fire spread. The process is basically one of applying the professional judgment of experienced fire protection engineers. Insofar as there are relatively few fire protection engineers in the world today, this represents a limitation of the approach.

The interpretation of compliance with the prescriptive building codes is facile and definite – it complies or it does not. In contrast, interpretation of probabilistic information is somewhat ambiguous. It is difficult to adjudge the significance of a probability value without some guidelines. In the absence of any such guideline, it would be necessary to assign costs or benefits to alternative levels of fire safety and to try to optimize the situation. Both the guidelines and costs may be elusive values.

16.4.7 SUMMARY OF THE MODIFIED APPROACH

In application, the modified approach offers several advantages over its predecessor. Of primary consequence is the theoretical basis, which is explicitly identified and applied in a standard fashion. This should create a more favorable acceptance by users familiar with the principles of probability theory. Similarly, the explicit identification of the underlying postulates of fire spread and other assumptions should make the modified approach intuitively acceptable to those who are in accord with these principles.

Finally, the application of the modified approach is facilitated by simplified input requirements and calculations. The primary inputs are four probability distributions, which are of a standard format and can be identified with available data or by experienced judgment. Discontinuities are handled by the reliability factors, which may similarly be either generated or estimated. Thus, the input is minimal and of a uniform nature. The calculation procedures are well defined and amenable to computerization. These characteristics of the modified approach contribute to the appropriateness of probabilistic measures of fire safety.

16.5 WPI engineering method

Since the advent of the GSA Goal-Oriented Systems Approach, WPI, in collaboration with others, has been developing engineering procedures for evaluating the performance of building fire safety systems (Fitzgerald, 1985a,b). The resultant method focuses on structuring building fire safety problems in a manner that is organized and consistent. It provides a descriptive framework for identifying specific problems and formulating solutions. The framework delineates functional fire and building systems such that their interactions and interdependencies can be identified.

Eventually, this method will develop into an integrated, calculation-based analysis, and design procedure for use by practicing professional engineers. The procedure is envisioned to function analogous to structural and mechanical engineering methods. Although the method has reached a level of maturity where it has been applied to real-world problems (e.g. Fitzgerald *et al.*, 1991), it has yet to evolve into a complete, evaluation-based procedure. Each new application provides additional experience toward the evolution of the method.

16.5.1 FRAMEWORK OF THE WPI ENGINEERING METHOD

Evaluation of building fire safety involves the integration of a large number of factors that comprise the complex system of fire safety. In development of the WPI engineering method,

Table 16.7. WPI engineering method framework

A. Performance identification and needs
  1. Establish performance criteria
    (a) People
    (b) Property
    (c) Continuity of operations
B. Building analysis
  2. Prevent ignition and established burning
    (a) Prevent ignition
    (b) Initial fire growth hazard potential
    (c) Special hazard automatic extinguishment
    (d) Occupant extinguishment
  3. Flame movement
    (a) Fire growth hazard potential
    (b) Automatic sprinkler extinguishment
    (c) Fire department extinguishment
    (d) Barrier effectiveness
  4. Smoke movement
    (a) Air volume generation
    (b) Smoke generation
      – Obscuration particulates
      – Toxicity
    (c) Air volume modifications
    (d) Barrier effectiveness
  5. Structural frame
    (a) Heat energy impact
    (b) Protection effectiveness
    (c) Deflection
    (d) Structural capability
  6. People movement analysis
    (a) Alert effectiveness
    (b) Path movement
    (c) Building design
  7. People protection
    (a) Evacuation
    (b) Areas of refuge
    (c) Defend in place
  8. Property protection
    (a) Move
    (b) Defend in place
  9. Continuity of operations

systematic procedures have evolved to structure fire safety problems and solutions. The complete method consists of nine major parts grouped into three categories. These parts and their components are shown in Table 16.7.

The five analysis components of Part B comprise an organized framework that identifies interrelationships of elements of the building fire safety system. Building parts and building code requirements can be associated with a specific analytical component of the system. For example, a door latch becomes a part of barrier effectiveness for flame or smoke movement, and the architectural layout is a factor in the analyses of various means of fire extinguishment.

Part B of Table 16.7 involves engineering procedures to predict the performance of an existing or proposed building and its fire safety system. Some parts of the framework are more fully developed than others. The organization and structure of the framework has been developed to address the functional engineering questions, rather than to conform to available data or computational models.

In using this methodology, it is important to differentiate between the analytical framework and the techniques of quantification. The analytical framework identifies components that must be evaluated, elements that make up the components, and procedures for combining components into a meaningful measurement. This is considered to be independent of the quantification of the elements.

Quantification of the elements of the methodology, means of expressing the results, and identification of levels of acceptability are separate considerations. Scientific research has established some values, but the state of the art is inadequate to quantify with confidence most elements of the methodology. However, lack of reliable scientific data has not prevented the methodology from being applied. Where data is lacking, codes, standards, experience, and engineering judgment have been utilized. The method indicates what information is needed and how it will be used.

16.5.2 EVENT TREES

An innovation of the WPI engineering method is the use of event trees to impute an element of time on the process. While the original GSA approach was based on logic tree representation of component interactions, WPI suggests there is a sequence for these events that facilitates conceptualization of the process.

For example, in the GSA approach, limiting flame movement is associated with the logic diagram of Figure 16.7. In the engineering method, it is assumed that these events occur in sequence. First, there is the possibility of self-termination of the fire before it gains enough headway to activate automatic fire protection. Thus, the automatic suppression event is considered to be conditional on the fire not self-terminating. Similarly, manual suppression is considered to be implemented after automatic suppression has failed (or if it is not installed).

These relationships are shown in the event tree of Figure 16.14. The event tree indicates that upon established burning (EB), a fire will either self-terminate (I) or not ($\bar{\text{I}}$). If the fire self-terminates, then limitation of flame movement (L) has been achieved. This is represented by the dotted line on the right-hand side of the tree.

In the same manner, the next event is automatic suppression (A) followed by manual suppression (M). For each of these events, if they are successful in controlling the fire, the process exits the event tree. If they do not limit flame movement, the method progresses to the next event.

16.5.3 QUANTIFICATION

State probabilities are obtained by multiplying the transition probabilities along the chain of events from the defined start of the analysis to the state being considered. In the WPI engineering method, the state probabilities that show flame termination are added. Their sum, up to any state, defines the probable limitations of flame movement up to that state.

The start of the building analysis begins with EB. Established burning is defined as some identifiable fire characteristic. For example, a flame height of 25 cm is easy to identify, and is about the size of flame that begins to develop strong radiative transfer.

To illustrate application of the method, assume that given established burning, the transition probability to room involvement, considering fuel and environment alone, is 0.3. $P(\text{I})$ is defined

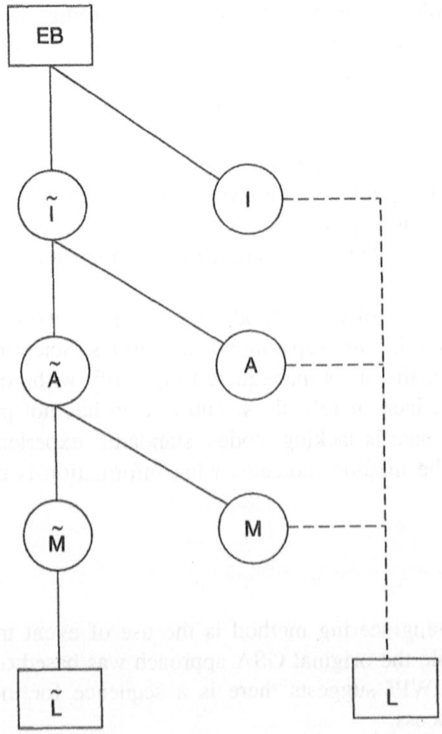

Figure 16.14. Event tree for room of origin

as the state probability of self-termination before room involvement, given established ignition; therefore,

$$P(\mathrm{I}) = 0.7$$

Given fires that do not self-terminate, $P(\mathrm{A})$ represents the state probability that automatic sprinklers will control the fire before full room involvement. Assume here that no automatic sprinklers are installed. Then,

$$P(\mathrm{A}) = 0.0$$

Given fires that have neither self-terminated nor been extinguished automatically, $P(\mathrm{M})$ expresses the probability of control by the fire department before full room involvement. Assume for illustration that,

$$P(\mathrm{M}) = 0.2$$

Within the WPI engineering method, the term "compartment" has particular meaning. It is a building space through which flames can move unimpeded and without barrier interference. Barriers, no matter how weak and no matter whether there are openings or not, are boundaries to the compartment. Doors and windows, for example, define the limits of a compartment regardless of whether they are open or closed. Barrier analysis addresses considerations of the barrier fire resistance and whether doors or windows are open.

Probability of full compartment involvement $P(\overline{\mathrm{L}})$ is produced by multiplying the probabilities in the chain of states (assumed to be independent) leading from "established burning" to "full room involvement." This is represented on the left side of the event tree in Figure 16.14. Given established burning, the probability of full compartment involvement is

$$P(\overline{\mathrm{L}}) = (0.3)(1.0)(0.8) = 0.24$$

Probability of termination before full compartment involvement $P(\mathrm{L})$ is calculated by summing separate values for each likelihood of termination. Therefore, given established burning, probability of termination before full room involvement is

$$P(\mathrm{L}) = 0.7 + 0.0 + (0.2)(0.3) = 0.76$$

The event tree process is extended to represent the compartment of origin, its barrier, the next compartment, and so forth as indicated in Figure 16.15 taken from an application to fire safety design of ships (Fitzgerald *et al.*, 1991).

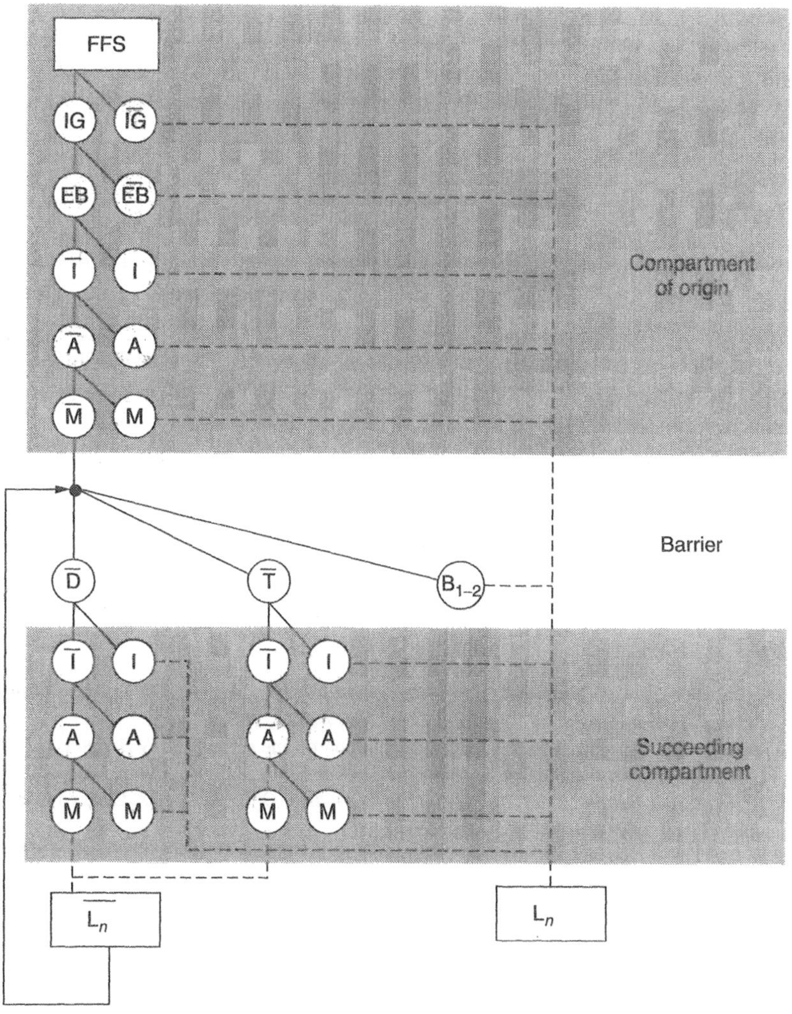

Figure 16.15. Flame movement analysis used for the PIR

16.5.4 LIMITATIONS OF THE WPI ENGINEERING METHOD

Present limitations of the WPI engineering method include lack of completeness, quantification, and documentation, and scope.

Analysis of smoke movement and people movement are presently crude. The framework for engineering evaluation of these components has not yet been developed.

Current applications require intuitive estimates of the probability of success for most fire safety building features. Support for this subjective probability judgment may include theoretical behavior based on scientific and engineering principles, interpretation of output from deterministic computer models, or deterministic relationships obtained from published literature. While the method encourages use of state-of-the-art resources, this can be time-consuming for the practitioner and expensive for the client.

A major deficiency of the WPI engineering method is the lack of adequate documentation. There is an acute need for peer-reviewed explanation of the theory, practices, and applications. While existing documentation has been evaluated to some extent by others, a rigorous, critical examination for accuracy, completeness, and appropriateness is required.

As more consistent techniques of quantification evolve, results will be compared numerically to other benchmarks, such as building codes and statistical data. However, at present, the method is limited to examining and evaluating the fire safety of unique structures and for risk management with quantifiable objectives.

16.5.5 SUMMARY OF THE WPI ENGINEERING METHOD

A major strength of the engineering method is its organizational structure, interrelating the many components of fire safety as an integrated system. Code officials, design professionals, fire officials, equipment manufacturers, and plant engineers are able to understand how the parts of this complex system interact and how to evaluate, on a subjective basis, expectations of building performance. The structure facilitates communication among these groups so that the problem and its solutions are understood.

The WPI engineering method has the capability to incorporate a variety of evaluation techniques. Quantification can be by engineering judgment utilizing theory, empirical results, or experience as the basis for the estimates. It is also possible to establish values and appropriate guidelines for selected conditions by computer simulation, loss experience statistics, Delphi procedures, or consensus.

References

American Society for Testing and Materials (1988). *Standard Method of Fire Tests of Building Construction and Materials*, ASTM-E-119, Annual Book of ASTM Standards, ASTM, Philadelphia.

Benckert, L-G (1962). The log-normal model for the distribution of one claim. *ASTIN Bulletin*, **II**, Part I.

Burros, R H (1975). Probability of failure of building from fire. *Journal of the Structural Division, ASCE*, **101**(9).

Culver, C G (1976). *Survey Results for Fire Loads and Live Loads in Office Buildings*, Center for Building Technology, National Bureau of Standards, Washington, DC.

Fitzgerald, R W (1985a). An engineering method for building fire safety analysis. *Fire Safety Journal*, **9**, 233–243.

Fitzgerald, R W (1985b). Risk analysis using the engineering method for building fire safety. *Fire Safety Science – Proceedings of the First International Symposium*, C E Grant and P J Pagni (Eds.), Hemisphere, Washington, DC, pp. 993–1002.

Fitzgerald, R W, Richards, R C and Beyler, C L (1991). Fire safety analysis of the polar ice breaker replacement design. *Journal of Fire Protection Engineering*, **3**(4), 137–150.

Ingberg, S H (1928). Tests of the severity of building fires. *NFPA Quarterly*, **22**(1).

Lie, T T (1972). Optimum fire resistance of structures. *Journal of the Structural Division, ASCE*, **98**(ST1).

National Fire Protection Association (1980). *Fire Safety Concepts Tree (Wall Chart)*, Quincy, MA.
National Fire Protection Association (1986). *Guide to the Fire Safety Concepts Tree*, NFPA 550, Quincy, MA.
Pape, R, Mavec, J, Kalkbrenner, D and Waterman, T (1976). *Semi-Stochastic Approach to Predicting the Development of a Fire in a Room From Ignition to Flashover*, IITRI Project J6367, IIT Research Institute, Chicago.
Ramachandran, G (1972). *Extreme Value Theory and Fire Resistance*, Fire Research Note 943, Fire Research Station, Borehamwood, UK.
Rennie, R A (1961). The measurement of risk. *Journal of Insurance*, **27**(1).
Watts, J (1977). *The Goal-Oriented Systems Approach*, NBS-GCR-77–103, National Bureau of Standards, Washington, DC.
Watts, J (1979). *A Theoretical Rationalization of Goal-Oriented Systems Approach to Building Fire Safety*, NBS-GCR-79–163, National Bureau of Standards, Washington, DC.
Watts, J (1983). A probability model for fire safety tree elements. *Hazard Prevention*, **19**(6), 14–15.

17 FIRE SAFETY ASSESSMENT IN THE PROCESS INDUSTRIES

17.1 Introduction

There are two characteristic features of the process industries, which occasion public concern. Firstly, there is the wide range of potentially hazardous substances that may be present either in storage or under stressful operating conditions such as high pressure, high temperature or controlled exothermic reaction. Secondly, if containment is lost, there may be quantities of dangerous materials that would not only endanger people directly involved in the process but also threaten those beyond the boundary of the enterprise. In the present context, we are concerned mainly with substances that give rise to major fire and explosion hazards. The Flixborough disaster may be quoted as an example of how the release of some tens of tonnes of a volatile flammable liquid (cyclohexane) caused not only havoc in the plant itself but also damage well beyond the boundary. However, the hazard that arises from the release of dangerous material extends to toxic and environmental hazards as well. Thus, the Bhopal disaster that took place in December 1984 in India resulted in the release of some 45 tonnes of methyl isocyanate to the atmosphere and this caused 2500 fatalities in an area round the plant. As an example of the environmental effect that release of material from a process plant can have, the incident that took place in Seveso in 1977 in Italy may be quoted. Dioxin was released from a relief vent that extensively contaminated the countryside, requiring the slaughter of many thousands of livestock to prevent contaminated meat getting into the food chain. Another example where environmental factors loomed large, was the fire that took place in one of the warehouses of a chemical company in Basle, Switzerland in 1986. The warehouse was used for storing agrochemicals. Roughly 10,000 m^3 of fire-fighting water drained into the nearby Rhine and with it about thirty metric tonnes of the chemicals stored in the warehouse. Among these were an estimated 150 kg of highly toxic mercury compounds. This brought about extensive contamination of the river.

17.2 Legislation for control of major hazards

The occurrence of these disasters in recent years, particularly Flixborough and Seveso, has given rise to extensive legislation for the control of these hazards. Thus a directive of the European Community (1982) requires member states to adopt provisions necessary to ensure that people

Evaluation of Fire Safety D. Rasbash, G. Ramachandran, B. Kandola, J. Watts and M. Law
 ISBN: 0-471-49382-1

in charge of such activities carry out specific precautions to identify major accident hazards and adopt appropriate safety measures to prevent such accidents. In the United Kingdom, this requirement has found expression in the Health and Safety Commission (1984) document entitled, "The Control of Industrial Major Accident Hazards" (CIMAH). More recently in the United States the Occupational Safety and Health Administration (OSHA) has issued a standard on process safety covering similar ground. As a first step to applying CIMAH regulations, it is necessary to identify substances handled in industry that can give rise to public concern. This is basically carried out in the document on the notification of installations handling hazardous substances regulations (HSC, 1982). CIMAH specifies quantities of the different materials that, if exceeded, bring in requirements for special design against major hazards. For flammable substances, the materials of concern may be identified as follows:

1. Flammable gases, substances that in gaseous state at normal pressures and mixed with air become flammable, the boiling point of which at normal pressure is 20 °C or below.
2. Highly flammable liquids, substances that have a flashpoint lower than 21 °C and a boiling point that at normal pressure is above 20 °C.
3. Flammable liquids, substances that have a flash point lower than 55 °C and which remain liquid under pressure where the particular processing conditions such as high pressure and high temperature may create major accident hazards.

The definition for explosive substances is substances that may explode under the effects of flame or which are more sensitive to shocks or friction than dinitrobenzene. For the three categories of flammable substances as indicated above, the amount specified requiring special steps to be taken are respectively: 200 tonnes, 50,000 tonnes, and 200 tonnes. For highly reactive materials, which are in general materials, with explosive properties, the amount specified is 50 tonnes although this is increased to 250 tonnes for sodium chlorate and 5000 tonnes for ammonium nitrate. For explosive substances themselves the amount specified is 50 tonnes, although there are a few materials for which the limiting amount is 10 tonnes. These figures contrast greatly with the specified amounts for materials with highly toxic properties, which can be as low as 1 kg.

The requirements that come into operation if the quantity of dangerous substance is exceeded firstly requires the preparation of a written report, which is submitted as a safety case to the Health and Safety Executive. The objectives of the safety case are as follows.

1. To identify the nature and scale of the use of the dangerous substances at the installation.
2. To give an account of the arrangements for safe operation of the installation, for control of serious deviations that could lead to a major accident and for emergency procedures at the site.
3. To identify the type, relative likelihood, and broad consequences of major accidents that might occur.
4. To demonstrate that management has appreciated the major hazard potential of a company's activities and has considered whether the controls are adequate.

Detailed guidance is given on the way these safety cases should be prepared. In addition, it is necessary to provide information to the authorities on significant modifications or to make periodic "no change" declarations. It is also necessary for them to prepare an on-site emergency plan, to provide information to the local authority to enable them to draw up an off-site emergency plan, and for information to be provided to the public about the major accident hazard that might threaten them. Following the Piper Alpha offshore disaster in UK waters (Chapter 3), many of these requirements, particularly the submission of a safety case, are now required for offshore installations handling large quantities of flammable liquids and gases.

Requirements for the safe design of processes handling hazardous substances, including submission of safety cases and emergency plans to the authorities, have given rise to a methodology for identification of hazard in process plants and for their assessment, analysis, and quantification, which will be briefly dealt with in the rest of this chapter. These are in addition to the extensive codes of practice of the process and petroleum industries that are a major feature in safe design and operation. A number of these quantification procedures are based on point systems similar to those described in Chapter 13. The main aim of the systems is to obtain an index of the hazard of the plant such that the greater the hazard the greater the index. However, no specific meaning is attached to the actual values of the index.

Hazard identification is associated mostly with the properties of the particular hazardous substances identified and mainly identifies the way they may be released from containment and how their hazard may be manifested. Guidance on this may be obtained from a scrutiny of the way in which dangerous accidents have occurred in the past. The Loss Prevention Bulletin of the UK Institution of Chemical Engineers periodically gives a summary of all major accidents within the Chemical and Process Industries worldwide. However, the major means for hazard identification is either a hazard and operability study (HAZOP) or a failure mode and effects analysis (FMEA). HAZOP is a systematic way of examining in depth the deviations that might occur in the operation of a process plant. FMEA is a "procedure by which each potential failure mode in a system is anticipated to determine the results or effects thereof on the system and to classify each potential failure mode according to its severity," (Dept. of Defense USA, 1980). Once a hazard is identified, analysis and quantification may be based on logic trees, particularly, fault trees and event trees. These trees would incorporate the effect of the safety procedure introduced to counter the hazard. The whole procedure of safety design in the process industry is sometimes called *hazard analysis*, although this term is usually confined to the quantification of identified hazards using logic trees.

17.3 Point schemes used in the process industry

A number of point schemes exist specifically for use within the Chemical and Process Industries. These schemes may be regarded as rapid assessment tools that serve particularly to identify areas that are in need of special attention. In order to apply these schemes, it is necessary to have access to detailed instructions. Many industrial and insurance companies have their own rapid ranking methods for which such details are not generally available. However, access is available to the two schemes mentioned below.

17.3.1 DOW CHEMICAL POINTS SCHEME

This is a partial points scheme put out by the Dow Chemical Company for the quantitative evaluation of risk on their various sites, which has also been extensively used elsewhere in the process industries (AIChE, 1973). It is intended to apply only to process units and not to auxiliary units and units such as power generation systems, control rooms, heaters and so on. The basic approach of the system is to determine "a fire and explosion index" that quantifies the risk factors in the process. Knowing this, there follows a selection of preventive and protective features. These are not ascribed a figure, but features are recommended for different values of the fire and explosion index and special features are recommended for specific items that appear in the index.

The first thing that is done is to divide the process plant into units, a unit being defined as the part of a plant that can be readily and logically characterized as a separate entity. The fire and explosion index is then determined for each of the units where a basic feature of the index is a material factor (MF) that can be identified either by the heat of combustion or the heat of reaction of the main material concerned. For combustible solids, liquids, and gases, the MF is defined as equal to the heat of combustion in BTUs per pound multiplied by 10^{-3}. Hence for methane, benzene, and ethylene the value of MF is 21.5, 17.3, and 25.1. The material factor is then increased by considering

(a) special material hazards

(b) general process hazards

(c) special process hazards.

Special material hazards consist of properties such as oxidizing capability, materials that react with water to produce combustible gas and materials subject to spontaneous heating, spontaneous polymerization, explosive decomposition or detonation. For each of these, the material factor can be increased by factors that can vary for each item up to 150%. General process hazards cover items such as handling and physical change only, continuous reactions, batch reactions, and multiplicity of reactions in the same equipment. Each of these may engender increases in the material factor up to 60%. Special process hazards include operation in or near the explosive range, operating temperatures or pressure, the difficulty of controlling the process of the reaction, the propensity for dust or mist explosions or explosion hazards generally. It needs to be shown that the process conditions would increase the hazard. The material factor can be increased for each item up to a value of 150%. In this sector, there is also a weighting factor for large quantities of flammable or combustible liquids that might be within the unit. For units of equipment containing more than 3000 gallons of flammable liquid the material factor is increased by 100%.

This information allows the fire and explosion index to be calculated. If it is between 0 and 20 the risk is classified as mild, between 20 and 40 as light, 40 and 60 as moderate, 60 and 75 as moderately heavy, 75 and 90 as heavy, and 90 and upwards as extreme.

All units are expected to have certain preventive and protective features that are listed. To these are added certain recommended "minimum preventive and protective features," which depend on the fire and explosion index. These can include such items as fire proofing (this means provision of fire resistance for structural supports), and protection from internal explosions, and provision of water spray, explosion relief, dyking, and blast and barrier walls. In general, these extra requirements are rated as being all "optional" if the fire and explosion index is low or "required" if the fire and explosion index is high.

Specific preventive features can provide the major form of protection. These are specific to certain items that have appeared in the evaluation of the fire and explosion index. Thus if the special process hazard is either a dust explosion hazard or operation is in or near the explosive range, then counteracting features are recommended. These could include the design of equipment to contain the explosion, design to relieve the explosion, provision of suppression, provision of dilution or inerting to keep the material out of the explosive range and provision of instrumentation with backup for process control. Similarly, where large quantities of flammable liquids are at risk, the recommended precautions could include the provision of instrumentation for remotely operated valves to minimize the flow of liquids, the provision of combustible gas monitors to raise an alarm below the lower flammable limit, provision of combustible gas monitors that automatically activate deluge systems or shut the system down, and provision of drainage and collection ponds to carry liquid spills away from process equipment.

There are drawbacks with the Dow Point scheme. There is no cover for the transport of materials from one unit to another and there is no assessment for management and housekeeping.

Also the special hazard of flashing following release of a flammable liquid is not specifically covered. Moreover the concentration on units tends to lose sight of the plant as a whole.

17.3.2 THE MOND FIRE, EXPLOSION AND TOXICITY INDEX

This is a development of the Dow Index (Lewis, 1979, 1980). The same basic approach has been used, particularly in dividing the process plant into units and defining a material factor for each unit (which for fire and explosion risks is the heat of combustion of the main item) and then moderating this factor according to special material hazards, general process hazards, and special process hazards. However, various additions and changes have been made to these material factor variations. Thus the special material hazard takes account of the manner in which the material, if released, mixes, and disperses with the atmosphere. It is general experience that, following leakage, hydrogen will escape quickly because of its buoyancy and ease of dispersion but a viscous material, although flammable, will present a small hazard relative to most flammable gases and vapors. A new factor has been introduced, which offsets the excessively high ratings given to hydrogen and certain other fuels on the basis only of their heat of combustion. Ignition sensitivity, explosive decomposition, condensed phase explosive properties, and gaseous detonation have also been more clearly defined as additional hazard factors in this section. Within the area of general process hazards, the risks of disconnecting pipework and the open transfer of liquids, as well as the use of transportable containers, have been included as risk factors. A number of extra items have also been added to the special process hazards, particularly corrosion and erosion effects, joint and packing leakages, vibration or support movements, ignition sensitivity, and electrostatic hazards. A toxicity hazard has also been included.

In addition to the fire and explosion index used in the Dow method, four indexes are estimated, namely, fire load, internal plant explosion, open flammable cloud explosion (called *aerial explosion* by the author), and toxicity index. There is also an overall risk rating that represents the potential size of an incident that would be likely to occur if all the safety and other offsetting features completely failed. If the estimated value of any of the indexes is unacceptable or borderline, a range of offsetting effects of safety control systems and other preventative features are considered and if relevant, may be introduced to reduce the overall risk. Unlike the Dow Index, these offsetting features are also quantified with factors less than one and are then used to offset the overall risk rating and the other hazard indexes.

17.4 Instantaneous fractional annual loss (IFAL)

This approach was evolved by the Insurance Technical Bureau in the United Kingdom (1979) and it specifically estimates the probability of a fire or explosion occurring in a chemical or process plant and the consequences that it may incur. The probability is based on an estimated frequency per annum of a fire or explosion occurring. The consequential damage that may result within the plant area is based on the amount of fuel involved and the type of fire or explosion incident. There are three elements to instantaneous fractional annual loss (IFAL), a process factor (p) which is calculated under the assumption that the process will be operated under good practice. Deviations from this good practice modify this process factor by an engineering adjustment factor (e) and a management factor (m). The IFAL of an operation is therefore given by

$$\text{IFAL} = p \cdot e \cdot m$$

The process properties need to be described in detail before a (p) factor can be obtained and these need to include design and operating standards that are chosen to represent good safety practice.

The hazards that are considered to contribute to the process factor on a typical chemical plant are P_1 liquid (pool) fires, P_2 vapor (flash) fires (mainly BLEVES), P_3 open flammable cloud explosions (termed by the authors as *percussive unconfined vapor cloud explosions*), P_4 vapor cloud explosions confined within buildings, and P_5 internal explosions within items of plant. P_5 is considered first, then P_1 to P_4 are evaluated by considering the sequence of events that lead to loss as follows:

c = loss of containment

i = ignition

s = spread of fire or explosion

d = damage

Loss of containment leads to emission of flammable material into the atmosphere. Both the frequency and the size of emission need to be specified. The frequency is obtained from historical data and depends upon the mode of failure that leads to emission, the type of equipment, its construction and the materials being handled. The amount of loss will also depend not only on these factors, but also on the holdup of the items concerned, and the conditions of pressure and temperature. Loss of containment can also take place as a result of pressure or missiles caused by internal explosions in plant P_1 and open flammable cloud explosions. The type of fire or explosion that is produced depends upon physical properties of the materials involved, the conditions to which they are subjected and the presence, siting, and intensity of ignition sources. The presence of an ignition source is based upon a monitoring of naked flames in the vicinity and an assessment of the amount of instrumentation, lighting, and maintenance required and various items that may produce ignition by sparks or hot surfaces in the area where the vapor may reach. Information on these factors is processed to give an estimate of the probability of ignition. An estimate is then made of items at risk within the vicinity that is calculated to be affected by the incident being considered. Knowing this the damage is estimated as the fraction of the total value of the investment that will need replacement as a consequence of the incident. This is then integrated for blocks in the plant to give the IFAL.

An algorithm based on more than 200 steps together with worksheets has been produced by the Insurance Technical Bureau to allow a systematic estimation of the p factor.

To estimate the possible influence of the engineering and management factors on the process, it is necessary to define the components that constitute good practice in engineering and management. Engineering includes hardware items particularly for protection against fire and explosions. Management would include steps that instill a good and continuing standard of safety practice and the extent to which people are trained to deal with hazardous incidents. Deviations from good practice can then be defined. Each of these deviations is then considered with regard to the effects on the subelements c, i, s, and d that are used in estimating the p factor, giving rise to a renewed value of the p factor.

The IFAL approach contains a great deal of statistical and physical information that allows both the frequency of fire and explosions and the damage that they may produce to be estimated. The approach therefore has a great deal more depth than the points schemes mentioned earlier. However, for any specific item of a plant, a quantitative approach of hazard in greater depth could be obtained from a fault tree and event tree analysis specific to the items concerned, since a great deal of the data used has been processed to allow application to a wide range of plant failure conditions. Also the types of fire incidents considered are not comprehensive since there is no special consideration of jet fires or running fires down vertical structures. Both these types of fire would be particularly important for offshore installations.

17.5 Hazard and operability studies

The methodology of hazard and operability studies was developed by the Imperial Chemical Industries in the United Kingdom round about 1970 (Lawley, 1974, Chemical Industries Association, 1977). The logic behind this study is that most hazards are not overlooked because of lack of knowledge but because they are hidden by the complexity of the plant. A small team of responsible persons involved in the design and operation of the plant carry out an in-depth, line-by-line and section-by-section analysis of different items in the plant looking for inadequacies in design. Variations in operating conditions during start up, shut down, and normal operation are visualized by guide and property words. Hazards can be identified by considering the cause and consequences of a deviation. Recommended solutions may involve changes in the design and/or operation of the process. The properties considered are flow, temperature, pressure, level concentration, heat, and cool. The guidewords are NO, NOT, MORE, LESS, AS WELL AS, PART OF, REVERSE, and OTHER THAN. The guide word "as well as" means that all the design and operating intentions are achieved together with some additional activity and "part of" means that only some of the intentions are achieved and some are not. The guide word "reverse" means that the logical opposite of the intention occurs and "other than" that no part of the original intention is achieved, something quite different happens.

The technique of using the guide words systematically generates a very large number of questions. It is essential that the team contains enough people with sufficient knowledge and experience to answer the majority of these questions without recourse to further expertise. However, the aim of the HAZOP is to identify the deficiencies in the design and operation of the plant. The team members must therefore be of sufficient seniority to authorize the changes recommended. When the HAZOP begins the design is frozen. Nothing should change during the study. Changes that are recommended by the team need to be implemented. Any additional changes would need to be submitted to the team for review before being authorized.

17.6 Failure mode and effects analysis (FMEA)

The safety and performance of a given system or plant depends on the reliability of equipment. Reliability of a system or of a component of the system is defined as the probability of the system or the component performing its design function satisfactorily. Many causes of hazards can be evaluated from deduced modes of equipment failure. Failure modes and effects analysis involves reviewing each element of a system by asking "What if?" type of questions to discover the modes of failure and then to examine the outcomes or consequences of each failure scenario. Reliability can be improved by eliminating or controlling some or all of the causes or modes of failure. As such, the FMEA process is more suited to examining equipment failures than process failures. It is a useful method for analyzing component failures contributing to the failure of the whole system. Being essentially an inductive method, this approach can also be used to examine the immediate effects of failure on other components of a system.

Lees (1996) identifies a number of objectives for undertaking an FMEA. They are as follows:

1. Identification of each failure mode of the sequence of events associated with it and of its causes and effects;

2. A classification of each failure mode by relevant characteristics, including the ability to detect, diagnose and test, the ability to replace an item and the compensating and operating provisions.

A failure is usually defined as an inherent state of a system component in which the item is unable to perform its design function (Wells, 1980). The definition of a failure scenario forms a basis for the identification of failure modes, causes, and possible outcomes or effects. This definition may involve collecting such information as the components of a system, modes of operation, the operating environment, and so on. Generally, two levels of failure modes are described: generic failure and specific failure. Generic failure modes may involve failure during operation, failure on demand, premature operation, and so on. Specific failures, on the other hand, pertain to the specific design or operating conditions under which a given component can fail, for example, fracture or exceeding design limits.

The effects of a failure can be manifest in several different ways, for example, by changes in the operation, function or status of a given system. Such effects can be local, that is, confined to the system component under consideration or global, affecting the system as a whole.

Information is collected about all the significant failure modes for each component in an FMEA analysis. Such information could be in the form of

- failure to open/close/start/stop/continue operation,
- spurious failure,
- degradation,
- erratic operation,
- scheduled service/replacement.

McCormick (1981) gives illustrative examples of FMEA.
The key steps in carrying out an FMEA are

- defining the objectives of the study and the output required,
- identifying the system under study,
- identifying separate subsystems and their interrelations,
- identifying failure modes and their causes and effects,
- identifying immediate effect on other components in the subsystem,
- identifying immediate effect on the subsystem,
- identifying effect on the whole system,
- identifying design and operating provisions against undesirable situations.

17.6.1 FAILURE MODE EFFECTS AND CRITICALITY ANALYSIS (FMECA)

The FMEA process combined with the Criticality Analysis is referred to as the *Failure Mode Effects and Criticality Analysis* (FMECA). It involves identifying critical areas and categorizing faults by their effect and some estimation of each failure mode frequency. Criticality is based on the reliability of a system or component, which is defined as the probability that a component will perform a required specified function. This may depend on a number of factors such as: commencing and continuing to operate on demand, not operating before demand and not continuing to operate after the demand has ceased. The reliability of multicomponent systems depends on the reliability of their components and the manner (series, parallel or a suitable combination of these) in which the components are connected.

Table 17.1. Severity categories and their resulting conditions

Severity category	Resulting conditions
Minor	(i) No effect on mission capability. (ii) Negligible effect on functional output at highest system level. (iii) Can be repaired by routine maintenance.
Major	(i) Negligible effect on mission capability. (ii) Degradation of functional output at high indenture level. (iii) Can be repaired within capability of organizational level maintenance.
Critical	(i) Some degradation of mission capability. (ii) Severe reductions in functional output of highest indenture level equipment. (iii) Cannot be repaired within immediate capability of organizational level maintenance.
Catastrophic	(i) Severe reduction in mission capability. (ii) Complete functional loss of items at highest indenture level. (iii) Requires maintenance outside immediate organizational level.

The categories of effect and the severity levels are specified by taking into account the following general characteristics:

- death or injury to the public or operating personnel;
- damage to other equipment;
- consequential economic loss.

Some of the severity categories are listed in Table 17.1. The failure rate frequencies or probabilities are obtained from the relevant component failure rate database. Equipment manufacturers or safety and reliability societies usually hold such information.

17.6.2 RELIABILITY OF FIRE PROTECTION SYSTEMS

Practically no investigation has so far been carried out in assessing the reliability of any fire protection device considering it as a system composed of components, although the reliability of the components has been evaluated in several research studies. Consider for example, a passive fire protection measure such as fire resistant compartmentation. Structural elements such as walls, floor, and ceiling are components of a compartment. These components constitute a "series" system since the compartment would "fail" to prevent the spread of heat and smoke to other compartments or areas in a building if any one of its components should "fail." Failure would occur if severity attained in a real fire exceeds the fire resistance of the structural elements. Failure can also be due to weakness caused by penetrations, doors or other openings in the structural boundaries of a compartment.

Likewise, the following main components of an automatic fire detection system installed in a building are in a "series" arrangement – the detector heads, zone control panels, central control panel, and the link to the fire brigade. The system would "fail" if a fire is not detected or detected but information about the fire is not communicated to the central control panel or the fire brigade. The system can trigger a "false alarm" in a nonfire situation, which is part of the unreliability of the system. The leading causes of failure are poor maintenance that leads to dust, insects and so on, clogging the heads, electrical and mechanical faults in the components, and unsuitable design or positioning of the heads and other components.

A sprinkler system would fail to operate and extinguish or control a fire if any one of its main components, which are in a "series" arrangement, should fail. These components are – sprinkler heads, pipework, and water supply from public mains, gravity tanks or other sources. The leading causes of failure are poor maintenance, system shut off for maintenance or repair, sprinkler top valves that are shut, defects in heads or other components, inadequate water supply, and blocked pipework.

17.7 Inherent safety

An alternative approach is to provide inherent safety in the design of fundamental features of the plant and its location so that the consequences of any accident is reduced (Kletz, 1984). The majority of accidents follow a loss of containment, so that a reduction of the amount of material being processed, particularly under stressful conditions of high temperature or pressure, will reduce the hazard. This may be achieved by converting a batch reaction to a continuous reaction, which may considerably reduce the quantity and time a material is held under stressful conditions. An alternative approach, particularly for exothermic reactions, is to use a semibatch method. One or more of the reactants is added over a period of several hours rather then mixing all the components at the start (Singh, 1993). The general reduction of hazardous materials inventory and the use, if possible, of processes employing safer materials under less stressful conditions are other approaches that need to be examined.

Jones (1992) has recommended the formulation of a systematic, disciplined approach to inherent safety with the acronym ISIN (Inherent Safety InPut), comparable in status to HAZOP. It would question elements of the intrinsic design that introduce a hazard or require the use of added protective systems.

17.8 Stages in safety design for new plant

A number of stages can usually be identified for safety design of a new plant in the process industry. These can vary from one firm to another, but the following sequence may be regarded as typical:

Stage 1. It is usual to identify the hazards at the project exploration stage, before the preliminary project assessment, based on the properties of the materials involved, particularly the fire, explosion, and toxic hazards. Data are gathered about the proposed processes and the nature of the site and previous experience with similar plant is assimilated. A decision is taken as to whether the hazards including the environmental hazard associated with the process are compatible with the location. As a first step, some organizations carry out a Dow points scheme assessment. At this stage the numerical criteria for the acceptable degree of hazard is laid down. This is usually based on the fatal accident frequency rate, FAFR (Chapter 8). The FAFR is 4 for the chemical industry in the United Kingdom and it is customary on the basis of this to allot a value of 0.4 as an upper limit to any one item of process plant.

Stage 2. At this stage one looks for major hazards, that is those that could radically change design through the possibility of large losses. When the flow sheet and the preliminary piping and instrumentation diagrams for the project become available a coarse hazard and operability study is carried out and major hazards thus identified are subject to hazard quantification. As far as fire and explosion are concerned, losses of containment that could lead to fatalities, particularly through a BLEVE or an open cloud flammable explosion, would be included in this process. Releases that could give rise to environmental problems caused by flares, and problems caused by toxicity, noise, dust, gaseous liquids, and solid effluents are also considered. Major hazards are compared with predetermined acceptable criteria and the design altered to meet the criteria. Protective systems are also broadly quantified. After this stage the project is sanctioned.

Stage 3. The major detailed hazard identification procedure is carried out in the form of a hazard and operability study and/or a failure modes and effects analysis. The information needed to carry out these studies comprises the piping and instrumentation diagrams, the engineering design of vessels and pipework, draft operating procedures, maintenance procedures, start up and shut down procedures, and emergency procedures. A study is made of the pressure systems including blow down and relief systems and includes an assessment of how the plant matches up to the criteria of the design.

The identification of major hazards is not expected in this stage, but minor hazards and operability problems are identified and remedied where necessary. Hazards uncovered by the identification processes are usually quantified using fault tree and cause/consequence analysis, such analysis being used to justify changes.

Having specified the probability of a flammable mixture being present, the plant is classified by area and electrical safety standards. If any design changes are made during this stage, then they would need to be resubjected to the hazard and operability study.

Stage 4. This takes place during the construction phase. Precommissioning checks are carried out in order to verify that all points arising from stage 3 have been implemented. These would include changes to the hardware and changes in operational procedures.

Stage 5. Before start up commences, the engineers check the operational safety of the plant, and also check that routine safety standards meet those laid down in safety legislation.

Stage 6. After the plant has been running some 12 to 18 months, the original hazard quantification predictions are updated in the light of operational experience. Account must be taken of any significant modifications that have taken place during the commissioning process. A report is prepared detailing design shortcomings and equipment failures and operational difficulties as they relate to the hazards in operating the plant. A final safety audit is carried out to check that the plant satisfies the original criteria for hazards and to provide comprehensive documentation for future reference.

17.9 Examples of logic tree analysis in process industries

Logic trees in Industry are discussed in Chapter 14.

References

AIChE (1973). Fire and explosion safety and loss prevention guide. *Hazard Classification and Protection*, CEP Technical Manual, American Institution of Chemical Engineers.

CIA (1977). *A Guide to Hazard and Operability Studies*, Chemical Industries Association Ltd.

Department of Defense (1980). Procedures for Performing a Failure Mode, Effects & Criticality Analysis, MIL-STD-1629A, USA.

EEC (1982). Council Directive on Major Accident Hazards of Certain Industrial Activities, 82/501/EEC European Economic Community.

HSC (1982). *Notification of Installations Handling Hazardous Substances Regulations*, Health & Safety Commission, HMSO, London.

HSC (1984). *Control of Industrial Major Accident Hazards, Regulations & Guidance*, Health and Safety Commission, HMSO, London.

Insurance Technical Bureau (1979). *Instantaneous Fractional Annual Loss (IFAL) Process Factor Algorithm for Petrochemicals*, Insurance Technical Bureau, London.

Jones, P (1992). Inherent safety: action by HSE. *The Chemical Engineer*, (526), 29.

Kletz, T A (1984). *Cheaper, Safer Plants or Wealth & Safety at Work*, Notes on Inherently Safer & Simpler Plants, Institution of Chemical Engineers, London.

Lawley, H G (1974). Loss prevention: operability studies and hazard analysis. *Chemical Engineering Progress*, **70**(4), 45.

Lees Frank, P (1996). *Loss Prevention in the Process Industries*, Vol. 1, 2nd Edition, Butterworth-Heinemann, Oxford, UK.

Lewis, D J (1979). *The Mond fire, explosion & toxicity index – a development of the Dow index*. AIChE Loss Prevention Symposium, Houston, TX, April 1979.

McCormick, N J (1981). *Reliability and Risk Analysis*, Academic Press, London.

Singh, B (1993). Assessing semi-batch reaction hazards. *The Chemical Engineer*, (537), 21–25.

Wells, G L (1980). *Safety in Process Plant Design*, George Goodwin Limited, London.

INDEX

Note: Page references in bold followed by 'f' represents a figure, 't' represents a table.

Abstract models 255
Accommodation methods 29, 30
Acronyms 151
Action time 302
Additive weighting 383
Addressable systems 233
AHP *see* analytic hierarchy process
Air Registry Board, England 143
Air transport, aviation standards 81
Airline Accident Digest Series 143
ALARM 1.0 (computer software) 368
Albini and Rand model 417–418
Alternative Approaches to Life Safety (NFPA) 367
Amelioration des Establissements Cantonaux d' Assurance Contre I' Incendie (Switzerland) 141
American Burn Association 142
American Institute of Chemical Engineers (AIChE) 72
American National Standards Institute (ANSI) 65
American Society of Civil Engineers (ASCE) 72
American Society of Mechanical Engineers (ASME) 72
American Society for Testing and Materials (ASTM) 70
Analytic hierarchy process (AHP) 371, 377, 382, 384, 386
Angle factors 266
Annual Abstract of Statistics, Central Statistical Office (UK) 145, 146
Area damage
  probability distribution in textile industry **202f**
  probability of 200–202
Arson, UCR estimates 148
ASKFRS (computer package) 179
Association of British Insurers (ABI) 139
Association of Canadian Fire Marshals and Commissioners 138
Atlanta Federal Building
  application of GSA approach 441–442
  L-curves using GSA and modified approaches **442f**
  probabilities of fire limitation by GSA and modified approaches **442t**
Attributes 363, 373–375, 386
  ratings 377–382
  weighting 375–377
Australia, fire statistics sources 138–139
Australian Assembly of Fire Authorities (AAFA) 139
Australian Fire Incident Reporting System (AFIRS) 138, 139
Australian National Fire Incident Statistics (1989–1990) 139
Austrian Zwenberg railway tunnel experiment 346, **347f**
Auto ignition temperatures 95
Automatic detectors 227–235
  average time and area damaged **230f**
  performance 227–228
  property protection effectiveness 228–231
  reliability 232–233
  response time and 248
Automatic fire detection (AFD) systems, hardware failure rates 232
Automatic flammable gas detection 122

Backdraft 106, 108
Barrier analysis 429–430, 446
Basic Building Code (US) 69
Basle disaster 451
Bedding, as material first ignited 180
Beverly Hills Supper Club, Kentucky disaster (1977) 51–53
Bhopal disaster (1984) 451
Blast walls 20

Evaluation of Fire Safety D. Rasbash, G. Ramachandran, B. Kandola, J. Watts and M. Law

ISBN: 0-471-49382-1

Boiling liquid compressed bubble explosion (BLCBE) 109
Boiling liquid expanding vapor explosion (BLEVE) 108, 109, 125
Boilover 108, 109
Boundary conditions 300–301
Bradford City Football Ground disaster (1985) 41–43, 187
British Insurance Association (BIA) 139
British Standards Institution 79
Building code organizations 69
Building codes 68–69
Building codes and standards, for existing and historic buildings 65
Building evaluation method (point system) 369
Building and Fire Research Laboratory (US) 74
Building fire safety systems, framework 444
Building fires, details for K433 and FDR1 forms 134–135
Building Officials and Code Administrators International (BOCA) 69, 369
Building regulations, effectiveness of 146
Building size, damage and **160f**
Building stock 146
Building type, fire growth rate **168t**
Buildings
  age of 146
  buoyancy smoke movement 265
  codes of practice and standards 79
  fire development 96
  leakage paths 267
  life safety 13**t**, 22
  population 190, 304–305
  quantitative fire safety design 32
  requirements in UK 78–79
  risk categories 146
  sudden massive flaming 106–108
Buildings Exits Code (US) 68
Bureau de normalisation du Québec (BNQ) 77
Burner fires 321
Burning
  and ignition 87–95
  mechanisms of 87–88
Burning rate, calculation 273
Burnout time, for rooms in office buildings 169
Burns 80, 178, 179

Canada
  building codes 74
  fire codes 74
  legal and fire safety requirements 74–78
  national fire statistics 138
Canadian Codes Centre (CCC) 75
Canadian Commission on Building and Fire Codes (CCBFC) 75–76
Canadian Construction Materials Centre 78
Canadian Gas Association 77
Canadian General Standards Board 77
Canadian National Codes 77
Canadian Standards 77
Canadian Standards Association 77
Carbon dioxide 121, 278
Carbon monoxide 109, 111, 112
Cardboard, as material first ignited 180
Casualties
  location of 177–178
  rate per fire 180–182
Casualty Return, Lloyds Register of Shipping annual publication 143
Causative agents 18
Ceiling jets 104
  comparison of predicted and measured radial profiles **347f**
  for confined and unconfined ceilings 345–346
Ceiling layer
  development of **261f**
  stratification 261–263
Cellular automata 305
Cellular automaton 306
Center for Fire Research, National Institute of Standards and Technology (US) 179
Central Government Department of Transport (UK) 141
Central Office Fire Risk Assessment *see* COFRA
Cerberus (Switzerland), fires casebook 144, 228
CFAST (multiroom zone model) 333–336
  four-room test with corridor **335f**
  gas species concentration 334–335
  upper-layer temperature and interface height 334
  validation 334
  vent flows 335–336
CFDS-FLOW3D field model 350–359
CFDS-FLOW3D solution method 300
CFDS-FLOW3D validation studies 340
Char 88, 94, 113
Chemical Industries Association 82
Chief and Assistant Chief Fire Officers Association (CACFOA Research Ltd) (UK) 136
Children, as cause of fires 177, 180, 187
Chimney fires 153
Chip or fat pan fires, television advertisement campaign 222–225
Chip pan safety, campaign in Exeter 222
CIFI (Circulation dans un Immeuble des Fumées de l'Incendie) model 268
Cigarettes, smoldering 180

City fires 16
Civil Aviation Authority (CAA) (UK) 81, 141
CLASP (Consortium of Local Authorities Special Programme) building 53, 54
Clothing, as material first ignited 180
Coconut Grove disaster 17
Code for Safety to Life from Fire in Buildings and Structures (US) 68
Codes
  adoption and enforcement in US 66
  consensus standards and 64–65
  limitations of existing US 65–66
  objective-based Canadian 76–77
  and standards 64
COFRA (Central Office Fire Risk Assessment) 371, 379
Combustion efficiency 259
Combustion products
  source terms 285–289
  travel time to escape route 184
Combustion toxicology 179
Comite Europeen des Assurances (CEA) 140
Commercial Fire Rating Schedule 140
Commonwealth Scientific Industrial and Research Organisation (CSIRO) 139
Compartment of origin, probabilistic evaluations 428
Compartment size, damage and **161f**
Compartmentation 243
  fire-resistant 154, 239–243
Compartments
  atmosphere layers 260
  fire resistance 241
Compliance alternatives, BOCA National Building Code 369
Computational fluid dynamics (CFD) modeling 295, **357f**
Computer fire models 295, 339, 367
  evacuation time calculation 305
  types of **256f**
Conduction, heat loss 267, 292–293
Cone calorimeter 110
Configuration factors 266
Conflagrations, prevention of 244
Consensus codes and standards 64–65
Conservation equations 272, 297
Consumer fire safety 80–81
Consumer Product Safety Commission (CPSC) (US) 73, 142, 144, 234
Consumer products, fire safety requirements 73
Control of Industrial Major Accident Hazard Regulations (CIMAH 1984) (UK) 79
Convection
  from flames 100–101
  from hot gas layers 103–105
  heat loss 291–292
Convective heat coefficient 294
Convective heat flux source terms 280
Convective heat transfer 100, 103, 125, 266, 278
Covers, as material first ignited 180
CPSC *see* Consumer Product Safety Commission
Critical events, outcome 25
Crown (government) properties, reports on fires 142
Crystal graphs 416
Cyclohexane hazard 451

Damage
  relationship between building size and 159
  severe structural 240
Danish State Fire Inspectorate 138
Data
  normalization of 378–379
  sources of 133–152
  transformation techniques 377
Databases 141–143
  quality control of 148
Deaths
  frequency distribution **188t**
  incendiary suicides 176
  probability of 185
  risk in different establishments 170
Decalin 36
Decision rules 379
Decision tables 379–382
Decision trees 26, 393–395
Definitions 7–9
Delamination, effect on flames 106, 107
Delphi method 374–375
Demographic Yearbook (UNO) 147
Denmark
  insurance organizations and fire protection associations 141
  sources of fire statistics 138
Density functions, fire loss 198, 200
Department of Health and Human Services (US) 72
Department of Housing and Urban Development (HUD) (US) 72, 144
Department of Labor, Occupational Safety and Health Administration (OSHA) 73
Department Store Exits Code (US) 68
Department stores, fire reports 143
Department of Transportation (DOT) (US) 73
Detector reliability, batteries and 232

Detectors
  application of exponential model of fire growth **229t**
  assessing economic value 229
  automatic 227–235, 248
  estimated average savings with detectors connected to fire brigade **230t**
  operation time of 183
Deterministic fire safety models 23, 164, 255–317, 399
Diffusion process 416
Dioxin disaster 451
Directorate for Fire and Explosion Prevention, Norway 138
Disaster-vulnerable people and daytime fire, fire death pattern in Japan 176
Discovery period 184
Discovery time, and fatal casualties **183t**
Discovery time and fatality rate vs number of deaths **183f**
Discrete probability distributions 432
Dispersion mechanisms 20
Distance maps **307f**, 308
Distribution function, of loss frequency for fires attended by fire brigades 200
Doors
  self-closing 187
  smoke-stop 4, 112, 117
Doorways
  centerline velocities and temperatures as function of height **341f**
  vertical openings 267
Dow Chemical Company, fire and explosion index 367
Dow Chemical Points Scheme 453–455, 460
Dow fire and explosion index 367
Dow Indices 367
Dry powders 121
Dupont Plaza Hotel, Puerto Rico disaster (1986) 45–51, 108
Dwelling fires, house-to-house visits 222
Dwellings
  fatal casualties **179t**
  spread of fire **178t**

Economic data 146–147
Eddy breakup model 300
Edinburgh model 370–371, 375–377, 383
Effective emissivity 293–294
EGRESS, evacuation model 306–307
Electric blankets 180
Electrical insulation, as material first ignited 180
Electricity, deaths due to 177
Elms and Buchanan, network model 408–410
Emergency equivalent network 412
Emergency lighting 184
Emergency services 20
Empirical models 320
Enclosure fires
  dynamics of 257–270
  field modeling 294–301
  forced ventilation 342–345
  large single-cell 356–359
  models 256–257
  with natural ventilation 340–342
Energy conservation 271–272, 277, 297
Engineering societies 71–72
Enhanced fire safety evaluation system for business occupancies 368
EPCOT Building Code (US) 68
Epidemic theory 417–418
Equivalent fire spread network 412–414
Escape facilities 184, 243–245
Escape methods 20, 22
Escape routes 122, 184
Escape time 302
Established burning 445
Established fires 17
Evacuation 243, 244
Evacuation models 184–186, 288, 302–308
  crowd movement 302–304
Evacuation time 184, 186
Event space method 409
Event tree analysis, gas release **396f**
Event trees 25, 125, 392, 393, 395–396, 447, 453
  room of origin **446f**
  WPI engineering method 445
Executive Agency of the Department of Health (UK) 142
Exit flow rate 302
EXITT, evacuation model 305–306
Explosion hazards, fire and 451
Explosions 5, 13, 19, 87, 109, 122–123
  fires associated with 54–61
  internal plant 455
  relief 20
Explosive substances, definition 452
Exposed building 154
Exposing building 154, 155
Extinction coefficient 110, 287, 288
Extreme order distributions 204, 210
Extreme order parameters
  estimation 204–206, 217–218
  standard normal distribution **211t**
Extreme order statistics, properties 217–220
Extreme order theory 203
Extreme value parameters **205t**
Extreme value theory 203–213, 215, 216

Fabrics, material testing 80
Factory Exits Code (US) 68
Factory Mutual Engineering and Research 140
Factory Mutual Engineering Risk Database 140
Factory Mutual Research Corporation (FMRC) 140–141
Factory Mutual Research (FMR) 72
Factory Mutual Research Laboratories, spreading fires growth rates 167
Failure mode and effects analysis (FMEA) 453, 457–460
Failure mode effects and criticality analysis (FMECA) 458–459
Failure modes
  definition 458
  and effects analysis 461
Failure rate 204
Fairfield Home, Nottinghamshire disaster (1974) 53–54
False alarms 144, 233–235, 459
FAST and HARVARD VI
  discussion of results 336–339
  experimental setup 336
  gas concentration 337
  layer height 338–339
  upper-layer temperature 337–338
FAST model 286, 305
  mechanical ventilation 285
  smoke obscuration 288
  toxicity assessment 289
FAST zone model 328–333
  comparison of predicted and measured upper-layer temperature in a three-room test facility **330f**
  comparison of predictions and measurements in a three-room test facility **332f**
  comparison of results 330–331
  corridor and lobby configuration **329f**
  experimental setup 330
  layer interface heights 331
  mass flow rates 331
  upper-layer temperatures 330–331
  validation study 331–333
Fatal Accident Frequency Rate (FAFR) 175, 190, 192, 460
Fatality, chance of 180
Fatality rates
  and discovery time 182, **183f**, **183t**
  estimate of increase in 184
  for fires discovered at ignition 182
Fault trees 25, 125, 142, 391, 392, 393, 397–398, 427, 453
Federal agencies (US) 72–74
Federal Aviation Administration (FAA) (US) 73, 81
Federal Emergency Management Agency (FEMA) (US) 73–74, 145, 148
Federation of Finnish Insurance Companies, Finland 141
Field models 256, 300, 320, 339–359
  enclosure fires 294–301
  validation 339, 359
Financial losses, FPA figures (UK) 149
FINDS (Fire Information National Data Service) (UK) 136
Finland, fire statistics sources 138
Fire
  interaction between people and 121–220
  interaction between structures and 114–117
Fire accommodation, consequential effects and 20–21
Fire Administration (US), fire statistics 164
Fire alarms 21, 22
Fire area and heat output, relationship between 168
Fire barriers
  effectiveness 438, 440
  failure modes 429
  stress–strength models 435–436
Fire brigades 119, 200
  automatic detectors and 248
  capability 160
  fire departments and 134–139
  sprinklers and 249
Fire casualties
  characteristics 176–177
  statistical data 176
Fire Certificates (Special Premises) Regulations (1976) (UK) 79
Fire Commissioner of Canada 138
Fire control 244
  calculating probability by lognormal stress–strength relationship **440t**
  calculating probability in a compartment by suppression **440t**
Fire damage
  estimated reduction with installed sprinklers **238t**
  room size and 159
Fire death rates, United States vs Japan 176
Fire deaths 12, 13, 32, 176
Fire Defense Agency of Ministry of Home Affairs, Japan 138
Fire detection 21, 22, 118–119

Fire disasters 5, 25
associated with explosions 54–61
Basle 451
Beverly Hills Supper Club 51–53
Bhopal 451
Bradford City Football Ground 41–43, 187
Coconut Grove 17
Dupont Plaza Hotel, Puerto Rico 45–51, 108
Fairfield Home, Nottinghamshire 53–54
Flixborough 451
Interstate Bank, Los Angeles 43–45, 107
involving dioxin 451
involving extensive spread of smoke 51–54
involving sudden rapid spread of fire 35–51
King's Cross Underground Station 41, 80, 82, 107, 187, 296
lessons learnt from 16–17
MV Betelgeuse, Ireland 54–56
and population experiencing them **191f**, 192
Ronan Point, London 56–60
Seveso 451
Stardust dance club fire 6, 38–40, 102, 107
Summerland Leisure Center, Isle of Man 35–38, 107, 108
Fire doors 117, 245–246
Fire engineering 9
Fire and explosion index 367, 453–454
Fire extinguishers 119, 120
effectiveness of 246–247
Fire Extinguishing Trades Association (UK) 144, 247
Fire fighting 22, 120
Fire frequency, age distribution of population and 172
Fire growth
combustible walls and ceiling tests on 107–108
rate of 165, **168f**
sprinklers and reduction in 239
square law 124
stages 400
within room of origin **166f**
Fire growth and suppression model, probabilities of extinguishment 406, **407t**
Fire hazards 3, 451
definition 7–8
evaluation 11
interaction between other hazards and 3–4
major areas 4–5
in mines 73
on vessels 73
Fire incidents 171
analyses of 134
family instability and 176
Fire load 169, 379, 437, 455
Fire locations 153
Fire loss
cumulative distribution curve **199f**
differences among nations 151
distribution 198
logarithms 206, **208t**
sprinklered vs. unsprinklered buildings 238
survivor probability distribution **214f**
Fire Loss in the United States (NFPA annual study) 137
Fire marshals 66
Fire modeling 300
techniques 320
variables 194
Fire plume source terms 276–277
Fire point temperatures 93, 94, 95, 96
Fire Precautions Act (UK) 78
Fire prevention 8, 21, 22
definition 9
fire occurrence and 17–18
measures 221–227
performance 148
Fire prevention campaigns 222
Fire Prevention (FPA journal) 139, 144
Fire protection 8, 9
cost effective strategy identification 394
cost effectiveness **31f**
methods 19
performance 148
points 363
Fire Protection Association (FPA) (UK) 28, 139, 150, 187
Loss Report Form 139, 140
Fire protection associations, insurance organizations and 139–141
Fire Protection Associations or Organizations 9
Fire Protection Engineering 9
Fire Protection Handbook (NFPA) 66
Fire protection systems
manufacturers fire reports 143
NFPA concepts 421
reliability 459–460
Fire Research Station (UK) 135, 139, 144, 179, 235, 236
Fire resistance 170
sprinklered compartments 249
testing structure for 114
Fire risk
assessment 193, 216
building groups and **157t**
components 391
definition 7, 147
due to severe weather conditions 147, 172

estimate of 170
evaluation 392
identification 392
management stages 392
probability of 154
reduction 392
target probabilities in buildings 32, **33t**
total cost of 21
transfer 392
Fire safety 9
acceptable level 30–33
assessment in process industries 451–462
audits and checklists 81–82
in the community 3–10
cost–benefit approach 30
data sources 16–17
decision-making level hierarchy **370t**
defining 370
equivalency 30
management 392
objectives 12–13
performance evaluation 65–66
prescriptive vs functional approach 5–6
public and private authorities requirements 63–83
quantitative evaluation approaches 22–26
statistics 14
subsystems 15–21
transport 81
in workplace 73
Fire safety attributes
beneficial 378
definition 373
detrimental 378, 379
nonmonotonic 378
ratings 377–382
weighting 375–377
Fire safety codes 14, 74, 154, 202, 244
model building and 68–70
vs building codes 66
Fire safety concepts tree 421–425
applications 423–425
checklist 424–425
communication 424
and derivative approaches 421–429
logic gates 422
major tree branches 422–423
OR gates 422
trade-offs 425
Fire safety design 8, 32
equivalent 30
fire scenarios 123–125
and management 21
steps in 13–16
trade-off 30
Fire safety engineering 9, 21–22
published sources of quantitative information **88t**
Fire safety evaluation 22–26, 65–66, 443
checklist 424–425
defensibility 365
multiattribute evaluation 365
point systems 363
steps **14t, 15f**
Fire safety evaluation system (FSES) 367–369
derivative applications 368–369
for health care occupancies 368
safety parameters 368
Fire safety management 8, 15, 22
Fire safety measures 21–22
effectiveness 221
performance of 221–252
reliability 221
Fire safety models, logic trees and 23
Fire safety related organizations in US, nonprofit 70–72
Fire safety strategies 376
Fire safety systems 11–34
Fire scenarios 19, 20, 123–125, 194
Fire services 121
performance of 148
Fire severity 169–170
calculation 369
estimation 430
models 437
Fire spread
among modules 433
calculating probability of limiting 438, **439f, 439t**
categories 162
circumstances favorable to rapid 106
containment 433
control 22
extent of 162–164
from buildings to adjoining buildings 243
limitations 433
massive energy transfer 433
mathematical model 274–275
postulates 433–434
probabilistic network **412f**
probability 417
quantification of 19
termination 433
Fire statistics 17, 32, 165
analysis by logit transformation 216
fires discovered first by smoke alarms 227

Fire statistics (*continued*)
loss of life 175
sources of 133–152
Fire Statistics, United Kingdom Home Office booklet 135, 148
Fire statistics
US Fire Administration 164
use of 147–148
Fire subsystems, direct detriment 20
Fire suppression 119–121
capability 432
effectiveness 438, 440
Fire Suppression Rating Schedule 72
Fire Test Standards Committee (E-5) 70, 71
Fire Test Standards (US) 71
"Fire in the United States"(FEMA/USFA fire statistics) 136
Fire-fighting equipment, provision of 159
Fire-generated flows 259–261
Fire-resistive buildings 140
Fireballs 125
Fireguards 80
Firemen
cause of death 192
safety of 123
Firemen's Advice Scheme, effect on dwelling fire frequencies 222
Fires
associated with explosions 54–61
average loss corrected for inflation 150
average loss per fire at 1966 prices **214t**
casualty rate vs delays in discovery 182
causes in dwellings 177
causes of incapacitation and death 179
conditional probability 158
damage to life 147
definition 7
development and control 19
direct damage 197
directional spread 418
estimation of time from ignition to discovery 135
expected area damage 162
fatality rate per **181t**
financial loss 5, 146, 149, 160, 162
frequency in industry 171
frequency of 146
frequency of spread beyond buildings 155
fuel-controlled 259
growth rate 164–169, 380
harmful effects 19–20
involving extensive spread of smoke 51–54
involving upholstered furniture in homes 193
multiple-death 186–190
nonfatal casualty rate **181t**
occurrence and growth of 153–174
occurrence probability 146, 190, 391
places of origins in spinning and doubling industry **156t**
probability damage and consequences 391
probability of one starting 147, 154, 155
probability vs building size 157
probable area damage 158–162
probable damage 146, 147
property damage 148
research studies 144–145
self-termination 438
severe winter weather and 172
severity expected 170
socioeconomic aspects in Britain 171–172
spread along surfaces 105
spread of 95–105
spread within a room 153, 154, 165–166
total cost to a community 5
ventilation-controlled 259
FireSEPC 372
Flame and burning objects, source terms 273–276
Flame cone, representation **273f**
Flame development, maximum extent **404t**
Flame movement analysis **447f**
Flame and smoke, spread of 263
Flame to object model 275–276
Flame volumes **98f**, 99
Flameover 104, 107, 108
Flames
buoyancy of 90
cine record of petrol fire **90f**
cocurrent 105, 106
convection from 100–101
countercurrent 105
diffusion 89, 121
dimensions 97
heat transfer measurements 101–102
limiting movement 445
luminous 89, 96, 101
profiles **90f**
radiation from 96–100
shapes for burning plastics 91
turbulent 90
Flaming, sudden massive 106–109
Flaming combustion, sudden increase in buildings 106
Flaming region 277
Flaming scenarios 194
Flammable clothing 80
Flammable cloud explosion 108, 109, 125
Flammable Fabrics Act 73

Flammable liquids
fire point temperatures 93
flash point temperatures 92, 93
Flammable substances, categories 452
Flash points 92, 93
Flashback 106, 108
Flashover 45, 48, 106, 107, 169, 194, 289
predicting 104
probability of 170
two-layer models and 262
Flixborough disaster 451
Flooding pressure 269
Flow boundaries 301
Flow rates
against density graph **303f**
comparison of calculated and measured mass **342t**
Flow Reynolds number 269
Flows
through narrow openings 269–270
through vertical vents source terms 280–285
Food fat, as material first ignited 180
Force of mortality 204
Forced convection 266
Forced ventilation fire, calculated and measured flow properties **343t**
Forced ventilation modeling 282–283
blower mode 284–285
extract mode 283
FPA *see* Fire Protection Association
Fractional effective dose (FED) 179, 288
France, fire statistics sources 138
Frequency distribution 198
Fuels
flame radiation properties of burning 96, **97t**
H/L values **89t**
heat of combustion 87
Furnishings, as material first ignited 180, 187
Furniture
as material first ignited 187
material testing 80
Furniture calorimeter 110
Furniture and Furnishings Fire Safety Regulations 1988 (UK) 80

Galbestos cladding 35
Gas
exposure to toxic 175
fires due to 171
Gas burner fire, mathematical model 274
Gas emissivity 294
Gas fatalities 178
Gas layers
convection from hot 102–103
radiation from hot 102–103
Gas-phase flaming combustion 87
Gases, radiation from 96
Gathering phase, evacuation model 184
General process hazards 454
General Services Administration 72
General Services Agency 11
Generalized least square method 212, 215
Generic failure modes 458
Geneva Association for the study of Insurance Economics 150
Geometric factors 266
Germany, national fire statistics 138
Glasgow Coma Scale 364
Global fire spread matrix 411
Glossary 386–387
Glowing combustion 87
Goal-oriented systems approach (GSA) 425–432
application 441
Atlanta Federal Building example 441–442
barrier analysis 429–430
compartment of origin 428–429
components 426–427
concept of time 432
fire safety approach 26
L-curves 429, 430
limitations 431–432
major branches **427f**
modified approach 432–443
objectives curve 427
probabilistic 421
probability curves 427–428
sensitivity analyses 432
summary 432
total probability theorem 436
Greek symbols 313–314
Gretener Method 366–367
computerized models 367
weighted product 384
Grinnell Corporation 143
Gross capital stock 146–147
Gross Fixed Capital Formation (GFCF) 147, 150
Gross Fixed Capital Stock (GFCS) 150
Gross national product (GNP) 146, 150
GSA *see* Goal-oriented systems approach
Guide to the Fire Safety Concepts Tree (NFPA) 421

Halon 121
Handbook of Fire Protection Engineering (SFPE) 71, 242, 366
Handbook for Reporting System of Fire Incident and Fire Death (Japan) 138

Hangar fire, analysis parameters **357t**
HARVARD V zone model (FIRST) 286, 287, 320–324
actual gas temperature profiles **323f**
comparison of model predictions with test data **322t**
comparison of results 321–324
comparison of several models **323f**
experimental fire test details **322t**
experimental setup 321
measured vs calculated species concentrations for MOD 20 case **324f**
measurements 321
types of fire installed 321
HARVARD VI, FAST and 336–339
HARVARD VI zone model 324–328
average temperature rise 326
corridor and lobby configuration **327f**
discussion of results 325–326
experimental setup 325
fire 325
fire test configuration **325f**
measurements 325
overall heat losses 326–328
pressure difference 326
vertical temperature profiles and layer depths 328
Hazard analysis 453
HAZARD (computer package) 179
Hazard function 204
HAZARD I (computer package) 192, 305
Hazard and operability study (HAZOP) 142, 453, 457, 461
Hazards
definition 7
fire and explosion 451
identification 461
industrial and process 79–80
legislation for control of major 451–453
as low as reasonable practicable (ALARP) 6, 32
Health and Safety Executive (HSE), Department of Employment (UK) 79
Heat conduction model 279
Heat conduction source terms 279–280
Heat detectors 118–119, 231
reliability 232
Heat exposure models 115, **116f**
Heat loss, from smoke by conduction 267
Heat release rate 257–259
from fire 291
typical computer work station 45
Heat transfer
and mass loss to surfaces 265–267
measurements from flames 101–102
rates 101, 114
Heated air, inhalation of 175
Heaters, safety of 80
High-rise buildings, point system using fuzzy system theory 371
Historic Fire Risk Index (US) 371
Home Office, United Kingdom 144, 187, 227
Horizontal openings 268–269
Hot gas layers, convective heat transfer rates **105t**
Hot gas temperature rise, calculation 290
Hot layer
calculation of depth 272
calculation of temperature 272–273
model 275
radiative flux 278
source terms 277–279
Human failure 18
HVAC (heating, ventilation, and air conditioning) models 256, 257
HVAC (heating, ventilation, and air conditioning) systems 266
Hydrogen cyanide 109

I-curves 428
Ignition 91–95
burning and 87–95
causes in different parts of a building 155
natural materials 180
sprinklers and 164
Ignition sources 18, 38, 153, 172, 456
human 153, 155
identification and power of 95
in multiple-death fires 187
nonhuman 153, 155
spinning and doubling industry **156t**
Imperial Chemical Industries (UK), methodology of hazard and operability studies 457
Industrial buildings, fire doors and property damage reduction 246
Industrial fire insurance, rebate calculating techniques 202
Industrial premises, representation of integrated fire hazards **27f**
Industry, sudden massive flaming 108–109
Inflation 146, 150, **205t**
Infrared detectors 118
Inherent safety 460
Injuries, nature of 178–179
Inlet velocities 301
Inspection, of equipment and structures 222
Inspectorate of Fire Services (Ministry of Interior) (The Netherlands) 138

Instantaneous fractional annual loss (IFAL) 455–457
Institute for Research in Construction (IRC) (Canada) 75, 77
Institution of Chemical Engineers (UK) 7, 79
  data on fires in chemical plants 142
Institution of Electrical Engineers 81
Institution of Structural Engineers 117
Insulation, in houses 4
Insurance Companies Association, Sweden 141
Insurance costs 30
Insurance industry, estimates of expected loss 29
Insurance Organisation Technical Committee (SKAFOR), Denmark 141
Insurance organizations 72, 139–141
Insurance Rating Bureau, Japan 141
Insurance requirements, United Kingdom 79
Insurance Services Office (ISO), New York 72, 140
  Commercial Fire Rating Schedule 140
Insurance Technical Bureau (UK) 455
  P factor 456
Integrated fire hazard, for industrial premises **27f**
Intensity function 204
Inter-governmental Maritime Consultative Organisation (IMCO) 81
Interior Bureau of Land Management, US 142
Intermittent flaming region 277
International Association of Electrical Inspectors and Underwriters Laboratories 143
International Association of Fire Safety Science 87
International Building Code (IBC) 70
International Chamber of Shipping 81
International Civil Aviation Organisation (ICAO) 81, 143
International Code Council (ICC) 70
International Conference of Building Officials (ICBO) 69
International databases 143
International Fire Code 70
International fire statistics, comparisons 150–151
International Maritime Organization (IMO) 81
Interstate Bank, Los Angeles disaster (1988) 43–45, 107
  estimated rate of heat release on 12th floor **46f**
  heat release rate of typical computer work station **46f**
  heat release rate vs ceiling height and floor area **47f**
Interval scale 378
Ionization smoke detectors 119, 228
  failure rate 232
ISIN (inherent safety InPut) 460
Japan
  fire death patterns in residences 176
  fire statistics sources 138
  preschool children death rate 176
  risk of deaths among older adults 177
JASMINE field model 296, 339–350
  for ceiling jet 345–346
  discussion of results 342, 345
  experimental setup 340, 343
  field model simulations 340–342

King's Cross Underground Station, London disaster (1987) 41, 80, 82, 107, 187, 296
Kolmogorov–Smirnoff goodness of fit test 437

L-curves 429, 430–431, 432, 433, 441
Laminar natural convection 266
Large losses
  behavior of 207–208
  fire frequency and **209f**
Large-Loss Fires in the United States (NFPA annual report) 137
Lawrence Livermore National Laboratory (LLNL) 320
Layering, smoke 118
Layers, mixing of 263–265
Legislation 33
  for control of major hazards 451–453
  literature 14
Life, loss of 154, 155, 175–196
Life risk 175
  impact of product choices 192–194
  measurements 190–192
Life safety 159–160
  in buildings 22
  effectiveness 231–232
Life Safety Code (NFPA) 65, 66, 68, 122, 367
Life safety and egress, selected variables influencing **24–25t**
Life safety objectives, for fires in buildings 13
Light scatter detectors 119
Likert scale 377–378, 379, 380, 387
Limiting oxygen index 338
Linear estimation, extreme order parameters 218
Ling and Williamson, network models 411–414
Liquid (pool) fires 456
Liquids, flaming combustion properties 88–91
Livermore test arrangement, sketch **321f**
Lloyds Register (LR) of Shipping 143
Logic gates 422
Logic trees 23, 379, 391–398, 453
  analysis in process industries 461
  ordinary combustibles **380f**
London Transport 141

Loss expectancy, statement used in insurance industry **29t**
Loss prevention bulletin, UK Institution of Chemical Engineers 453
Loss Prevention Council (LPC) (UK) 79, 139
Loss size, using logit model to quantify 217
Losses
  average and total 209–213
  inflation corrected natural logarithms of large **205t**

Main safety subsystem 20
Major disasters, in buildings involving sudden rapid spread of fire 35–51
Major explosion disasters 35, 54–60
Major fire disasters 35–53
Major hazards 8
Man-caused events, frequency in United States **189f**
Management Strategy for Fire (UK Fire Protection Association) 28
Manual fire fighting 120
Markov chains 402
Markov models
  for fire spread within a room or compartment 400–404
  mathematical representation 400–401
  state transition model 404–407
  weaknesses 404
Markov process 402–404
Mass conservation 270–271, 297
Mass flow boundaries 301
Mass loss model 179
Mass transfer to surfaces 267
Material factor, definition 454
Materials, first ignited 180, 187
Mathematical models, fire safety 22, 23, 255
Maximum likelihood estimation 211, 212
  extreme order parameters 218
  parameters of log normal distribution **213t**
Methyl isocyanate 451
Mine health and safety administration 73
Ministry of Interior, Finland 138
Mixed-is-burnt model 300
Mobile home fires, studies 144
Model code writing organizations 69
Model codes 64
Model validation 319–361
Modeling, interpretations of 255
Models
  enclosure fires 256–257
  uncertainty of 255
  validation 319–361
Modified Delphi exercise 375, 376
Modified event space tree **410f**
Modified goal-oriented systems approach 438
Moment estimation, extreme order parameters 218
Momentum conservation 297
Mond fire, explosion and toxicity index 455
Monte Carlo simulation 407
Mortality Statistics (Office of Population Censuses and Surveys, OPCS) 148
Motorcar manufacturers, fire reports 143
Multiattribute decision analysis 372
Multiattribute evaluation 372–384
  analytic hierarchy process (AHP) 377
  attribute examples 373–374
  attribute generation 373
  attribute ratings 377–382
  attributes 373–375
  characteristics 372
  examples 369–372
  scoring methods 382–384
  weight assessment techniques 375
Multiattribute utility theory 372
Multicomponent systems, reliability 458
Multicriteria decision making 372
Multimode detectors 233
Multiple regression 215, 216
  analysis 197, 215–216
Multiple-death fires 186–190
Multiple-death fires in the United States (NFPA annual report) 137
Multiple-occupancy dwellings
  average number of deaths per fire 188
  installation of automatic detectors 185
  number of fires and fire deaths 182
Multiroom enclosure fires with natural ventilation 353–356
  ceiling jet temperatures **355f**
  discussion of results 355–356
  experimental setup 354–355
  temperature profiles for thermocouple tree **356f**
MV Betelgeuse, Ireland disaster (1979) 54–56

N-Gas model 288
Narcotic gases, incapacitation caused by 179
National Accounts Statistics Yearbook 147
National Association of State Fire Marshals (NASFM) 66
National Board of Fire Research, Sweden 167
National Building Code (Canada) 74, 75, 77
National Building Code (US) 69
National Bureau of Standards (NBS) *see* National Institute of Standards and Technology (NIST)
National Center for Health Statistics (US) 148

National Code Documents (Canada), development of 76
National Electronic Injury Surveillance System (NEISS) (US) 142
National Fire Academy (NFA), Maryland 74
National Fire Code (Canada) 74, 75
National Fire Codes (NFPA) 66, 69
National Fire Incident Reporting System (NFIRS) 136
National Fire Protection Association (NFPA) (US) 11, 66–68, 136, 140, 145, 150, 176, 186
  annual survey 137
  building code 68
  codes and standard-making process 67
  Fire Incident Data Organisation (FIDO) 137, 140
  fire protection systems concepts 421
  fire safety approach 26, 28
  fire safety concepts tree 376, 385
  incendiary fire estimates 148
  NFPA 103 fire safety evaluation system for business occupancies 368
  NFPA 103 *Life Safety Code* 367
  performance options 67–68
National Fire Service (UK) 134
National Fire statistics, gaps in 149–150
National Health Service (NHS) (UK) 141
National Highway Traffic Safety Administration (NHTSA) (US) 73, 142
National Institute of Standards and Technology (NIST) (US) 74, 145, 179, 192, 239, 437
National Model Code (Canada) 75
National Research Council of Canada 74–75
National Smoke Detector Project, reasons for nuisance alarms 234
National Statistics, limitations in use of 148–149
National Transportation Safety Board (NTSB) (US) 142
Naval Safety Center, Norfolk, VA 143
Netherlands, fire statistics sources 138
Network models 408–414
New Zealand, fire statistics sources 138
New Zealand Fire Commission 138
NFPA *see* National Fire Protection Association
NHS Estates 142
Nomenclature 125–127, 309–313, 387–388
Nominal systems 378
Non-fire-resistive buildings 140
Nondeterministic models 399
Nondisaster-vulnerable people and daytime fire 176
Normalization, of data 378–379
Norway, fire statistics sources 138
Nuclear power stations, risk tolerability 32
Nuclear Regulatory Commission 72
Numerical solution procedure assumptions 293–294
Numerical solutions 300

Occupant types, specification **305t**
Occupational Safety and Health Administration (OSHA) 452
Occupied buildings, fires in 153
Office buildings, evacuation time design 186
Office of Population Censuses and Surveys (OPCS) 148
Oil, fire frequency 171
Oil heaters 187
Oil pool fire model 295
Oil Tanker and Terminal Safety 81
One-layer models 256, 257
One-zone modeling, postflashover fires 289–294
Ontario Housing Corporation, Canada 145, 232
Opacity 110
Open coal fires 80
Open flammable cloud explosions 108, 109, 125, 455, 456
Open flash fires 125
Optical density, smoke measurement 110
Ordinal systems 378
Ordinary combustible attributes
  equipment maintenance 380
  fire growth rate 380
  fire load 379
  fuel ignitability 380
  growth 379
  housekeeping 380
  ignition 379
  ignition sources 380
  transfer processes 380
Organization for Economic Co-operation and Development (OECD) 147
Oroglas (poly methyl methacrylate) 35, 36, 38
Outdoor fires 153
Output losses 147

Paper, as material first ignited 180
Parameter interaction matrix 371
Pareto distribution 199, 415
  area damage 202, 203
  use in fire damage 415
Particle optical density 288
People
  definition of individual types **304t**
  interaction between fire and 121–122
Perception time 302
Percolation process 416–417
Petrol fire, cine record of flames **90f**

PHOENICS computer model 295, 296, 339
Photoelectric type smoke detectors, bulb lifetime 232
Physical data 87–132
Physical models 255
Piper Alpha Platform, North Sea disaster (1988) 60–61, 79, 114, 452
Planes of symmetry 301
Plastics pool fires, flame diameter vs height above fuel surface **92f**
Platt, network model 410–411
Plume region 277
Plumes, maximum height 260
Point schemes 22–23, 33
Point smoke detectors, European standard 119
Point system fire safety evaluation, effectiveness of decision tables 382
Point systems 363–390
  advantages 365, 368
  application 366
  attributes 363
  concepts 364–366
  defensibility 365
  Dow fire and explosion index 367
  Edinburgh model 370–371
  examples 366–372
  Gretener Method 366–367
  guidelines 385–386
  multiattribute evaluation 365
  multiattribute evaluation examples 369–372
  purpose 384
  timber-frame apartment buildings 372
Polymers, fire points **94t**
Pool fires 321
  mathematical model 274
Population, affect on frequency of fires and fire casualties 145
Population data 145
Portable heating appliances 177
Portable kerosene heaters, legislation 80
Postflashover fires
  analysis 289
  in buildings 112–114
  models 257, 262
  one-zone modeling 289–294
Preflashover enclosure fires, zone modeling 270–289
Preflashover fires
  heat release rate 437
  models 257, 262
  sprinkler evaluation by stress–strength model 436
Preschool children, fire death rate 176
Pressure boundaries 301
Pressure pulses 123
Principal Component Analysis 170
Private authorities, fire safety requirements 63–83
Probabilistic models 23, 256, 391
Probability curves 427–428
Probability distributions 399, 437
  fire damage 197–203
  model verification 437
  modified approach 437–441
  overview 438
  a priori analysis 437
  procedures 438–440
Probability levels, values of $y_{(m)}$ **207t**
Probability trees **163f**, 164
  textile industry **406f**
Process industries
  fire and explosion hazards 451
  fire safety assessment 451–462
  point schemes used 453–455
Process plants, identification of hazard 453
Products standards and testing 70–71
Property damage 148, 154, 197–220
Property protection 12
  effectiveness 237–239
Public authorities, fire safety requirements 63–83
Public education programmes 222
Public lands statistics, US 143
Pulmonary complications 179
Pyrolysate vapors 88
Pyrolysis rate 273, 274

Quantitative attributes 377
Quantitative risk assessment (QRA) 392

Radiant heat transfer 125
Radiation
  estimates 99
  from flames 96–100
  from hot gas layers 102–103
  heat loss 293
  heat transfer models 275
Radiative heat flux source terms 280
Radiative heat transfer 100, 266, 278, 279
Railway Inspectorate (Health and Safety Executive) (UK) 141
Random walk 199, 401, 414–416
Rapid phase transition (RPT) explosion 109
Rapid vapor explosion 109
Ratio scale 378
Realm transition descriptors **403f**
Recognition time 184
Recommended National Building Code (US) 69
Reliability theory 434
Residual differences 359

Retail premises, Pareto distribution of area damage 203
Return period 209
Reynolds stresses 298
Richardson numbers 261, 262
Risk, definition 7
Risk agent 8
Risk assessment 18
Risk management 28–29
Risk premiums 202
Risk protection 29
Risk reduction 29
Road transport, UK safety requirements 81
Ronan Point, London disaster (1968) 56–60
Room corner test, flameover production 108
Room fire, significant features of wall effect **264f**
Royal Society for the Prevention of Accidents (ROSPA) 142

Safety, definition 8
Safety design for new plant, stages 460–461
Safety of Life at Sea Conventions (SOLAS) 81
Safety in Mines Research Establishment 141
Scaling systems 378
School Exits Code (US) 68
SCOPES (Specific Commercial Property Evaluation Schedule) 72
Scoring methods 382–384
  additive weighting 383
  weighted product 383–384
Seveso disaster 451
SFPE Engineering Guide to Performance-Based Fire Protection Analysis and Design of Buildings 71
Shape factors 266
Ship to Ship Transfer of Liquefied Gases 81
SIMULEX, evacuation model 304, 307–308
Single-enclosure fires, with natural ventilation 350–353
Single-occupancy dwellings, number of fires and fire deaths 182
Smoke 19
  defenses against 117–118
  growth rate 167
  visibility 111–112
  visual obscuration 179
Smoke alarms, fires discovered by 227
Smoke control systems 20, 118, 184
"Smoke Detector Operability Survey - Report on Findings"report 234
Smoke detectors 118, 119, 145, 231
  effectiveness study 164
  sensitivity 111
  standards 119
Smoke fatalities 178
Smoke and fumes, travel to escape routes and occupied areas 177
Smoke layer, absorption coefficient 103
Smoke obscuration 287–288
Smoke particles 175
  deposition of 267
Smoke and toxic gases
  hazard levels 111
  production and movement of 109–112
Smoke volume, quantification models 118
Smoke-related casualties in homes, increase 180
Smokers' materials 180, 187
Smokiness 110, 111
Smoldering fire scenarios 194
Smoldering ignition 95
Society of Fire Protection Engineers (SFPE) 71
Solid fuels
  fire point measurement 95
  fire point temperature 94
Solid-phase smoldering 87
Solids, flaming combustion properties 88–91
Solution methods 300
Soot
  absorption coefficient 102
  heat radiation and 266
  radiation from 96
Southern Building Code Congress International (SBCCI) 69
Space heater fires
  television advertisement campaign 225–227
  vs. temperature **226f**
Space heating 180
Special material hazards 454
Special process hazards 454
Species conservation 297
Specific failures 458
Spontaneous ignition temperatures 95
Spray fires 321
Sprinkler alternative probability tree **163f**
Sprinkler systems, failure to operate 460
Sprinklers 30, 119, 120, 159, 164, 235–240
  estimated reduction in fire damage **238t**
  life safety effectiveness 239–240
  in multiple-occupancy dwellings 185
  operation time in experimental fires 235
  performance 143, 235–238
  property protection effectiveness 238
  reasons for failure 236
  requirements 235
  size and degree of loss 216
  success in controlling fire spread 236–238
  water damage from 237
Standard Building Code (US) 69

Standard Fire Code (SBCCI) 69
Standard fire resistance test 242
Standards
  Canadian 77
  codes and 64
  consensus codes and 64–65
Standards Council of Canada 77
Stardust dance club fire, Dublin disaster (1981) 6, 38–40, 102, 107
State transition model
  spread from room to room 407–408
  spread within a room 404–407
Statistical Abstract, Bureau of Census, USA 145
Statistical Bureau, Stockholm 141
Statistical data, sources of 133–152
Statistical Yearbook (UNO) 147
Statistics, analysis of 134
Steel and concrete structures, performance in fires 117
Stochastic fire risk modeling 399–419
Stochastic models 23
Stress–strength models 434–436
Stress–strength relationship 438, 440
Structural damage, probability of 169
Structural failure 240, 241
Structural fire protection, sprinklers and 249–250
Structural models 115, **116f**
Structural (passive) fire protection 240–246
Structural stability 22
Subattribute growth, decision table **380t**
Subattribute ignition, decision table **381t**
Success trees. 26
Summerland Leisure Center, Isle of Man disaster (1973) 35–38, 107, 108
Surfaces, heat transfer and mass loss to 265–267
Survivor function 190
Switzerland
  fire statistics 144
  insurance organizations and fire protection associations 141
Symbols 9, 33, 172–173, 195, 359
Synthetic materials, ignition of 180

Tail functions 190
Temperature–time curves **115f**, 116, 117
TENAB (computer package) 179
Textile industry
  analysis 207–209
  extent of fire spread and average area damaged **162t**
  sprinklers in 159
Textiles, as material first ignited 180
"The Control of Industrial Major Accident Hazards"(CIMAH) 452
Theater fires 16
Thermoplastic fibers 180
Town gas, dealing with 59
Toxic hazards 288–289
Toxic products 19, 179
Toxicity index 455
Transition descriptors **403t**
Transport, fire safety 73, 81
Travel time 302
Tunnel fires
  calculated and measured results **349t**
  comparison of predictions with forced ventilation measurements **348f**
  comparison of predictions with natural ventilation measurements **348f**
  experimental setup 346
  field model simulations 346–350
Turbulence, stationary and nonstationary **297f**
Turbulence modeling 299
Turbulent flow, wall functions 301
Turbulent mixing layer, formation **265f**
Turbulent natural convection 266
Turbulent viscosity 298, 299
Two-layer models 256, 257, **258f**

Ultraviolet detectors 118
Unconfined vapor cloud explosions 109
Underwriters Laboratories of Canada 77
Underwriters' Laboratories (UL) (US) 71, 119
Uniform Building Code (US) 69
Uniform Crime Reports (UCR) 148
Uniform Fire Code (US) 69
United Kingdom
  building requirements 78–79
  case studies of multiple-death fires 187
  fire departments and brigades fire statistics 134–139
  FPA financial losses figures 149
  insurance organizations and fire protection associations 139–140
  insurance requirements 79
  legal fire safety requirements 78–82
  nonfatal casualties 177
  special databases 141–142
United Nations Organization (UNO), New York statistical office 147
United States of America
  Coast Guard 142
  fire fighter deaths (NFPA annual report) 137
  fire fighter injuries (NFPA annual report) 137
  fire risk assessment method 192
  fire statistics 144
  fire statistics sources 136–137
  Forest Service 142

frequency of man-caused events **189f**
heating appliance deaths 177
insurance organizations and fire protection associations 140–141
legal fire safety requirements 63–66
special databases 142–143
United States Fire Administration (USFA) 74, 136
Upholstered furniture
fire loss data 145
US risk model deaths **194t**
Upholstery, material first ignited 180
Urethane foam cushioning materials 180
US National Fire Prevention and Control Administration, National Fire Data Center 150–151

Vapor (flash) fires 456
Vehicle crashes, ignition source 187
Velocity against density graph **303f**
Vent flow rates 269
Ventilation, natural 280–282
Ventilation factor 259
Ventilation pool fire results, analysis of forced 324
Ventilation systems
effectiveness of 247–248
interaction between sprinklers and 250
Venting 40, 247–248
Vertical openings 267–268
Vertical temperature profiles, in compartment corner fire **342f**
Veterans Administration 72
View factors 266

Walking velocity against interperson distance **304f**
Walls to object, model 275
Water, in sprinklers 235
Water mist systems 121
Weather, fire risk and severe 147, 172
Weibull distribution 201
White Book on Fire Service in Japan 138
Wildlife Statistics and National Forest Fire Report (US) 142
Window vertical openings 267
Wisconsin Administrative Code 369
Wood crib fire, sizes at different heat output **91f**
World Airline Accident Summary 143
World Fire Statistics Center 150
World Health Organization, Statistical Annual 143
WPI (Worcester Polytechnic Institute) engineering method 421, 443–448
compartments 446
event trees 445
framework 443–445
limitations 448
quantification 445–447
strengths 448
summary 448

Yearbook of Industrial Statistics (UNO) 147

Zone models 256, 270–289, 320, 339
stratification 261–263
theory and concept of 257
validation 320, 359

Printed and bound by CPI Group (UK) Ltd, Croydon, CR0 4YY
16/09/2026
14963479-0008